D1657273

Biorefinery of Inorganics

Wiley Series in Renewable Resources

Series Editor:

Christian V. Stevens, Faculty of Bioscience Engineering, Ghent University, Belgium

Titles in the Series:

Wood Modification: Chemical, Thermal and Other Processes
Callum A. S. Hill

Renewables-Based Technology: Sustainability Assessment
Jo Dewulf, Herman Van Langenhove

Biofuels
Wim Soetaert, Erik Vandamme

Handbook of Natural Colorants
Thomas Bechtold, Rita Mussak

Surfactants from Renewable Resources
Mikael Kjellin, Ingegärd Johansson

Industrial Applications of Natural Fibres: Structure, Properties and Technical Applications
Jörg Müssig

Thermochemical Processing of Biomass: Conversion into Fuels, Chemicals and Power
Robert C. Brown

Biorefinery Co-Products: Phytochemicals, Primary Metabolites and Value-Added Biomass Processing
Chantal Bergeron, Danielle Julie Carrier, Shri Ramaswamy

Aqueous Pretreatment of Plant Biomass for Biological and Chemical Conversion to Fuels and Chemicals
Charles E. Wyman

Bio-Based Plastics: Materials and Applications
Stephan Kabasci

Introduction to Wood and Natural Fiber Composites
Douglas D. Stokke, Qinglin Wu, Guangping Han

Cellulosic Energy Cropping Systems
Douglas L. Karlen

Introduction to Chemicals from Biomass, 2nd Edition
James H. Clark, Fabien Deswarte

Lignin and Lignans as Renewable Raw Materials: Chemistry, Technology and Applications
Francisco G. Calvo-Flores, Jose A. Dobado, Joaquín Isac-García, Francisco J. Martin-Martínez

Sustainability Assessment of Renewables-Based Products: Methods and Case Studies
Jo Dewulf, Steven De Meester, Rodrigo A. F. Alvarenga

Cellulose Nanocrystals: Properties, Production and Applications
Wadood Hamad

Fuels, Chemicals and Materials from the Oceans and Aquatic Sources
Francesca M. Kerton, Ning Yan

Bio-Based Solvents
François Jérôme and Rafael Luque

Nanoporous Catalysts for Biomass Conversion
Feng-Shou Xiao and Liang Wang

Thermochemical Processing of Biomass: Conversion into Fuels, Chemicals and Power, 2nd Edition
Robert C. Brown

Chitin and Chitosan: Properties and Applications
Lambertus A.M. van den Broek and Carmen G. Boeriu

The Chemical Biology of Plant Biostimulants
Danny Geelen, Lin Xu

Forthcoming Titles:

Waste Valorization: Waste Streams in a Circular Economy
Sze Ki Lin, Chong Li, Guneet Kaur, Xiaofeng Yang

Process Systems Engineering for Biofuels Development
Adrián Bonilla-Petriciolet, Gade Pandu Rangaiah

Biobased Packaging: Material, Environmental and Economic Aspects
Mohd Sapuan Salit, Rushdan Ahmad Ilyas

Biorefinery of Inorganics

Recovering Mineral Nutrients from Biomass and Organic Waste

Edited by

ERIK MEERS
Department of Green Chemistry & Technology, Ghent University, Belgium

GERARD VELTHOF
Wageningen Environmental Research, The Netherlands

EVI MICHELS
Department of Green Chemistry & Technology, Ghent University, Belgium

RENÉ RIETRA
Wageningen Environmental Research, The Netherlands

WILEY

This edition first published 2020
© 2020 John Wiley & Sons Ltd

All rights reserved. No part of this publication may be reproduced, stored in a retrieval system, or transmitted, in any form or by any means, electronic, mechanical, photocopying, recording or otherwise, except as permitted by law. Advice on how to obtain permission to reuse material from this title is available at http://www.wiley.com/go/permissions.

The right of Erik Meers, Gerard Velthof, Evi Michels and René Rietra to be identified as the authors of the editorial material in this work has been asserted in accordance with law

Registered Offices
John Wiley & Sons, Inc., 111 River Street, Hoboken, NJ 07030, USA
John Wiley & Sons Ltd, The Atrium, Southern Gate, Chichester, West Sussex, PO19 8SQ, UK

Editorial Office
The Atrium, Southern Gate, Chichester, West Sussex, PO19 8SQ, UK

For details of our global editorial offices, customer services, and more information about Wiley products visit us at www.wiley.com.

Wiley also publishes its books in a variety of electronic formats and by print-on-demand. Some content that appears in standard print versions of this book may not be available in other formats.

Limit of Liability/Disclaimer of Warranty
In view of ongoing research, equipment modifications, changes in governmental regulations, and the constant flow of information relating to the use of experimental reagents, equipment, and devices, the reader is urged to review and evaluate the information provided in the package insert or instructions for each chemical, piece of equipment, reagent, or device for, among other things, any changes in the instructions or indication of usage and for added warnings and precautions. While the publisher and authors have used their best efforts in preparing this work, they make no representations or warranties with respect to the accuracy or completeness of the contents of this work and specifically disclaim all warranties, including without limitation any implied warranties of merchantability or fitness for a particular purpose. No warranty may be created or extended by sales representatives, written sales materials or promotional statements for this work. The fact that an organization, website, or product is referred to in this work as a citation and/or potential source of further information does not mean that the publisher and authors endorse the information or services the organization, website, or product may provide or recommendations it may make. This work is sold with the understanding that the publisher is not engaged in rendering professional services. The advice and strategies contained herein may not be suitable for your situation. You should consult with a specialist where appropriate. Further, readers should be aware that websites listed in this work may have changed or disappeared between when this work was written and when it is read. Neither the publisher nor authors shall be liable for any loss of profit or any other commercial damages, including but not limited to special, incidental, consequential, or other damages.

Library of Congress Cataloging-in-Publication Data

Names: Meers, Erik, 1976- editor.| Velthof, Gerard, 1964- editor. | Michels, Evi, 1980- editor. | Rietra, Rene, 1967- editor.
Title: Biorefinery of inorganics : recovering mineral nutrients from biomass and organic waste / edited by Erik Meers, Faculty of Bioscience Engineer, Laboratory of Analytical Chem, Gerard Velthof, Evi Michels, Rene Rietra, Wageningen University.
Description: First edition. | Hoboken, NJ : John Wiley & Sons, Inc., [2020] | Series: Wiley series in renewable resources | Includes bibliographical references and index.
Identifiers: LCCN 2020005302 (print) | LCCN 2020005303 (ebook) | ISBN 9781118921456 (hardback) | ISBN 9781118921463 (adobe pdf) | ISBN 9781118921470 (epub)
Subjects: LCSH: Sewage–Purification–Nutrient removal. | Factory and trade waste–Purification. | Nutrient pollution of water
Classification: LCC TD758.5.N58 .B56 2020 (print) | LCC TD758.5.N58 (ebook) | DDC 631.8/69–dc23
LC record available at https://lccn.loc.gov/2020005302
LC ebook record available at https://lccn.loc.gov/2020005303

Cover Design: Wiley
Cover Image: © Natascha Kaukorat/Shutterstock, Education globe © Ingram Publishing/Alamy Stock Photo

Set in 10/12pt TimesLTStd by SPi Global, Chennai, India

Printed and bound by CPI Group (UK) Ltd, Croydon, CR0 4YY

10 9 8 7 6 5 4 3 2 1

Contents

Section III State of the Art and Emerging Technologies in Nutrient Recovery from Organic Residues 65

Section V Agricultural and Environmental Performance of Biobased Fertilizer Substitutes: Overview of Field Assessments 189

List of Contributors

Fabrizio Adani Gruppo Ricicla, Lab. - Università degli Studi di Milano, DISAA, Agricoltura e Ambiente, Milano, Italy

Falentijn Assinck Wageningen Environmental Research, Wageningen University & Research, Wageningen, The Netherlands

Zhaohai Bai Center for Agricultural Resources Research, Institute of Genetic and Developmental Biology, CAS, Shijiazhuang, Hebei, China

Evangelina Belia Primodal Inc., Québec, Canada

Enrico Benetto Luxembourg Institute of Science and Technology (LIST), Environmental Research and Innovation (ERIN), Luxemburg

Jayanta Kumar Biswas Department of Ecological Studies, University of Kalyani, West Bengal, India
International Centre for Ecological Engineering, University of Kalyani, West Bengal, India

Gabriele Boccasile Regione Lombardia DG-Agricoltura, Milan, Italy

Aleksandra Bogdan Department of Green Chemistry & Technology, Faculty of Bioscience Engineering, Ghent University, Ghent, Belgium

Claudio Brienza Department Green Chemistry and Technology, Faculty of Bioscience Engineering, Ghent University, Ghent, Belgium

Sander Bruun Department of Plant and Environmental Sciences, University of Copenhagen, Denmark

Fridtjof de Buisonjé Wageningen Livestock Research, Wageningen University & Research, Wageningen The Netherlands

Jeroen Buysse Department of Agricultural Economics, Faculty of Bioscience Engineering, Ghent University, Belgium

Juan Tur Cardona Department of Agricultural Economics, Faculty of Bioscience Engineering, Ghent University, Belgium

Lies De Clercq Department of Physical and Analytical Chemistry, Ghent University, Ghent, Belgium

Giuliana D'Imporzano Gruppo Ricicla, Lab. - Università degli Studi di Milano, DISAA, Agricoltura e Ambiente, Milano, Italy

Marie-Line Daumer Institut national de recherche en sciences et technologies pour l'environnement et l'agriculture, Rennes, France

Kimo van Dijk Wageningen Environmental Research, The Netherlands
Soil Quality Department, Wageningen UR, Wageningen, The Netherlands

Renjie Dong Key Laboratory of Clean Utilization Technology for Renewable Energy in Ministry of Agriculture, College of Engineering, China Agricultural University, Beijing, PR China

Phillip Ehlert Wageningen Environmental Research, Wageningen University & Research, Wageningen, The Netherlands

Ania Escudero Glasgow Caledonian University, Glasgow, Scotland

Wim van Geel Wageningen Plant Research, Wageningen University & Research, Wageningen, The Netherlands

Sam Geerts Aquafin n.v., Aartselaar, Belgium

Katarzyna Golkowska Luxembourg Institute of Science and Technology (LIST), Environmental Research and Innovation (ERIN), Luxemburg

Piet Groenendijk Wageningen Environmental Research, Wageningen University & Research, Wageningen, the Netherlands

Marius Heinen Wageningen Environmental Research, Wageningen University & Research, Wageningen, the Netherlands

Paul Hoeksma Wageningen Livestock Research, Wageningen University & Research, Wageningen The Netherlands

Marieke ten Hoeve Department of Plant and Environmental Sciences, University of Copenhagen, Denmark

Gertjan Holshof Wageningen Livestock Research, Wageningen University & Research, Wageningen, The Netherlands

Yong Hou Soil Quality Department, Wageningen University & Research, Wageningen, The Netherlands
College of Resources and Environmental Sciences, China Agricultural University

Yong Hou College of Resources and Environment Sciences, China Agricultural University, Beijing, China

Lars Stoumann Jensen Department of Plant and Environmental Sciences, University of Copenhagen, Denmark

Daniel Klein Lippeverband, Essen, Germany

Daniel Koster Luxembourg Institute of Science and Technology (LIST), Environmental Research and Innovation (ERIN), Luxemburg

Chad Kruger Center for Sustaining Agric & Natural Resources, WA, USA

Viooltje Lebuf Flemish Coordination Centre for Manure Processing, Belgium

Evert Leijenhorst BTG Biomass Technology Group B.V. Enschede, The Netherlands

Jan Peter Lesschen Wageningen Environmental Research, Wageningen University & Research, Wageningen, The Netherlands

Qian Liu Soil Quality Department, Wageningen University & Research, Wageningen, The Netherlands

Lin Ma Center for Agricultural Resources Research, Institute of Genetic and Developmental Biology, CAS, Shijiazhuang, Hebei, China.

Adrien Marchi Aquafin n.v., Aartselaar, Belgium

Johan te Marvelde Lippeverband, Essen, Germany

Erik Meers Department of Green Chemistry & Technology, Faculty of Bioscience Engineering, Ghent University, Ghent, Belgium

Evi Michels Department of Green Chemistry & Technology, Faculty of Bioscience Engineering, Ghent University, Ghent, Belgium

Jantine van Middelkoop Wageningen Livestock Research, Wageningen University & Research, Wageningen, The Netherlands

Anders Nättorp School of Life Sciences FHNW, Muttenz, Switzerland

Myles Oelofse Department of Plant and Environmental Sciences, University of Copenhagen, Denmark

Oene Oenema Soil Quality Department, Wageningen Environmental Research, Wageningen University & Research, Wageningen, The Netherlands

Valentina Orzi Gruppo Ricicla, Lab. - Università degli Studi di Milano, DISAA, Agricoltura e Ambiente, Milano, Italy

Ole Pahl Glasgow Caledonian University, Glasgow, Scotland

Hervé Paillard Véolia Environnement, Aubervilliers, France

Marie Edith Ploteau Lippeverband, Essen, Germany

Reinhart Van Poucke Department of Green Chemistry & Technology, Faculty of Bioscience Engineering, Ghent University, Ghent, Belgium

Wei Qin Soil Quality Department, Wageningen University, The Netherlands
College of Resources and Environmental Sciences, China Agricultural University

Sukanta Rana International Centre for Ecological Engineering, University of Kalyani, West Bengal, India

Gregory Reuland European Biogas Association, Renewable Energy House, Belgium
Department of Green Chemistry & Technology, Faculty of Bioscience Engineering, Ghent University, Ghent, Belgium

Carlo Riva Gruppo Ricicla, Lab. - Università degli Studi di Milano, DISAA, Agricoltura e Ambiente, Milano, Italy

Ana Robles Department of Green Chemistry & Technology, Faculty of Bioscience Engineering, Ghent University, Ghent, Belgium

Bart Saerens Aquafin n.v., Aartselaar, Belgium

René Schils Wageningen University & Research, Wageningen Plant Research, Wageningen University & Research, Wageningen, The Netherlands

Karl-Georg Schmelz Emschergenossenschaft, Essen, Germany

Oscar Schoumans Wageningen Environmental Research, Wageningen University & Research, Wageningen, the Netherlands

Jaap Schröder Wageningen Plant Research, Wageningen University & Research, Wageningen, The Netherlands

Brajendra K. Sharma Illinois Sustainable Technology Center, University of Illinois at Urbana-Champaign, Champaign, Illinois, USA

Ivona Sigurnjak Department of Green Chemistry & Technology, Faculty of Bioscience Engineering, Ghent University, Ghent, Belgium

Luc Sijstermans Slibverwerking Noord-Brabant, Moerdijk, the Netherlands

Fulvia Tambone Gruppo Ricicla, Lab. - Università degli Studi di Milano, DISAA, Agricoltura e Ambiente, Milano, Italy

Paul Thiers Department of Political Science, Washington State University–Vancouver, Vancouver USA

Ian Vázquez-Rowe Pontificia Universidad Católica del Perú, Department of Engineering, Peruvian LCA Network, Peru
University of Santiago de Compostela, Department of Chemical Engineering, Santiago de Compostela, Spain

Céline Vaneeckhaute BioEngine, Research Team on Green Process Engineering and Biorefineries, Chemical Engineering Department, Université Laval, Quebec, Canada

Peter Vanrolleghem BioEngine, Chemical Engineering Department, Université Laval, Québec, Canada

Nicolas De La Vega European Biogas Association, Brussels, Belgium

Gerard Velthof Wageningen Environmental Research, Wageningen University & Research, Wageningen, The Netherlands

Jingmeng Wang Soil Quality Department, Wageningen University & Research, Wageningen, The Netherlands

Marjoleine Weemaes Aquafin n.v., Aartselaar, Belgium

Shubiao Wu Key Laboratory of Clean Utilization Technology for Renewable Energy in Ministry of Agriculture, College of Engineering, China Agricultural University, Beijing, PR China

Georgine Yorgey Center for Sustaining Agric & Natural Resources, WA, USA

Frank Zepke EuPhoRe GmbH, Telgte, Germany

Wei Zheng Illinois Sustainable Technology Center, University of Illinois at Urbana-Champaign, Champaign, Illinois, USA

Kor Zwart Wageningen Environmental Research, Wageningen University & Research, The Netherlands

Series Preface

Renewable resources, their use and modification are involved in a multitude of important processes with a major influence on our everyday lives. Applications can be found in the energy sector, paints and coatings, and the chemical, pharmaceutical, and the textile industry, to name but a few.

The area interconnects several scientific disciplines (agriculture, biochemistry, chemistry, technology, environmental sciences, forestry, ...), which makes it very difficult to have an expert view on the complicated interaction. Therefore, the idea to create a series of scientific books that will focus on specific topics concerning renewable resources has been very opportune and can help to clarify some of the underlying connections in this area.

In a very fast changing world, trends are not only characteristic for fashion and political standpoints; science is not free from hypes and buzzwords. The use of renewable resources is again more important nowadays; however, it is not part of a hype or a fashion. As the lively discussions among scientists continue about how many years we will still be able to use fossil fuels - opinions ranging from 50 to 500 years - they do agree that the reserve is limited and that it is essential not only to search for new energy carriers but also for new material sources.

In this respect, renewable resources are a crucial area in the search for alternatives for fossil-based raw materials and energy. In the field of energy supply, biomass and renewable-based resources will be part of the solution, alongside other alternatives such as solar energy, wind energy, hydraulic power, hydrogen technology and nuclear energy. In the field of material sciences, the impact of renewable resources will probably be even bigger. Integral utilization of crops and the use of waste streams in certain industries will grow in importance, leading to a more sustainable way of producing materials.

Although our society was much more (almost exclusively) based on renewable resources centuries ago, this disappeared in the Western world in the nineteenth century. Now it is time to focus again on this field of research. However, this should not mean a 'retour à la nature', but it should be a multidisciplinary effort on a highly technological level to perform research towards new opportunities, to develop new crops and products from renewable resources. This will be essential to guarantee a level of comfort for a growing number of people living on our planet. It is 'the' challenge for the coming generations of scientists to develop more sustainable ways to create prosperity and to fight poverty and hunger in the world. A global approach is certainly favoured.

This challenge can only be dealt with if scientists are attracted to this area and are recognized for their efforts in this interdisciplinary field. It is, therefore, also essential that consumers recognize the fate of renewable resources in a number of products.

Furthermore, scientists do need to communicate and discuss the relevance of their work. The use and modification of renewable resources may not follow the path of the genetic engineering concept in view of consumer acceptance in Europe. Related to this aspect, the series will certainly help to increase the visibility of the importance of renewable resources.

Being convinced of the value of the renewables approach for the industrial world, as well as for developing countries, I was myself delighted to collaborate on this series of books focusing on different aspects of renewable resources. I hope that readers become aware of the complexity, the interaction and interconnections, and the challenges of this field and that they will help to communicate on the importance of renewable resources.

I certainly want to thank the people of Wiley's Chichester office, especially David Hughes, Jenny Cossham and Lyn Roberts, in seeing the need for such a series of books on renewable resources, for initiating and supporting it, and for helping to carry the project to the end.

Last but not least, I want to thank my family, especially my wife Hilde and my children Paulien and Pieter-Jan, for their patience and for giving me the time to work on the series when other activities seemed to be more inviting.

Christian V. Stevens,
Faculty of Bioscience Engineering
Ghent University, Belgium

Series Editor, '*Renewable Resources*'
June 2005

Preface

The Recovery and Use of Nutrients from Organic Residues

In the transition from a fossil-based to a circular and biobased economy, it has become an important challenge to maximally close the nutrient cycles and migrate to a more sustainable resource management, both from an economical perspective and from an ecological one. Nutrient resources are rapidly depleting, significant amounts of fossil energy are used for the production of chemical fertilizers, and the costs of energy and fertilizers are increasing. Biorefinery (i.e. the refining of chemicals, materials, energy, and products from biobased [waste] streams) is gaining more and more interest.

The aim of this book is to present the state of the art regarding the recovery and use of mineral nutrients from organic residues, with a focus on use in agriculture. Contributors come from Europe, the United States, Canada, India, and China. The target groups are students (MSc and PhD), scientists, policy advisors, and professionals in the fields of agriculture, waste treatment, environment, and energy.

The introductory section (Section I) gives insights into the global nutrient flows in food systems, with a focus on nitrogen (N) and phosphorus (P) flows and balances.

The transition from a fossil-based to a circular and biobased economy is difficult, due to obstacles in legislative systems, lack of integration of institutional and governance structures, and lack of coordination between the different stakeholders. The legislative, socioeconomic, and technical constraints to the use of recovered nutrients in the European Union, United States, and China are described in Section II.

A series of emerging technologies in the recovery of nutrients from livestock manure, wastes, wastewater, and human excreta are presented in Section III. Cases of nutrient recovery processes in (agro-industrial) practice are illustrated in Section IV, including struvite recovery from domestic wastewater, mineral concentrates from manures, biochar from agro-digestates, and the use of minerals from the ash produced by incineration of chicken manure.

Sections V–VII deal with agronomic, environmental, and economic assessments of the recovery of nutrients and the use of biobased products. Field assessments of agricultural and environmental performance of biobased fertilizer substitutes are described in Section V. An economic assessment of biobased products in comparison to current fossil-based counterparts is given in Section VI. Environmental impact assessments of the production and use of biobased mineral fertilizers are described in Section VII.

Mathematical models have become important tools for the design and optimization of waste-treatment facilities. Moreover, models are used to extrapolate results obtained under controlled conditions to a larger scale. Section VIII deals with the modeling of nutrient recovery from waste(water) flows and dynamic modeling of nutrient cycling in soils.

We gratefully acknowledge all the authors for their contributions to this book.

Erik Meers
University of Ghent, Belgium

Gerard Velthof
Wageningen University & Research, The Netherlands

Evi Michels
University of Ghent, Belgium

René Rietra
Wageningen University & Research, The Netherlands
November 2019

Section I

Global Nutrient Flows and Cycling in Food Systems

1

Global Nutrient Flows and Cycling in Food Systems

Qian Liu[1,3], Jingmeng Wang[1,3], Yong Hou[1,3], Kimo van Dijk[1,2], Wei Qin[1,3], Jan Peter Lesschen[2], Gerard Velthof[2], and Oene Oenema[1,2]

[1] *Soil Quality Department, Wageningen University, Wageningen, The Netherlands*

[2] *Wageningen Environmental Research, Wageningen, The Netherlands*

[3] *College of Resources and Environmental Sciences, China Agricultural University, Beijing, China*

1.1 Introduction

Plants require 14 nutrient elements (N, P, K, Mg, Ca, S, Fe, Mn, Zn, Cu, B, Mo, Cl, and Ni) in specific amounts for growth and development, in addition to carbon dioxide (CO_2), water (H_2O), and photosynthetic active radiation (sunlight). The growth and development of plants are distorted when the supply of one or more of these nutrient elements is sub-optimal [1]. For growth, development, reproduction, and maintenance of body functions, animals and humans require some 22 nutrients (N, P, K, Mg, Ca, S, Fe, Mn, Zn, Cu, Mo, Cl, Co, Na, Se, I, Cr, Ni, V, Sn, As, and F) in specific amounts in addition to water, carbohydrates, amino acids (protein), and vitamins [2–4]. The health, growth, production, and reproduction of animals are distorted when the supply of one or more of these nutrients is suboptimal. Over-optimal supply of nutrients may lead to toxic effects and imbalances in nutrient supply, and thereby also to malfunctioning.

The supply of nutrients in most soils to plant roots is limited, and that is why farmers apply animal manures and fertilizers to cropland. The most limiting elements are nitrogen (N), phosphorus (P), and potassium (K), but other elements may also limit plant growth and

Biorefinery of Inorganics: Recovering Mineral Nutrients from Biomass and Organic Waste, First Edition.
Edited by Erik Meers, Gerard Velthof, Evi Michels and René Rietra.
© 2020 John Wiley & Sons Ltd. Published 2020 by John Wiley & Sons Ltd.

development. Excessive or inappropriate use of nutrients in crop and animal production leads to large nutrient losses to the wider environment and creates a range of unwanted environmental effects and possible negative human health effects [5]. Excess P in surface waters is associated with water pollution, eutrophication, and biodiversity loss. Losses of N create a cascade of threats to water, air, and soils, affecting biodiversity, climate, and potentially human health [6]. The current losses of N and P to the atmosphere and surface waters exceed the so-called safe planetary boundaries [7]. Transgressing a boundary increases the risk that human activities could inadvertently drive the Earth System into a much less hospitable state, damaging efforts to reduce poverty and leading to a deterioration of human wellbeing in many parts of the world, including wealthy countries. This indicates that losses of N and P to the environment have to be decreased by proper management. Nutrient management is generally defined by a coherent set of activities with the objective of achieving both agronomic targets (food production and quality, as well as income for farmers) and environmental ones (minimal losses of nutrients to the wider environment).

Nutrient management has to consider also that the geological reserves (e.g. phosphorus, potassium, magnesium, copper, zinc, selenium-rich rocks) used to manufacture fertilizers and nutrient supplements are finite and may become depleted within a few generations [8, 9]. This indicates that actions are needed to reduce nutrient losses, to realign current nutrient use, to recover and recycle nutrients from wastes, and possibly to redefine food systems [10]. This is also the background to the chapters presented in this book: which techniques, technologies, and managements are available and can be used to improve the utilization of nutrients from animal manures, residues, and wastes?

This introductory chapter briefly summarizes the driving forces of nutrient cycling and the changes in global nutrient flows and balances in agricultural systems and food systems during the last couple of decades. The emphasis is on N and P in food production–consumption systems; these nutrients most limit global food production among essential nutrient elements, while losses of N and P to the wider environment have significant negative human health and ecological effects [5, 11].

1.2 Primary and Secondary Driving Forces of Nutrient Cycling

In terrestrial systems, nutrients cycle between soils, plants, and animals and then back to soils again. Some nutrients also cycle through the atmosphere (e.g. N and S), while fractions of basically all nutrients are transported to groundwater bodies or the sea and then to sediments.

Four primary *energy* sources are distinguished in natural systems (Figure 1.1). These fuel a number of secondary driving forces, which subsequently fuel nutrient transformation and transport processes. Sunlight fuels photosynthesis, the hydrological cycle (evapotranspiration), and wind and water currents (in combination with gravitational energy and internal particle energy). Natural gravity and the internal energy of particles govern the earth motion (seasonal and diurnal cycles), the physical interaction between elementary particles (including diffusion), and the physical transport of particles, following the laws of thermodynamics. The heat (energy) in the core of the earth governs tectonic uplift and volcanic activity [12].

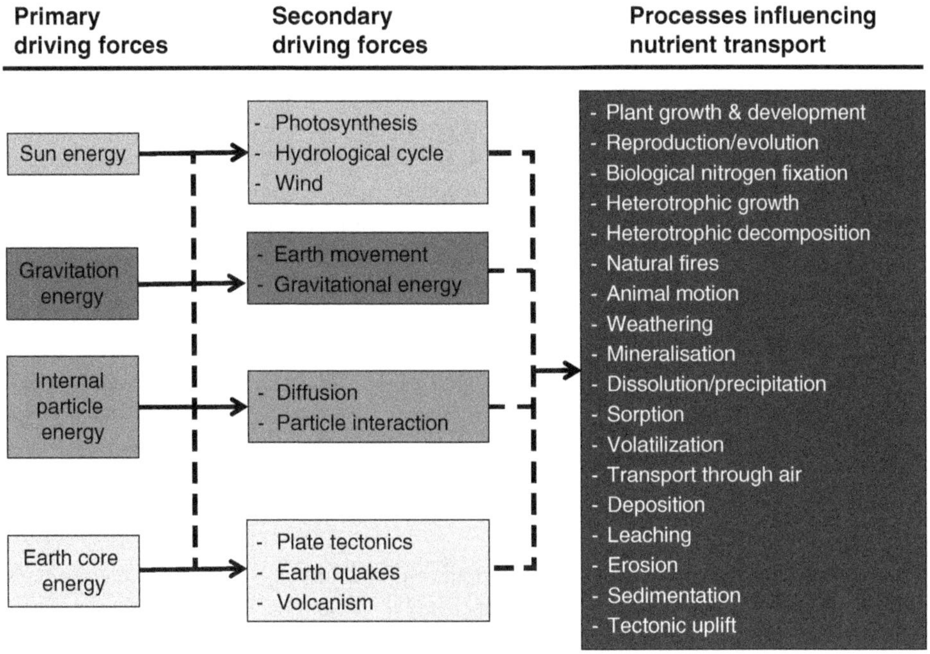

Figure 1.1 *Driving forces of nutrient transformations and transport in natural systems. There are four primary energy sources (first column), which fuel the secondary driving forces and subsequently the nutrient transformation and transport processes in nature.*

Wind, rain, and evapotranspiration are considered secondary driving forces for nutrient transformation and transport, because they are fueled by the energy from the sun, gravitation, and the internal particle energy. This holds as well for the orbital motion of the earth and the other so-called secondary driving forces. These secondary driving forces, alone or in combination, drive a large number of nutrient transformation and transport processes. This includes heterotrophy: that is, utilization by bacteria, animals, and humans of the energy derived from the sun and found in the carbohydrates, proteins, and nutrients in plants for growth, development, and (re)production [13]. Biological N_2 fixation is indirectly also fueled by the sun; microorganisms utilize carbohydrates from plants and other source to cleave the triple bond in atmospheric N_2 and to produce ammonia and amines [14].

The transport and cycling of nutrients in ecosystems depend also on the reactivity and mobility of the nutrient elements (related to the internal particle energy). Some nutrient elements (e.g. N, S, Fe, Mn) are involved in reduction–oxidation processes, and thereby to changes in valence, reactivity, and acidity. Nitrogen and sulfur (S) are termed "double mobile" because they are mobile both in gaseous phase (e.g. NH_3, N_2, N_2O, NO, NO_2, H_2S, and SO_2) and in aqueous phase (e.g. NO_3, NH_4, and SO_4), while some N and S species occur in non-mobile solid phase (organically bound N, elemental sulfur [S°]). Other nutrients (e.g. Ca, Mg, K, Na, Cl, and B) are soluble in aqueous phase, while most metals (e.g. Cu, Zu, Cr, and Ni) and oxyanions of P, Mo, and Se are easily adsorbed to soil particles and therefore have low solubility [15].

1.3 Anthropogenic Influences on Nutrient Cycling

Since the invention of agriculture some 10 000 years ago, humans have transformed natural systems, first in agricultural systems and later partly also in urban, infrastructural, and industrial areas. Agriculture encompasses all activities related to crop and animal production. An agricultural or farming system is commonly defined as a population of individual farms that have broadly similar resource bases, enterprise patterns, household livelihoods, and constraints, and for which similar development strategies and interventions would be appropriate. Depending on the scale of the considerations, a farming system can encompass a few dozen or millions of households. A main distinction is commonly made between (i) specialized crop production systems, (ii) specialized animal production systems, and (iii) mixed production systems.

A food system includes all activities involving the production, processing, transport, and consumption of food and the recycling of wastes. It also includes the inputs needed and outputs generated at each of these steps. Food systems operate within and are influenced by social, political, economic, and environmental contexts. They have developed over time, under the influence of changes in demography and culture, science and technology, markets and governmental policy.

The change from natural systems to agricultural and food systems has greatly altered nutrient cycles. The main driving forces for this change are the increasing and agglomerating human population and the increasing prosperity (of at least part of the population), which lead to an increasing food demand and to changes in food choices and nutrient cycling (Figure 1.2). The change has been greatly facilitated by scientific and technological

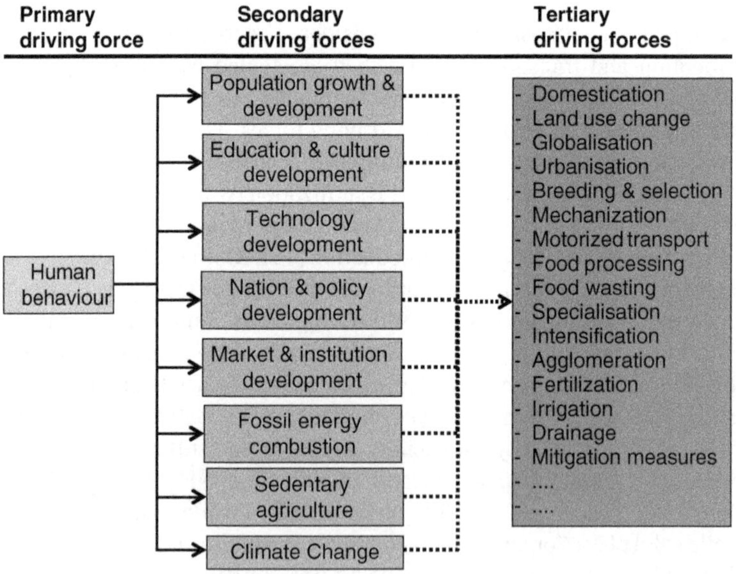

Figure 1.2 Anthropogenic driving forces for changes in nutrient transformations and transport in agriculture and food systems. A distinction is made between primary, secondary, and tertiary driving forces, which all influence the nutrient transformation and transport processes indicated in Figure 1.1.

developments, such as the invention of inorganic fertilizers (the Haber–Bosch process [14, 16]), high-yield crop and animal breeds, improved pest and disease management, mechanization, irrigation, processing, storage, and transport. These developments have contributed much to increased crop and animal productivity and to improvements in the processing, storage, and marketing of food, feed, fiber, and fuel products, especially during the last century.

Liberalization and internationalization of markets combined with relatively cheap transport facilities have opened up and enlarged markets, increased competition, and decreased prices. As a result, nutrients embedded in food and feed products are transported across the world in all sorts of transport vessels [16–18]. Governmental policies have supported agricultural production in many countries, through facilitation of the build-up of a good knowledge and physical infrastructure, through market and product support, and through subsidies on inputs (including fertilizers). The alterations in global nutrient cycles through land use change, agriculture, food processing, and transport have increased especially from the twentieth century onward [14, 19]. Finally, climate change is also increasingly seen as a driver of changes in the global cycling of nutrients, as changes in temperature regimes and rainfall patterns affect biological processes and the transport of nutrients.

1.4 The Global Nitrogen Cycle

The N cycle is a most complex nutrient cycle. Nitrogen exists in different forms, most of which are biologically, photochemically, or radiatively active; these forms are therefore termed reactive N or Nr. Forms of Nr can be transformed into other forms, depending on environmental conditions. These transformations are often mediated by microorganisms [14]. The most reduced Nr forms have a valence state of −3 (as in NH_3 and amines) and are energy-rich, while the most oxidized Nr forms (NO_3^-) have a valence state of +5. Nitrifying bacteria (e.g., Nitrosomonas, Nitrobacter) can utilize the energy in reduced N forms, and thereby increase the valence state up to the level of that in NO_3^- through a process called nitrification. Conversely, heterotrophic bacteria can utilize the oxidative power (electron-accepting ability) of oxidized Nr forms and thereby reduce nitrate (NO_3^-) and nitrite (NO_2^-) to nitrous oxide (N_2O) and di-nitrogen (N_2) through a process called denitrification. Molecular N (N_2) may also be formed through anaerobic ammonium oxidation (anammox; $NH_4^+ + NO_2^- \rightarrow N_2 + 2H_2O$), by chemoautotrophic bacteria under specific conditions. Most of the N on earth exists in the form of molecular nitrogen or di-nitrogen (N_2) in the atmosphere. This N_2 is unavailable to most plants and organisms because of the strength of the triple bond (N≡N) between the two N atoms. Only a few microorganisms, either free-living or symbiotic in leguminous plant root systems, have the capability to utilize (fix) N_2, converting it to organically bound N. The fertilizer industry uses the Haber–Bosch process to convert N_2 into ammonia/ammonium (NH_3/NH_4^+) with the use of methane (H_2 source), external heat, and pressure [14]. The NH_3/NH_4^+ thus formed and the NO_3^- from nitrification can be taken up by plants (assimilation). Following the senescence of plants and organisms, the organic N is transformed again into NH_3/NH_4^+ (through mineralization), which may then be nitrified to NO_3^- and subsequently denitrified to N_2.

Figure 1.3 presents a summary of the global N cycle. The largest pool of N is N_2 in the atmosphere. Sediments and rocks are also large pools of N, but this is locked up. Most of

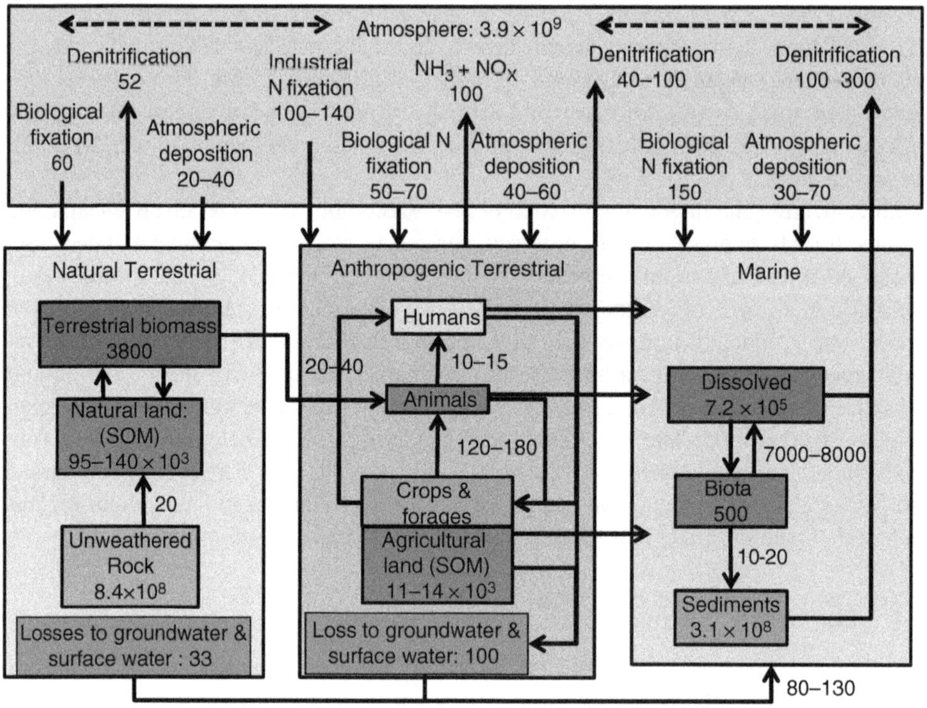

Figure 1.3 The global nitrogen cycle, with four main compartments: the atmosphere, the natural terrestrial biosphere, the anthropogenic terrestrial system, and the marine system. Boxes represent pools, arrows represent annual fluxes. Numbers in pools are N stocks in Mt, numbers near arrows are N fluxes in Mt per year. Industrial N fixation includes 100 Mt chemical N fertilizer and 36 Mt for other industrial usage (1 Mt = 1 million tons = 10^{12} g = 1 Tg). The direct exchange between the natural terrestrial and anthropogenic terrestrial systems is considered negligible. Source: After Smil [20], Galloway et al. [16, 21], Schlesinger and Bernhardt [13], and Fowler et al. [22].

the N in soils exists in organic forms, and the soil organic matter content is a good predictor of the N content of soils. Cropland soil contains some 11–14 × 10^3 Mt organic N [23]. The pool of inorganic N (NH_4^+ and NO_3^-) in soils is very small, because these forms are easily taken up by plants and microorganisms. The kinetics of the N cycling between pools differs greatly. Large amounts of N enter the anthropogenic systems via industrial and biological N_2 fixation and via atmospheric Nr deposition (Figure 1.3).

The global N cycle is strongly influenced by anthropogenic activities. The changes in human diets toward more animal-derived protein have increased the total amount of N required in Europe (to deliver the food of one person) to more than 100 kg per person per year [24, 25]. More than half of the food eaten by humans is now produced using N fertilizers via the Haber–Bosch process [26]. Industrial N_2 fixation is now as large as or larger than biological N_2 fixation in the terrestrial system. In addition, large-scale deforestation and soil cultivation have increasingly mobilized N from the organic N pools, which have subsequently contributed to the increased Nr flux from the terrestrial system to the aquatic/marine system and to the atmosphere [16].

Commonly, not more than 40–50% of the applied fertilizer N is recovered in harvested crop [27, 28]. The amount of N in harvested crop (from cropland) was estimated at 74 Tg in 2010 [28]. About 30–50% of this was utilized by humans as plant food. The other 50–70% was utilized by domestic animals (mainly cattle, pigs, and poultry). In addition, domestic animals used some 90–125 Tg N in herbage and forages. Humans utilized about 10–15 Tg N from animal production in the form of milk, meat, and eggs in 2010. Most of the N in food and feed is not retained by humans and animals, but is excreted via urine and feces. These excreta are a large source of Nr, which can be used as fertilizer. However, a large fraction is lost to air and waterbodies, because of poor management [5, 29].

The total loss of N from the anthropogenic system to the aquatic/marine system was about 100 Tg in 2010 (Figure 1.3). In addition, some 100 Tg was lost as NH_3 and NO_X to the atmosphere, and another 40–100 Tg was denitrified into N_2. All these estimates are rather uncertain, which is in part related to the large spatial and temporal variations of all N losses. Losses of NH_3 and NO_X to the atmosphere and of NO_3^-, NH_4^+, and dissolved and particulate organic N have greatly increased during the last couple of decades. Main sources are agriculture, households (sewage), fossil fuel combustion (mainly NO_X), and industry. These losses have created a range of unwanted effects, including human health effects, eutrophication and pollution of surface waters, biodiversity loss, and climate change [6, 16]. The impacts are evident at regional and global scales, especially in countries with intensive crop- and animal-production systems and in urban and industrial areas with little or no environmental regulation. There is a strong need to increase the N use efficiency in food systems (crop and animal production, food processing, consumption, fossil energy combustion, and waste recycling [5]).

1.5 The Global Phosphorus Cycle

Phosphorus is a reactive, non-gaseous, and therefore rather immobile element; the P cycle may serve therefore as an example for the cycling of other "immobile" elements, such as metals, Ca, and Mg. The P cycle is relatively simple (Figure 1.4) compared to the cycle of the "double mobile" nutrient elements N and S.

Most P is locked up in rocks and sediments. Relatively large amounts are also stored in soils and ocean waters. Much smaller amounts are found in natural biomass, crops, animals, and humans. Annually, between 6 and 11 Mt P enters households in crop- and animal-derived food, detergents, and other products. Following their use, only a tiny amount is retained in human biomass and households; some is returned to cropland as compost or sewage, but most is landfilled or discharged. Many of the pools and fluxes shown in Figure 1.4 have relatively large uncertainty [30]. This holds especially for the quantifications of regional distributions of pools and fluxes, for which data availability is limited.

Phosphorus in soils, sediments, and rocks is associated with oxides, carbonates, (heavy) metals, and uranium (actinide elements), mainly because of the reactivity of phosphate anions (PO_4^{3-}). Cadmium (Cd) and uranium (U) are natural components of phosphate rock, and the use of mineral P fertilizer can thus lead to soil pollution by cadmium and uranium. Cadmium in soil can be taken up by plants and can enter the food chain, where it may contribute to kidney damage. Other toxicological effects are skeletal damage and carcinogenicity. The European Food Safety Authority (EFSA) has set the tolerable weekly intake

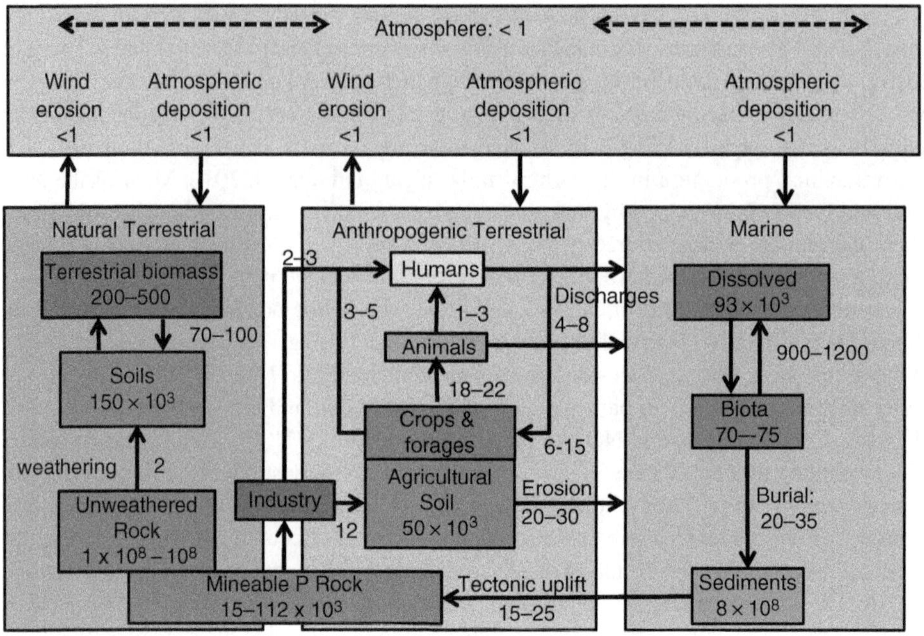

Figure 1.4 *The global phosphorus (P) cycle. Boxes represent pools, arrows represent annual fluxes. Numbers in pools are P stocks in Mt, numbers near arrows are P fluxes in Mt per year (1 Mt = 1 million tons = 10^{12} g = 1 Tg). Source: After Smil [30] and Schlesinger and Bernhardt [13].*

of Cd at 2.5 μg per kg of body weight. Phosphate rock from Morocco and Western Sahara, the largest rock P reserves, has particularly high Cd levels. The occurrence of low-cadmium phosphate rock (1–20 mg Cd per kg of P_2O_5) is small, however, indicating that Cd may have to be removed during the processing of P fertilizer from phosphate rock [31].

The soil-solution pool and the surface layers of oceans waters have relatively low P concentrations. These pools are depleted by P uptake through plant roots and phytoplankton and by adsorption and precipitation processes. They are replenished several times a year by mineralization, desorption, and weathering processes [15]. The cycling of P in terrestrial biomass, including crop residues, manures, and wastes in soil, has turnover times of 1–10 years. The mining of rock P for the manufacture of P fertilizers, and human-induced soil erosion and discharge to the oceans (where the P eventually sinks into sediments), occur at civilization times scales (i.e. 10^3 years) [30]. On the time scale of hundreds of millions of years, these sediments are uplifted and subject to rock weathering, completing the global cycle [13].

The availability of P in most soils limits crop production around the world. That is why farmers apply animal manures, composts, and P fertilizers. Phosphorus fertilizers are derived from minable rock P, which occurs in a limited number of locations [8, 9]. The 2007/2008 price peak of P fertilizers provoked an intensive debate about the rate of depletion of minable P rock reserves [8, 9, 32, 33]. The concept of "peak phosphorus" was introduced, by analogy with peak oil, to describe the point in time when the global use of P rock is maximal and will start to decline because of increasing cost to mine the diminishing reserves. Estimates of P rock reserves show a wide uncertainty range, from 50 to 400 years. Wellmer and Scholz [34] recently argued that a peak cannot be predicted with the present base of knowledge. Though there is debate about the magnitude of the P rock reserves, scientists and policy makers agree that they are finite, and that P is essential for life and cannot be substituted by other elements.

The availability of P in surface waters often also limits phytoplankton growth [35–37]. Pollution of surface waters with P from agriculture and urban activity (including industry) has greatly increased phytoplankton productivity in receiving waters, and has led to a series of ecological impacts, including algal blooms, hypoxia, biodiversity loss, and fish kills [38, 39]. The leaching and runoff of P from agricultural land, farmyards, and households is called non-point or diffuse pollution, because it involves widely dispersed activities. These pollution sources are difficult to measure and regulate due to their dispersed origins and because they vary with the seasons and the weather. The discharge of P from households and wastewater treatment plants has decreased in many countries, following the introduction of regulations on P in laundry and dishwashing detergents and the implementation of improved wastewater treatment. Bans on waste discharge from P fertilizer industries and other industries have greatly reduced "point-source pollution." The leaching and runoff of P from agricultural land and farmyards requires a whole range of activities, including improved P fertilization recommendations, improved manure management, erosion control, and "end-of-pipe" solutions [40].

Both the possible depletion of easily accessible P rock reserves and the eutrophication of surface waters caused by P losses from agriculture and households necessitate that the efficiency of P use in food systems be greatly increased. This is especially important for the European Union (EU) with its large population, intensive agriculture, and essentially zero P rock reserves. In 2005, the EU-27 imported 2.4 Tg P via P fertilizers, food, and feed, but only a quarter (0.8 Tg) reached households. Almost 40% of the total P input accumulated in agricultural soils (0.9 Tg), while about half was lost as waste (1.2 Tg). Hence, a total of 4 kg of P was needed in order to obtain 1 kg of P in food and other consumer products in EU-27 countries (Figure 1.5; [41]). Through increased recycling of P from residues and wastes and via P loss mitigation measures, this ratio can be decreased to 2 : 1; that is, a P use efficiency in the whole food production–consumption chain of 50%.

Europe is strongly dependent on net P imports via mineral P fertilizers, food, feed, and detergent components from foreign countries [10]. Withers et al. [10] proposed a so-called 5R stewardship framework (Realign P inputs, Reduce P losses, Recycle P in bioresources, Recover P in wastes, and Redefine P in food systems) to make Europe less dependent on P import and to reduce P losses. There is a need for a better quantitative understanding of the P flows and cycling in the food production/consumption/waste management chain across Europe, and for actions and business models aimed at achieving a P-efficient Europe. The need for more efficient P utilization and a reduction of P losses holds for many countries, especially rapidly developing ones like China [42, 43].

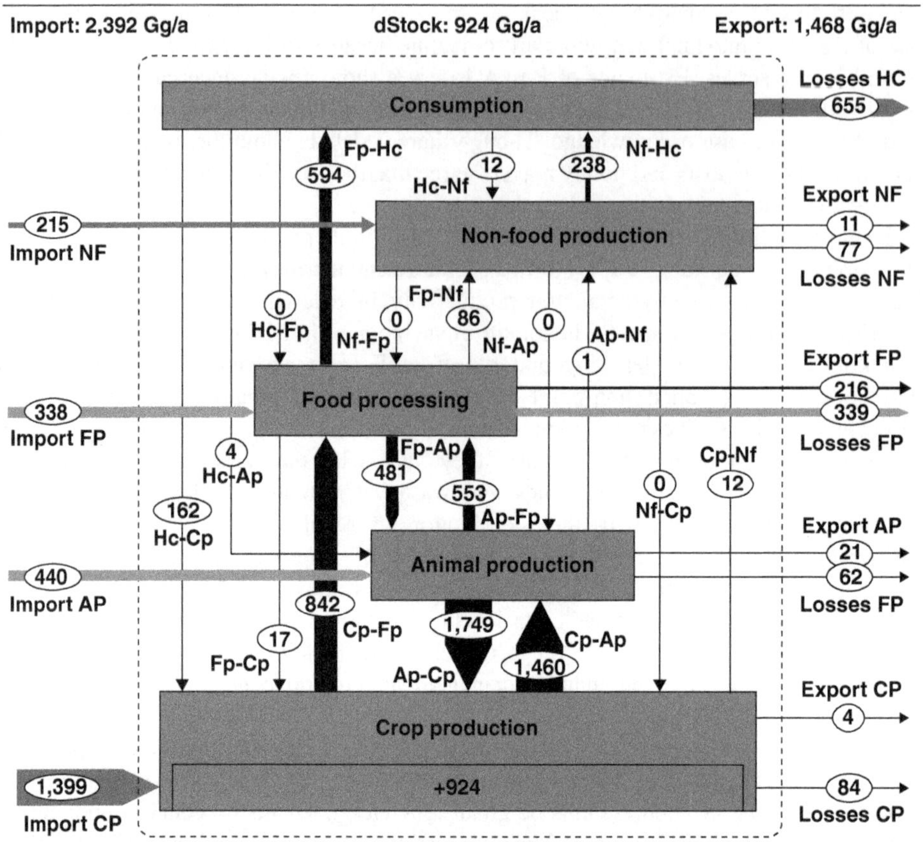

Figure 1.5 Phosphorus (P) use in the EU-27 in 2005 (Gg P/year), aggregated at the food and non-food production–consumption–waste chain. Boxes in the middle indicate the main sectors: crop production (CP), animal production (AP), food processing (FP), non-food production (NF), and consumption (HC). Arrows indicate the size of the P flows; arrow thickness indicates the relative flow sizes. Imports are shown on the left-hand side, P export and P losses are shown on the right-hand side. Internal upward/downward flows are shown in black [41].

1.6 Changes in Fertilizer Use During the Last 50 Years

The use of fertilizers greatly increased in the second half of the twentieth century, when fertilizer manufacturing technologies became more mature, transport facilities became cheaper, and the knowledge among farmers regarding how and how much fertilizers to use greatly increased. Figure 1.6 shows the changes over time in fertilizer N, P, and K use in the world and per continent.

There are large differences in the changes over time between continents. Fertilizer use started relatively early in Europe and North America, but slowed from the 1980s and, in Europe, decreased due to saturation of markets and the implementation of environmental policies. Asia became the largest user from the 1980s; it has the largest area of cropland (580 million ha in 2012) and the largest population. Fertilizer application rates are often very high in Asia, because of the rapidly increasing food demand among the increasing

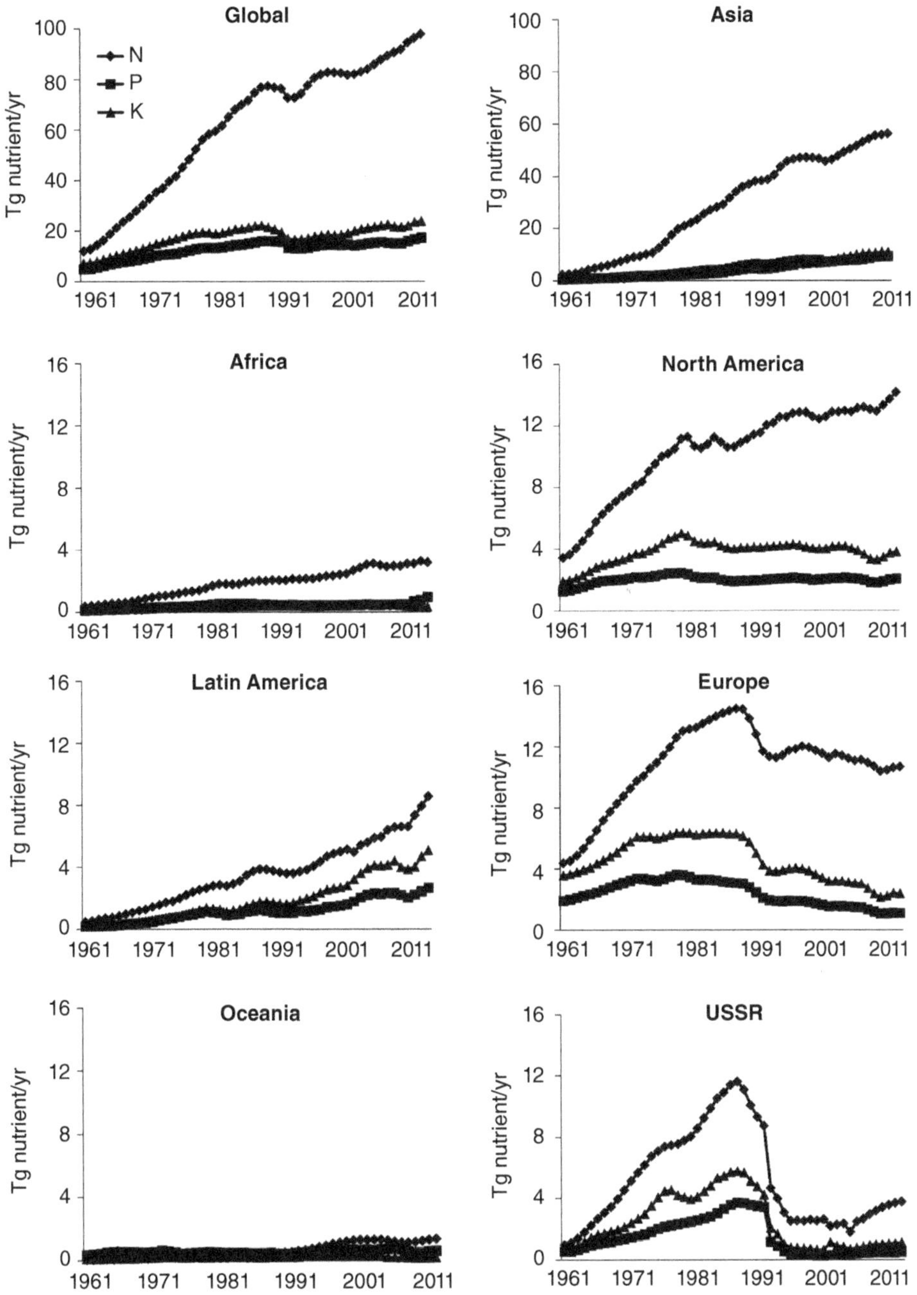

Figure 1.6 *Consumption of synthetic nitrogen (N), phosphorus (P), and potassium (K) fertilizers globally (upper left panel) and per continent or region between 1961 and 2012. Note the differences on the y-axis between the upper two panels and the lower six panels. Europe does not include the European part of the Russian Federation. The Russian Federation is included separately as the USSR, the former Union of Soviet Socialist Republics. Source: FAOSTAT [44].*

human population and the subsidies on fertilizers. Fertilizer use is still very low in Africa relative to its large population and large agricultural area (220 million ha), mainly because of market and infrastructure constraints.

Fertilizer use in Europe decreased drastically from the early 1990s due to (i) political changes in Central European countries and the removal of subsidies on synthetic fertilizers and (ii) the implementation of environmental policies in the EU countries. The effect of political changes on fertilizer use in the 1990s was even stronger in the former countries of the Union of Soviet Socialist Republics (USSR). The implementation of the Nitrates Directive in the EU in 1991 increased the utilization of nutrients from animal manures and thereby decreased the need for synthetic fertilizer input [45]. Forecasts indicate that fertilizer NPK use may decrease further and then stabilize during the next few decades in the EU countries.

1.7 Changes in Harvested Crop Products and in Crop Residues During the Last 50 Years

Following the increased use of synthetic fertilizers (Figure 1.6), the introduction of high-yield crop varieties, and the improved pest and disease management and crop husbandry practices from the early 1960s, crop yields and the amounts of N and P in harvested crops and crop residues greatly increased (Figure 1.7). Increases in yield were especially large in Asia. Total production of most cereals increased via an increase in the yield per unit surface area, but total production of oil crops like soybean, oil palm, and

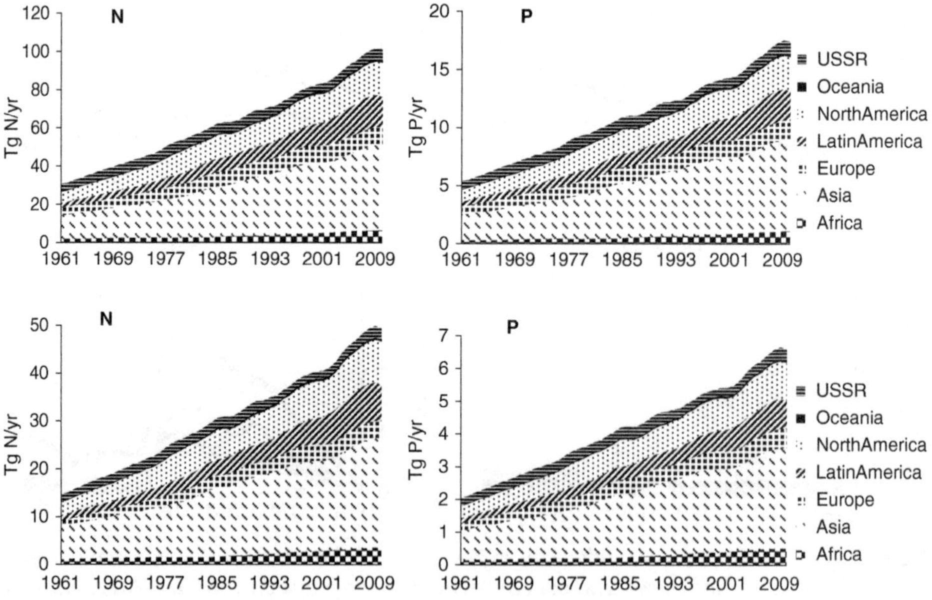

Figure 1.7 *Changes in the total amounts of nitrogen (N) and phosphorus (P) in harvested crop products (upper panels) and in aboveground crop residues (lower panels) per continent between 1961 and 2010. Source: FAOSTAT [44].*

Table 1.1 *Amounts of co-products, by-products, and wastes from agriculture in EU-28.*

Types		Dry matter (Tg)	Nitrogen (Gg)	Phosphorus (Gg)
Straw	Wheat	95	600	50
	Barley	48	300	20
	Maize	39	400	20
	Other straws	56	600	40
Prunings	Vineyard	2	20	< 10
	Olives	2	20	< 10
	Fruit trees	2	20	< 10
Manure	Liquid	35	3100	530
	Solid manure	69	3100	600
Slaughterhouse residues		20	500	330
Food processing residues		9	300	40
Total		377	8960	1660

Source: Based on data from Eurostat, FAOSTAT, and the authors' calculations.

rapeseed increased through a combination of increases in both cultivated area and yield per unit of cultivated area [46].

The amounts of N and P in harvested crop globally in 2012 were roughly similar to the total amounts of fertilizer applied in that same year (~100 Tg N and nearly 20 Tg P). This may suggest that all fertilizer N and P is recovered in harvested crop, but this is not the case. On average, not more than 40–50% of applied fertilizer N is recovered in harvested products, while apparent recovery percentages for applied P fertilizers are commonly not more than 20%. A large fraction of the N and P comes from sources other than fertilizers, such as the soil, biological N_2 fixation, animal manure, and atmospheric deposition. Currently, about half of the food produced in the world is derived from fertilizer N, while the other half comes from other sources [47].

Approximately half of the N and P in harvested crop ends up in crop residues (Figure 1.7). Some of these residues are harvested and used as animal feed or biofuel. Most, however, are returned to the soil and thereby contribute to the maintenance of the soil organic carbon pool and to soil fertility. Stubble and roots (not shown in the figure) also contribute to the build-up of the soil organic matter pool, and to recycling of N and P.

The amount of N in harvested crops in EU-28 countries is about 10 Tg per year (Table 1.1). In addition, there is some 3–4 Tg N in crop residues. Approximately 2 Tg N is found in the straw of cereals (Table 1.1). The uncertainties in these estimates are relatively large, especially because yields of grassland, forages, and crop residues on farmers' fields are not recorded on a regular basis for statistical purposes.

1.8 Changes in the Amounts of N and P in Animal Products and Manures

Approximately two-thirds of the agricultural area in the world is used to feed livestock [24, 25]. This includes about 3 billion ha of grassland and about 1 billion ha of cropland (mainly maize, soybean, and wheat). In addition, livestock consume residues from the food industries, feed on kitchen residues, utilize crop residues, and scavenge farmyards, roadsides, and "unmanaged" lands. The total amounts of ingested biomass and of N and P are not accurately known, however.

Livestock retain only a small percentage of the ingested carbon and nutrients in live-weight gain, milk, and eggs. The remainder is excreted in feces and urine. For N and P, the percentage retained ranges from about 10% in ruminants used for beef and mutton production to 20–30% for dairy production and to 30–45% for pork and poultry production. The total amount of N in milk, meat, and eggs produced globally increased from about 7 Tg in 1961 to slightly more than 20 Tg in 2011. The total amount of P in milk, meat, and egg increased from about 1 Tg in 1961 to about 3 in 2011 [29]. The total amounts of N and P in manure increased from about 60 and 12 Tg, respectively, in 1961 to about 120 and 24 Tg, respectively, in 2011. The larger increase in the amounts of N and P in milk, meat, and eggs than in animal manure reflects the increase in animal productivity and the stronger increase in the number of monogastric animals than in less productive ruminant animals [29]. The total amounts of N and P in manure are larger than the total use of fertilizer N and P in the world. However, the increase in fertilizer N and P use has been larger than the increase in the production of animal-manure N and P.

1.9 Changes in the Trade of Food and Feed

In the second half of the twentieth century, agricultural systems became more specialized and agglomerations of specialized systems developed. At the same time, more and more people started to live in urban areas. As a result, food and feed products needed to be transported over longer distances. The specialization and agglomeration of food production systems was further facilitated by transnational corporations (TNCs), which increasingly influence the food production–consumption chain [48], as well as contributing to the diversification of food products in supermarkets, increasing the length of the food chain, and adding value to food products. TNCs are increasingly powerful and rich; for example, the value of the 10 largest retailers in the world in 2005 was roughly similar to that of the total production of all farmers [49].

The urbanization, specialization, and separation of crop and livestock production systems have contributed to a rapid increase in cross-border transport of crop and animal products. Expressed in amounts of N, the total sum of imported crop products was about 20 Tg in 2010 [18], which is equivalent to about 20% of the total amount of N in the harvested crop. Similarly, the amount of P in imported animal products was 3.5 Tg in 2010, which is again about 20% of the total amount of P in livestock products produced. Europe and Asia were relatively large importers of both crop products and animal products in 2010. The biggest exporters were North America, Latin America, and Europe. The EU-28 was a net importer of crop products, while its imports and exports of animal products were roughly balanced [46].

1.10 Changes in Nutrient Balances

Surpluses of input–output balances of food production–consumption systems strongly increased during the second half of the twentieth century; N and P inputs increased more than N and P outputs. Surpluses are indicators for potential losses (especially for N) or for accumulation in the system (especially for P).

Overviews of the N and P balances of agriculture in the 35 Organisation for Economic Co-operation and Development (OECD) members show large differences between countries [50]. In 2009, mean N surpluses at country level ranged from about 0 to more than 200 kg ha^{-1}, and mean P surpluses from −15 to 50 kg ha^{-1} [50]. Most OECD countries had increases in N and P surpluses between 1950 and 1990, and a decrease during the last 20 years. The increased awareness of the ecological and human health impacts of N and P losses and the introduction of good agricultural practices and regulations have increased N and P use efficiency in the agriculture of many countries. In the United States, P surpluses started to decrease from the 1970s–1980s following an increased awareness of the build-up of soil P levels and of the effects of P losses on the eutrophication of surface waters. However, N surpluses have slowly but steadily increased during the last four decades. In EU countries, N and P surpluses increased during the period 1950–1980/1990. Nitrogen surpluses started to decrease from the 1980s and P surpluses from the 1970/1980s, following an increased awareness of soil P test values and of the fertilizer value of animal manures, as well as the implementation of agri-environmental policies. Most of the surplus P has accumulated in agricultural soils, and the challenge is to utilize this so-called legacy soil P more effectively and to efficiently recycle P from wastes and residues [10, 51].

Bouwman et al. [52, 53] explored the N and P balances of crop and animal production systems at continental levels between 1900 and 2050. In the early twentieth century, N and P surpluses were relatively small. Between 1900 and 1950, the global N surplus almost doubled (to 36 Tg per year) and the P surplus increased by a factor of 8 (to 2 Tg per year). Between 1950 and 2000, the N and P surpluses increased further to 138 Tg per year for N and to 11 Tg for P. The relatively low N and P surpluses in 1900 also follow from the relatively large recovery of N and P in crop products at that time (Figure 1.8). Nutrient recovery in crop production is defined here as the amounts of N and P in harvested crop products, as per cent of total N and P input, respectively. Nutrient recovery in crop products decreased between 1900 and 1970/2000 and increased again (or is expected to increase) between 1970/2000 and 2050, depending on the region (Figure 1.8). Nutrient recovery in animal production is defined here as the amounts of N and P in milk, meat, and eggs, as a per cent of the total feed N and P intake by the animals, respectively. Recovery of N and P in animal products tended (or is expected) to increase steadily between 1900 and 2050, due to (i) an increase of animal productivity (a result of breeding and improved feeding), (ii) a relative increase in the proportion of monogastric animals relative to ruminant animals, and (iii) improved herd, health, and feed management [29].

1.11 General Discussion

This introductory chapter has briefly presented the main changes in the global nutrient flows in food production–consumption systems during the last 50 years, resulting from anthropogenic influences (Figure 1.2). The changes in nutrient flows and cycling were very large during this period. These changes have had huge impacts on the terrestrial and aquatic biosphere. The focus in this chapter is on N and P, as these nutrients are limiting crop production across the world, and because excess N and P in the environment creates a range of ecological effects, which severely limit the sustainability of nutrient management practices that cause high N and P losses [5, 7, 54].

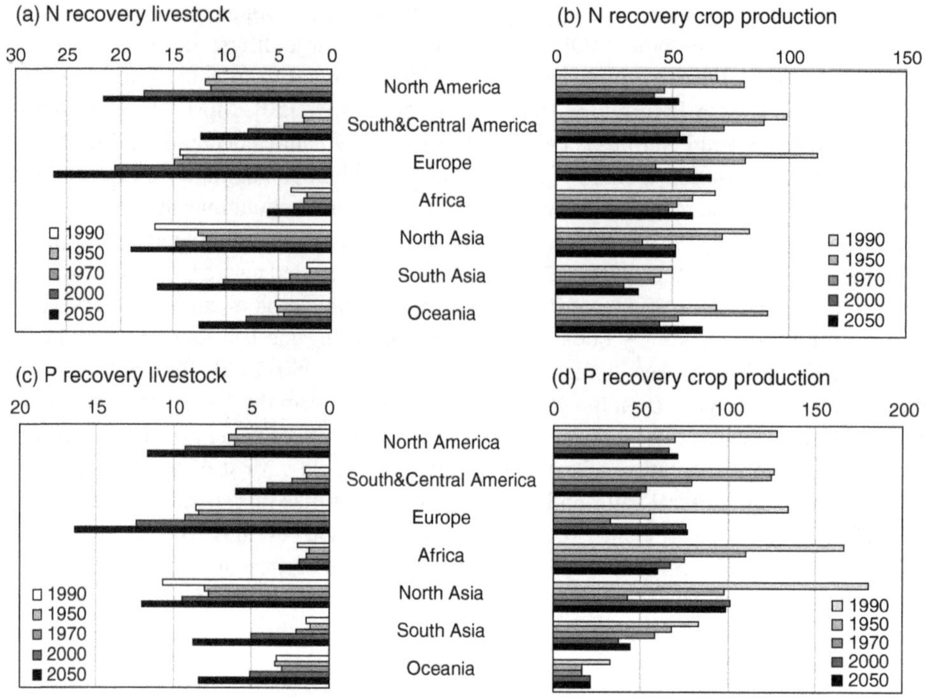

Figure 1.8 Global recovery of N (a and b) and P (c and d) from livestock and crop production per continent for the years 1900, 1950, 1970, 2000, and 2050. Source: Based on Bouwman et al. [53].

There are significant differences between continents in the changes over time in the production and consumption of plant- and animal-derived food, and in the use of fertilizers, residues, manures, and wastes as nutrient sources. These differences are related to socioeconomic conditions (e.g. GDP, demographics, education, land area, and tenure), environmental conditions (e.g. climate, geomorphology, soil), and cultural factors (e.g. religion, customs, values). Yet, the patterns discussed also indicate commonalities and phases in development. During the first stage of development (phase 1), there are significant increases in the use of fertilizers, increased production and consumption of animal food, and increases in nutrient surpluses and nutrient losses to the environment. In the second stage (phase 2), there is first a stabilization and subsequently a decrease in the use of fertilizers, a stabilization/decrease in the consumption of animal-source food, and a decrease in nutrient losses. The change from phase 1 to phase 2 mainly depends on socioeconomic conditions, changes in the animal-production sector, and governmental regulations [28].

The progress of phase 2 depends on the utilization of nutrients from crop residues, animal manures, and wastes. Global crop residues and animal manures contained some 50 and 120 Tg N and 7 and 24 Tg P, respectively, in 2010. These large nutrient resources are often not utilized in an effective manner, because of a poor match between N and P supply among the residues and manures and the N and P demand by the growing crop, both in absolute amounts and in terms of timing. As discussed later in this book, processing and recovery of the residues, manures, and household wastes (sewage) is then a possible gateway to enhancing the utilization of the nutrients from these sources. The biofertilizers will replace synthetic fertilizers if they come to be accepted by farmers. This will depend on their price and form, the presence of unwanted substances, their spreading characteristics, and the degree to which a nutrient ratio's in the biofertilizers can be tuned to the site-specific demands of a crop.

Wastes from households are also a significant source of nutrients, which are currently not utilized effectively. The average total consumption of plant- and animal-derived protein in the world is about 28 kg per capita per year, which is equivalent to about 4.5 kg N per capita per year [25]. Similarly, the amount of P in the food used by households is about 0.8 kg per capita per year [41]. This would suggest that the global amounts of N and P in human wastes were about 32 Tg N and 6 Tg P in 2010.

Table 1.1 presents an overview of the estimated amounts of N and P in various crop residues, animal manures, and wastes from slaughterhouse residues and food-processing residues in the EU-28. This table does not yet include wastes from households, which we estimate at 3 Tg N and 0.5 Tg P. Altogether, there are approximately 12 Tg N and 2.1 Tg P in crop residues, animal manures, and wastes from the food-processing industries and from households in EU-28 countries available for recycling and re-use. These amounts are roughly similar to the fertilizer N and P use in Europe (Figure 1.6).

The nutrients from crop residues, animal manures, and wastes are often not utilized effectively in agriculture because of various practical, cultural, and legal barriers. Yet, there is an increasing need to do so [10]. The main reasons/barriers to their use include the following: (i) the availability of the nutrients is often unknown and unpredictable, while the ratio of the nutrient elements in the supply often does not match that of the nutrient elements demanded by the crop; (ii) residues, manures, and wastes are often voluminous (bulky) and thus expensive to store, transport, and apply to land, while the production on specialized farms is often too large to be utilized effectively on the farm (or on nearby farms); and (iii) the possible presence of pollutants, pathogens, odor nuisances, and so on makes the transport, handling, and use elsewhere of residues, manures, and wastes less attractive or even prohibitive. Legislation may also be a barrier for recycling in some countries [10].

The processing and treatment of residues, animal manures, and wastes may contribute to the removal of some of these barriers. A range of possible techniques and technologies is available to process and treat the residues, manures, and wastes and to recover and utilize the nutrients in a form that allows their much more effective utilization [55]. Thus, the input via mineral fertilizers, and hence the fertilizer cost to farmers, may decrease, and the nutrient losses to the wider environment may also decrease. This is the main perspective underlying the processing and treatment of residues, animal manures, and wastes.

This book presents a range of possible techniques and technologies for the treatment of residues, manures, and wastes in order to improve the recovery and utilization of nutrients. The applicability of these techniques and technologies depends on the amount, form, and availability of residues, manures, and wastes within a region, and on the socioeconomic conditions. The need for such processing and treatment techniques and technologies will likely increase in the future as the production of residues, animal manures, and wastes increases, especially in regional agglomerations of intensive crop- or animal-production systems and near urban areas.

References

1. Marschner, H. (2012). *Mineral Nutrition of Higher Plants*. London, San Diego, New York, Boston, Sydney, Tokyo: Academic Press.
2. McDonald, P., Edwards, R.A., Greenhalgh, J.F.D. et al. (2010). *Animal Nutrition*, 7e. Harlow: Prentice Hall.
3. Suttle, N.F. (2010). *Mineral Nutrition of Livestock*, 4e. Wallingford: CABI.
4. Thompson, B. and Amoroso, L. (2011). *Combating Micronutrient Deficiencies : Food-Based Approaches*. Wallingford: CABI.
5. Sutton, M.A., Bleeker, A., Howard, C.M. et al. (2013). *Our Nutrient World: The Challenge to Produce More Food and Energy with Less Pollution. Global Overview of Nutrient Management*. Nairobi: Centre for Ecology and Hydrology, Edinburgh & United Nations Environment Programme.
6. Sutton, M.A., Howard, C.M., Erisman, J.W. et al. (2011). *The European Nitrogen Assessment*. Cambridge: Cambridge University Press.
7. Steffen, W., Richardson, K., Rockström, J. et al. (2015). Planetary boundaries: guiding human development on a changing planet. *Science* **347** (6223): 1259855.
8. Van Kauwenbergh, S.J. (2010). *World Phosphate Rock Reserves and Resources*. Muscle Shoals, AL: International Fertilizer Development Center (IFDC).
9. Reijnders, L. (2014). Phosphorus resources, their depletion and conservation, a review. *Resources, Conservation and Recycling* **93**: 32–49.
10. Withers, P.J.A., van Dijk, K.C., Neset, T.S.S. et al. (2015). Stewardship to tackle global phosphorus inefficiency: the case of Europe. *Ambio* **4**: 193–206.
11. Mueller, N.D., Gerber, J.S., Johnston, M. et al. (2012). Closing yield gaps through nutrient and water management. *Nature* **490**: 254–257.
12. Smil, V. (2017). *Energy and Civilization: A History*. Cambridge, MA: MIT Press.
13. Schlesinger, W.H. and Bernhardt, E.S. (2013). *Biogeochemistry. An Analysis of Global Change*. Durham, NC: Duke University.
14. Smil, V. (2001). *Enriching the Earth*. Cambridge, MA: MIT Press.
15. Pierzynsky, G.M., McDowell, R.W., and Sims, J.T. (2005). Chemistry, cycling and potential movement of inorganic phosphorus in soils. In: *Phosphorus: Agriculture and the Environment*, Agronomy Monograph 46, 53–86. Madison, WI: ASA, CSSA, and SSSA.
16. Galloway, J.N., Townsend, A.R., Erisman, J.W. et al. (2008). Transformation of the nitrogen cycle: recent trends, questions, and potential solutions. *Science* **320** (5878): 889–892.
17. Lassaletta, L., Billen, G., Grizzetti, B. et al. (2014). 50 year trends in nitrogen use efficiency of world cropping systems: the relationship between yield and nitrogen input to cropland. *Environmental Research Letters* **9**: 105011.
18. Lassaletta, L., Billen, G., Grizzetti, B. et al. (2014). Food and feed trade as a driver in the global nitrogen cycle: 50-year trends. *Biogeochemistry* **118** (1–3): 225–241.
19. Vitousek, J.P.W., Aber, J.D., Howarth, R.W. et al. (1997). Human alteration of the global nitrogen cycle: sources and consequences. *Ecological Applications* **7** (3): 737–750.
20. Smil, V. (1999). Nitrogen in crop production: an account of global flows. *Global Biogeochemical Cycles* **13** (2): 647–662.
21. Galloway, J.N., Dentener, F.J., Capone, D.G. et al. (2004). Nitrogen cycles: past, present and future. *Biogeochemistry* **70** (2): 153–226.
22. Fowler, D., Coyle, M., Skiba, U. et al. (2013). The global nitrogen cycle in the twenty-first century. *Philosophical Transactions of the Royal Society B: Biological Sciences* **368** (1621).
23. Lal, R. (2004). Soil carbon sequestration to mitigate climate change. *Geoderma* **123** (1–2): 1–22.
24. Smil, V. (2013). *Should We Eat Meat? Evolution and Consequences of Modern Carnivory*. Chichester: Wiley.

25. Westhoek, H., Lesschen, J.P., Rood, T. et al. (2014). Food choices, health and environment: effects of cutting Europe's meat and dairy intake. *Global Environmental Change* **26**: 196–205.
26. Smil, V. (2001). *Enriching the Earth: Fritz Haber, Carl Bosch, and the Transformation of World Food Production*. Cambridge, MA: MIT Press.
27. Mosier, A.R., Syers, J.K., and Freney, J.R. (eds.) (2004). *Agriculture and the Nitrogen Cycle. Assessing the Impacts of Fertilizer Use on Food Production and the Environment*. SCOPE Vol. 65. Chicago, IL: Bibliovault OAI Repository, the University of Chicago Press.
28. Zhang, X., Davidson, E.A., Mauzerall, D.L. et al. (2015). Managing nitrogen for sustainable development. *Nature* **528**: 51–59.
29. Liu, Q., Wang, J., Bai, Z.H. et al. (2017). Global animal production and nitrogen and phosphorus flows. *Soil Research* **55** (6): 451–462.
30. Smil, V. (2000). Phosphorus in the environment: natural flows and human interferences. *Annual Review of Energy and the Environment* **25** (1): 53–88.
31. Science Communication Unit (2013). *Sustainable Phosphorus Use. Science for Environment Policy*. Report produced for the European Commission DG Environment. Bristol: University of the West of England.
32. Cordell, D., Drangert, J., and White, S. (2009). The story of phosphorus: global food security and food for thought. *Global Environmental Change* **19**: 292–305.
33. Scholz, R. and Wellmer, F.W. (2013). Approaching a dynamic view on the availability of mineral resources: what we may learn from the case of phosphorus? *Global Environmental Change* **23**: 11–27.
34. Wellmer, F.W. and Scholz, R.W. (2016). Peak minerals: what can we learn from the history of mineral economics and the cases of gold and phosphorus? *Mineral Economics* **30** (2): 73–93.
35. Correll, D.L. (1998). The role of phosphorus in the eutrophication of receiving waters: a review. *Journal of Environmental Quality* **27** (2): 261–266.
36. Carpenter, S.R., Caraco, N.F., Correll, D.L. et al. (1998). Nonpoint pollution of surface waters with phosphorus and nitrogen. *Ecological Applications* **8** (3): 559–568.
37. Carpenter, S.R. (2008). Phosphorus control is critical to mitigating eutrophication. *Proceedings of the National Academy of Sciences of the United States of America* **105** (32): 11039–11040.
38. Diaz, R.J. and Rosenberg, R. (2008). Spreading dead zones and consequences for marine ecosystems. *Science* **321** (5891): 926–929.
39. Schindler, D.W., Carpenter, S.R., Chapra, S.C. et al. (2016). Reducing phosphorus to curb lake eutrophication is a success. *Environmental Science & Technology* **50** (17): 8923–8929.
40. Schoumans, O.F., Chardon, W.J., Bechmann, M.E. et al. (2014). Mitigation options to reduce phosphorus losses from the agricultural sector and improve surface water quality: a review. *Science of the Total Environment* **468–469**: 1255–1266.
41. Van Dijk, K.C., Lesschen, J.P., and Oenema, O. (2016). Phosphorus flows and balances of the European Union Member States. *Science of the Total Environment* **542 B**: 1078–1093.
42. Ma, L., Velthof, G.L., Wang, F.H. et al. (2012). Nitrogen and phosphorus use efficiencies and losses in the food chain in China at regional scales in 1980 and 2005. *Science of the Total Environment* **434**: 51–61.
43. Ma, L., Wang, F., Zhang, W. et al. (2014). Environmental assessment of management options for nutrient flows in the food chain in China. *Environmental Science & Technology* **47** (13): 7260–7268.
44. FAOSTAT (2015) Production and consumption data per country and year. Available from: http://faostat3.fao.org/home/E (accessed December 23, 2019).
45. Velthof, G.L., Lesschen, J.P., Webb, J. et al. (2014). The impact of the Nitrates Directive on nitrogen emissions from agriculture in the EU-27 during 2000–2008. *Science of the Total Environment* **468–469**: 1225–1233.
46. Wang, J., Liu, Q., Hou, Y. et al. (2018). International trade of animal feed and its relationships with livestock density and N and P balances at country level. *Nutrient Cycling in Agroecosystems* **110** (1): 197–211.

47. Erisman, J.W., Sutton, M.A., Galloway, J. et al. (2008). How a century of ammonia synthesis changed the world. *Nature Geoscience* **1** (10): 636–639.
48. UNCTAD (2009). *World Investment Report 2009: Transnational Corporations, Agricultural Production and Development*. Geneva: United Nations Publications.
49. Von Braun, J. and Díaz-Bonilla, E. (2008). *Globalization of Food and Agriculture and the Poor*. Oxford: Oxford University Press.
50. OECD (2013). Nutrients: nitrogen and phosphorus balances. In: *OECD Compendium of Agri-Environmental Indicators*. Paris: OECD Publishing.
51. Rowe, H., Withers, P.J.A., Baas, P. et al. (2016). Integrating legacy soil phosphorus into sustainable nutrient management strategies for future food, bioenergy and water security. *Nutrient Cycling in Agroecosystems* **104** (3): 393–412.
52. Bouwman, A.F., Beusen, A.H.W., and Billen, G. (2009). Human alteration of the global nitrogen and phosphorus soil balances for the period 1970–2050. *Global Biogeochemical Cycles* **23** (4).
53. Bouwman, A.F., Klein Goldewijk, K., van Der Hoek, K.W. et al. (2013). Exploring global changes in nitrogen and phosphorus cycles in agriculture induced by livestock production over the 1900–2050 period. *Proceedings of the National Academy of Sciences of United States of America* **110** (52): 20882–20887.
54. Tilman, D., Cassman, K.G., Matson, P.A. et al. (2002). Agricultural sustainability and intensive production practices. *Nature* **418**: 671–677.
55. Hou, Y., Velthof, G.L., Lesschen, J.P. et al. (2017). Nutrient recovery and emissions of ammonia, nitrous oxide, and methane from animal manure in Europe: effects of manure treatment technologies. *Environmental Science & Technology* **51**: 375–383.

Section II

The Role of Policy Frameworks in the Transition Toward Nutrient Recycling

2.1

Toward a Framework that Stimulates Mineral Recovery in Europe

Nicolas De La Vega[1] and Gregory Reuland[1,2]

[1]*European Biogas Association, Renewable Energy House, Brussels, Belgium*

[2]*Department of Green Chemistry & Technology, Faculty of Bioscience Engineering, Ghent University, Ghent, Belgium*

2.1.1 The Importance of Managing Organic Residues

The European Union (EU) has all the makings of an excellent testing ground for mineral recycling from organic material. To start with, the region has a state-of-the-art industrialized agricultural sector with a strong focus on livestock farming. In addition, Europe is mostly an urbanized continent with vast amounts of organic waste in the form of municipal solid waste and sewage sludge, as well as considerable waste streams from the industrial and commercial sectors (food scraps and leftovers, catering and beverage waste, slaughterhouse waste, feed) and agroindustry (crop residues, manure).

On the other hand, as shown in Figure 2.1.1, the EU is highly dependent on imported protein-rich plants. In 2016/2017, it used 27 million tons of crude proteins, of which 17 million were imported, overwhelmingly to supply the feed market [1]. In 2017, the EU-28 agricultural sector consumed an estimated 11 588 236 tons of nitrogen and 1 343 870 tons of phosphorous in the form of inorganic fertilizers. Phosphorous being a finite resource of which Europe has only very limited amounts, most of the phosphate rock used for fertilizers must be imported, which results in high production and transportation

Biorefinery of Inorganics: Recovering Mineral Nutrients from Biomass and Organic Waste, First Edition.
Edited by Erik Meers, Gerard Velthof, Evi Michels and René Rietra.
© 2020 John Wiley & Sons Ltd. Published 2020 by John Wiley & Sons Ltd.

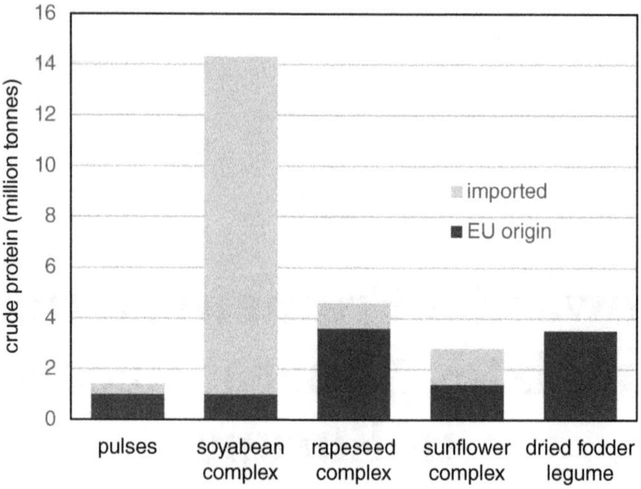

Figure 2.1.1 *Use of proteins and their sources (in million tons of crude protein) in the EU in 2016/2017 [1].*

costs. Nitrogenous fertilizers, such as ammonium nitrate and urea, are produced from the burning of natural gas – a fossil fuel – and are therefore strongly linked to oil prices. These activities result in significant concentrations of nutrients, particularly nitrogen, and increasingly phosphorous, in large areas throughout Europe. This nutrient accumulation has caused certain parts of Europe to become nitrogen "hotspots," leading to soil, water, and atmospheric pollution [2].

Other undesired effects related to inadequate treatment of organic material include the risk of spreading plant and human pathogens, intense odor, and greenhouse gas emissions. Methane is a particularly powerful greenhouse gas, which is rated by the UN Intergovernmental Panel on Climate Change (IPCC) as having a global-warming potential (GWP) of 25 over the course of a century. In uncontrolled conditions, organic matter such as manure, food waste, and sewage sludge will emit such greenhouse gases into the atmosphere [3].

In Europe, the two organic streams with the biggest adverse effects on climate are livestock manure, kept in non-enclosed storages (i.e. from agriculture), and the organic biodegradable fraction of solid municipal waste: in the EU, agriculture accounts for 53% of all methane (CH_4) emissions, while waste represents 28% [4]. Optimizing the management of organic materials can thus simultaneously address several environmental issues: water and soil pollution, nutrient recycling, and greenhouse gas emission mitigation.

2.1.2 The Rise of Nutrient and Carbon Recycling

To tackle these problems, a series of innovative techniques were developed over the past few decades that treat organic materials with varying levels of deployment in the EU. Today, anaerobic digestion (AD) and composting are among the most common options for successfully treating significant volumes of organic streams. Both techniques are based on well-tested and proven technology with reasonable investment costs, and both provide the possibility of recycling nutrients and organic carbon in the form of organic fertilizers or soil improvers.

Nutrients in organic material can also be recovered separately, in a more concentrated form, using innovative techniques to produce struvite, membrane-filtered concentrates, ash, biochar, and similar products. Several nutrient recovery techniques are already tested at full-scale in Europe and provide marketable products, such as recovered phosphorous from polluted sewage sludge and ammonia stripping/scrubbing from manure/digestate in areas with high nutrient surpluses. Combining two or more nutrient recovery techniques can increase the quality and value of the recycled material.

2.1.3 The European Framework for Nutrient Recovery and Reuse (NRR)

This chapter will center on the EU's current legislative framework and the ongoing reforms that aim to increase nutrient recycling. The EU is the most integrated region in the world, with its origins going back to the post-World War 2 Europe of the early 1950s. Since then, the regional economic block has developed a complex decision-making system with a defined distribution of competences between European supranational institutions and national (as well as regional) governments. Nutrient recovery and reuse (NRR) technologies fit well within the EU's current areas of competence, particularly in relation to environmental and agricultural policies. While the general focus lies on nutrient recycling as a whole, this chapter will refer mainly to EU legislation relevant to organic material and novel recovery techniques.

In addition to providing fertilizing products and renewable energy, recovery of organic material plays well into Europe's Bioeconomy Strategy, aimed toward a more sustainable use of renewable biological resources. The EU's Circular Economy Action Plan (2015) [5] and the subsequent fertilizer regulation revision [6] also provide a favorable platform for NRR technologies. Under the slogan "closing the loop," the Circular Economy aims to maintain the value of products, materials, and resources in the European economy for as long as possible, while reducing the generation of waste. This ambitious plan spans across many sectors and will be instrumental in reaching the EU's Sustainable Development Goals by 2030.

The emphasis given to waste management (organic and inorganic) in fostering the recycling of high-quality waste and to encourage valuable materials to find their way back into the economy can be of particular interest to NRR technologies. The Circular Economy also aims to boost the market for secondary raw materials, water reuse, and biobased products. Though recycled nutrients have been recognized by European authorities as an important component of the bioeconomy, new measures are required for the establishment of an EU-wide market for organic and waste-based fertilizers.

2.1.4 EU Waste Legislation

Waste management is a policy that testifies to the interplay between various levels of governance. At the EU level, there are framework directives that set common definitions and waste-management targets. National and regional authorities establish detailed laws adapted to country specificities while also integrating the provisions of the EU framework. Then, municipalities and cities are responsible for implementing them in practice at the local level. Two waste legislations are pivotal for nutrient recycling: Directive 2018/850 on the landfill of waste (amending Landfill Directive 1999/31/EC) and the Waste Framework Directive (WFD) 2018/851 (amending Directive 2008/98/EC).

The Landfill Directive sets progressive diversion targets for biodegradable waste. It states that Member States shall take the necessary measures to ensure that by 2035, the amount of landfilled municipal waste is reduced to 10% (or less) of the total amount of municipal waste generated (by weight). This is a positive measure toward nutrient circularity, as it is very costly to recover nutrients from landfills. During the last two decades, the EU has substantially reduced its landfilling rate, as well as greenhouse gas emissions from existing sites, with the help of landfill gas-recovery systems.

The amended WFD sets the European ground rules for waste management. It is based on a set waste hierarchy that favors resource-efficient techniques such as waste prevention and recycling over more linear approaches like energy recovery (incineration) and disposal (landfilling) (Figure 2.1.2). It sets a minimum target of 65% for the re-use and recycling of municipal waste by Member States by 2035. Digesting and composting biodegradable waste can also be counted toward the recycling target as long as the resulting product is used for soil fertilization or improvement, as specified in the Calculation Methods 2011/753/EU of the WFD. This measure is favorable toward nutrient recycling as it encourages national authorities to support such techniques in order to fulfill EU mandatory targets.

While the WFD generally favors nutrient recycling, it has weak provisions for biodegradable waste in two important instances. First, it does not require the separate collection of biodegradable waste, whereas this is already mandatory in municipalities for paper, metal, plastic, and glass. This means that nutrients in organic material often land in the mixed fraction of solid municipal waste, making their recovery much harder due to the presence of contaminants. Second, biodegradable waste is the last major municipal waste fraction that lacks criteria under the WFD defining when a recycling process can end its legal waste status. Consequently, the waste status persists after treatment, with the recycled material still falling under waste legislation instead of being considered a product. This has created big differences between waste streams, where those with European separate collection duties and end-of-waste status have developed higher recycling rates, while biodegradable waste recycling has increased more slowly and only in specific regions. These shortcomings have had a negative impact on nutrient recycling and need to be addressed.

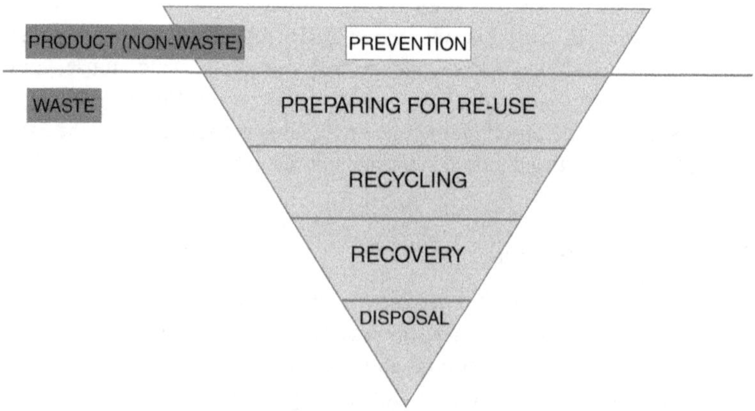

Figure 2.1.2 *The waste hierarchy as it is set under the Waste Framework Directive 2008/98/EC (http://ec.europa.eu/environment/waste/framework).*

Another point of concern is the need to better define by-products, a category between waste and product, which enables the agro-industry and farmers to avoid waste status when handling leftovers from their primary production (e.g. spent grain after brewing beer). The by-products provision within the WFD is a flexible tool that may boost nutrient recycling from homogeneous feedstock. Nonetheless, most national and regional authorities have not implemented this provision adequately since its entry into force in late 2010, where most EU countries still retain a binary legislative system divided into "waste" and "products." Clearer wording and the duty to recognize by-products are necessary in order to improve the integration of this important provision within national legislation.

When it comes to material recovery, a growing issue in several EU countries is the increasing incineration capacity across the EU and the lack of political resolve to control this expansion. This technology has several advantages, including the recovery of energy and certain metals; more importantly, it can handle residual and hazardous waste that cannot be recycled, and which otherwise would go to a landfill. Nonetheless, there is a high risk that while biodegradable waste is easily recyclable, it may still land in the mixed fraction (rather than be separately collected) and go directly to incinerators. This would constitute a significant loss of nutrients, which could easily be recovered via well-established technologies such as composting or AD. Though biodegradable waste has high water content and therefore a low energy output when combusted, there are two factors that significantly increase the likelihood of its being incinerated: (i) incinerators usually receive a gate fee based on weight in fresh mass; and (ii) several EU countries count the energy produced in incinerators from biowaste as part of their national renewable energy targets in EU Directive 2009/28/EC. This incentivizes public authorities to prioritize straight energy recovery rather than more resource-efficient recycling techniques such as NRR (as with AD). To address this challenge, it is crucial to establish a diversion target for biodegradable waste from incinerators, as is already done for landfilling under the Landfill Directive.

2.1.5 Moving from Waste to Product Legislation and the Interplay with Other EU Legislation

Once waste is processed as a valuable secondary resource (i.e. upcycled) it is essential it finds its way onto the market as a product. This might be in the form of energy, materials, or chemicals derived from organic material following the biorefinery principle.

The focus of this section will be on fertilizers, as it is the most common use for recycled organic material and due to the imperative need to mitigate agriculture's impact on the environment. In short, organic fertilizers and soil improvers provide three critical nutrients that are essential for agriculture: nitrogen, phosphorous and potassium. Unlike conventional mineral fertilizers, organic fertilizers also contain significant amounts of organic carbon and trace nutrients. The carbon compounds promote soil fertility, soil structure, and the biological and physical health of the soil, and counter erosion. Additionally, mineral fertilizing products can be derived from organic products such as digestate. This "new generation" of fertilizers, based on NRR techniques, possesses the same agronomic properties as inorganic (mineral) fertilizers.

Generally, organic fertilizers such as digestate are used locally, so as to avoid transporting large volumes of material. Nonetheless, there are several instances where having the ability to trade across EU borders can be crucial: that is, in areas with intensive livestock farming activity (e.g. from Belgium to France or from the Netherlands to Germany), which tends to concentrate large volumes of nutrients and runs the associated risk of soil and water pollution. These areas, under agricultural pressure, must displace the excess nutrients to areas where soils are lacking organic matter and nutrients.

Until now, only industrially manufactured mineral fertilizers from non-organic origins could be traded freely across the EU as "EC fertilizer." In other words, organic fertilizing products were not included and therefore were regulated under national legislation (i.e. national end-of-waste), though not all EU countries have a clear or even viable legal framework for these products. As a result, organic fertilizers remained confined to national markets within Member States, hindering nutrient flows as well as economies of scale. This legal vacuum at the EU level has resulted in counterproductive developments where countries established a heterogenous panel of testing methods and requirements for organic fertilizers.

Fortunately, on November 20, 2018, the European Commission, the Council, and the Parliament agreed on a new Fertilizer Regulation, which will come into force in January 2022 and which aims to facilitate the access of organic and waste-based fertilizers to the EU Single Market. Important changes include limit values for toxic contaminants in EU fertilizing products, such as a 60 $mg\,kg^{-1}$ limit for cadmium in phosphate fertilizers. The regulation goes further in addressing the end-of-waste and end-of-animal by-product status for digested organic waste and animal by-products. More scientific work has to be done before the requirements for new techniques to recycle nutrients in mineral form can be defined; the EU's Joint Research Centre is currently evaluating struvite, ashes, and biochar to be added to the Regulation in coming years.

All recovered fertilizing materials treated under the new Fertilizer Regulation will follow common rules on quality, safety, and labeling requirements in order to be traded freely across the EU. To ensure compliance with the requirements set in the new regulation and to avoid cases of fraud, fertilizer producers and distributors will be subject to quality controls by national notified bodies under the scrutiny of national competent authorities.

2.1.6 Complying with Existing Environmental and Health & Safety Legislation

Having adequate product legislation is paramount to boosting nutrient recycling. However, in a regulatory system as complex as the EU's, several pieces of legislation must be taken into account in order to ensure a smooth integration of new products. For digestate, compost, and other products derived from organic materials, the end-of-waste status and end-of-animal by-product status are on the right track under the amended Fertilizer Regulation.

AD, compost, and several other nutrient recovery techniques play an important role in reducing nitrogen leaching from animal manure. For example, AD and other techniques can significantly increase nutrient availability for plants as compared to raw animal manures and slurries. Due to technological advances in digesting manure and the possibility of post-treatment of digestate, nitrogen availability in mineral form can be increased significantly, opening the path to a new generation of biobased fertilizers. In combination with good agricultural practices (fertilization plan, time of application), the use of recycled fertilizing products, both in organic and in mineral form, could prove providential for a more effective dosage of quick- and slow-release nutrients, thus providing farmers with an efficient tool for a reasoned and environmentally sustainable use of fertilizers.

In response to increasing amounts of nitrates of agricultural origin being found in water bodies in the early 1990s, the EU put in place the Nitrates Directive to protect the quality of Europe's surface and groundwaters. In setting out steps for Member States to keep track of nitrate runoff and leaching into water bodies, the Nitrates Directive established good agricultural practices – to prevent and reduce nitrate pollution – and designated Nitrate Vulnerable Zones (NVZs), which correspond to areas that are more susceptible to agricultural nitrate pollution. In a nutshell, in the designated NVZs, up to 170 kg N per ha from livestock manure can be applied in each calendar year.

The current WFD 2008/98/EC [7] states that animal by-products are excluded from the scope of the Directive, including processed products, except those which are destined for incineration, landfilling, or use in a biogas or composting plant. In other words, manure that is applied to the field (agricultural use) is not considered waste, whereas that which is digested in a biogas plant is (with all the ensuing paperwork and obligations associated with the Directive). This state of affairs is massively counterproductive and highly unsatisfactory as it hinders political, technical, and organizational efforts to increase the energetic use of manure in biogas plants.

Products derived from animal manure/digestate, via NRR technologies, cannot be defined as inorganic fertilizers regardless of the resulting properties (in several cases resembling inorganic fertilizers rather than manure). The maximum allowable level of nitrogen to be spread under the legal status of "animal manure" stemming from the Nitrates Directive puts manure-based products or those partly based on manure at a disadvantage compared to inorganic fertilizers, which do not face this restriction. This creates an imbalance in the market, as one type of fertilizer is disadvantaged against another. A case in point is ammonium sulfate, which is a 100% mineral fertilizer recovered by extracting ammonia-rich air from stables or digestate, upgrading it, and capturing it in a "scrubber" solution; despite its mineral composition, this upgraded fertilizer must comply with the 170 kg N per ha associated with manure-based fertilizers according to the Nitrates Directive. The Directive can also be a hindrance in NVZs, where the quantity of N fertilizer exceeds the 170 kg N annual limit.

This is contradictory to the aim of including organic fertilizers within the Fertilizer Regulation, in the sense that the material in question does not stop being manure (no "end-of-waste"/"end-of-manure" proposition regardless of processing). Though the Fertilizer Regulation defines product standards, the use of the fertilizer is not regulated by the Regulation. In order to solve this issue, either the Nitrates Directive should be revised to specify when manure can lose this status and be considered a resource or product after it is adequately processed or the relevant EU authorities should give clear guidelines on when processed manure products derived from manure (ammonium sulfate, struvite, magnesium calcium, etc.) can be exempt, based on the latest scientific knowledge and in cooperation with national experts, working groups, and authorities. Consistency between the Nitrates Directive and the new Fertilizer Regulation will be necessary to create a level playing field between chemical fertilizers and the manure-based alternatives.

The Fertilizer Regulation includes a small but important amendment to the Animal By-Product Regulation, which enables a wide range of animal by-products (except those with higher risks) to exit their legal status as such if digested, thereby reducing administrative burdens to producers. This implies that this material is no longer subject to the Animal By-Products Regulation and therefore can fall into product legislation. In other words, animal by-products can have a regulatory framework that is compatible with product legislation for compost and digestate in particular. Other techniques for processing animal by-products are likely to be added in the future following the scientific evaluation of the Joint Research Centre.

2.1.7 Conclusion

Finding a framework for nutrient recycling is challenging, yet essential. It touches upon legislation in many policy areas, which can either act as a catalyst or as an obstacle to the reutilization of minerals and carbon from organic material. Questions and doubts concerning legislation can hinder the development of nutrient recycling from organic materials, because operators/investors are hesitant to engage in routes where the regulatory context is unclear. A coherent and ambitious European policy for a circular economy for nutrients is therefore essential.

While adapting legislation to this end is imperative, this has to be done in a coherent way that is in line with consumer expectations and respectful of the environment. Moreover, EU legislation must be inclusive but not prescriptive, so that it can be adapted to the different realities of regions and countries across Europe. Much progress has been made in terms of adapting existing legislation and creating new provisions that are more conducive to nutrient recovery. Big efforts are being made to harmonize requirements among EU countries in order to create a bigger market for recycled products. As innovation continues apace, the years leading to 2020 and beyond are likely to be decisive for the nutrient recovery sector in Europe.

References

1. European Union. Report from the Commission to the Council and the European Parliament on the development of plant proteins in the European Union. COM/2018/757. Brussels: European Commission; 2018.
2. Erisman, J.W., van Grinsven, H., Grizzetti, B. et al. (2011). The European nitrogen problem in a global perspective. In: *The European Nitrogen Assessment: Sources, Effects and Policy Perspectives* (eds. M. Sutton, C. Howard, J.E. Erisman, et al.), 9–31. Cambridge: Cambridge University Press.
3. European Environmental Agency. Trends and Projections in Europe 2018: Tracking Progress Towards Europe's Climate and Energy Targets. No. 16/2018. Luxembourg: Publications Office of the European Union; 2018.
4. European Environmental Agency. Air Quality in Europe – 2018. Report No. 12/2018. Luxembourg: Publications Office of the European Union; 2018.
5. European Commission. Communication from the Commission to the European Parliament, the Council, the European Economic and Social Committee and the Committe of the Regions. Closing the Loop – An EU Action Plan for the Circular Economy. Brussels: Publications Office of the European Union; 2015.
6. European Union (2019). Regulation (EU) 2019/1009 of the European Parliament and of the Council of 5 June 2019 laying down rules on the making available on the market of EU fertilising products and amending Regulations (EC) No. 1069/2009 and (EC) No. 1107/2009 and repealing Regulation (EC) No. 2003/2003. *Official Journal of the European Union* **62**: 1–114.
7. European Commision (2008). EC Directive 2008/98/EC of the European Parliament and of the Council of 19 November 2008 on waste and repealing certain directives. *Official Journal of the European Union* **L312** (3).

2.2

Livestock Nutrient Management Policy Framework in the United States

Georgine Yorgey and Chad Kruger

Center for Sustaining Agriculture & Natural Resources, Washington State University Northwestern Washington Research and Extension Center Mount Vernon, Mount Vernon, WA, USA

2.2.1 Introduction

Nutrient recovery efforts in the United States are shaped largely in response to federal legislative mandates that set a minimum standard for the protection of surface- and groundwater quality and, to a lesser extent, air quality – rather than a framework that explicitly encourages nutrient recovery or recycling. It is also important to recognize that in the United States, much of the implementation and enforcement occurs at the state level, where state legislative standards may be higher than the minimum federal standard or where local conditions lead to unique implementation. Disputes relating to states' or other jurisdictions' implementation and enforcement of federal requirements are adjudicated – and the associated financial risk makes legal precedent a powerful force. Finally, in some cases, companies may impose even stricter supply-chain requirements that impact nutrient management, with these most often occurring in response to consumers' desire for sustainability attributes associated with the items they purchase.

The Federal Clean Water Act (CWA) [1] establishes the structure for regulating discharges of pollutants into US waters and for regulating surface-water quality standards. Under the CWA, the US Environmental Protection Agency (EPA) has developed national water-quality criteria recommendations for pollutants in surface waters and has

Biorefinery of Inorganics: Recovering Mineral Nutrients from Biomass and Organic Waste, First Edition.
Edited by Erik Meers, Gerard Velthof, Evi Michels and René Rietra.
© 2020 John Wiley & Sons Ltd. Published 2020 by John Wiley & Sons Ltd.

implemented pollution-control programs such as setting wastewater standards for industry. The EPA's National Pollution Discharge Elimination System (NPDES) permit program controls discharges of pollutants from industrial, municipal, and other point sources into navigable waters. Meanwhile, water quality in groundwater and other public water systems is protected through legally enforceable standards for contaminants including nitrates and nitrites, as required by the Safe Drinking Water Act [2].

While the CWA provides the overarching legal authority for protecting surface-water quality, additional federal and non-federal regulations provide additional detail relevant to specific nutrient sources. Thus, for example, the treatment of residuals from wastewater treatment processes (biosolids) is managed under the NPDES permit program, as well as the Code of Federal Regulations, Title 40, Part 503, *Standards for the Use or Disposal of Sewage Sludge*.

Because the policy context relevant to nutrient recovery varies by state as well as between and even within nutrient sources (e.g. between municipally-generated and agriculturally-generated nutrients), we provide a case study of nutrient management for livestock producers, and more specifically within the dairy industry, to provide an understanding of the financial, logistical, and technical challenges that individual farmers face in managing manure nutrients, and the resulting interest in – and barriers to – more widespread adoption of advanced nutrient-recovery approaches.

2.2.2 The Legal-Regulatory Framework for Manure Nutrient Management

The EPA is the authorized federal regulatory agency for implementing the requirements of the federal CWA and the NPDES Concentrated Animal Feeding Operation (CAFO) permit, which applies to all livestock operations that meet the regulatory definition of a concentrated animal feeding operation [3]. Complementing federal legislation and rules, many states have legislative and regulatory frameworks for livestock nutrient management that include both provision of technical support and administration of permitting, inspection, and enforcement.

A look at several specific states gives a sense of how this varies across the United States. In Washington State (in the Northwest), legislation (Chapter 90.48 and Chapter 90.64 Revised Code of Washington) designates the Washington State Department of Ecology as the official regulatory authority for the enforcement of federal and state CWAs, as well as CAFO rules and permits. As detailed in a Memorandum of Agreement [4], the Washington State Department of Ecology coordinates with the Washington State Department of Agriculture to work toward a shared goal of assuring water quality compliance related to livestock activities.

As reflected in the Memorandum, the approach in Washington relies heavily on working cooperatively with the dairy industry and other government agencies such as state and local conservation agencies to protect water quality while supporting a viable private-sector dairy industry – an approach that is common across many US states. This principle is captured in the preamble of the Revised Code of Washington 90.64 related to *Dairy Nutrient Management*, which states:

> *The legislature finds that there is a need to establish a clear and understandable process that provides for the proper and effective management of dairy nutrients that affect the quality of surface or ground waters in the state of Washington. The legislature finds that there is a need*

for a program that will provide a stable and predictable business climate upon which dairy farms may base future investment decisions.

(90.64.005 RCW)

Within this context, a key activity that livestock operations in Washington State carry out to show their compliance with nutrient management requirements is the completion and maintenance of a Nutrient Management Plan (NMP), detailing how the operation will ensure nutrients are fully used by crops and do not degrade water quality. The goal of this approach is to match the application of manure nutrients to projected plant uptake on available cropland, but to leave the strategy at the discretion of the farm enterprise (e.g. purchase sufficient cropland or export nutrients to distant fields).

Similar approaches relying on an NMP of some sort are used in many other states, though which livestock operations are required to complete NMPs and which can complete them voluntarily varies by state. In practice, however, many livestock facilities that are not required to complete NMPs do, as they feel that participation provides some guarantee that they will be in compliance with CWA, limiting risks for enforcement or legal action and making them eligible for cost-share programs aimed at implementing nutrient management technology and practices.

Across the United States, preparation of NMPs is normally carried out using technical resources developed by the United States Department of Agriculture (USDA) Natural Resource Conservation Service (NRCS). The NRCS maintains a set of reference materials, design standards, and technical guidance tools to help farmers, conservation districts, and private consultants plan, design, build, and manage manure-management and nutrient-management systems. These references and tools include the Comprehensive Nutrient Management Plan (CNMP), a Field Office Technical Guide (FOTG) specific to the local county, the Agricultural Waste Management Field Handbook (AMFH), and the Animal Waste Management (AWM) planning software [5–9]. NRCS design and management criteria and standards are often cited in legislation and rules as the minimum standards to which manure-management systems and plans must be designed and managed for compliance with CWA and state laws. In many states, local Conservation Districts and private technical service providers directly assist dairies with preparing their NMP (using NRCS tools and standards). They also provide other forms of technical support for manure management, such as annual nutrient score cards.

2.2.3 Current Manure-Management Practices

Most manure in the United States is land-applied, with policy and regulation supporting a goal of applying manure at a rate that is consistent with plant uptake. However, a number of characteristics of manure combine to create farm-level incentives for maximizing nutrient loading, despite the policy context. These include the expense of transporting manure, the variability in nutrient form and content, the difference between the nitrogen, phosphorus, and potassium (NPK) ratio of manure and the ratio required by crops, a tendency to target nutrient application toward high-yield goals rather than average yields, and food safety concerns that largely limit manure applications to field crops [10–15]. An analysis of nitrogen applications to croplands in 2006 estimated that 93% of cropland acreage receiving manure did not meet best management practices for rate, timing, and application method, compared with 62% of acres receiving nitrogen as non-manure fertilizer [14]. Based on author review of data from the Washington State Department of Agriculture, evidence suggests that the Washington Dairy Nutrient Management Program has been increasingly effective at reducing excess nitrogen application to croplands over its duration, suggesting improvements in at least some areas of the country since the mid-2000s.

Given the costs of hauling and land-applying manure, CAFOs with liquid manure systems have long used first-generation nutrient recovery systems such as screens and settling basins to recover large solids and fibers. Composting of solids is also used in some cases. However, to date, adoption of more advanced nutrient recovery technologies has been fairly limited.

Adoption of anaerobic digestion, a technology that reduces methane emissions from manure treatment and generates renewable energy but does not on its own reduce or recover nutrients, has also been modest. As of the end of 2018, 282 anaerobic digestion facilities were operating or in the process of being constructed on livestock operations in the United States [16]. To give a sense of penetration rates, in 2018 digesters were estimated to serve approximately 6% of the US dairy herd. Inconsistent, non-binding, and oscillating incentives for renewable energy are one important cause of limited adoption, alongside other barriers that include economic and financing hurdles, impediments to accessing energy markets in order to sell the energy generated, underdeveloped markets for solid and liquid digestate co-products, and the lack of a federal climate policy [17, 18].

2.2.4 Public Investments for Improvement of Manure-Management Practices

While current manure-management practices are driven largely by the need to comply with applicable rules and regulations relating to water quality, there are also several public sector investments designed to catalyze innovation and improved adoption of best manure management practices, relying on a voluntary rather than a regulatory approach. NRCS manages two key grant programs, the Environmental Quality Incentives Program (EQIP) and the Conservation Innovation Grants (CIGs) [19, 20]. EQIP is a voluntary Farm-Bill program that provides cost-share grants to eligible producers to incentivize investments in technology and management practices that improve environmental outcomes. Priorities for EQIP investment are generally made at the local level through Conservation District Boards of Supervisors, but there are also national-level EQIP initiatives. To be eligible, farmers must have a current Conservation Activity Plan and must request cost share for approved technologies or management practices. The CIG Program is a subset of EQIP targeting the development of new environmental protection technologies and management practices that are based in existing research but which have not had significant on-farm adoption. The CIG program has provided a clear investment opportunity for deploying new dairy manure- and nutrient-management technology [21].

Another key public-sector investment supporting improved dairy manure nutrient management is the research and extension system of the Land Grant University system and the USDA Agricultural Research Service (ARS). Scientists within these public-sector research institutions are the primary resource for publishing peer-reviewed research that is used by the regulatory and technical support agencies to develop and update technology and practice standards and guidance tools. Targeted funding from federal agencies including the USDA and EPA provides opportunities for public-sector research scientists to carry out research projects that advance knowledge and technology for managing manure nutrients. As one example, in 2016, the EPA solicited proposals as part of a Nutrient Recycling Challenge to facilitate the transfer of technology for next-generation nutrient recovery.

Meanwhile, within some states, state-level policy has also provided incentives for adoption of improved manure management, with implications for nutrient recovery. California, which accounts for about a fifth of the nation's milk, has implemented state-wide climate policy, including a 2016 law that requires the livestock industry to cut methane emissions to 40% of 2013 levels by 2030 and establishes targets for reducing organic waste in landfills [22]. To achieve livestock-related reductions, the state is currently implementing a voluntary approach, using funds from fees collected under its cap-and-trade program to invest in both digester and non-digester methane-reduction technologies. Among the non-digester technologies included are solid–liquid separators with drying or composting, scrape and vacuum collection of manure with drying or composting, and pasture-based practices. By September 2018, the program had committed an estimated 31.5 million USD in alternative manure-management practices that reduce methane emissions and likely also nutrient impacts to groundwater [23]. There is ongoing discussion about the need to better understand the implications of these practices for nutrients, particularly air emissions such as volatile organic compounds (VOCs), nitrous oxide, and ammonia [23].

In Washington State, the legislature recently implemented a 3.8 million USD pilot program for dairy manure nutrient recovery technology through the Washington State Conservation Commission, in order to test, evaluate, and demonstrate next-generation nutrient recovery technology [24]. This program was developed in response to concerns around water quality impacts from the storage and application of dairy manure.

2.2.5 The Role of the Judicial Process and Consumer-Driven Preferences

While agencies and private industry utilize a mix of mandatory and voluntary activities to address water quality concerns, the final authority for compliance and enforcement is held by the judicial system, as is the case for all potential conflicts in the United States. Anyone with standing who believes that the actions of another party has caused them injury under the law has the right to file a civil lawsuit to enforce their rights, bringing the activity to a stop or seeking damages from the defendant. In many cases, such litigation has resulted in evolving interpretations of how the CWA applies to dairy manure nutrient management. Impacts from litigation can be national in scope.

In 2015, in Washington State, a summary judgment was issued in the United States District Court Eastern District of Washington for *Community Association for Restoration of the Environment* v. *Cow Palace* (2015). The judgment has the potential to have far-reaching implications for manure management for livestock facilities because it is the first decision to apply the regulations of the Resource Conservation and Recovery Act (RCRA), a federal law governing the disposal of solid waste and hazardous waste, to manure [25, 26]. In applying the RCRA, the ruling found that the manure could be considered a "solid waste" rather than a beneficial product, because of the way in which it was managed. Beyond the fact that this decision applies the RCRA to manure, the agricultural sector has been unsettled by the fact that the dairy in this case had held a state-approved NMP since 1998, and had followed NRCS guidelines (which allow for permeability) in constructing its manure lagoon [25–27].

While this case was settled out of court, the findings hold significant potential impact both within Washington State and nationally. In the wake of the settlement, the Washington Department of Ecology evaluated and issued a new CAFO Permit rule and the Washington State Department of Agriculture initiated a stakeholder Dairy Nutrient Advisory Committee to review the Dairy Nutrient Management Program with recommendations focused on improving implementation of the program and possibly requesting legislative updates to RCW 90.64. It has also provided some motivation nationwide for individual dairies to consider and pilot implementation of nutrient recovery technology, as livestock facilities re-evaluate the level of risk associated with their current manure-management practices.

Meanwhile, social pressures from consumers continue to increase, and have the potential to significantly change the landscape in which livestock facilities in the United States operate. This is felt particularly in the form of economic leverage from companies that purchase a significant amount of dairy products. Starbucks, which is a major purchaser of dairy, particularly fluid milk, has had a preference in North America since 2009 to purchase from firms using industry best practices for animal husbandry and processing for dairy, egg, and meat production [28]. Responding to similar desires by many other large and small companies throughout the milk-related value chain, the Innovation Center for US Dairy has created a US Dairy Stewardship Commitment [29]. Companies that adopt the US Dairy Stewardship Commitment meet defined criteria relating to animal care, the environment, and food safety, and report on their impact. Environmental aspects include metrics related to both nutrient management and resource recovery.

2.2.6 Limitations of the Current Framework

The framework for manure management in the United States has largely focused on concern for the impairment of surface- and groundwater quality, an approach that has potentially significant limitations. While agricultural operations that emit large quantities of air pollutants may be subject to Clean Air Act regulation and permits (as well as, in some cases, other regulations), there has been ongoing disagreement about whether or not existing data provide a sufficient basis for regulating and managing air emissions from animal feeding operations [30].

Within this context, tighter regulation to protect water quality during land application of manure has the potential to cause changes to manure management that reduce losses of nitrogen-nitrate, by trading them for losses of nitrogen-ammonia [31], given high costs for transporting manure. There may thus be implicit incentives to reduce the nitrogen content of manure through strategies that promote the creation of ammonia and its volatilization into the atmosphere, such as by storing the manure in uncovered lagoons, aerating the manure, or surface-applying slurry rather than injecting it. Available anecdotal evidence suggests that all of these strategies are used at least to some extent by livestock producers in the United States [31–34]. Because volatilization addresses water-quality regulations, producers have little incentive to recover the nitrogen through ammonium sulfate, struvite, or other costly technologies. The addition of air-quality regulations may change this equation in the future. Ammonia volatilization creates a negative impact on air quality, which is becoming an increasing concern in some areas of livestock concentration. For example, air quality is a significant concern in the San Joaquin Valley, California, where Federal PM 2.5 standards are being exceeded [35], and in the Yakima Valley, Washington, where meeting air-quality standards remains an ongoing concern [36]. Future federal climate policy, increasing numbers of states with state-level policies, and companies with carbon-based purchasing standards could also provide additional policy contexts that would impact nutrient management and nutrient recovery, albeit not directly.

Though more comprehensive regulatory approaches may be one solution to this problem, ongoing technology development is also essential to bringing down costs and to providing farmers with viable options to monitor and improve nutrient applications, as well as in some cases to concentrate and export nutrients. The development of economically viable technologies also depends crucially on paying more attention to the market for products generated via nutrient recovery [37].

Some of the need related to products is technical in nature; for example, in the case of ammonium sulfate, the composition of the product recovered from manure is not of the same quality or consistency as commercially available ammonium sulfate, so it does not compete well in the market. In other cases, policy barriers also need to be addressed to ensure a clear marketplace and consistent markets. For example, pathogens can be a concern with soil-amendment products derived from manure, though processing that includes composting or anaerobic digestion can reduce (but not eliminate) these concerns. Clarifying when and how manure-derived products can be safely used on food crops would provide more certainty for producers, and could expand the number of acres available to accept these materials, aiding in overall nutrient-management efforts.

As of the end of 2018, the US Food and Drug Administration (FDA), as part of the Food Safety Modernization Act, which applies to produce normally eaten raw, was conducting a risk assessment and research on the number of days that should be left between applications of raw manure as a soil amendment and harvesting to minimize the risk of contamination [15]. In the meantime, the FDA encouraged produce farmers to observe specific waiting periods: 90 days for crops whose edible parts do not touch the soil, 120 days for those whose edible parts do (CFR Title 7). While the Food Safety Modernization Act specifically addresses composting – establishing limits on detectable amounts of bacteria and allowing stabilized compost prepared using approved methods to be applied without a specified waiting period as long as application methods minimize the potential for contact with produce – it does not specifically address anaerobic digestion or other manure-processing technologies; as additional approaches become more common, this may be required.

Ultimately, the solution to the concentration of nutrients in dairy manure is cost-effective technology options that enable distribution of those nutrients across a larger land base. Advanced nutrient-recovery technologies are necessary in order to improve the distribution of nutrients, but scientists and others need to be more attentive to the marketplace needs and barriers to accepting manure-derived nutrients.

2.2.7 Conclusion

In light of these recent increases in environmental, legal, and social concerns, nutrient management is becoming an increasing concern of many livestock producers in the United States. In a recent survey, dairy farmers in the Northwest identified nutrient management as the lead concern that would cause them to adopt new manure-management technologies [38]. A handful of livestock operations are also experimenting with advanced nutrient recovery (beyond solids removal), with a handful of large commercial dairies using or piloting technologies that include dissolved air flotation for removal of fine solids, vermi-biofiltration, distillation, and membranes. Recent analyses have identified cost, complexity, and the lack of markets for recovered products as major barriers [37, 39, 40].

However, given limited public willingness to support governmental incentives for expensive nutrient recovery technology on CAFOs and economic constraints on the livestock industry, viable technologies will most likely need to either successfully minimize expenses or generate concentrated nutrient products that can be sold to offset costs, or both. Additionally, it is plausible that increased regulatory burdens and legal concerns will exacerbate the trend toward increasing consolidation and concentration of livestock facilities in an effort to aggregate sufficient capital for investment in additional pollution control through nutrient recovery or by other means.

Given so far limited commercial penetration, there has been considerable ongoing interest in public–private partnerships to spur investment in technology innovation in order to contribute solutions to this issue across all livestock sectors. The future social and policy framework for dairy nutrient management in the United States will likely be a continuation of the existing mixed strategy of mandatory and voluntary activities spurred by both legal and market pressure, and one that continues to prioritize and incentivize innovation and the development of cost-effective nutrient-management technology.

References

1. Clean Water Act. 33 USC §1251–1387. 1972.
2. Safe Drinking Water Act. 42 USC §300f et seq. 1974.
3. US-EPA. Water Permitting: NPDES: NPDES Home. US Environmental Protection Agency. Available from: http://water.epa.gov/polwaste/npdes (accessed December 23, 2019).
4. Memorandum of Understanding Between the Washington State Department of Agriculture and the Washington State Department of Ecology Related to The State of Washington's Efforts to Protect Water Quality Related to Livestock Activities Under the Authority of Chapter 90.48 RCW, Water Pollution Control Act and Chapter 90.64 RCW, Dairy Nutrient Management Act. Available from: https://ecology.wa.gov/DOE/files/6f/6f30de07-feb0-463a-958e-cf48df3a43bf.pdf (accessed December 23, 2019).
5. USDA-NRCS. Manure and Nutrient Management: United States Department of Agriculture. Available from: https://www.nrcs.usda.gov/wps/portal/nrcs/main/national/technical/ecoscience/mnm (accessed December 23, 2019).
6. USDA-NRCS. Agricultural Waste Management Field Handbook: United States Department of Agriculture, Natural Resources Conservation Service. Available from: http://www.nrcs.usda.gov/wps/portal/nrcs/detailfull/national/technical/ecoscience/mnm/?cid=stelprdb1045935 (accessed December 23, 2019).
7. USDA-NRCS. Comprehensive Nutrient Management Plan: United States Department of Agriculture, Natural Resources Conservation Service. Available from: http://www.nrcs.usda.gov/wps/portal/nrcs/detail/wi/farmerrancher/?cid=nrcs142p2_020843 (accessed December 23, 2019).
8. USDA-NRCS. Animal Waste Management: United States Department of Agriculture. Available from: http://go.usa.gov/Zcrm (accessed December 23, 2019).
9. USDA-NRCS. Agricultural Waste Management Field Handbook. United States Department of Agriculture, Natural Resources Conservation Service. Available from: https://www.nrcs.usda.gov/wps/portal/nrcs/detailfull/national/water/?&cid=stelprdb1045935 (accessed December 23, 2019).

10. Davis, J., Iversen, K., and Vigil, M. (2002). Nutrient variability in manures: implications for sampling and regional database creation. *Journal of Soil and Water Conservation* **57** (6): 473–478.
11. Eghball, B., Wienhold, B.J., Gilley, J.E., and Eigenberg, R.A. (2002). Mineralization of manure nutrients. *Journal of Soil and Water Conservation* **57** (6): 470–473.
12. Ribaudo, M., Kaplan, J.D., Christensen, L.A. et al. (2003). Manure management for water quality costs to animal feeding operations of applying manure nutrients to land. *SSRN Electronic Journal* https://doi.org/10.2139/ssrn.757884.
13. USDA-ERS. Manure Use for Fertilizer and for Energy. Report to Congress. Washington, DC: United States Economic Research Service; 2009.
14. USDA-ERS. Nitrogen in Agriculture Systems: Implications for Conservation Policy. Washington, DC: United States Economic Research Service; 2011.
15. US Food and Drug Administration. Food Safety Modernization Act: Final Rule on Produce Safety. 2015.
16. US-EPA. Livestock Anaerobic Digester Database, January 2019. Washington, DC: United States Environmental Protection Agency; 2019. Available from: https://www.epa.gov/agstar/livestock-anaerobic-digester-database (accessed December 23, 2019).
17. Gloy, B.A. and Dressler, J.B. (2010). Financial barriers to the adoption of anaerobic digestion on US livestock operations. *Agricultural Finance Review* **70** (2): 157–168.
18. Edwards, J., Othman, M., and Burn, S. (2015). A review of policy drivers and barriers for the use of anaerobic digestion in Europe, the United States and Australia. *Renewable and Sustainable Energy Reviews* **52**: 815–828.
19. USDA-NRCS. Environmental Quality Incentives Program: United States Department of Agriculture, Natural Resources Conservation Service, Financial Assistance. Available from: http://www.nrcs.usda.gov/wps/portal/nrcs/main/national/programs/financial/eqip (accessed December 23, 2019).
20. USDA-NRCS. Conservation Innovation Grants: United States Department of Agriculture, Natural Resources Conservation Service, Financial Assistance, Environmental Quality Incentives Program. Available from: http://www.nrcs.usda.gov/wps/portal/nrcs/detail/national/programs/financial/eqip/?cid=nrcs143_008205 (accessed December 23, 2019).
21. USDA-NRCS. Conservation innovation grants 2017. Available from: https://www.nrcs.usda.gov/wps/portal/nrcs/main/national/programs/financial/cig (accessed December 23, 2019).
22. CA SB-1383. Senate Bill 1383 Chapter 395. Short-lived climate pollutants: methane emissions: dairy and livestock: organic waste: landfills 2016. Available from: https://leginfo.legislature.ca.gov/faces/billNavClient.xhtml?bill_id=201520160SB1383 (accessed December 23, 2019).
23. California Air Resources Board. Recommendations to the State of California's Dairy and Livestock Greenhouse Gas Reduction Working Group. November 26, 2018. Available from: https://ww3.arb.ca.gov/cc/dairy/dairy_subgroup_recommendations_to_wg_11-26-18.pdf (accessed December 23, 2019).
24. Washington State Conservation Commission. $3.8 million awarded for innovative dairy nutrient management projects. June 15, 2018. Olympia, WA. Available from: https://scc.wa.gov/dairynutrientprojects-0618 (accessed December 23, 2019).
25. Bruderer LC. Federal court decides manure can be a solid waste. January 29, 2015. Available from: https://www.downeybrand.com/legal-alerts/federal-court-decides-manure-can-be-a-solid-waste/ (accessed December 23, 2019).
26. Dumas C. Case could impact dairies, livestock operations nationwide. January 21, 2015. Available from: https://perma.cc/4QLT-M6VC?type=image (accessed December 23, 2019).
27. Merlo C. Washington dairy decision could set national nitrogen management precedent. February 5, 2015. Available from: http://www.agweb.com/article/washington-dairy-decision-could-set-national-nitrogen-management-precedent-NAA-catherine-merlo (accessed December 23, 2019).

28. Starbucks. Animal welfare-friendly practices. Updated December 18, 2014. Available from: http://globalassets.starbucks.com/assets/313ef95924754048b3ca8cea3cc2ff90.pdf (accessed December 23, 2019).
29. Innovation Center for US Dairy. US Dairy Stewardship Commitment. Available from: http://commitment.usdairy.com (accessed December 23, 2019).
30. Copeland C. Air quality issues and animal agriculture: EPA's air compliance agreement. August 18, 2014. Congressional Research Service. Available from: https://nationalaglawcenter.org/wp-content/uploads/assets/crs/RL32947.pdf (accessed December 23, 2019).
31. Aillery, M., Gollehon, N., Johansson, R. et al. (2005). *Managing Manure to Improve Air and Water Quality*. Washington, DC: United State Economic Research Service.
32. Gay, S.W. and Knowlton, K.F. (2005). *Ammonia Emissions and Animal Agriculture*. Ettrick, VA: Virginia Cooperative Extension, Virginia Tech, and Virginia State University.
33. Jiang, A., Zhang, T., Zhao, Q. et al. (2010). Integrating ammonia recovery technology in conjunction with dairy anaerobic digestion. In: *Climate Friendly Farming: Improving the Carbon Footprint of Agriculture in the Pacific Northwest* (eds. C. Kruger, G. Yorgey, S. Chen, et al.). Pullman, WA: Washington State University.
34. Sweeten, J., Erickson, L., Woodford, P. et al. (2000). *Air Quality Research and Technology Transfer White Paper and Recommendations for Concentrated Animal Feeding Operations*. Washington, DC: USDA Agricultural Air Quality Task Force.
35. US-EPA (2012). *The Green Book Nonattainment Areas for Criteria Pollutants*. Washington, DC: United States Environmental Protection Agency.
36. Pruitt, G. (2013). *Yakima Regional Clean Air Agency: Message to the Public from the Director*. Yakima, WA: Yakima Regional Clean Air Agency.
37. Frear, C., Ma, J., and Yorgey, G. (2018). *Approaches to Nutrient Recovery from Dairy Manure*. Extension Publication EM112E. Pullman, WA: Washington State University.
38. Bishop, C.P. and Shumway, C.R. (2009). The economics of dairy anaerobic digestion with coproduct marketing. *Review of Agricultural Economics* **31** (3): 394–410.
39. Ziobro, J. and Bonnelycke, N. (2015). *The Importance of Markets for Co-products and Innovations for Farms of all Sizes. Waste to Worth: Spreading Science and Solutions*. Seattle, WA: US Environmental Protection Agency.
40. Yorgey, G., Frear, C., Kruger, C.E., and Zimmerman, T. (2014). *The Rationale for Recovery of Phosphorus and Nitrogen from Dairy Manure*. Pullman, WA: Washington State University Extension.

2.3

Biomass Nutrient Management in China: The Impact of Rapid Growth and Energy Demand

Paul Thiers

School of Politics, Philosophy and Public Affairs, Washington State University Vancouver, Vancouver, WA, USA

2.3.1 Introduction

In China, biomass management has been shaped far less by intentional regulatory action (as in Europe) or litigation (as in the United States) and far more by broad economic and forces and policies in other areas. The driving factors have been a decline in the availability of agricultural labor, the rapid industrialization of rural areas, the ever-increasing demand for energy, and, recently, a national policy decision to promote renewable energy. This has led to ongoing conflict over the use of biomass beyond the regulatory or legal battles of the Western democracies. The Chinese case shows how policy changes that have nothing to do with nutrient management in their intent can, nonetheless, drive dramatic changes in the nutrient cycle. The inability of policy makers to establish an effective regulatory environment also illustrates the additional challenges rational nutrient management faces outside of the wealthy, democratic states of the First World.

2.3.2 The Impact of Economic Liberalization Policy in the 1980s and 1990s

When the Chinese government introduced economic reforms in the early 1980s, moving away from decades of a closed, purely socialist economy, it began with reforms in the agricultural sector. Large, inefficient People's Communes were broken up into household-scale

Biorefinery of Inorganics: Recovering Mineral Nutrients from Biomass and Organic Waste, First Edition.
Edited by Erik Meers, Gerard Velthof, Evi Michels and René Rietra.
© 2020 John Wiley & Sons Ltd. Published 2020 by John Wiley & Sons Ltd.

units of production. This household responsibility system gave peasant families direct market incentives to maximize production, especially the intensive production of vegetables that could be sold directly to newly permitted local markets once state-established grain production quotas had been met. At the same time, the government-enforced household registration system made it illegal for rural people to migrate to cities in search of employment. With no off-farm opportunities for return on labor and virtually no capital available, labor intensity in agriculture increased rapidly. Under these policies, labor-intensive forms of nutrient management (composting, cover-cropping, etc.) increased dramatically. Production per acre also increased, and the total value of agricultural output rose from just under 100 billion Yuan in 1978 to 500 billion Yuan in 1990, primarily due to intensification of on-farm labor [1, 2].

By the late 1980s, however, changes in policy and accelerated economic growth began to reverse this trend. Government policies were implemented to stimulate industrial production both in the cities and in local government-owned rural enterprises, creating a demand for off-farm labor that would continue to increase for the next 20 years. Simultaneously, the accumulation of rural wealth in the early stage of reform created opportunities to invest capital in agricultural production through the purchase of external inputs. These changes in opportunity costs to labor and the availability of capital drove a dramatic change in the nutrient cycle in China [1].

While reliable data concerning changes in all on-farm nutrient management during this period are not available, interviews with agricultural researchers indicated that the shift from labor-intensive agriculture to a more capital-intensive form reflected a move away from composting and other traditional practices as farmers bought chemical fertilizers and herbicides in ever-increasing amounts. The use of chemical inputs to replace labor accelerated until, by the mid-1990s, chemical application rates in some eastern provinces were among the highest in the world [3]. By most accounts, these extremely high rates of application have continued until very recently. Bai et al. [4] document that the rapid and unsustainable increase in mineral phosphorus use continued from the beginning of the agricultural reform policy in the 1980s right up until 2014. Official data from the Chinese Statistical Bureau should always be treated with skepticism, but it is worth noting that officially, national consumption of chemical fertilizer increased every year from 1952 to 2015 and then declined slightly in 2016 and 2017 [5].

While rapidly changing policies and economic forces were not specifically intended to change nutrient-management practices, it seems clear that they had the effect of reducing the on-farm recycling of nutrients. This created a surplus in biomass. In the North China Plain, a major agricultural region, more and more field residue was diverted to paper production. Many small-scale paper mills along the Huai River used rice and wheat straw as a feedstock. These small mills, many operating without any regulatory supervision, were highly inefficient and polluting, leading to a serious degradation of water quality in the Huai [6].

2.3.3 Environmental Protection Efforts and Unintended Consequences

After a number of highly publicized fish kills and local protests about water quality, the central government initiated a campaign to close down dozens of these mills [6]. As with so many policy initiatives, this campaign had unintended consequences for biomass management. Closing paper mills reduced the market for field straw, forcing famers to go back to on-farm management. But another decade of rapid economic growth and a loosening of restrictions on rural migrants finding employment in growing cities had further reduced the availability of on-farm labor. As a result, burning field straw has become increasingly prevalent since 2000. This is an effective way to save labor while removing biomass, but it is also a major source of air pollution in North China. While the central government has decreed that field burning is not allowed, a 2017 study using satellite imaging and other data sources found that in-field burning of corn, rice, and wheat straw is still widespread and is the major source of nitrous oxides air pollution in the country [7].

Rapid economic growth and increasing affluence in urban areas have had another profound influence on nutrient cycling in China. Consumer demand for meat and dairy products has expanded very quickly [8]. This has led to the emergence of two production types, "specialty households" raising a few dozen to a few hundred animals, and large concentrated animal feeding operations (CAFOs), raising hundreds of thousands or even millions. Ma et al. [9] report that generated manure increased from 400 Tg (fresh weight) in 1949 to 2670 Tg in 2009. Both production types are significant sources of surface and water pollution that have proved extremely difficult to regulate. Specialty households are disbursed and largely outside of the limited capacity of the central government to regulate. Large CAFOs are often owned by, or connected to, local government agricultural bureaus, making regulation politically difficult. While some manure from these operations is disposed of on local farm fields, the percentage disposed in this way has substantially decreased as chemical fertilizer has become more available, meaning that at least half is disposed of directly into waterways or landfills [9].

The Chinese government has made significant and repeated efforts to decrease pollution from agriculture. The need to promote ecologically sound agriculture has been mentioned in every Five-Year Plan since 1996. The Environmental Protection Law of 2014 established policies and subsidies for the diversion of animal waste toward the production of renewable energy (see later). Efforts to rationalize the use of chemical fertilizer have primarily been coordinated through the Ministry of Agriculture. The most recent significant effort is the Ministry's 2015 Action to Achieve Zero Growth of Chemical Fertilizer Use by 2020, which was given central Party/State legitimacy in the 13th Five-Year Plan in 2016. The Action Plan goes beyond macro-level policy statements to focus specifically on improving monitoring of the quantity and composition of applied chemical fertilizer and its replacement, where possible, with organic fertilizer [10]. Shuqin and Fang [10] conclude that the first year of the plan made some but insufficient progress. They identify several barriers to success, most specifically the lack of reliable data and the conflicts of interest among technical extension workers and local government agricultural bureaus, both of whom frequently profit from fertilizer sales [10].

As noted earlier, the Chinese National Bureau of Statistics reported that 2016 and 2017 saw the first official declines in chemical fertilizer use in more than 60 years [5]. If this decline, or even a plateau, is indeed taking place, it is a significant change that may be the result of the Ministry of Agriculture action. Even as China has become wealthier and more concerned about pollution, inadequate enforcement of environmental policy, particularly in rural areas, has continued to be pervasive [11]. So reductions in chemical fertilizer use are likely to be a result of the extension of better practices through programs like the Ministry of Agriculture's. On the manure-management side, however, it seems that changes in nutrient management have come, not from the regulation of agriculture, but from the promotion of alternative energy production. Decades of rapid economic growth have led to equally rapid increases in energy demand. Most of this demand has been met through increases in coal consumption. But urban air pollution from coal consumption has become a major political issue. Coal consumption has also caused China to become the world's largest emitter of greenhouse gases. In the last decade, the Chinese government has launched a number of national policy initiatives to change the country's energy profile. These initiatives have caused a renewed interest in biomass, now as a potential source of energy.

2.3.4 Renewable Energy Policy and Its Impact on Biomass Management

While reforms since the 1980s have greatly increased the importance of market signals, the Chinese economy is still heavily influenced by central planning. The state issues Five-Year Plans, representing intense negotiations among government ministries and interests, which set production goals and lead to specific policies and allocations. The 11th Five-Year Plan (2005–2010) and 12th Five-Year Plan (2011–2015) set a new direction in energy policy, including the promotion of renewable energy [1]. The 13th Five-Year Plan (2016–2020) sets the ambitions goal of reaching 90 GWh of electricity from biomass and 8 Billion cubic meters of bionatural gas per year by 2020 (http://Asiapacificenergy.org). As a legal foundation to this energy transition, the Chinese government issued a Renewable Energy Law outlining specific policies to promote alternative energy sources including the production of energy from biomass [12].

While a number of policy initiatives have impacted biomass use, three national policies and one international initiative have proven particularly important. Nationally, the government established pilot projects for the production of fuel ethanol, a feed-in tariff to reward electricity produced form biomass, and a subsidy for the construction of biogas digestion facilities. Internationally, China embraced the Clean Development Mechanism (CDM) of the Kyoto Protocol, linking Chinese biomass management to international carbon markets. The original design of each of these policies, and the unintended consequences due to implementation problems, illustrate how policy decisions around energy production and greenhouse gas emissions impact biomass management both intentionally and unintentionally.

With the 11th Five-Year Plan, the central government began to commit significant amounts of funding to the promotion of renewable energy, expected to total 800 billion USD between 2009 and 2020 [13]. As part of this spending, the government substantially increased subsidies for ethanol fuel production to about 302 USD per ton and increased funding for four large fuel ethanol plants that had been established in 2006 [14]. These plants produced fuel ethanol from both grain and cellulosic material (wheat straw and corn stover), the latter of which is heavily subsidized, as pilot projects intended to develop and demonstrate cellulosic ethanol technology. But, in combination with state financial support, they were also quite profitable and high-status. The low profitability of food grain agricultural production made these plants an attractive customer for farmers. According to plant managers interviewed in 2010, one cellulosic plant was consuming about one-third of the corn stover within a 25-mile radius just 2 years after they began production. However, the real profitability for these and other plants lay in the use of grain for ethanol production, particularly considering that the cost of cellulosic production, once the value of the government subsidy was removed, was still 30% higher than the market price of ethanol.

The inevitable preference for grain as a feedstock for ethanol production led to a very rapid increase in the amount of food grain being diverted to energy production. While it was technically illegal to divert land dedicated to food production into production of energy crops, several stakeholders interviewed in 2010 acknowledged that clandestine diversion of land and grain was widespread [14]. Unable to fully enforce its restrictions, the central government cut fuel ethanol subsidies from 302 USD per ton in 2009 to 244 USD per ton in 2010 [14]. Even so, some local governments continued to promote ethanol production as a feedstock for local production of fuel and even more lucrative alcohol production, illustrating the lack of enforcement capacity in the face of economic incentive.

The government also took steps to promote biomass production of electricity, primarily through a fee-in tariff (FIT). A FIT is a type of subsidy where the producer of electricity is paid an additional amount at the point of sale to the electric power grid. The Chinese Renewable Energy Law set a FIT for electricity produced from renewable sources, including electricity produced from biomass, at 0.25 RMD (0.04 USD) per kilowatt hour above the prevailing price paid for coal-generated electricity within the region [15]. This was a significant incentive that led to a rapid expansion of renewable energy, especially wind power. It also increased the use of biomass for electricity, including both the biodigestion of animal waste (see later) and the direct combustion of field straw.

Plans to construct power plants generating electricity through the direct combustion of large amounts of field straw began as soon as the presence of a FIT in the Renewable Energy Law became known. While no systematic research is available, it is clear that these plants made use of the large quantities of field residue that had become available with the decline of on-farm composting and small-scale paper production. One such plant is the Hebei Jinzhou 24 MW Straw-Fired Power Project, which received project approval from the Chinese National Development and Reform Commission in late 2006 [16]. The plant is designed to combust about 176 000 tons of corn and wheat straw in two large boilers to generate about 132 GWh of power per year. That is about 44% of the total corn and wheat residue estimated to be produced in the Jinzhou region, giving some idea of the potential significance of these plants to biomass management. It is likely that these new combustion plants produce less local air pollution than field burning. The energy generation is also perceived as a benefit to China's goals of reducing coal consumption. The impact on soil fertility of diverting large quantities of biomass to energy generation has not been systematically assessed.

China's renewable energy policy has also driven changes in the management of animal manure on large CAFOs. The Chinese Ministry of Agriculture has a tradition of promoting household- and village-scale anaerobic digestion of animal manure to produce methane for farmhouse cooking and heating, dating back to before the economic reforms. Millions of digesters of about three cubic meters were installed, though most fell out of use with lack of maintenance and rising opportunity costs to labor. With the growth of energy demand and government support for renewable energy, anaerobic digestion is undergoing a renaissance, focused on the new, very large CAFOs. In 2008, the National Development and Reform Commission announced that it would pay a substantial portion of construction costs for medium-sized and large biogas plants in rural areas: 45% in the poorest regions, falling to 25% in rich parts of the country [17]. Because upfront costs are a major consideration in new power plant construction, this was a significant subsidy, allowing many projects to go forward.

China's participation in the CDM of the Kyoto Protocol on climate change also spurred activity on biomass to energy projects. The CDM is an international trading scheme established to allow wealthy countries to meet some of their greenhouse gas reduction commitments by buying certified emission-reduction credits from poorer countries that have made no such commitments. Credits are generated when an enterprise adopts behavior that reduces or avoids emitting greenhouse gases. But the behavior must be carefully documented and certified by a third party and registered with both national and international agencies. China, as a rapidly industrializing country with considerable inefficiencies in energy use, offers a variety of low-cost opportunities for Western governments to fund reductions in greenhouse gases. Since the Kyoto Protocol entered into force in 2005, China has been the world's leading producer of certified emission-reduction credits, accounting for more than half of the world's total [18].

While most of the CDM-funded projects in China have focused on changing chemical industry processes to reduce emissions of specific climate changing chemicals or on constructing large wind farms to replace coal power, some projects have directly sought to change biomass management. Biomass-relevant projects have centered around two categories of CDM-approved methodologies: the use of biomass to generate electricity and changes in agricultural waste management to reduce the uncontrolled emission of methane. A large CAFO can apply for credits under both. By moving from open lagoon storage of waste to controlled anaerobic digestion, methane previously emitted into the atmosphere can be flamed, converting emissions to carbon dioxide, a far less potent greenhouse gas. If the flamed methane is used to generate electricity, the electricity sold to the grid can earn credits equivalent to the coal that would be required to generate the same volume of power.

As the CDM became a hot topic in China, several enterprises generated proposals for greenhouse gas emissions reduction. Several were simply claiming the coal-reduction benefits of direct combustion of field residue, including the Jinzhou Straw Fired Power Project discussed earlier, which claims to have generated sufficient electricity to replace coal consumption equivalent to more than 178 thousand tons of CO_2 emissions per year since it registered with the CDM and began generating power in 2007 [16].

The potential for CAFOs that install anaerobic digestion to produce and sell certified emission-reduction credits using both methodologies (reducing methane emissions from lagoons and burning methane to generate electricity) could reasonably have been expected to play a significant role in promoting anaerobic digestion of manure with subsequent benefits for nutrient recycling and local pollution reduction. In the end, only four projects registered with and went through the complex validation procedures for anaerobic digestion plants during the 8 years of CDM activities in China (2005–2012), and only one of these actually completed the process and sold certified emission-reduction credits.

Site visits to three of these projects [19, 20] highlighted some of the problems with using carbon trading to promote anaerobic digestion in a developing-world context. First and foremost, the extremely high accounting and validation costs required to certify the emission reduction credits greatly reduced the profitability and attractiveness of the process. All of these CAFOs were very large, constituting a best-case scenario for economies of scale. The one fully successful project was the largest: a five-million-chicken CAFO in Shandong Province. But even at these very high economies of scale, managers reported that transaction costs were prohibitive; in the case of a 10 000-head dairy CAFO in Inner Mongolia Province, they were high enough to lead the managers to abandon the CDM process even as it was nearing completion. Another CAFO was denied certification of emission-reduction credits during the validation process, though the reason for this denial was considered confidential [19].

These problems point to the relative failure of carbon pricing to promote anaerobic digestion in rural China, and bioenergy has declined as an element of China's climate-change policy. Bioenergy is not a component of China's new domestic cap-and-trade program, coming on line now (Ba 2018), and it received only passing mention in China's national commitments under the Paris Accord in 2015 [21]. On a more promising note, research and development money for biological nitrogen fixation does appear in the document, indicating, perhaps, an appreciation for the contribution that direct nutrient management policy could make to greenhouse gas emission. But the CDM experience did at least create a foothold for large-scale digestion technology as a step toward manure management in China. All of these large CAFOs did, in fact, install biodigesters. Managers indicate that the biogas construction subsidy and the feed-in tariff were far more significant than the potential profits offered by the CDM in their decision to build and operate these digesters. This indicates that, with the right mix of policies, anaerobic digestion can become reality on at least the very largest of CAFOs despite the lack of enforcement of environmental regulations. One of the CAFOs has begun to process methane into natural gas for sale to local natural gas-powered vehicles, another transition promoted by government energy policy and the high market demand for energy. Because energy demand and transition to renewable energy are such driving forces in China, it seems likely that domestic renewable energy policy, rather than pollution prevention or nutrient recycling, will continue to be primary path promoting this technology.

This is not to say that implementation of energy policy is always reliable. The same problems that plague enforcement of environmental policy are also present in the energy sector. There have long been reports that electric grid managers refuse to buy energy produced by renewable energy sources because they do not trust decentralized sources and do not want to pay the higher price. This is a direct violation of the requirement in the renewable energy law that grid operators guarantee grid access to renewable sources [22]. The manager of the biogas plant in Inner Mongolia confirmed that the grid frequently refused to purchase its electricity, meaning that staff had to burn off the methane. This problem seems to be most prominent in provinces such as Inner Mongolia where coal producers have a great deal of economic and political power [20]. While one of the large CAFOs is experimenting with processed natural gas as an alternative energy product, the reliability of grid access is essential if energy production is to drive this technology forward. A lack of grid access and long-term grid contracts is cited as one reason why biogas production has lagged behind the 12th Five-Year Plan target of 30 GW of installed capacity by 2020, reaching only 6.7 GW by 2012 [22].

2.3.5 Conclusion

Biomass management in China has changed dramatically since the late 1970s. These changes have been driven, not by environmental regulation or a direct interest in nutrient recycling for fertility, but by rapid economic growth and unintentional effects of policies in unrelated areas such as labor, energy, and climate change. At present, it is the energy potential of field and animal waste, more than pollution or soil fertility issues, that will hold the attention of policy makers and enterprise managers. This interest in energy recovery may unintentionally be leading to some better outcomes for local environments. However, real progress will depend on stakeholders making nutrient recycling for its own sake a priority. If the central government's new emphasis on air and water pollution prevention can overcome China's longstanding lack of regulatory enforcement, there may be increasing opportunities to improve nutrient management in the near future.

References

1. Ash, R. and Kueh, Y.Y. (1996). *Agricultural Development in China Since 1978*. Oxford: Clarendon Press.
2. Colby WH, Crook FW, Webb S-EH. Agricultural Statistics of the People's Republic of China, 1949–90. Statistical Bulletin 154783, United States Department of Agriculture, Economic Research Service. 1992. Available from: https://ideas.repec.org/p/ags/uerssb/154783.html (accessed December 23, 2019).
3. Thiers, P. (1997). Pesticides in China: policy and practice. *Pesticide Outlook* **8** (1): 6–10.
4. Bai, Z., Ma, L., Ma, W. et al. (2016). Changes in phosphorus use and losses in the food chain of China during 1950–2010 and forecasts for 2030. *Nutrient Cycling in Agroecosystems* **104** (3): 361–372.
5. CEIC. China Chemical Fertilizer Consumption 1952–2017. CEIC data 2019. Available from: https://www.ceicdata.com/en/china/consumption-of-chemical-fertilizer/cn-chemical-fertilizer-consumption (accessed December 23, 2019).
6. Economy, E.C. (2010). *The River Runs Black: The Environmental Challenge to China's Future*. Ithaca, NY: Cornell University Press.
7. Zhou, Y., Xing, X., Lang, J. et al. (2017). A comprehensive biomass burning emission inventory with high spatial and temporal resolution in China. *Atmospheric Chemistry and Physics* **17** (4): 2839.
8. He, Y., Yang, X., Xia, J. et al. (2016). Consumption of meat and dairy products in China: a review. *Proceedings of the Nutrition Society* **75** (3): 385–391.
9. Ma, L., Zhang, W., Ma, W. et al. (2013). An analysis of developments and challenges in nutrient management in China. *Journal of Environmental Quality* **42** (4): 951–961.
10. Shuqin, J. and Fang, Z. (2018). Zero growth of chemical fertilizer and pesticide use: China's objectives, progress and challenges. *Journal of Resources and Ecology* **9** (1): 50–59.
11. Economy EC. China's environmental enforcement glitch. Available from: http://thediplomat.com/2015/01/chinas-environmental-enforcement-glitch (accessed December 23, 2019).
12. Qiu, X. and Li, H. (2012). Energy regulation and legislation in China. *Environmental Law Reporter* **42**: 10678–10693.
13. Zhang J. Green energy: Development and investment opportunities in China.Vice Chairperson of Energy Research and Development Center of the China Investment Association and CEO of the China Energy Investment Net International Conference on Biomass Energy Technologies. August 19, 2010. Beijing, China.
14. Nesbitt, E.R., Thiers, P., Gao, J. et al. (2011). China's vision for renewable energy: the status of bioenergy and bioproduct research and commercialization. *Industrial Biotechnology* **7** (5): 336–348.
15. Jiang, X., Sommer, S.G., and Christensen, K.V. (2011). A review of the biogas industry in China. *Energy Policy* **39** (10): 6073–6081.
16. CDM. Project Design Document, Hebei Jinzhou 24MW Straw-fired Power Project. Clean Development Mechanism of the UNFCC2006.
17. General Office of the National Development and Reform Commission and the General Office of the Ministry of Agriculture. Rural Biogas No. 2519, Construction of Rural Biogas Project 2009. 2008.

18. UNFCCC. Data for CDM Project Activities. United Nations Framework Convention on Climate Change 2017. Available from: http://cdm.unfccc.int/Statistics/Public/CDMinsights/index.html#iss (accessed December 23, 2019).
19. Thiers P. Assessing Policies to Promote Biogas Energy on Concentrated Animal Feeding Operations in China: Lessons from the Clean Development Mechanism. Presentation. PRC Environmental Tradeoffs: Modern China's Environment, Sciences and Landscapes; Richmond, CAN. September 21, 2015.
20. Pierce HC. Assessing the clean development mechanism in rural China: Can ecological modernization function in an authoritarian context? Master's thesis: Washington State University; 2013.
21. UNFCCC. Enhanced Actions on Climate Change: China's Intended Nationally Determined Contributions. United Nations Framework Convention on Climate Change 2015. Available from: https://www4.unfccc.int/sites/ndcstaging/PublishedDocuments/China%20First/China%27s%20First%20NDC%20Submission.pdf (accessed December 23, 2019).
22. Gutzke M. Challenges for Biogas in China: The Effectiveness of Chinese Policy Mechanisms to Address Discriminating Barriers for the Environmentally Sound and Financially Viable Production of Biogas in the Chinese Husbandry Sector. Unpublished master's thesis. Kings College London; 2012.

2.4

Nutrient Cycling in Agriculture in China

Lin Ma[1], Yong Hou[2], and Zhaohai Bai[1]

[1] *Center for Agricultural Resources Research, Institute of Genetic and Developmental Biology, CAS, Shijiazhuang, Hebei, China*

[2] *College of Resources and Environment Sciences, China Agricultural University, Beijing, China*

2.4.1 Introduction

China's population is the largest in the world, jumping from 1.01 billion in 1980 to 1.44 billion in 2017, an increase of 43% [1]. The economy has also developed rapidly; the average gross domestic product (GDP) per capita rose from 195 to 8759 USD between 1980 and 2017, an increase of a factor of 44 [2]. Tilman et al. [3] indicate that there is positive correlation between food consumption (on energy and protein basis) and average income level. China has also experienced a strong increase in urbanization rate, from 19% in 1980 to 58% in 2017. Diets differ between rural and urban residents, with the latter consuming more fruits and animal products [4]. Overall, these factors have led to higher food consumption and an increasing demand for animal products. Nutrient inputs to agricultural production systems have strongly increased to fulfill the food demand. Nutrient management has also changed, and nutrient cycling between crop and animal production has been broken. This has resulted in severe environmental issues at the national and regional scales, including airborne pollution, eutrophication of waters, and high nutrient accumulation in cropland. In response to the severe problems with nutrients from agriculture, the Chinese government has implemented several acts since 2013 aiming to improve nutrient management at both the national and the regional scale.

Biorefinery of Inorganics: Recovering Mineral Nutrients from Biomass and Organic Waste, First Edition.
Edited by Erik Meers, Gerard Velthof, Evi Michels and René Rietra.
© 2020 John Wiley & Sons Ltd. Published 2020 by John Wiley & Sons Ltd.

2.4.2 Nutrient Cycling in China

The total production of N in animal manure increased from 6.8 Tg in 1980 to 17 Tg in 2010 (Figure 2.4.1), due to the rapid development of livestock production [7]. Manure P production increased from 0.96 Tg to 3.7 Tg during the same period. Bai et al. [5] estimated that the actual manure N and P production is even higher when all the backup/breeding animals and the protein and P contents in these animals are considered: 23 Tg N and 4.6 Tg P. Most of the manure is derived from landless industrial cattle and pig production systems [5]. The amount of N and P in straw (mainly of wheat, rice, and maize) increased in the period 1980–2010 from 3.4 to 6.5 Tg N and from 0.10 to 0.31 Tg P (Figure 2.4.1). Livestock and crop production were decoupled following the livestock transition between 1980 and 2010 [7]. Traditional and grazing systems were mixed production systems, as most feed came from own land and manure was applied to own land. However, the rapid intensification

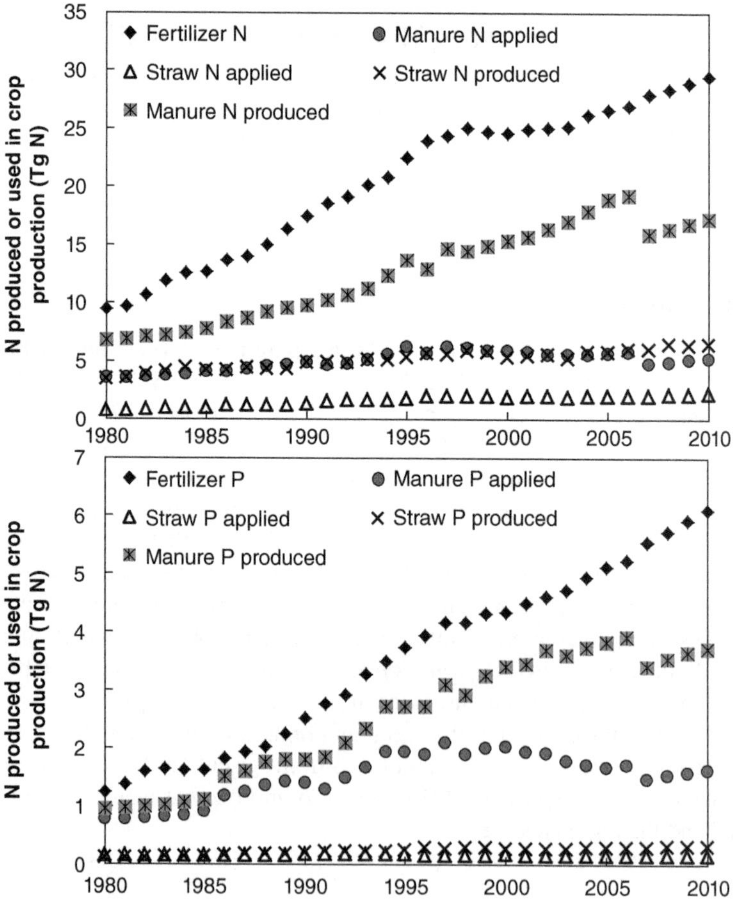

Figure 2.4.1 *Nitrogen (upper panel) and phosphorus (bottom panel) fertilizer use, production and application of manure, and production and utilization of straw in crop production in China between 1980 and 2010 [5, 6].*

of agricultural production led to the decoupling of livestock and crop production. The recycling rate of manure N decreased from 50 to 30%, and that of manure P from 80 to 40% between 1980 and 2010 (Figure 2.4.1), with the remaining manure either lost to atmosphere via ammonia emission and denitrification or lost to watercourses via direct discharge without treatment. The highest recycling rate for manure in the period 1980–2010 was shown for systems with ruminant animals and mixed systems, but the manure production in these systems was relatively small compared to other livestock systems [7]. The rapid decrease in recycling of manure was also partially related to the use of mineral fertilizer. In 1980, manure was the most important source of nutrients in agricultural production; only small amounts of mineral fertilizer were used. However, the amount of mineral fertilizer use has increased greatly, due to the existence of native P rock reserves in China, as well as governmental subsidies for the production, distribution, and use of mineral N and P fertilizers [8, 9]. Due to the oversupply of fertilizer N and P, there has been less demand regarding manure N and P, leading to the decoupling of crop–livestock production.

China plays a key role in global N and P fertilizer use, being the largest mineral N and P fertilizer producer and consumer (28 and 32% of global amounts) in the world in 2016 [1]. The large N and P use has resulted in relatively low N and P use efficiencies compared to many other regions in the world [10]. In crop production, N use efficiency has decreased from 52 to 42%, and P use efficiency from 79 to 49%, between 1980 and 2010. The low nutrient use efficiency in crop production is mainly attributed to overfertilization. In addition, the increased use of N and P fertilizers in China has been ascribed to the increased food and feed needs of the increasing human and animal populations [9, 11, 12]. Moreover, the increased consumption of vegetables and fruits since the 1990s has contributed to a higher N and P fertilizer use, because of the relatively large nutrient supply to vegetable and fruit production [5, 13]. Both N and P use efficiencies for livestock production have steadily increased at the herd level. However, N and P use efficiencies at the whole food system have decreased greatly [5, 6] because of the increased contribution of livestock production to the total agriculture production. The N and P use efficiencies of livestock production are still much lower than for crop production, despite improvements. Hence, the increasing contribution of livestock production will continue the low N and P use efficiency at the food system level.

2.4.3 Effects on the Environment

There is increasing concern about water pollution in China. High concentrations of nitrate (NO_3^-) have been shown in the groundwater, especially in the intensively managed agricultural regions of northern China. In 2016, the Ministry of Water Resources of China reported that 80% of 2103 monitored wells were not suitable for the provision of drinking water, mostly because NO_3^- concentrations exceeded the World Health Organization (WHO) quality standard for drinking water of 11 mg NO_3-N L^{-1} or 50 mg NO_3 L^{-1} [14]. Most of these monitored wells are shallow groundwater, not widely used as a drinking water source by the urban population, but still an important water source for rural people in China. Drinking water in rural areas is mainly used directly, without sufficient testing for quality. It has been reported that 31% of the rivers and six of nine major coastal bays in China are suffering from eutrophication. In addition, more than 60% of the monitored drinking water wells are severely contaminated, classified as level IV (scale I to V, where V is the worst) or worse, with NO_3^- concentrations exceeding $130\,mg\,l^{-1}$ – far beyond the WHO quality standard of $50\,mg\,l^{-1}$ [15–17]. Loss of N and P from agriculture is the main cause of the poor water quality. Han et al. [16] assessed available data sets on NO_3^- concentrations and showed that groundwater and surface-water pollution are ubiquitous throughout China, with the most serious contamination in the intensively managed agricultural regions of the north. Strokal et al. [18] concluded that livestock production systems contribute significantly to N losses to the surface and groundwater. Isotope analysis of NO_3^- in the groundwater also indicated that most originates from fertilizer and manure [16, 19].

Heavy air pollution has reduced Chinese urbanities' sense of happiness [20]. In fact, the poor air quality has serious impacts on human health and life expectancy. In 2010, it was estimated that 1.77 million people aged over 65 years in China had premature mortality related to long-term exposure to high fine particle ($PM_{2.5}$) concentrations >$30\,\mu g\,m^{-3}$ [21]. There are greater regional differences in deaths related to high $PM_{2.5}$ concentration, most of which are concentrated in the North China Plain and Yangtze River Delta [21]. The mechanisms of formation of high levels of $PM_{2.5}$ in extreme weather conditions across China are complicated, however, with many studies indicating that ammonia (NH_3) emissions contribute greatly to $PM_{2.5}$ formation [22]. The contribution of NH_3 to $PM_{2.5}$ formation was on average ~30% during a severe haze event in 2015. Ammonia not only poses a risk to human health because of $PM_{2.5}$ formation, but also decreases biodiversity and increases soil acidification [23]. Agricultural production, especially livestock production, contributes to more than 90% of NH_3 emission in China [24]. Total NH_3-N emissions increased from 5.6 Tg in 1980 to 13.6 Tg in 2010. Most of the NH_3 emission was from traditional, mixed production systems, but a rapidly increasing amount comes from landless systems [7]. Livestock production in China emitted 6.7 Tg NH_3-N in 2010, equivalent to 49% of the total NH_3 emissions from agriculture (Figure 2.4.2). Housing systems and manure storage are the major sources of NH_3 emission, representing up to 73% of total NH_3 emissions from livestock production. Inadequate manure collection and storage are the main causes of release (Figure 2.4.2). Manure treatment and application are also important, releasing 0.8 and 1.0 Tg NH_3-N per year, respectively (Figure 2.4.1a). In addition, urea is also a major source of ammonia emission, as it is the most popular N fertilizer in China.

Direct discharge of manure into surface waters accounts for over two-thirds of the load of N and P in the northern rivers and for 20–95% in the central and southern rivers [18]. In 2010, 5.5 Tg of manure N entered surface-water systems, 97% of which originated from livestock manure seepage and direct discharge from housing and manure storage, and only 3% from manure application through runoff and erosion (Figure 2.4.1a,c). The direct discharge of manure occurs mainly as a result of (i) lack of enforcement of regulations for manure storage capacity, (ii) lack of obligations or incentives to recycle manure back to crop production, (iii) lack of appropriate monitoring and control, and (iv) poor appreciation of the fertilizer value of manures. An analysis of soil cores from the edge of a 20-year-old layer hen manure store showed that NO_3-N contents were $50–130\,mg\,kg^{-1}$ in the top 100 cm of the soil – higher than the contents in the soil of a nearby 30-year-old fertilized wheat–maize rotation system. Similar high soil NO_3-N concentrations were found near a 12-year-old dairy manure store (Figure 2.4.1b). Clearly, current livestock housing and manure storage in China pose a great threat to groundwater and air quality.

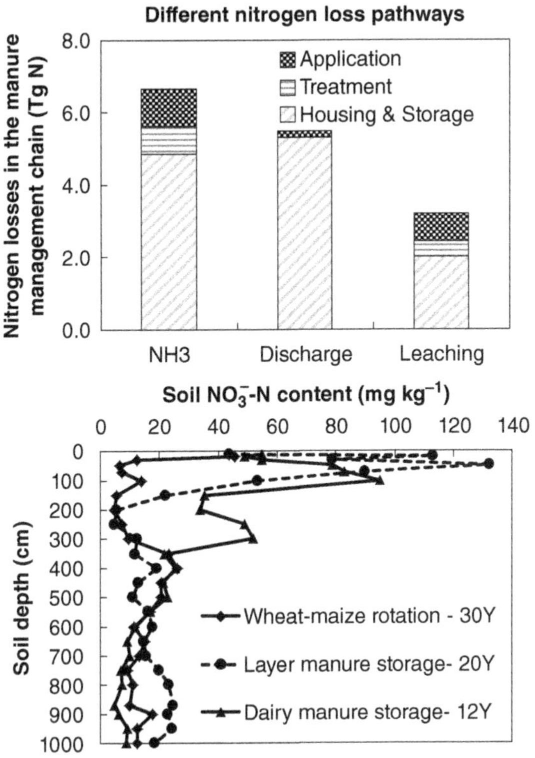

Figure 2.4.2 *(a) Total nitrogen losses from the manure management chain in China in 2010. (b) NO_{3}-N contents of different soil depths from different management systems in the North China Plain. Source: (a) Derived from Bai et al. [5]; (b) Derived from on-farm sampling and laboratory analysis in 2016.*

2.4.4 Nutrient Management Policies

To address the environmental issues raised by livestock production, there is a need for governmental regulations and policy instruments. Regulations on manure management have been implemented in many countries since the 1990s (e.g. the Nitrates Directive in the European Union). These regulations define how manure must be stored, treated, and applied, and how it can be applied to crops. In contrast, policy instruments regarding manure management are not yet or only poorly implemented in developing countries.

Since the policy reform of 1978, the scale and number of livestock farms in China have increased under market-originated driving forces, as described in the previous section. In the early 2000s, Chinese livestock policies gradually shifted from an objective of economic development to a more integrative one, including environmental protection. In 2001, management measures for pollution control and standards of pollutant discharge were issued in the livestock sector, though these were subjected only to large-scale industrialized livestock producers, such as the Discharge Standard of Pollution for Livestock Production (GB 18596-2001; National Environmental Protection Administration). However, environmental policies so far have performed poorly in rural China, due to inadequate regular monitoring systems operated by local environmental agencies and insufficient instruments on management measures [25, 26].

In more recent years, there have been increasing attempts and policy pressures to address environmental pollution in the livestock sector and the recycling of livestock waste. In January 2014, China enacted its Intensive Livestock Farming Pollution Prevention Regulations, which encourage and support the disposal and utilization of livestock manure. In May 2015, the Chinese government released an Action Plan for Prevention and Control of Water Pollution (also known as Ten Water Protection Action). An important measure of this plan is the establishment of non-livestock production regions (NLPRs) near vulnerable water bodies, natural scenic places, and human residential areas. The main aim of the NLPR policy is to reduce the pollution of water courses. The NLPRs were implemented within 2.5 years, and by the end of 2017 there were already >90 000 of them, with a total area of around 0.82 million km^2, equivalent to 16% of the total area of agricultural land in China. The NLPRs have forced 0.26 million pig farms to shut down or move to other regions [4], decreasing the number of slaughtered pigs by 46 million head per year between 2014 and 2017 [4], or 7% of the total number of pigs slaughtered per year in China; this is equivalent to 80% of the pigs slaughtered per year in Germany, the third largest pig producer in the world [1].

This spatial planning policy of the Ministry of Agricultural and Rural Affairs appears at first sight to be very effective, as (potential) pollution sources are removed from vulnerable areas. As the consumption of pork is forecast to grow by 50% between 2010 and 2050 [7], it is important to address the question: Which farms and regions will take over the market share in the future? The Chinese government has assigned relatively poor provinces in the north and west as development regions for pig production. The main rationales for this are the current relatively low pig population density in these regions, the relatively high pig feed production potential, and the relatively large land availability for the application of pig manure. However, the pollution burden will be transferred to these provinces, especially if no strict emission mitigation technologies are implemented. It is as yet unclear whether the northern and western provinces of China can accommodate the additional environmental burden; they have a semi-arid climate, fragile natural grasslands, and already high NO_3 leaching and NH_3 emissions.

Some recent policies have been developed in order to increase the recycling or reuse of livestock manure in China, following the national actions for Agricultural Green Development [27]. For examples, the livestock manure policy target for manure reuse is 75% by 2020 [28], which has led to the implementation of 100 demonstration county projects where local government supports innovative management practices to improve the sustainable reuse of manure nutrients. In addition, in 2017, the Ministry of Agriculture (MOA) initiated the national action of Replacing Fertilizer by organic manure in 100 counties, with emphasis on cash crops including fruits, vegetables, and tea; this has since been extended to 150 counties. In 2015, the MOA introduced an Action to Achieve Zero Growth of Chemical Fertilizer Use by 2020, the first objective of which is to achieve annual growth rates of chemical fertilizer use of less than 1% from 2015 to 2019 and to realize zero growth of chemical fertilizer use for the principal crops by 2020. To achieve this objective, four key technical roadmaps are suggested: promotion of the use of organic fertilizers in crop production, promotion of soil testing and formulated fertilization, improvement of fertilization methods, and implementation of new fertilizers and new techniques.

Meanwhile, a number of environmental policies regarding renewable energy and air pollution have been introduced to manage livestock pollution. For example, the central government began offering biogas subsidies in 2005, and this is now one of the major instruments used to promote household biogas production in rural China [29] . The Blue Sky Act (2018) has set targets to reduce concentrations of SO_2 and NO_x by 15% between 2015 and 2020, with an emphasis on increasing manure recycling, though without a clear target [30]. This is less comprehensive than the current legislation in the EU, which requires member states to reduce emissions of SO_2, NO_x, non-methane volatile organic compounds, and NH_3 to agreed targets.

Overall, China has been increasingly concerned with the recycling of nutrients in agricultural waste, especially livestock manure, and more restrictive environmental policies are expected. However, the current policies have many limitations, including a lack of effective instrumentation, an unclear target of mitigation, and limited information and services provisioning. Future research on the policy instruments aimed at better nutrient cycling in China's agricultural systems should seek to reduce these limitations.

2.4.5 Future Perspectives

2.4.5.1 National Nutrient Management Strategy

This review highlights developments in nutrient management over the last several decades, and the challenges that must be overcome if China is to achieve the triple result of food security, resource use efficiency, and environmentally sound food production and consumption. Though various promising nutrient-management concepts and technologies have been developed and tested in research, especially in crop production, their adoption in practice is still negligible, because of some complex barriers and constraints. A coherent national strategy is needed to overcome the most persistent limitations to the widespread adoption of new approaches, including:

(i) a greater emphasis on knowledge and technology transfer from research to practice, through education, training, demonstration, and extension services;
(ii) a change in focus from merely increasing animal production to improving animal production and manure management, and linking the livestock-production and crop-production sectors;
(iii) a shift in focus from nutrient management almost exclusively in crop production to nutrient management across the whole food production–consumption chain;
(iv) a shift in focus in national policies from purely food security to an integrated approach that emphasizes food security, resource use efficiency, and environmentally sound production and consumption; and
(v) a shift in research from largely disciplinary, separately focused studies in the crop and animal sectors to both disciplinary and multidisciplinary studies, scenario analyses, and integrated assessments of whole food systems at different spatial levels.

2.4.5.2 Challenges of Technology Transfer in Manure Management

Though there has been a tradition of recycling manure nutrients for centuries, there is little scientific knowledge in Chinese research about the effectiveness of applied manure nutrients as soil amendments for crop production. Hence, there is a need to quantify the fertilizer replacement value of different animal manures in practice and to establish sound recommendations for their use. In addition to differences in fertilizer value between animal species, information is needed on how the agronomic value, and potential environmental impact, of "farmyard" manures (smallholders) may differ from manures and wastewaters produced by intensive livestock production systems.

Equally important is the implementation of incentives for recycling manure nutrients in crop production. Current subsidies on fertilizers discourage the recycling of manure nutrients. Even more important perhaps is the reintegration of animal-production systems with crop-production systems. Animal production in China is rapidly growing, and basically all growth is centered upon intensive landless livestock-production systems, which import feed from elsewhere and have no nearby land for proper manure disposal. This is not a sustainable system. Agglomeration of animal production systems near consumers and food processors is economically attractive, but discourages recycling of manure nutrients in cropland because of the large infrastructure and transportation costs. The central government should reconsider current national policies and redirect development pathways for animal production that fully integrate animal manure management as a valuable resource rather than a useless waste. Opportunities for sound recycling of manure nutrients must receive high priority when designing new industrial livestock production systems and locations for animal production. Such opportunities can be obtained through spatial planning (i.e. limiting livestock density within an area) or high-tech manure processing and subsequent transport and distribution of processed manures to cropland. When processed, manure nutrients can be less bulky and thus cheaper to transport. However, manure processing has considerable economic cost.

Apart from the proposed interventions by the central or regional governments in the spatial planning and design of animal production system, there is a need for the establishment of multidisciplinary teams of animal scientists, veterinarians, agronomists, engineers, economists, and environmental scientists to develop a long-term strategy for sustainable animal production systems with sound manure-management practices. Currently, these scientists often operate in isolation, and as a consequence, governments, policy makers, businesses, farmers, and society at large receive conflicting messages. Research in animal production has mostly focused on increasing animal productivity via breeding and high-quality feed formulations. The results of this research are adopted by intensive livestock-production systems (and not by the traditional smallholder mixed farming systems), but efficient use of the resulting manure production is neglected. Here, multidisciplinary research teams must develop integrated options for both improved animal production and sound manure management.

2.4.5.3 Environmental Protection

The United States and European Union also experience significant contamination of water bodies by N and P from agricultural sources. The Clean Water Act was introduced in the United States in 1972 and the Nitrates Directive in the EU in 1991, to protect water resources. All EU Member States have to designate Nitrate Vulnerable Zones (NVZs), defined as areas of land draining to waters that are affected by NO_3^- pollution or that could be affected if action is not taken to decrease NO_3-leaching. Designation of such areas vulnerable to N and P pollution could increase the effectiveness of water quality protection in China. Criteria for the design of such NVZs (for N and P losses to water) should include: (i) information about the current eutrophication status of waters, based on monitoring data for NO_3^- and P concentrations in surface waters and groundwater; (ii) a comprehensive evaluation of the risks of NO_3^- and P leaching from agriculture at regional levels; (iii) a comprehensive evaluation of the risk of N and P runoff in relation to proximity of surface water bodies; and (iv) a risk analysis regarding manure leaching and discharge to surface waters. The regions where N and P concentrations are already excessive should be termed Vulnerable Zones, where immediate policies and actions should be implemented, while the regions with a high risk of N and P loss are deemed as potential Vulnerable Zones, where further measurements of N and P in the watercourses should be taken as soon as possible. Scientists also need to develop action program for each NVZ, which should include measures to reduce N and P leaching and runoff, such as by calculating the maximum N and P application rates based on balanced fertilization in order to avoid overfertilization, increase recycling of manure, and decrease the risk of water pollution. A series of field studies needs to be conducted to quantify optimal nutrient application rates and the nutrient availability in manures, depending on crop type, soil type, climate, and yield level. Bai et al. [7] used the most recent monitoring data for NO_3^- concentrations in groundwater and surface waters and a comprehensive evaluation of the risks of NO_3^- and P leaching and runoff in more than 2000 counties in 2012 to identify the first N and P Vulnerable Zones (NPVZs) in agricultural land in China [31]. The total area of NPVZs covered around 68 million ha arable land (51% of the agricultural land), of which 20% was potential NPVZs. A combination of balanced N fertilization, precision fertilization and irrigation techniques, and a decrease in direct manure discharge into watercourses decreased N losses from the area of potential NPVZs by nearly 50% compared to 2012 values [32].

Mitigation of NH_3 emission from agricultural sources may be more effective if NH_3 emission ceilings are introduced at a regional level, comparable to those set in the EU National Emission Ceilings (NEC). The NH_3 ceiling should be adjusted to the risk of $PM_{2.5}$ formation and protection of biodiversity in the region. If emission exceeds the NH_3 ceiling, measures within the region must be taken to decrease emissions.

2.4.6 Conclusion

Four options, and their integration, have been identified as pathways to improve P use efficiency and decrease P losses in the food chain in China. Each single option is less effective than the integration of all four. Balanced P fertilization (S1) in crop production decreased mineral fertilizer P input and increased P use efficiency in crop production (PUEc; Figure 2.4.3). Precision animal feeding (S2) decreased the P input in animal production and thereby the amounts of P in animal manure and the P losses from animal production. Improved animal manure management (S3) greatly decreased P losses from animal production and at the same time decreased the need for mineral fertilizer P input. Compliance with the recommendations for a "healthy diet" (S4) will lead to more vegetable and fruit production and less animal production; as a result, manure P production, mineral fertilizer P use, and P loss will decrease. The integration of the four options reduced mineral

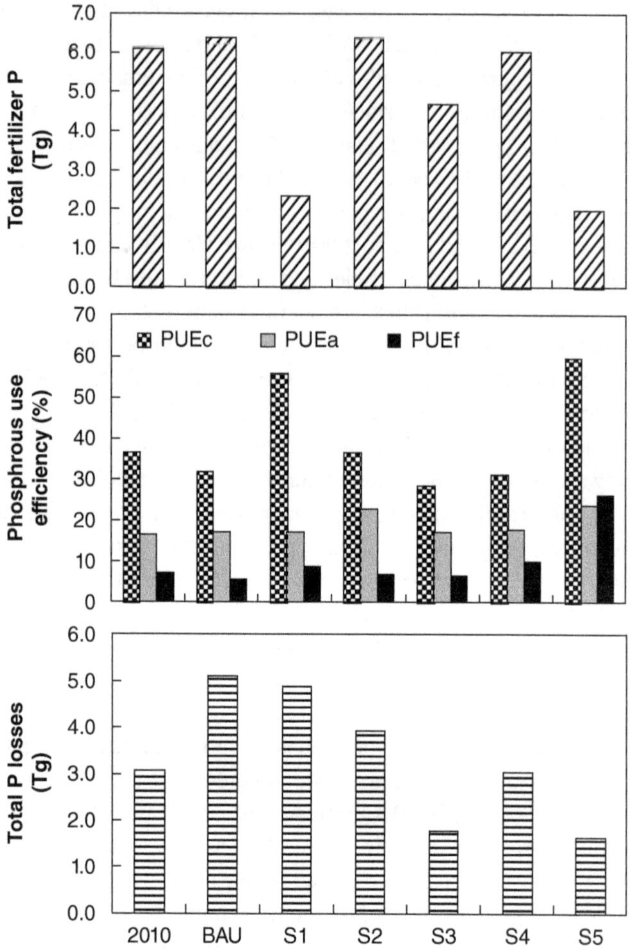

Figure 2.4.3 *Forecasts of (a) annual P fertilizer consumption, (b) P use efficiency in crop production (PUEc), animal production (PUEa), and the food chain (PUEf), and (c) total P losses in 2030, according to six scenarios: BAU, business as usual; S1, balanced P fertilization in crop production; S2, precision animal P feeding (i.e. lowering the P supplementation of animal feed by 20%); S3, improved manure management (i.e. 95% of manure P is collected and applied to cropland); S4, compliance to Chinese dietary recommendations; S5, integration of S1–S4. Source: Results derived from Bai et al. [5].*

P fertilizer requirements by 69% and total P losses by 68% (Figure 2.4.3). Also, P use efficiency in crop production increased by 88%, that in animal production by 38%, and that of the whole food chain by 350% compared to business as usual (BAU) (Figure 2.4.3) [5]. Similar achievements in terms of fertilizer N savings and N loss reduction could be achieved under the same scenarios.

References

1. Food and Agricultural Organization (FAO) database 2019. Available from: http://www.fao.org/faostat/en/#data (accessed December 23, 2019).
2. World Bank database 2019. Available from: https://data.worldbank.org (accessed December 23, 2019).
3. Tilman, D., Balzer, C., Hill, J., and Befort, B.L. (2011). Global food demand and the sustainable intensification of agriculture. *Proceedings of the National Academy of Sciences* **108** (50): 20260.
4. National Bureau of Statistics of China (NBSC) 2019. Available from: http://www.stats.gov.cn/tjsj/ndsj/ (accessed December 23, 2019).
5. Bai, Z., Ma, L., Ma, W. et al. (2016). Changes in phosphorus use and losses in the food chain of China during 1950–2010 and forecasts for 2030. *Nutrient Cycling in Agroecosystems* **104** (3): 361–372.
6. Hou, Y., Ma, L., Gao, Z.L. et al. (2013). The driving forces for nitrogen and phosphorus flows in the food chain of China, 1980 to 2010. *Journal of Environmental Quality* **42** (4): 962–971.
7. Bai, Z., Ma, W., Ma, L. et al. (2018). China's livestock transition: driving forces, impacts, and consequences. *Science Advances* **4** (7): eaar8534.
8. Zhang, W., Ma, W., Ji, Y. et al. (2008). Efficiency, economics, and environmental implications of phosphorus resource use and the fertilizer industry in China. *Nutrient Cycling in Agroecosystems* **80** (2): 131–144.
9. Li, Y., Zhang, W., Ma, L. et al. (2013). An analysis of China's fertilizer policies: impacts on the industry, food security, and the environment. *Journal of Environmental Quality* **42** (4): 972–981.
10. MacDonald, G.K., Bennett, E.M., Potter, P.A., and Ramankutty, N. (2011). Agronomic phosphorus imbalances across the world's croplands. *Proceedings of the National Academy of Sciences* **108** (7): 3086.
11. Ma, L., Wang, F., Zhang, W. et al. (2013). Environmental assessment of management options for nutrient flows in the food chain in China. *Environmental Science & Technology* **47** (13): 7260–7268.
12. Ma, L., Zhang, W.F., Ma, W.Q. et al. (2013). An analysis of developments and challenges in nutrient Management in China. *Journal of Environmental Quality* **42** (4): 951–961.
13. Yan, Z., Liu, P., Li, Y. et al. (2013). Phosphorus in China's intensive vegetable production systems: overfertilization, soil enrichment, and environmental implications. *Journal of Environmental Quality* **42** (4): 982–989.
14. Minister of Water Resources of the People's Republic of China (MOWR) 2016. Available from: http://www.mwr.gov.cn/sj/tjgb/dxsdtyb/ (accessed December 23, 2019).
15. Gu, B., Ge, Y., Chang, S.X. et al. (2013). Nitrate in groundwater of China: sources and driving forces. *Global Environmental Change* **23** (5): 1112–1121.
16. Han, D., Currell, M.J., and Cao, G. (2016). Deep challenges for China's war on water pollution. *Environmental Pollution* **218**: 1222–1233.
17. World Health Organization (1993). *Guidelines for Drinking-Water Quality*. Geneva: World Health Organization.
18. Strokal, M., Ma, L., Bai, Z. et al. (2016). Alarming nutrient pollution of Chinese rivers as a result of agricultural transitions. *Environmental Research Letters* **11** (2): 024014.
19. Wang, S., Zheng, W., Currell, M. et al. (2017). Relationship between land-use and sources and fate of nitrate in groundwater in a typical recharge area of the North China plain. *Science of the Total Environment* **609**: 607–620.
20. Zheng, S., Wang, J., Sun, C. et al. (2019). Air pollution lowers Chinese urbanites' expressed happiness on social media. *Nature Human Behaviour* **3** (3): 237–243.

21. Li, T., Zhang, Y., Wang, J. et al. (2018). All-cause mortality risk associated with long-term exposure to ambient PM2·5 in China: a cohort study. *The Lancet Public Health* **3** (10): e470–e477.
22. An, Z., Huang, R.-J., Zhang, R. et al. (2019). Severe haze in northern China: a synergy of anthropogenic emissions and atmospheric processes. *Proceedings of the National Academy of Sciences* **116** (18): 8657.
23. Sutton, M.A., Howard, C., and Erisman, J.W. (2011). *The European Nitrogen Assessment: Sources, Effects and Policy Perspectives*. Cambridge: Cambridge University Press.
24. Zhang, X., Wu, Y., Liu, X. et al. (2017). Ammonia emissions may be substantially underestimated in China. *Environmental Science & Technology* **51** (21): 12089–12096.
25. Swanson, K.E., Kuhn, R.G., and Xu, W. (2001). Environmental policy implementation in rural China: a case study of Yuhang, Zhejiang. *Environmental Management* **27** (4): 481–491.
26. Liu H. Underestimation for pollution emissions from livestock production: 45% of COD emissions in China come from livestock production. 2013 Available from: http://china.caixin.com/2013-01-04/100479637.html (accessed December 23, 2019).
27. MOA. Implementation of five actions towards agricultural green development. 2017. Available from: http://jiuban.moa.gov.cn/zwllm/tzgg/tz/201704/t20170426_5584189.htm (accessed December 23, 2019).
28. MOA. Recycling of livestock manure policy the target for manure recycling is 75% by the end of 2020. 2017. Available from: http://jiuban.moa.gov.cn/zwllm/tzgg/tz/201707/t20170710_5742847.htm (accessed December 23, 2019).
29. Qu, W., Tu, Q., and Bluemling, B. (2013). Which factors are effective for farmers' biogas use?–evidence from a large-scale survey in China. *Energy Policy* **63**: 26–33.
30. The State Council of China. 2018. Available from: http://www.gov.cn/zhengce/content/2018-07/03/content_5303158.htm (accessed December 23, 2019).
31. Bai, Z., Lu, J., Zhao, H. et al. (2018). Designing vulnerable zones of nitrogen and phosphorus transfers to control water pollution in China. *Environmental Science & Technology* **52** (16): 8987–8988.
32. Lu J, Bai Z, Chadwick D, et al. Mitigation options to reduce water pollution by nitrate from crop and livestock production in China. Submitted.

Section III

State of the Art and Emerging Technologies in Nutrient Recovery from Organic Residues

3.1

Manure as a Resource for Energy and Nutrients

Ivona Sigurnjak[1], Reinhart Van Poucke[1], Céline Vaneeckhaute[2], Evi Michels[1], and Erik Meers[1]

[1]*Department Green Chemistry and Technology, Faculty of Bioscience Engineering, Ghent University, Ghent, Belgium*

[2]*BioEngine, Research Team on Green Process Engineering and Biorefineries, Chemical Engineering Department, Laval University, Quebec, QC, Canada*

3.1.1 Introduction

Intensification of livestock production has been widely advocated as a solution to meet the increasing demands for livestock products, which are expected to double in the next 10–15 years as a response to population growth, urbanization, economic progress, and changing consumer preferences [1–3]. If not done in a sustainable manner, meeting this demand will place increased pressure on the environment. Over the last several decades, intensification of livestock production has led to an excess of animal manure in certain world regions, such as North America, Europe, and South and South East Asia. If not managed properly, manure excess can lead to detrimental effects on the environment due to ammonia and nitrate loss from animal excretion. In order to prevent and mitigate these potential environmental problems, Member States of the European Union (EU) have enacted regulations on manure management. One of the most important regulations is that in areas at risk for agricultural nitrate pollution – so-called Nitrate Vulnerable Zones (NVZs) – application of N from animal manure must not exceed 170 kg N $ha^{-1}y^{-1}$ [4]. This limitation has led to a nutrient paradox, where in manure-surplus regions (e.g. Flanders

Biorefinery of Inorganics: Recovering Mineral Nutrients from Biomass and Organic Waste, First Edition.
Edited by Erik Meers, Gerard Velthof, Evi Michels and René Rietra.
© 2020 John Wiley & Sons Ltd. Published 2020 by John Wiley & Sons Ltd.

(Belgium), the Netherlands, Denmark, Brittany (France), Po Valley (Italy), Ireland, Aragon and Catalonia (Spain)), synthetic N fertilizer is used to satisfy crop nutrient requirements despite the N present in the manure surplus; in minor cases, only synthetic N fertilizer application is possible (e.g. application of N in several dressings, application on wet soil in early spring).

In the meantime, the world is faced with the depletion of fossil fuels and nutrients such as phosphorus (P) and potassium (K). Estimates of current P and K reserves are highly uncertain, but based on population growth and future nutrient demand, it is predicted that depletion will occur within 93–291 years for P and 235–510 years for K [5]. Moreover, awareness of fossil-fuel depletion is reflected in the volatile price of natural gas, and hence in the price of synthetic N fertilizers, whose production via the Haber–Bosch process uses approximately 2% of the world's energy [6]. These findings have triggered the development of nutrient and bioenergy recovery techniques and their subsequent implementation into the manure management chain (Figure 3.1.1).

This chapter reviews the most common bioenergy and nutrient recovery techniques for which manure can be used as a source for energy and nutrients. Bioenergy recovery techniques are defined here as techniques that convert animal manure for the production of power and heat, such as anaerobic digestion (AD), combustion, gasification, and pyrolysis. Nutrient recovery techniques are defined as techniques that (i) create an end-product with higher nutrient concentrations than the raw digestate/manure or (ii) separate the envisaged nutrients from organic compounds, with the aim of producing an end-product that is fit for use in the chemical or fertilizer industry or as a mineral fertilizer replacement [5]. Nutrient recovery options include phosphorus precipitation, ammonia stripping and scrubbing, membrane filtration, and phosphorus extraction from ashes.

3.1.2 Energy Production from Animal Manure

The energy content of animal manure can be estimated from its heating value, expressed as higher heat value (HHV) or lower heat value (LHV). HHV is defined as the total heat generated when a substance is combusted, including the latent heat generated from water vapor condensation [7]. LHV is determined by subtracting the latent heat of water vapor from the HHV. The HHV of animal manures can differ depending on the type of livestock manure and its characteristics (Table 3.1.1). The moisture and ash contents are crucial parameters, since they are negatively correlated with HHV and LHV. Therefore, if the ash or moisture content of manure is high, less energy will be recovered.

Ro et al. [9] estimated the annual energy content of the 35 million dry tons of manure produced in United States to be approximately 0.43 EJ, providing renewable energy with an approximate worth of 0.7 billion USD per year. In the EU-28, with an estimated 104 million dry tons of manure available for recycling and reuse (Chapter 1.1), the financial benefits could even be higher. These findings demonstrate that effective utilization of livestock waste as a renewable energy source can have significant impact on a country's energy budget and economy.

The renewable energy from animal manure can be extracted via biological and thermochemical conversion (TCC) processes. Biological processes convert biomass by utilizing

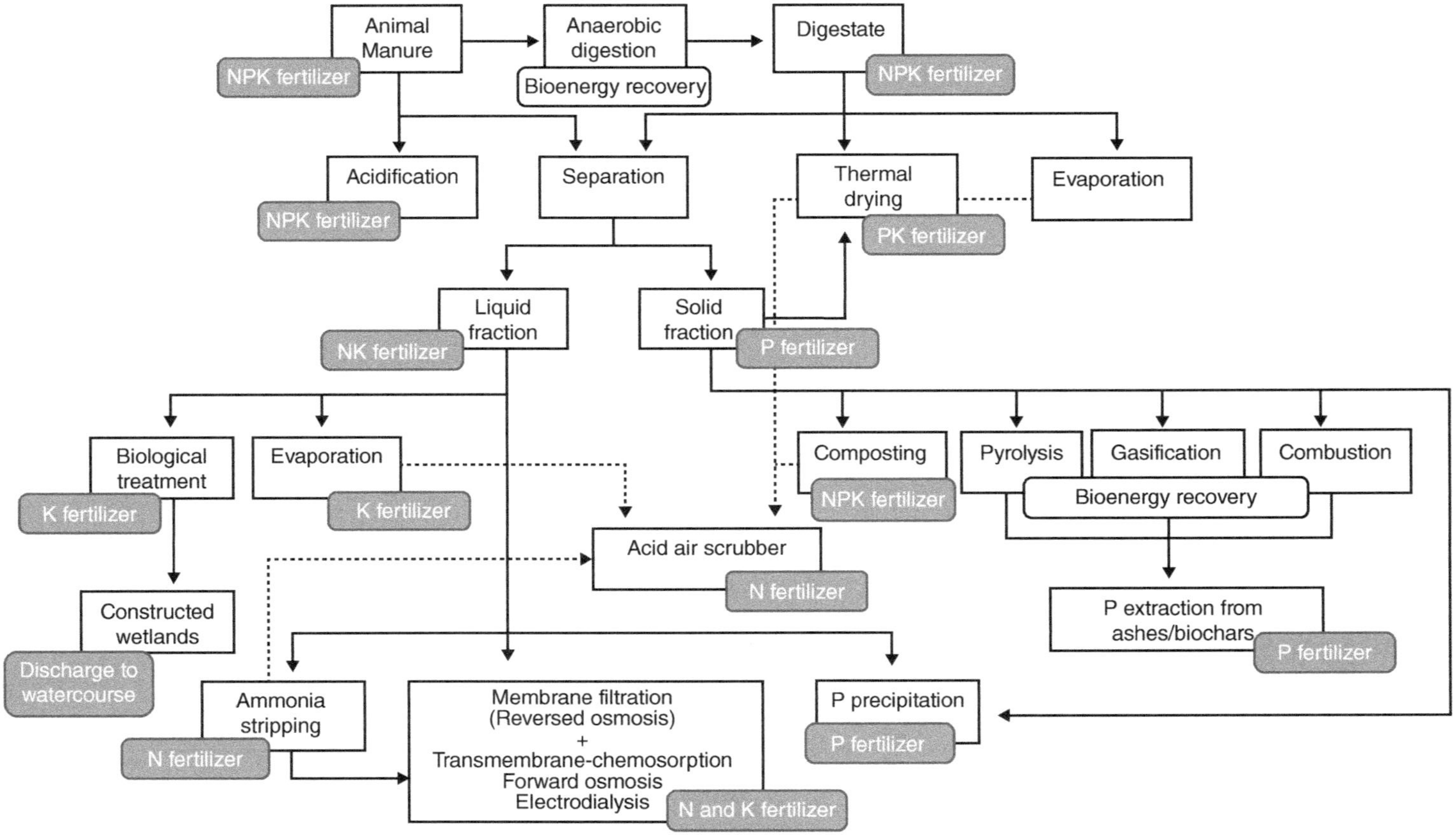

Figure 3.1.1 *Systematic overview of manure and digestate bioenergy and nutrient processing techniques. Bioenergy processing techniques are anaerobic digestion, pyrolysis, gasification and combustion. Nutrient recovery techniques are acid air scrubbers, ammonia stripping, P precipitation, P extraction from ashes/biochars, membrane filtration (reversed osmosis) – including transmembrane-chemosorption, forward osmosis and electrodialysis. The grey boxes present the by- or end- product of the respective technique that can be applied on the agricultural land. The square dot dashes indicate the gaseous flow of N recovery via acid air scrubber.*

Table 3.1.1 *Higher heat value (HHV) and composition of different manures as compared to common fossil fuel sources on dry basis.*

Feedstock	Proximate analysis				Ultimate analysis (wt% dry)				
	HHV (MJ kg^{-1} dry)	M (wt%)	FC (wt% dry)	Ash (wt% dry)	C	H	O	N	S
Cattle manure	17.36	13.88	13.86	15.87	45.39	5.35	36.82	0.96	0.29
Cattle manure	11.30	24.60	4.56	42.34	29.15	3.55	20.75	2.25	0.61
Beef manure	16.09	75.56	19.58[b]		37.64	5.06	28.64	1.87	0.18
Dairy manure	16.56	73.28	13.73[b]		38.06	5.18	28.15	1.85	0.25
Beef manure mixture[a]	14.87	29.54	24.24[b]		37.92	4.39	29.18	1.93	0.33
Dairy manure mixture[a]	14.22	29.85	31.86[b]		32.51	4.41	24.83	2.12	0.48
Sheep manure	16.02	47.80	13.90	20.90	40.60	5.10	30.70	2.10	0.60
Pig manure	19.44	91.26	34.43[b]		36.70	4.82	31.55	3.43	0.91
Pig manure	13.79	92.10	13.30	35.40	34.96	4.38	21.30	2.79	–
Pig manure	17.84	12.70	21.90	15.20	43.66	4.18	33.51	2.47	0.48
Chicken manure	17.14	39.70	–	17.55	41.18	6.25	28.19	6.09	0.50
Chicken manure	15.49	15.80	–	17.05	38.54	5.29	31.86	6.82	0.34
Layer manure	11.92	68.61	44.70[b]		26.77	3.33	30.52	2.25	0.49
Broiler manure mixture[a]	17.86	18.54	15.93[b]		37.84	5.17	35.58	5.44	0.59
Duck manure mixture[a]	12.42	51.84	36.35[b]		29.66	4.00	27.98	2.12	0.61
Digestate	17.84	7.20	20.20	14.90	44.38	5.57	35.90	2.22	0.48
Coal, lignite	24.63	13.40	46.97	3.83	63.54	4.52	28.95	0.70	0.31
Natural gas, Groningen	42.13	0.00	0.00	0.00	58.17	18.72	1.55	21.56	0.00
Methane 100%, Groningen	55.55	0.00	0.00	0.00	74.87	25.13	0.00	0.00	0.00

M moisture, *FC* fixed carbon, *wt* weight.

[a] Mixture means that mentioned manure was collected together with bedding material such as sawdust or rice husk.

[b] Numbers indicate fixed solids, which in this case present the sum of fixed carbon and ash.

Source: Modified from Choi et al. [7] and the ECN Database for Biomass and Waste [8].

microorganisms, while TCC processes utilize heat with or without the presence of oxygen. Both of these technologies have a dual function: to reduce the organic waste and to produce energy out of it.

3.1.2.1 Anaerobic Digestion

AD is a bioenergy recovery technique that utilizes microorganisms or enzymes to convert animal manure into valuable biogas, with nutrient-rich digestate as a side product. The digestion process begins with bacterial hydrolysis of the input materials in order to break down insoluble and complex organic polymers such as carbohydrates and make them available for other bacteria [10, 11]. Second, acidogenic bacteria convert the sugars and amino acids into carbon dioxide, hydrogen, ammonia, and organic acids. Third, acetogenic bacteria convert the resulting organic acids (i.e. the propionic acid and butyric acid) and alcohols into acetic acid, along with additional ammonia, hydrogen, and carbon dioxide. Finally, methanogens convert acetogenesis products to CH_4 (55–70% of biogas composition) and CO_2 (30–45% of biogas composition). Other minor compounds (1–2%) that can be found in biogas are H_2S, NH_3, and H_2, while CO, N_2, and O_2 can be found in trace amounts [12]. The biogas yield of manure is significantly lower than that of other feedstocks. Weiland [13] report a biogas yield of 25 and 30 m^3 t^{-1} fresh matter (FM) for cow and pig manure, respectively. Agricultural raw materials such as grass, maize, and corn yield 102, 200, and 630 m^3 biogas t^{-1} FM, respectively. Despite the low yield, manure is often digested because of its low price and abundancy in many regions of the world. AD of animal manure, as a single substrate (monodigestion), can inhibit the process of methanogens due to the low organic loads and the high N concentration in animal manure. Adding one or more additional substrates in so-called anaerobic co-digestion can overcome the drawbacks of monodigestion, while also improving the economic viability of AD plants due to higher methane production [14]. As a major biogas component and the combustible part of biogas, the percentage of methane will determine the HHV of biogas. For example, biogas with 60% of CH_4 will have an HHV of 33.33 MJ kg^{-1}, which corresponds to 60% of methane HHV (Table 3.1.1). The actual HHV and biogas composition depend on what has been decomposed, and how.

During AD, about 20–95% of the feedstock organic matter (OM) is degraded (depending on feedstock composition and residence time) and transformed into CH_4 and CO_2 [12, 15]. This implies that the OM and dry matter (DM) content decrease in the digestate. However, only easily degradable OM is decomposed, while complex OM, such as lignin, remains in the digestate [16]. Furthermore, the AD process converts a high percentage of the N in manure to NH_4^+-N (Table 3.1.2), especially for feedstocks with a high degradability, producing a digestate with an NH_4^+-N/N total proportion of above 80% [15, 18] and enhancing the digestate N fertilizer efficiency. Also, more than 90% of the volatile fatty acids (VFA) are decomposed, which leads to significantly lower odor emissions during field application of digestate in comparison to pig slurry [5]. On the other hand, decomposition of VFA results in a pH increase, which causes a higher risk for ammonia volatilization. This volatilization during fertilization can be reduced by injecting or incorporating digestate into the soil. The AD process does not influence P content, which means that the P content of digestate is entirely defined by the ingoing streams. Similarly, the AD process does not alter

Table 3.1.2 *Range of main characteristics of digestate (untreated, liquid, and solid fractions from solid/liquid separation) in comparison with undigested animal slurries.*

Parameter	Digestate[a]		Liquid fraction of digestate[b]		Solid fraction of digestate[b]	
	Absolute values	Difference with raw manure	Absolute values	Difference with liquid undigested	Absolute values	Difference with solid undigested
DM (%)	1.5–13.2	−1.5 to −5.5	1.6–6.6	−0.6 to −0.9	13.4–24.7	−0.3 to +0.3
Total C (% DM)	36–45	−2 to −3	33–48	−0.7 to −10.7	40–43	+0.8 to +1.0
Total N ($g\,kg^{-1}$ FM)	1.20–9.10	≈0	2.0–5.1	≈0	4.2–6.5	≈0
NH_4^+/N (%)	44–81	+10 to +33	40–80	+6 to +13	26–49	+3 to +5
C : N ratio	3.0–8.5	−3 to −5	2.4–4.8	−1.6 to −3.1	11–19	−2.9 to +0.1
Total P ($g\,kg^{-1}$ FM)	0.4–2.6	≈0	0.2–1.0	−0.24	1.7–2.5	+0.4 to +0.8
Total K ($g\,kg^{-1}$ FM)	1.2–11.5	≈0	2.6–5.2	−0.13 to −0.17	2.4–4.8	+0.5 to +0.6
pH	7.3–9.0	+0.5 to +2	7.9–8.4	+0.66 to +1.19	8.5–8.7	+0.5 to +0.7

DM, dry matter; FM, fresh matter

[a]Data from Möller and Müller [15].

[b]Data from Möller and Müller [15] and Monaco et al. [17].

Source: European Commission figures.

the heavy metal content. However, during digestion, DM decreases, consequently increasing the concentration of P and heavy metals in the digestate. Impurities such as weed seeds and pathogens can be killed off during the AD process. The extent to which this inactivation is sufficient depends entirely on temperature (mesophilic or thermophilic), residence time in the digester, and the type of organism [5].

3.1.2.2 Thermochemical Conversion Process

In contrast to AD, which requires a processing time of days to weeks, TCC technologies require treatment times in the span of seconds to hours. TCC technologies are high-temperature chemical reforming processes in which the bonds of OM are broken down and intermediates are formed into gas and hydrocarbon fuels [19]. Similar to the AD process, a by-product rich in minerals and fixed carbon, commonly referred to as char, is formed.

Feedstock characteristics have a significant effect on the conversion efficiency of TCC processes, since they can influence process parameters such as total solids (TS), volatile solids (VS), and ash contents. Animal manure from dairy and swine production is typically a diluted solid waste stream composed primarily of discharged wash water but also manure, urine, and undigested feed. The discharged manure characteristics are highly dependent on the growth stage of the animals, the type of manure handling, the collection system, and the amount of added water [19]. Compared to poultry- and cattle-based manure, swine manure has a high VS/TS ratio (69–84%) [11, 19], which indicates that more organic material is available for conversion, resulting in higher HHV. This was confirmed by Ro et al. [20] in hydrothermal gasification analysis, where swine manure as a feedstock produced gases with the highest energy per kilogram of dry TS out of five major types of animal manure (swine manure, dairy manure, poultry litter, unpaved feedlot manure, and paved feedlot manure).

Among the TCC processes, combustion, pyrolysis, and gasification are the most widely used biomass conversion techniques (Table 3.1.3). A general prerequisite for all TCC processes is an input stream with a DM content of at least 85% [27]. Therefore, manure or digestate needs to be dried prior to being subjected to the TCC process. Drying is more easily performed on solid fraction (SF) after mechanical separation, after which SF of manure or digestate can be pelletized and subjected, for example, to the combustion process. **Combustion** refers to the complete oxidation of biomass to produce heat, which can be converted into electricity as a final product. The HHV value of digestate pellets has been reported to vary between 16.4 and 17.3 MJ kg^{-1}, and to be similar to that of wood [28, 29]. Kuligowski and Luostarinen [30] report that 50% of the HHV of the digestate pellets is consumed during the drying process. In general, animal manure or digestate has the potential to be combusted if its moisture content is less than 60%, but the combustion is rarely efficient due to the presence of high ash content in these products [31, 32], which could obstruct brazier holes and partially choke combustion after a few minutes [29]. Also, production of impurities (i.e. greenhouse gases, heavy metals, acidic gases, polyaromatic hydrocarbons [PAHs]) leads to lower energy efficiency, similar to the energy efficiency of combusting municipal solid waste (MSW) (c. 15%; [33]). Under these conditions, pig manure and digestate with an energy content of 13.79 and 17.84 MJ kg^{-1} DM, respectively (Table 3.1.1), would yield 2.07 and 2.68 MJ of electricity per kg DM.

Table 3.1.3 *Main characteristics of process conditions and products of combustion, pyrolysis, and gasification.*

	Combustion	Pyrolysis	Gasification
Process characteristics			
Temperature (°C)	800–1450	250–900	550–900 with air 1000–1600 with other agents
Atmospheric pressure	1	1	1–45
Stoichiometric air-to-fuel ratio	>1, typically 1.3 for solid fuels	0	0–1, typically 0.2–0.4
Carbon conversion (%)	>99	≈75	80–95
Oxidant	Mostly air, and less pure oxygen	None	Air, pure oxygen, steam of their combinations
HHV of product ($MJ\ kg^{-1}$)	0	16–19	5–20
Process products			
Gas phase	CO_2, H_2O, O_2, N_2 **70–75%** gas yield value	H_2,CO, H_2O,N_2, hydrocarbons **13%** gas yield value in fast pyrolysis **35%** gas yield value in slow pyrolysis	<1000 °C CO, H_2, CH_4, CxHy, aliphatic hydrocarbons, benzene, toluene, tars, and CO_2, H_2O, and N_2 in the case of gasification in air (so-called producer gas) >1200 °C H_2, CO, CO_2, CH_4, H_2O, and N_2 in the case of gasification in air (so-called syngas) **85%** gas yield value
Solid phase	Sometimes ash in the case of lower-quality feedstock as MSW **20–30%** bottom (heavier solid) ash **1–6%** is fly (dust) ash	Ash, char **12%** char yield value in fast pyrolysis **35%** char yield value in slow pyrolysis	Ash, char **10%** char yield value
Liquid phase	–	Pyrolysis oil and water **75%** oil yield value in fast pyrolysis **30%** oil yield value in slow pyrolysis	Sometimes oil **0–5%** oil yield value

HHV, higher heating value; MSW, municipal solid waste.
Source: Modified from EC [21], IAE [22], Helsen and Bosmans [23], DEFRA [24], UNEP [25], and Shakorfow [26].

Due to these issues, it is often desirable to mix manure with dry fuel such as coal in so-called co-firing. This arrangement not only reduces the use of fossil fuels, but also lowers emissions of NOx and SO_2, which are released during coal combustion [31, 34]. A reduction in NOx is observed due to the release of NH_3 from animal manure, which subsequently combines with NOx to produce harmless N_2 and water [31], whereas SO_2 reduction is attributed to SO_2 capture by alkali ash of the manure feedstock [35]. Finally, flue (exhaust) gas is formed during combustion process, which requires cleaning (c. 15–35% of the total capital investment [23]) prior to its release into the atmosphere, making small-scale combustion not viable [5].

Pyrolysis and gasification are considered to be more efficient techniques for energy recovery from animal manure as they result in end-products with HHV [36], leading to a cleaner product for combustion compared to the original feedstock. **Pyrolysis** is defined as a non-oxygen thermochemical decomposition of solid waste [37] at a high temperature (around 500 °C), into the following products [38]: liquid (bio-oil), solid (biochar), and gaseous, non-condensable fractions. Due to the process precondition of high DM for input streams [39], poultry litter is considered as preferable feedstock for pyrolysis as compared to other manure types [36, 40]. In the case of digestate pyrolysis, several studies [41, 42] report an additional increase in energy yield as compared to AD alone, presenting a combination AD and pyrolysis system as an interesting option. This approach is relevant for low degradable feedstock that results in carbon-rich digestate whose SF after mechanical separation can be subjected to pyrolysis or gasification. The combination of AD and TCC has several advantages over combustion of manure. Pyrolysis and gasification will increase the energy density of the products and improve their combustion characteristics. Further, condensation is not required. The pyrolysis vapors can be combusted as such in the combined heat and power (CHP) of the biogas plant and the hygienized char can be amended to improve the soil quality and nutrient status. Leijenhorst et al. [43] obtained a yield of 16 wt% gas, 57 wt% oil, and 27 wt% char after pyrolysis of SF of digestate at 500 °C. These results are similar to that reported by Monlau et al. [42], where 9 wt% gas, 58 wt% oil, and 33 wt% of char was recovered under similar experimental conditions. The same study reported heating values of 23.5 MJ kg^{-1} and 15.7 MJ Nm^{-3} for the oil and gas, respectively. Moreover, the energy balance of the study showed that by utilizing these products for electricity, an increase of electrical production of 42% can be expected as compared to standalone AD [42].

Gas, bio-oil, and biochar as energy intermediates/products of pyrolysis have a high added value, and can therefore be used for energy production, either for CHP generation or for production of liquid fuels. In order to maximize the quantity of intermediates that are derived from pyrolysis, it is possible to apply different types of pyrolytic techniques [25], such as slow pyrolysis, fast pyrolysis, and flash pyrolysis. All of these reactions differ by temperature, residence time, and quantity of derived products. For example, fast pyrolysis maximizes liquid production rather than the production of biochar. In general, the value of pyrolysis is recognized not only in the industrial, but also in the agricultural sector. While bio-oil can be used for heating purposes [40], biochar can be applied to soil [37] as an improver of soil fertility and structure. Besides energy recovery, pyrolysis can also be applied for P recovery (see later). To our knowledge, however, pyrolysis as a part of manure management chain can at present only be found at the laboratory and pilot/demo scale.

Gasification is a TCC process at high temperature (>800 °C), aimed at decomposing the OM of dry or wet biomass [37] into gas. The yield, composition, and heating value of gas all depend on several parameters, including operational temperature, pressure, residence time, composition of biomass, amount of oxidizing agent [37, 38], and type of gasification process (wet or dry). As a result, two types of gasification gas can be distinguished: producer gas and syngas. The producer gas is formed at lower temperatures (<1000 °C), whereas syngas is produced at higher ones (>1200 °C). The main components of syngas are CO, H_2, CO_2, and CH_4. It contains fewer other gases as compared to producer gas, which makes it chemically more similar to the gas derived from fossil sources. Nevertheless, syngas can be made from producer gas by heating, causing thermal cracking and catalytic reforming [25]. Syngas as a product of gasification can be injected into the natural gas grid or used as a bioenergy feedstock for fuel, heat, and electricity production [37]. Whether syngas purification is needed beforehand depends on the content of impurities [44], which can minimize its value. Another important characteristic of the gasification process is that it occurs where there is a limited supply of an oxidizing agent (air, O_2, or steam), which prevents complete combustion [30]. Consequently, char/ash from gasification has a relatively high content of P and K as compared to the char from combustion, which is usually more contaminated with Zn, Cu, and Cr [44]. As a result, ashes from gasification may be of interest in P and K fertilizer recovery (Section 3.1.3.4). With regard to manure gasification, this technology as pyrolysis has not yet been applied on a commercial scale.

3.1.3 Nutrient Recovery Techniques

All above described bioenergy recovery techniques result in nutrient-rich by-products such as digestate, char, or ash, with energy as the main output. Further nutrient recovery processes occur mostly on digestate and ash streams. Figure 3.1.1 shows a diverse range of techniques suitable for digestate processing, but not all of them may be considered as nutrient recovery techniques. According to the definition given earlier, techniques that make it possible to reuse nutrients and close the nutrient cycle are phosphorus precipitation, ammonia stripping and scrubbing, membrane filtration, and phosphorus extraction from ashes. To operate efficiently, these techniques must consume a certain amount of energy, depending on the process requirement for pre-treatment (e.g. separators), chemicals (e.g. acids), heat, electricity, and so on [45, 46]. The most promising techniques are probably those that can combine energy generation and nutrient recovery, resulting in nutrient recovery coupled to AD or TCC technology.

As mentioned earlier, legislation is one of the main drivers for the introduction of bioenergy and nutrient recovery techniques. Next to N, the Nitrates Directive indirectly also limits P fertilization rates. For some European regions (i.e. Flanders, Estonia, Brittany, Germany, Ireland, Northern Ireland, Norway, Sweden, and the Netherlands) at risk of soil with a high P status, this limitation is not sufficient. Depending on the crop type and the soil P, total P fertilization rates range from 0 to 125 kg P ha y^{-1} [47]. In order to tackle the P issue, digestate is often separated into liquid fraction (LF) and SF. Due to its high P content, the SF is usually dried or composted and exported out of NVZ. As mentioned earlier, SF of digestate (and manure) is not yet subjected to thermochemical processes on a full commercial scale.

LF of digestate can be subjected to similar valorization pathways as LF of animal manure. The further valorization pathways of both product fractions are described hereinafter.

3.1.3.1 Phosphorus Precipitation

P precipitation is a reaction in which, at pH 8.3–10, phosphate salts are formed by addition of several ions (e.g. Mg, Ca, Fe, or K) to a solution containing soluble P (ortho-P). In the case of Mg (as $MgO/MgCl_2$) addition, Mg ion will bind with ammonium and phosphate in a liquid stream and form $MgNH_4PO_4$ (MAP or struvite) precipitates, which can be further separated by means of sedimentation [48]. This process captures 80–90% of soluble P and 10–40% of the NH_4-N initially present in the treated stream [49]. Instead of Mg addition, Ca, Fe, or K can be used to recover P as calcium phosphate, iron phosphate, or $K_2NH_4PO_4$ (potassium struvite), respectively. This P recovery strategy has been implemented at full scale for wastewater, (digested) sludge, and manure treatment and at pilot scale for the treatment of digestate [50].

With regard to manure or digestate treatment, P can be recovered from the manure or digestate as whole, from its SF or LF. Though the highest P concentrations can be found in SF, a certain part is still present in LF at lower concentrations. To precipitate struvite from LF, both P (as H_3PO_4) and Mg need to be added, due to the high N/P ratio of LF, which does not allow the $Mg:NH_4:PO_4$ ratio of 1 : 1 : 1 under which struvite precipitates [51]. As a result of H_3PO_4 and Mg addition, 90% of N and P can be recovered from LF [51]. For whole manure or digestate and its SF, most of the P is incorporated in OM and thus not easily available. This P can be recovered via a similar "acid–base" approach, where the tested stream is first acidified and separated, and P precipitation is induced on the LF of the tested stream [52]. In the acidification step, the pH of manure is reduced from its usual 7–8 to a minimum of 5. In this way, almost all of the mineral P becomes soluble and can be more easily retrieved from manure. With this approach, 80–100% of the total P in the tested stream can be recovered, but removal efficiencies of 50–60% are more typical [50, 52]. Though, worldwide, some utilities have installed these systems, the uptake of this technology has not been widespread due to market, regulatory, and site-specific conditions. Also, important technical challenges remain in the further reduction of chemical requirements, the guarantee of a pure product, and the stable and controlled production of struvite [50].

3.1.3.2 Ammonia Stripping and Scrubbing

Ammonia (NH_3) removal from N-rich waste streams (e.g. LF of digestate or of animal manure) usually involves two steps. First, NH_3 is removed (stripped) by blowing air or steam through the waste stream in a packed bed tower. As a result, NH_3 is transferred from the aqueous phase to a gas phase, after which stripgas (charged with NH_3) is removed (scrubbed) by washing it with a strong acidic solution in the scrubbing system. To obtain optimal removal, the pH and temperature of the waste stream are increased, often to 10 and 70 °C, respectively, thus shifting the NH_4^+/NH_3 equilibrium toward free NH_3 [48]. This technique can reach an NH_3 removal efficiency of 99% [53, 54].

For economic reasons, H_2SO_4 is normally used as an acidic solution. Ammonia can also be removed with hydrochloric (HCl), nitric (HNO_3), and phosphoric (H_3PO_4) acid. The reaction of NH_3 with H_2SO_4 results in ammonium sulfate $(NH_4)_2SO_4$ (also known as air scrubber water [ASW]: the term includes all NH_4^+-N-rich waters obtained after scrubbing NH_3-saturated air), which can be used as an NS fertilizer. The end-product is characterized by acidic pH and a high salt content. In the system, deposition of the formed $(NH_4)_2SO_4$ can take place when the maximum solubility of the salt is exceeded. This leads to clogging and subsequently increases energy requirement during the production phase. To prevent this effect, in Flanders (Belgium) and the Netherlands, the N content in the ASW is not legally allowed to exceed 58.8 g N l^{-1}, which is about three times lower than the maximum solubility of the salt (164 g N l^{-1}; [54]). Next to the removal of NH_3 from waste streams, NH_3 can also be recovered from livestock or manure operations such as housing, separation, composting, and drying units. The ASW from livestock or manure operations contains N completely in mineral N form and as such is recognized in Flanders [55] via national derogation as a substitute for synthetic N fertilizer. On the other hand, the ASW obtained by stripping and scrubbing ammonia from animal waste streams is still seen as animal manure despite having the same product characteristics. The "animal manure" status is currently assigned to this material because its use has not been regulated on the European level (VCM, pers. comm.).

3.1.3.3 Membrane Filtration

In contrast to the end-product of ammonia stripping and scrubbing, the LF of animal manure or digestate still contains high amounts of organic N. In order for this product to be accepted as a mineral N fertilizer, an increase of NH_4^+/Ntotal ratio is therefore desirable. This can be achieved by subjecting LF to a membrane filtration process, where the waste stream is forced through membranes by pressure. The material that is retained on the membrane is called retentate or mineral concentrate, and is known for containing N (NH_4^+-N/N_{total} = 0.9–1.0) almost completely in NH_4^+-N form [56, 57]. There are several types of membranes used in manure/digestate processing: microfiltration (MF; pores >0.1 μm, 0.1–3 bar), ultrafiltration (UF; pores > nm, 2–10 bar), and reverse osmosis (RO; no pores, 10–100 bar) membranes. In an MF concentrate, suspended solids are retained; in a UF concentrate, macromolecules are retained as well. Both filtration steps can be used as a pre-treatment for RO, in order to prevent either suspended solids or macromolecules from blocking the RO membrane [5]. Another technique that can be used prior to RO is dissolved air flotation (DAF), where small air bubbles are blown through the LF, entraining suspended solids to the surface, where they form a crust. This crust is then scraped off. When using DAF, coagulants (e.g. Fe(III)Cl_3, polyaluminium chloride [PAC]) and flocculants (e.g. Chitosan) are often added [58]. The permeate of RO, which consists mainly of water and small ions, can be discharged, if necessary after a "polishing" step, or used as process water. The biggest problem reported in membrane filtration is the blocking of the membrane during MF and UF. This is mainly caused by suspended solids, which form a cake on the surface of the membrane. Most installations reduce the blocking of the membrane pores by continuously dosing acid solution to the RO system, which is the most efficient way to reduce scaling and fouling [5]. In the Netherlands, a large research project has been ongoing since 2008, looking at the RO concentrate of eight different manure/digestate processing installations, using a UF or DAF step as pre-treatment [59]. Next to RO membrane filtration, there has been an increased interest in forward osmosis, transmembrane chemosorption, and electrodialysis. However, these techniques are still under development, and as such are not yet applied at full scale [58].

3.1.3.4 Phosphorus Extraction from Ashes

The ashes remaining after combustion or gasification of digestate/manure contain P-, K-, Al-, and Si-compounds and possibly also some heavy metals such as Cu, Zn, and Cd [60]. P extraction has been tested extensively for dried or dewatered sludge and ashes from sludge incineration. However, tests on SF, ashes, and biochar from manure/digestate are absent from the literature. Ash from the TCC process is usually used as a construction material for roads or for concrete production. In the light of nutrient depletion, P can also be recovered from the ash by thermochemical and wet chemical techniques. The wet chemical treatment involves ash redissolution by addition of acid or base. Since most metals are less soluble in alkaline than in acid solutions, leaching with base results in a P product with a lower metal contamination. However, alkaline agents are usually more expensive [61], and during leaching with base the Ca in sludge binds P as calcium phosphate, which reduces the degree of P release [62]. In contrast to base leaching, acid will result in higher release of P, but also higher metal contamination, since metals are more soluble in acid. Once the insoluble compounds are removed, P can be extracted from the liquid stream through precipitation, ion exchange, nanofiltration, and so on [63]. In thermochemical treatment, ashes are exposed to chlorine salts, such as potassium chloride or magnesium chloride, and thermally treated at temperatures above 1000 °C [60]. Simultaneously, a large fraction of the heavy metals is turned into heavy metals chlorides, which vaporize and subsequently are captured by flue gas treatment [63]. Any N that might still be present in the dry SF of manure or digestate is converted to gaseous N, and none is leached to ground and surface waters. As such, the char from the gasification process could be considered as a soil improver in agriculture.

3.1.4 Conclusion

Struvite precipitation, NH_3 stripping/scrubbing, and membrane filtration can be selected as the best available technologies for nutrient recovery from manure/digestate. They have already been implemented at full scale and have the ability to produce marketable end-products. On the other hand, TCC processes are less applied for energy recovery from manure/digestate as compared to AD. The main reason is the low DM content of manure/digestate, and the subsequent energy requirement in order to dry these materials prior to the TCC process. Nevertheless, these processes are currently under investigation, though they are not yet available on full scale. Future research should further explore, verify, and improve the fertilizer characteristics and marketing value of these products toward industrial and agricultural end-users.

References

1. Thornton, P.K. (2010). Livestock production: recent trends, future prospects. *Philosophical Transactions of the Royal Society B: Biological Sciences* **365** (1554): 2853–2867.
2. Udo, H.M.J., Aklilu, H.A., Phong, L.T. et al. (2011). Impact of intensification of different types of livestock production in smallholder crop-livestock systems. *Livestock Science* **139** (1): 22–29.

3. Sakadevan, K. and Nguyen, M.L. (2017). Livestock production and its impact on nutrient pollution and greenhouse gas emissions. In: *Advances in Agronomy*, vol. **141** (ed. L.S. Donald), 147–184. Cambridge, MA: Academic Press.
4. European Commision (1991). Directive of the Council of 12 December 1991 concerning the protection of waters against pollution caused by nitrates from agricultural sources (91/676/EC). *Official Journal of the European Communities* **L375**: 1–8.
5. Lebuf V, Accoe F, Van Elsacker S, et al. Techniques for Nutrient Recovery from Digestate. Report within the ARBOR Project. Available from: https://www.researchgate.net/publication/286862877_Inventory_Techniques_for_nutrient_recovery_from_digestate (accessed December 23, 2019).
6. Sutton MA, Bleeker A, Howard CM, et al. Our Nutrient World: The Challenge to Produce Moore Food and Energy with Less Pollution. Global Overview of Nutrient Management. Report. Centre for Ecology and Hydrology, Edinburgh (UK on behalf of the Global Partnership on Nutrient Management and the International Nitrogen Initiative). 2013.
7. Choi, H.L., Sudiarto, S.I.A., and Renggaman, A. (2014). Prediction of livestock manure and mixture higher heating value based on fundamental analysis. *Fuel* **116**: 772–780.
8. ECN Database for Biomass and Waste. Available from: https://phyllis.nl/ (accessed December 23, 2019).
9. Ro, K.S., Cantrell, K.B., Hunt, P.G. et al. (2009). Thermochemical conversion of livestock wastes: carbonization of swine solids. *Bioresource Technology* **100** (22): 5466–5471.
10. Bhatia, S.C. (2014). *Advanced Renewable Energy Systems: Part I*. New Delhi: Woodhead Publishing India.
11. Jensen, L.S. and Sommer, S.G. (2013). Manure organic matter – characteristics and microbial transformations. In: *Animal Manure Recycling – Treatment and Management* (eds. S.G. Sommer, M.L. Christensen, T. Schmidt and L.S. Jensen). Chichester: Wiley.
12. Jørgensen, P.J. (2009). *Biogas – Green Energy*. Aarhus: Faculty of Agricultural Sciences, Aarhus University.
13. Weiland, P. (2010). Biogas production: current state and perspectives. *Applied Microbiology and Biotechnology* **85** (4): 849–860.
14. Mata-Alvarez, J., Dosta, J., Romero-Güiza, M.S. et al. (2014). A critical review on anaerobic co-digestion achievements between 2010 and 2013. *Renewable and Sustainable Energy Reviews* **36**: 412–427.
15. Möller, K. and Müller, T. (2012). Effects of anaerobic digestion on digestate nutrient availability and crop growth: a review. *Engineering in Life Sciences* **12** (3): 242–257.
16. Chynoweth, D.P., Owens, J.M., and Legrand, R. (2001). Renewable methane from anaerobic digestion of biomass. *Renewable Energy* **22** (1): 1–8.
17. Monaco, S., Sacco, D., Borda, T., and Grignani, C. (2010). Field measurement of net nitrogen mineralization of manured soil cropped to maize. *Biology and Fertility of Soils* **46** (2): 179–184.
18. Sørensen, P. and Jensen, L.S. (2013). Nutrient leaching and runoff from land application of animal manure and measures for reduction. In: *Animal Manure Recycling:Treatment and Management* (eds. S.G. Sommer, M.L. Christensen, T. Schmidt and L.S. Jensen), 195–208. Chichester: Wiley.
19. Cantrell, K., Ro, K., Mahajan, D. et al. (2007). Role of thermochemical conversion in livestock waste-to-energy treatments: obstacles and opportunities. *Industrial & Engineering Chemistry Research* **46** (26): 8918–8927.
20. Ro, K.S., Cantrell, K., Elliot, D., and Hunt, P.G. (2007). Catalytic wet gasification of municipal and animal waste. *Industrial and Engineering Chemistry Research* **46**: 8839–8845.
21. European Commision (1995). *Combustion and Gasification of Agricultural Biomass – Technologies and Applications*. Geneva: European Commission, Directorate-General for Energy.
22. International Energy Agency. Annual Report 2006. Available from: http://www.globalbioenergy.org/uploads/media/0707_IEA_-_Bioenergy_annual_report.pdf (accessed December 23, 2019).

23. Helsen L, Bosmans A, eds. Waste-to-energy through thermochemical processes: matching waste with process. 1st Int Symposium on Enhanced Landfill Mining (October 4–6, 2010). Available from: http://energy.cleartheair.org.hk/wp-content/uploads/2016/05/FILE_772F83CC-9C54-437C-B222-39ECC79108EB.pdf (acccessed December 23, 2019).
24. Department for Environment, Food and Rural Affairs (2013). *Incineration of Municipal Solid Waste*. London: DEFRA.
25. United Nations Environment Programme (2013). *Technologies for Converting Waste Agricultural Biomass to Energy*. Nairobi: UNEP.
26. Shakorfow, A.M. (2016). Biomass. Incineration, pyrolysis, combustion and gasification. *International Journal of Sceince and Research (IJSR)* **5** (7): 13–25.
27. Prins W, Dahmen N, eds. Processes for thermochemical conversion of biomass. 10th European Conference on Industrial Furnaces and Boilers; 2015; Porto Gaia, Portugal.
28. Kratzeisen, M., Starcevic, N., Martinov, M. et al. (2010). Applicability of biogas digestate as solid fuel. *Fuel* **89** (9): 2544–2548.
29. Monlau, F., Sambusiti, C., Ficara, E. et al. (2015). New opportunities for agriculutral digestate valorization: current situation and perspectives. *Energy and Environmental Science* **8**: 2600–2621.
30. Kuligowski K, Luostarinen S. Thermal gasification of manure. Project partly financed by EU: Baltic manure – WP6 Energy potential. 2011.
31. Foged HL. Livestock manure to energy: status, technologies and innovation in Denmark. Available from: https://inbiom.dk/admin/public/download.aspx?file=/Files/Files/ecom/products/Biogas-Go-Global/Livestock-Manure-to-Energy.pdf (accessed December 23, 2019).
32. Triolo, J.M., Ward, A.J., Pedersen, L., and Sommer, S.G. (2013). Characteristics of animal slurry as a key biomass for biogas production in Denmark. In: *Biomass Now – Sustainable Growth and Use* (ed. M.D. Matovic). Vienna: InTech.
33. Malkow, T. (2004). Novel and innovative pyrolysis and gasification technologies for energy efficient and environmentally sound MSW disposal. *Waste Management* **24** (1): 53–79.
34. Otero, M., Sánchez, M.E., and Gómez, X. (2011). Co-firing of coal and manure biomass: a TG–MS approach. *Bioresource Technology* **102** (17): 8304–8309.
35. Shah, Y.T. (2015). *Energy and Fuel Systems Integration*. Boca Raton, FL: CRC Press.
36. Hussein, M.S., Burra, K.G., Amano, R.S., and Gupta, A.K. (2017). Temperature and gasifying media effects on chicken manure pyrolysis and gasification. *Fuel* **202**: 36–45.
37. Cantrell, K.B., Ducey, T., Ro, K.S., and Hunt, P.G. (2008). Livestock waste-to-bioenergy generation opportunities. *Bioresource Technology* **99** (17): 7941–7953.
38. Wang, L., Shahbazi, A., and Hanna, M.A. (2011). Characterization of corn Stover, distiller grains and cattle manure for thermochemical conversion. *Biomass and Bioenergy* **35** (1): 171–178.
39. Azuara, M., Kersten, S.R.A., and Kootstra, A.M.J. (2013). Recycling phosphorus by fast pyrolysis of pig manure: concentration and extraction of phosphorus combined with formation of value-added pyrolysis products. *Biomass and Bioenergy* **49**: 171–180.
40. Mante, O.D. and Agblevor, F.A. (2011). Parametric study on the pyrolysis of manure and wood shavings. *Biomass and Bioenergy* **35** (10): 4417–4425.
41. Li, Y., Zhang, R., He, Y. et al. (2014). Anaerobic co-digestion of chicken manure and corn stover in batch and continuously stirred tank reactor (CSTR). *Bioresource Technology* **156**: 342–347.
42. Monlau, F., Sambusiti, C., Antoniou, N. et al. (2015). A new concept for enhancing energy recovery from agricultural residues by coupling anaerobic digestion and pyrolysis process. *Applied Energy* **148** (Suppl. C): 32–38.
43. Leijenhorst EJ, Meers E, Van Poucke R, Raymaekers F. Pyrolysis of digestate. Working Paper INEMAD. 2015.
44. Olwa, J. (2011). *Investigation of Thermal Biomass Gasification for Sustainable Small Scale Rural Electricity Generation in Uganda*. Stockholm: KTH School of Industrial Engineering and Management.

45. Schoumans O.F., Ehlert P.A.I., Regelink I.C., et al. Chemical Phosphorus Recovery from Animal Manure and Digestate; Laboratory and Pilot Experiments. Wageningen Environmental Research Report 2849. 2017.
46. de Vries JW. From animals to crops – environmental consequences of current and future strategies for manure management. PhD thesis: Wageningen University; 2014.
47. Amery, F. and Schoumans, O. (2014). *Agricultural Phosphorus Legislation in Europe*. Merelbeke: Institute for Agricultural and Fisheries Research (ILVO).
48. Lemmens B, Ceulemans J, Elsander H, et al. Beste Beschikbare Technieken (BBT) voor mestverwerking. 2007.
49. Le Corre, K.S., Valsami-Jones, E., Hobbs, P., and Parsons, S.A. (2009). Phosphorus recovery from wastewater by struvite crystallization: a review. *Critical Reviews in Environmental Science and Technology* **39** (6): 433–477.
50. Vaneeckhaute, C., Lebuf, V., Michels, E. et al. (2017). Nutrient recovery from digestate: systematic technology review and product classification. *Waste and Biomass Valorization* **8** (1): 21–40.
51. Melse, R.W., Buisonjé, F.E., Verdoes, N., and Willers, H.C. (2004). *Quick scan van be- en verwerkingstechnieken voor dierlijke mest*. Wageningen: Wageningen University.
52. Schoumans O.F., Ehlert PAI, Nelemans JA, et al. Explorative study of phosphorus recovery from pig slurry; Laboratory experiments. Wageningen, Alterra Wageningen UR (University & Research Centre), Alterra Report. 2014.
53. Melse, R.W. and Ogink, N.W.M. (2005). Air scrubbing techniques for ammonia and odor reduction at livestock operations:review of on-farm research in the Netherlands. *Transactions of the ASAE* **48** (6): 2303–2313.
54. Van der Heyden, C., Demeyer, P., and Volcke, E.I.P. (2015). Mitigating emissions from pig and poultry housing facilities through air scrubbers and biofilters: state-of-the-art and perspectives. *Biosystems Engineering* **134**: 74–93.
55. Aanwenden van specifieke meststoffen. Flemish Land Agency. Available from: https://www.vlm.be/nl (accessed December 23, 2019).
56. Schröder, J.J., De Visser, W., Assinck, F.B.T. et al. (2014). Nitrogen fertilizer replacement value of the liquid fraction of separated livestock slurries applied to potatoes and silage maize. *Communications in Soil Science and Plant Analysis* **45** (1): 73–85.
57. Velthof GL. Mineral concentrate from processed manure as fertiliser. Wageningen, Alterra Wageningen UR (University & Research Centre), Alterra Report 2650. 2015.
58. Vaneeckhaute C. Nutrient recovery from bio-digestion waste: from field experimentation to model-based optimization. PhD Thesis: Faculty of Bioscience Engineering, Ghent Univeristy/Laval University; 2015.
59. Velthof G. Synthesis of the research within the framework of the Mineral Concentrates Pilot. Wageningen, Alterra, Alterra Report 2224. 2011.
60. Schoumans O.F., Rulkens WH, Oenema O, Ehlert PAI. Phosphorus recovery from animal manure; Technical opportunities and agro-economical perspectives. Wageningen, Alterra, Alterra Report 2158. 2010.
61. Jensen, L.S. (2013). Animal manure residue upgrading and nutrient recovery in biofertilisers. In: *Animal Manure Recycling: Treatment and Management* (eds. S.G. Sommer, M.L. Christensen, T. Schmidt and L.S. Jensen), 271–294. Chichester: Wiley.
62. Levlin, E., Löwén, M., and Stark, K. (2005). *Phosphorus Recovery from Sludge Incineration Ash and Supercritical Water Oxidation Residues with Use of Acid and Base*. Stockholm: Royal Institute of Technology.
63. Cornel, P. and Schaum, C. (2009). Phosphorus recovery from wastewater: needs, technologies and costs. *Water Science and Technology* **59** (6): 1069–1076.

3.2

Municipal Wastewater as a Source for Phosphorus

Aleksandra Bogdan, Ana Alejandra Robles Aguilar, Evi Michels, and Erik Meers

Department of Green Chemistry and Technology, Faculty of Bioscience Engineering, Ghent University, Ghent, Belgium

3.2.1 Introduction

Phosphorus is one of the key elements for all biochemical processes of living organisms. Recognition of its importance in agriculture drove the application of mineral fertilizers to become a common practice. However, the overuse of phosphorous fertilizer over the years has resulted in serious environmental problems such as eutrophication. Since 1991, European legislation has made important steps toward solving the problem, starting with restrictions on phosphorous discharge into streams.

In the last century, scientists made predictions regarding the depletion of global phosphate deposits due to extensive mining of non-renewable rock phosphate. According to those predictions phosphate deposits are not expected to last through the twenty-first century, with the global peak in phosphorus production being reached around the year 2030 [1–3]. The European Commission declared phosphate rock a critical raw material in the 2014 [4]. The most optimistic estimates are that supplies will last from one up to a few hundred years [5]. Nevertheless, phosphate mines are found only in a few places on earth, with 75% of known reserves located in Morocco, and none in European Union, which thus relies heavily on imports [6].

It is thus apparent that finding optimal resources and technologies that allow more sustainable production and utilization of phosphorous is imperative. An important step

Biorefinery of Inorganics: Recovering Mineral Nutrients from Biomass and Organic Waste, First Edition.
Edited by Erik Meers, Gerard Velthof, Evi Michels and René Rietra.
© 2020 John Wiley & Sons Ltd. Published 2020 by John Wiley & Sons Ltd.

in that direction has been made by estimating that at least 15% of phosphorus could be recovered from wastewater streams [7]. Wastewater contains phosphorus in the forms of ortho-, poly-, and organic phosphate [8]. However, the recovery of P from these sources is often case-specific. Factors of interest are population equivalent, type of P recovery within a wastewater treatment plant (WWTP), P status on land, and social and legislative barriers. Optimal solutions will differ between countries, as well as between the regions within a country. In order to make the right choice of technology, it is first necessary to understand the phosphorous market and its specific demands on a particular level of interest (national, regional, etc.). This is still at a starting phase in most of Europe. The second main consideration should be the adoption or development of the most suitable technology. Some of the first P recovery technologies are already operating at full industrial scale (see later), while many others are at the beginning of their development. As a final point, it is important to bring solutions to the public and legal acceptance level. In Europe, there is currently a new Fertilizer Regulation and the EU-created STRUBIAS group reports, which are developing and bringing a representative act to fulfill the demands of the novel P recovery technologies and their valuable P products market.

3.2.2 Phosphorus Removal from Wastewater

In an effort to define the most suitable technology for phosphorous recovery within a given area, it is important to begin with an overview of the current state of wastewater treatment. The focus should be on the number and proportion of WWTPs, the phosphorous content in the wastewater, phosphorus removal technology, phosphorous-rich sludge production within the WWTPs, and sludge processing methods (sludge disposal).

Phosphorus removal methods may differ depending on the size of the WWTP. Physical treatment such as filtration can lower the particulate P content. The second most widely used method is chemical precipitation (CP) via various iron and aluminum salts (and, rarely, calcium salts), which is suitable for both large and small scales, and offfers easy maintenance and a low cost accompanied with high efficiency. Next is enhanced biological phosphorus removal (EBPR), designed typically for large WWTPs, which ensure the stable P flow necessary for the functionality of the biological process [9, 10]. According to the urban water treatment directive data viewer (EC, 2013–2014), there are 26 812 WWTPs in Europe, of which 1664 implement exclusively EBPR and 9649 operate CP (Figure 3.2.1). In addition, more novel P removal technologies such as "Nereda," based on biological aerobic granular sludge P removal, often combined with conventional chemical precipitation, are expanding (Nereda WWTP, The Netherlands, pers. comm.). Physical techniques such as membrane technologies can also significantly lower P, but their price and maintenance may be more suitable for smaller-scale, extremely low discharge limits and higher P-rich streams [12]. Moreover, several running investigations (e.g. the RAVITA project pilot, commenced in 2016) are focusing on forming a closed loop system within the WWTP itself, without discharging the P as waste but instead formulating directly valuable products and moving in the direction of nutrient recovery [13].

3.2.3 Sludge Management

P removed via EBPR or CP ends up in the form of sewage sludge, which is further collected and processed. Typical phosphorus concentration in sewage sludge is around 2–4%

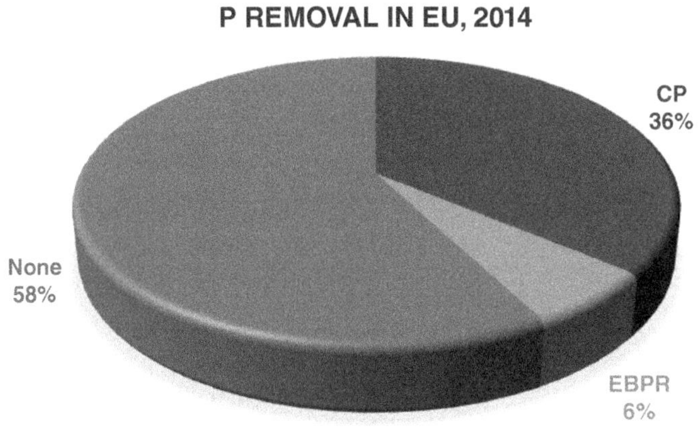

Figure 3.2.1 *Phosphorous removal in the European Union [11]. CP, chemical precipitation; EBPR, enhanced biological phosphorus removal.*

dry matter. EUROSTAT records that the sludge production in Europe (32 countries) reaches around 8.7 million tons (up to 300 million when including Turkey) [14]. It is expected that this number will further increase with the increasing population and expansion of urban WWTPs, mainly in the new EU Member States. Sludge handling can be achieved by incineration, composting, direct application, land reclamation, and landfilling. Direct application used to be the favored process, but its use is decreasing with more stringent legislation (some EU countries have already banned it) in favor of composting and incineration.

What is certain is that the concentrated phosphorus flows in sludge present an opportunistic raw material for the production of secondary phosphorus fertilizers. In Europe, it has been estimated that recovery potential for phosphorus from sewage sludge accounts for a minimum of 15% [7].

3.2.4 Current State of P Recovery Technologies

At the present moment, very few or no recycled mineral phosphorus products are processed and used worldwide. The first steps toward choosing the correct P recovery technique are based on the stage of the wastewater process from which the technology is built. P can be recovered directly from the wastewater, after centrifugation, drying, or incineration. Technologies suitable for P recovery in smaller-scale WWTPs include algal technologies, absorption, and precipitation, whereas larger-scale WWTPs can provide a good source for ash technologies, precipitation, and various metalo- and thermochemical processes (Table 3.2.1).

3.2.4.1 Phosphorus Salts Precipitation

Struvite is the most often produced form of the P salt formed via recovery techniques. It is a mineral, chemically formulated as magnesium ammonium phosphate hexahydrate or MAP ($MgNH_4PO_4 \times 6H_2O$). Several studies on struvite have focused on improving the recovery from municipal wastewaters, swine and dairy wastewaters, semiconductor water, anaerobic digesters, urine, and manure, examining which physicochemical conditions and technologies are the most efficient for in terms of the source of input [20, 21]. Depending

Table 3.2.1 P recovery: technology and product overview [15–19].

Technology	P recovery, %	Product	P_2O_5, %
Seaborn/Gifhorn	49	Struvite	28
Stuttgart	45	Struvite	27
AirPrex	7	Struvite	21
Crystalactor	40	Magnesium phosphate, calcium phosphate, struvite, or potassium magnesium phosphate	29–39
Phospaq		Struvite	29–40
Pearl	12	Struvite	28
LeachPhos	70–90	Calcium/aluminum phosphate	20–40
Struvia	11	Struvite, calcium phosphate	29
Pasch	60–80	Struvite, calcium phosphate	20–40
NyReSys	>30	Struvite	23–31
REM-NUT	50–70	Struvite	26–28
EcoPhos	97	H_3PO_4, dicalcium phosphate	-
TetraPhos	85	H_3PO_4	-
RecoPhos	89	Elemental P_4, P_2O_5, H_3PO_4	38
Mephrec	81	P slag	10–25
Kubota	80–90	P slag	17
AshDec	98	P ash	15–25

on the source material, struvite quality may differ. The most distinct variation can be observed between the struvites produced from sludge liquor (rather pure) and directly from the sludge (more impure).

The relevance and impact of the different ions, pH, P : N : Mg ratio, and mixing on struvite precipitation is thoroughly explained elsewhere [22–25]. From these reports, it is evident that the presence of Ca ions can affect crystallization meaningfully, and that Mg : Ca ratio should be kept at a value lower than 2. The optimal P : N : Mg molar ratio was selected to be around 1 : 1 : 1, and optimal pH around 9. Various kinetics models were employed in order to define and explain the crystal growth. Most researchers obtained satisfying results when applying simple first-order kinetics [26, 27]. On the other hand, the complexity of all of the ion interactions in the precipitation process needs further research. The different reactors types can have a significant effect on precipitation efficiency. The most used reactors are the fluidized bed reactor (FBR) and continuous stirred tank reactor (CSTR), alongside specific Rem Nut technology [28] and attempts to use microbial electrolysis cells (MECs) [3].

Struvite's commercial sale as a fertilizer supplement commenced in Japan in 1998 [29]. By 2001, Nawamura [30] reported that high struvite precipitation levels had been achieved at the Hiagari sewage treatment plant in Kitakyushu City, where 70% of phosphates were precipitated in a pilot FBR using environmentally friendly seawater as a source of Mg and air for pH adjustment. At the same time, a pilot FBR at Brisbane Water, Australia managed to precipitate 94% of P, by using $Mg(OH)_2$ as a source of Mg [31].

In the last 10 years, a breakthrough in P salt technologies let to the development of several processes:

The Airprex® process from CNP-Technology Water and Biosolids GmbH is a struvite precipitation technology using sewage sludge as source material. Struvite is formed by the addition of magnesium chloride salts ($MgCl_2$) in an air-lift reactor, followed by a sand washer.

Waterschap Velt en Vecht is the first Dutch water board to implement struvite technology. At its WWTP, struvite is formed by submitting sewage sludge to a stream of carbon dioxide (CO_2) stripped by bubble aeration, in which magnesium is added.

The STRUVIA™ process, a modification of the phosphorus recovery technology Phostrip (Showo Kankyo Systems K.K.) (SKS) by Véolia Water, recovers P in the form of struvite from sludge liquor. The technology consists of a CSTR containing special mixing technology (Turbomix®) and a lamella settler. Mg is added in the form of $MgCl_2$ and sodium hydroxide (NaOH) is used to keep the pH between 8 and 9.

The Pearl® process combined with WASSTRIP® is a successful solution from OSTARA Nutrient Recovery Technologies Inc. (Vancouver, Canada). OSTARA's technology treats the sludge liquor via a crystallization reactor installed after the dewatering unit to form struvite by dosing with $MgCl_2$ and increasing the pH with NaOH. The specific design of the mixing and harvesting reactor enables production of a uniform crystal size. Struvite is dried in a fluidized bed dryer, allowing the stability of the pellets (e.g. moisture content) to remain at the same level over long-term use [15].

The NuReSys can be applied to both centrate and digestate. It combines stripper and crystallizer (CSTR). Struvite is formed with the addition of $MgCl_2$.

The Crystalactor is developed by the Dutch consultancy and engineering company DHV. The main reactions are situated in the fluidized bed-type crystallizer, in which is placed a cylindrical reactor, partly filled with seed material with up-flow. The process can lead to various products, such as pellets of magnesium phosphate, calcium phosphate (CaP), struvite, and potassium magnesium phosphate, based on the chemical adjustment and process optimization.

The Phospaq technology from Paques recovers struvite from dewatering reject streams. The process is performed in a CSTR by aeration (pH adjustment) and addition of magnesium oxide (MgO) as a magnesium source [16–19].

3.2.4.2 Phosphorus Recovery Via Wet-Chemical Processes

Wet-chemical processes typically consist of two phases, the first of which uses acid (mainly sulfuric acid) to release the favorable P-rich matter from the sludge. Several processes have been established:

The STUTTGART has been developed at University of Stuttgart by the Institute for Sanitary Engineering (ISWA). It consists of leaching and precipitation phases. First, acidic extraction of P from digested sludge occurs at a pH of 4 with the addition of H_2SO_4. Second, after solid/liquid separation, struvite precipitation is favored by dosing of MgO and adjusting the pH to 8 with NaOH.

The LeachPhos, developed by BSH Umweltservice GmbH, extracts P from sewage sludge ash (SSA) in two stages. The first reaction is performed by addition of diluted sulfuric acid, where P is transferred into the leachate and filtered. In the second, the addition of sodium hydroxide or lime increases pH to form CaP sludge, by which filtration the final product CaP is collected [15].

The EcoPhos® process from the EcoPhos group is a leaching technology based on the use of hydrochloric acid (HCl) on ashes or low-grade rocks. It has a highly efficient heavy metal removal step and can result in the production of H_3PO_4 or dicalcium phosphate (DCP). By using sulfuric acid, it is possible to regenerate HCl while producing gypsum (construction material).

The Seaborne process from Seaborne Environmental Laboratory AG, later optimized as the Gifhorn process, uses digested sludge to produce P salts in two stages. The first takes place at pH 4.5 by the addition of sulfuric acid (H_2SO_4), followed by the dosing of sodium sulfide (Na_2S) and sulfide precipitation at pH 5.6, adjusted with NaOH. After decanting, the second phase occurs with the dosing of magnesium hydroxide (Mg(OH)) to form struvite/CaP at pH 9 (adjusted with NaOH). Addition of Mg below the stoichiometric ratio finalizes calcium precipitation, prevents scaling in the stripping reactor, and produces hydroxylapatite along with the struvite.

The PASCH® technology was developed at RWTH Aachen University. It is an acid-based leaching process, followed by a lamella separator and filter. The filtrate is subjected to extraction with alamine and tributylphosphate (TBP) for heavy metal removal, giving the final precipitates, such as CaP or struvite, depending on the chemical reagent used (i.e. lime or magnesium compounds).

The TetraPhos® process, developed by Remondis, uses phosphoric acid (H_3PO_4) to leach the ash [16–19]. Leaching is followed by the filtration step. The liquid fraction is further treated with the sulphuric acid and filtration step to separate calcium in the form of gypsum. The residing liquid fraction is processed via ion exchange to remove and eventually recover the metals (such as Al, Fe). The formed P acid is further concentrated and purified depending on the product quality demand.

3.2.4.3 Phosphorus Recovery Via Thermal Processes

Thermal processes allow good separation of the P-rich product and the heavy metals by use of high temperatures. We can differentiate between metallurgical and thermochemical processes. Thermal processes are slowly gaining huge interest. The most used technologies are incineration, pyrolysis, and gasification. The most screened processes are as follows:

The Mephrec process by Ingitec is a metallurgical treatment of dry sewage sludge or ash in order to recover P. Reaction temperatures in a shaft furnace are above 1450 °C. The end-product is silicophosphate (analogous to "thomas phosphate") [15].

The AshDec is a thermochemical treatment of SSA (FP6 project SUSAN, 2005–2008), developed by Outotec and BAM Federal Institute for Materials Research and Testing. The thermal reaction takes place at 900–1000 °C for approximately 20 minutes in a directly heated rotary kiln in counter flow. The ash is mixed with sodium sulfate (Na_2SO_4) or magnesium chloride ($MgCl_2$). Heating induces metals/metalloids to react with the salts, become gaseous, and evaporate. The end-product is the calcined ash with $CaNaPO_4$ phase [15, 19].

The RecoPhos is a thermochemical process formulated in the course of the RecoPhos project [32]. The process uses sewage sludge, meat and bone meal, and sewage sludge ashes as input materials [33]. The thermal process implies electromagnetically induced heating of a reactor bed consisting of coke or graphite at a temperature of 1200–1400 °C.

The reaction leads to high-grade elemental P in the form of vapor, which can be further condensed and harvested as P_4, additionally oxidized to P_2O_5, or transformed into H_3PO_4 [16–19].

The thermochemical (sludge melting) process from Kubota converts municipal solid waste, sewage sludge, landfill waste, and ashes at temperatures of 1300 °C into the valuable P slag. The melting furnace is a double cylinder, between which is placed the input material. Combustion takes place in the inner part, after which the ash component melts and flows down. By controlling the air and rotating the cylinder, optimal conditions can be achieved. Organic components are a source of fuel, whereas P and heavy metal volatility are controlled by additives such as Fe compounds. The slag produced is cooled in a water tank under the furnace and granulated [34].

3.2.4.4 Choice of Phosphorus Technologies Today

The present infrastructure for sludge handling and population size are important factors determining the application of several technologies.

Struvite precipitation from sludge liquor or sludge proves to be the most promising approach from the point of view of economy as well as of energy and climate change impact. Its slow-release properties and high quality are what distinguish struvite from other fertilizers. However, this technology has its restrictions. First, it has a low phosphorous recovery yield compared to other methods. Second, it demands the use of sewage sludge from enhanced biological phosphorous removal or phosphorous precipitation using green coagulants instead of the commonly used iron. In addition, struvite has not yet been recognized as a fertilizer (except in a few countries), and restrictions in fertilization laws thus further mitigate against the implementation of this technology. Besides struvite, this technology can also recover CaP, a commonly used fertilizer. Application of CaP or other fertilizers obtained from sludge precipitation demands further adjustments in the law.

Sludge/ash leaching technology has moderate phosphorus recovery and good separation of heavy metals as strong attributes. However, extremely high demands for chemicals and energy make this technology unfavorable. It could be considered as beneficial only where a country's high chemical production (e.g. of sulfuric acid) could compensate for its high consumption.

Metallurgical sludge treatment appeared to be one of the best technologies from the perspective of phosphorous recovery potential. Nevertheless, its high energy demands and reasonably high costs strongly reduce its applicability.

Ash recovery has an obvious advantage as an upgrade to the mono-incineration process, especially in the case of EcoPhos. However, in places without existing mono-incineration, it is rather unfavorable due to its high capital cost [15–19].

In the near future, if there is no existing mono-incineration plant in place, the recovery of phosphorus from sludge rather than ash appears the best way forward. However, knowing that rules on sludge application in agriculture may become stricter, as they already are in several EU countries, it is advisable to set a long-term goal of building ash recovery and metallurgical treatment plants, especially in highly populated areas.

3.2.5 Future P Recovery Technologies

Successful stories of the first P recovery technologies, as well as continuously arising P demands, are driving development to a great extent. Many new works are being produced on the laboratory scale and slowly progressing to higher technology readiness level (TRL). A few examples from the ongoing Phos4You project are presented in the following subsections.

3.2.5.1 Phosphorus Salt Recovery Upgrades

The PULSE process is a modification of the PASCH technology currently under development at the University of Liege. Fully dried sewage sludge is processed in three major steps 1) acid leaching aimed at P release from the sludge 2) reactive-extraction with various chemicals (e.g. polymers) to reduce the metal content, and 3) P salt precipitation to recover the valuable P product. The technology is equipped with a sludge dryer, a leaching vessel, solid/liquid separation, a mixer–settler cascade for reactive extraction, various storage, precipitation, and supply vessels, pumps, piping, and required sensors. The use of the strong acids demands that the equipment is made of highly corrosion-resistant materials. The resulting precipitate is a fertilizer-grade calcium or magnesium phosphate. In comparison with simple precipitation from wet sludge, this process enables greater P recovery (not only of dissolved P) while effectively removing Fe, Al, and heavy metals [38].

Biological phosphorus dissolution as a pre-step to P precipitation is a two step process that combines 1) bio-acidification, sludge phosphorus release and 2) precipitation, formation of struvite (using Struvia™ technology). This process should allow for a significant increase in the P recovery yield from sewage sludge liquor (up to 75% of the influent P of a WWTP). The bio-acidification is led by addition of an easily degradable carbohydrate source under strictly anaerobic conditions. Two mechanisms play a role here: (i) P release by the microorganisms that have accumulated the P during wastewater treatment; and (ii) released P maintenance by the bacteria in solution and dissolution of the mineral P combined with cations by developing bacteria, producing lactic acid in situ (decreasing the pH down to 4–5). Additional information about this technology can be found in the section 4.6.3.2 [39].

With the expanding development of **materials science (polymeric, composite, and smart)**, the precipitation of struvite is expected to be achieved faster and more efficiently.

It is already known that polymeric ligand-exchange resins (PLEs), such as PhosX^{np} iron nanoparticle resins, have the ability to remove phosphate originating from reverse osmosis concentrate [35]. Similarly, alginate-like exo-polysaccharides, isolated from aerobic granular sludge, were found to be able to capture potassium and magnesium ions, creating the supersaturation conditions suitable for struvite (potassium or ammonium magnesium phosphate) formation [36].

Currently, within the scope of a circular economy approach, work is being done on finding optimal, preferably biobased, sorption materials. One such is being upgraded to Véolia's P adsorption reactor (FiltraPHOS™) for small wastewater discharges (i.e. from septic tanks, individual dwellings, or business premises). The FiltraPHOS™ employs an enhanced mechanism combining a rapid gravitational filtration of raw water with a granular medium and its continuous self-backwashing. The filtering media is composed of single or dual media layers. Use of local resources as the adsorption material should be employed when feasible. Additional details on these technology developments are described in Section 4.6.3.3.

3.2.5.2 Thermal Processes

The EuPhoRe® process from EuPhoRe GmbH is a thermal treatment of sewage sludge. After the drying process, additives are provided to the input material. The reactor used is rotary kiln. The process consists of two steps (anoxic, oxic) pyrolysis/dry carbonation and an additional incineration step. Pyrolysis occurs during the reduction regime at temperatures of 650–750 °C, after which carbon is fixed by post-combustion at temperatures of 900 and 1100 °C. The flue gas from the oxidation process, the pyrolysis gas, and the vapor known as process gas are extracted out of the kiln at the sludge-feeding side and treated in two burning chambers. The part is combusted and reused in the rotary kiln. The hot flue gas is introduced at the side where the ash is discharged due to the counter-current system. Released flue gas is treated via a fine-particle filter and adsorbents to comply with air-emission regulations. The resulting product is P ash free of organic matter (and therefore pathogens), low in heavy metals, and P-available [40]. More details on this technology are given in Section 4.6.3.2.

The REMONDIS TetraPhos process adapted to the Emscher and Lippe regions, which produces sewage sludge ashes with a low P load (8%–11% P_2O_5), is set to enable the recovery of phosphate from ashes. The ashes are first treated with phosphoric acid and, after separation of acid-insoluble residue, the resulting leachate is purified with the addition of sulfuric acid and capacity of ion exchange and selective nano-filtration to generate an industrial-quality phosphoric acid (brand name RePacid®) [41]. More insight into this particular P technology development is provided in Section 4.6.3.1.

3.2.5.3 Natural Process for the Recovery of Phosphorus

The microalgal potential for P recycling and the production of algae-based biofertilizers has already been recognized and affirmed [37]. In the presence of adequate nutrients and sunlight, algae can perform luxury P uptake and form biomass, which can be used for biofertilizer production. In small WWTPs, nutrients such as P are usually not recovered and therefore are discharged into the environment, leading often to eutrophication problems. Recovering P from these decentralized small WWTPs requires systems offering robustness, low maintenance, and sustention of highly variable P load. The capacity of the extremophile microalga *Chlamydomonas acidophila* to acquire high phosphorus and nitrogen rates (up to 90%) while operating at low temperatures makes it a valid candidate for P recovery at those specific sites. Though light can be a limiting factor for this type of technology, it is not for *C. acidophila*, which requires a very low light intensity (40–113 μmol photons $m^{-2}\,s^{-1}$) for its growth. Additionally, *C. acidophila* is able to maintain its functionality even in the presence of 1000-times higher concentrations of pharmaceuticals than are found in effluents [42]. A detailed description of this technology is given in Section 4.6.3.4.

3.2.6 Conclusion

With the development of the first-full scale P recovery technologies, it became obvious that the greatest interest in recycled phosphorus was coming from fertilizer producers. However, there are concerns about the recycling, mainly regarding its price and the inability of production to fulfill demand. The main limitations on the development of phosphorus recycling are a lack of research, social barriers, and a lack of governmental pressure. Even when larger distributors recognize its potential, most products end up being sold as blends instead of as the final product, with its true value. Therefore, the successful entrance of recycled products into the market will be tightly connected with the implementation of novel legislation. Customers' suspicions related to recycled products, as in the case of every newly implemented technology, can be eliminated by a good marketing strategy supported by successful case studies.

It is expected that the tremendous passion for circular economy development and thus phosphorus recovery from wastewater will drive the formulation of fertilizers to a novel level, with a more stable composition, easier handling, and greater safety for the whole ecosystem.

References

1. Steen, I. (1998). Phosphorus availability in the 21st century: management of a non-renewable resource. *Phosphorus & Potassium* **217**: 27–31.
2. Cordell, D., Drangert, J.-O., and White, S. (2009). The story of phosphorus: global food security and food for thought. *Global Environmental Change-Human and Policy Dimensions* **19** (2): 292–305.
3. Cusick, R.D. and Logan, B.E. (2012). Phosphate recovery as struvite within a single chamber microbial electrolysis cell. *Bioresource Technology* **107**: 110–115.
4. European Commission (2014). *Report on Critical Raw Materials for the EU*. Geneva: European Commission.
5. Fixen, P.E. and Johnston, A.M. (2012). World fertilizer nutrient reserves: a view to the future. *Journal of the Science of Food and Agriculture* **92** (5): 1001–1005.
6. Schoumans, O.F., Bouraoui, F., Kabbe, C. et al. (2015). Phosphorus management in Europe in a changing world. *Ambio* **44**: S180–S192.
7. van Dijk, K.C., Lesschen, J.P., and Oenema, O. (2016). Phosphorus flows and balances of the European Union Member States. *Science of the Total Environment* **542**: 1078–1093.
8. Tchobanoglous, G., Burton, F.L., and Stensel, H.D. (2004). *Wastewater Engineering: Treatment and Resource Recovery*, 4e. New York: McGraw-Hill Higher Education.
9. Henze, M., Loosdrecht, v., M.C.M., Ekama, G.A., and Brdjanovic, D. (2008). *Biological Wastewater Treatment: Principles, Modelling and Design*. London: IWA Publishing.
10. Tarayre, C., De Clercq, L., Charlier, R. et al. (2016). New perspectives for the design of sustainable bioprocesses for phosphorus recovery from waste. *Bioresource Technology* **206**: 264–274.
11. Urban water treatment directive (UWWTD) data viewer: Council Directive 91/271/EEC of 21 May 1991 concerning urban waste-water treatment. 2013–2014.
12. Jayawardana HMCM, Dissanayaka DMSH, Kumarasinghe PPU, Mowjood MIM. Phosphorus Removal from Wastewater using Soil as an Adsorbent. Water Professionals' Day 2015.
13. RAVITA_project. 2012–2016. Available from: https://www.hsy.fi/ravita/en/results/Sivut/default.aspx (accessed December 23, 2019).
14. EUROSTAT. Sewage sludge production and disposal from urban wastewater (in dry substance (d.s)). 2017. Available from: http://ec.europa.eu/eurostat/web/products-datasets/-/ten00030 (accessed December 23, 2019).

15. P-REX_project. 7th Europen Framwork program. Main P-REX project deliverables 2012–2015. Available from: http://doi.org/10.5281/zenodo.242550 (accessed December 23, 2019).
16. Egle, L., Rechberger, H., and Zessner, M. (2015). Overview and description of technologies for recovering phosphorus from municipal wastewater. *Resources Conservation and Recycling* **105**: 325–346.
17. Egle, L., Rechberger, H., Krampe, J., and Zessner, M. (2016). Phosphorus recovery from municipal wastewater: an integrated comparative technological, environmental and economic assessment of P recovery technologies. *Science of the Total Environment* **571**: 522–542.
18. Ye, Y., Ngo, H.H., Guo, W. et al. (2017). Insight into chemical phosphate recovery from municipal wastewater. *Science of the Total Environment* **576**: 159–171.
19. Huygens D, Saveyn H, Eder P, Delgado Sancho L. DRAFT STRUBIAS Technical Proposals. DRAFT nutrient recovery rules for recovered phosphate salts, ash-based materials and pyrolysis materials in view of their possible inclusion as Component Material Categories in the Revised Fertiliser Regulation. Interim Report. Circular Economy and Industrial Leadership Unit DB-GaI, Joint Research Centre; 2017.
20. Battistoni, P., De Angelis, A., Pavan, P. et al. (2001). Phosphorus removal from a real anaerobic supernatant by struvite crystallization. *Water Research* **35** (9): 2167–2178.
21. Zarebska, A., Nieto, D.R., Christensen, K.V. et al. (2015). Ammonium fertilizers production from manure: a critical review. *Critical Reviews in Environmental Science and Technology* **45** (14): 1469–1521.
22. Le Corre, K.S., Valsami-Jones, E., Hobbs, P., and Parsons, S.A. (2005). Impact of calcium on struvite crystal size, shape and purity. *Journal of Crystal Growth* **283** (3–4): 514–522.
23. Pastor, L., Mangin, D., Ferrer, J., and Seco, A. (2010). Struvite formation from the supernatants of an anaerobic digestion pilot plant. *Bioresource Technology* **101** (1): 118–125.
24. Jordaan, E.M., Ackerman, J., and Cicek, N. (2010). Phosphorus removal from anaerobically digested swine wastewater through struvite precipitation. *Water Science and Technology* **61** (12): 3228–3234.
25. Huang, H., Xu, C., and Zhang, W. (2011). Removal of nutrients from piggery wastewater using struvite precipitation and pyrogenation technology. *Bioresource Technology* **102** (3): 2523–2528.
26. Nelson, N.O., Mikkelsen, R.L., and Hesterberg, D.L. (2003). Struvite precipitation in anaerobic swine lagoon liquid: effect of pH and Mg: P ratio and determination of rate constant. *Bioresource Technology* **89** (3): 229–236.
27. Rahaman, M.S., Ellis, N., and Mavinic, D.S. (2008). Effects of various process parameters on struvite precipitation kinetics and subsequent determination of rate constants. *Water Science and Technology* **57** (5): 647–654.
28. Liberti, L., Petruzzelli, D., and De Florio, L. (2001). REM NUT ion exchange plus struvite precipitation process. *Environmental Technology* **22** (11): 1313–1324.
29. Ueno, Y. and Fujii, M. (2001). Three years experience of operating and selling recovered struvite from full-scale plant. *Environmental Technology* **22** (11): 1373–1381.
30. Nawamura Y. Struvite recovery using seawater as a magnesium source. SCOPE Newsletter 2001.
31. Munch, E.V. and Barr, K. (2001). Controlled struvite crystallisation for removing phosphorus from anaerobic digester sidestreams. *Water Research* **35** (1): 151–159.
32. RecoPhos_project. 2012–2015. Available from: http://www.recophos.org (accessed December 23, 2019).
33. Steppich D. The thermo-reductive RecoPhos Process Leoben:RecoPhos – Demonstration Event2015. Available from: http://www.recophos.org/inc/mod/downloads/download.php?id=8 (accessed December 23, 2019).
34. Hosho F, Yoshioka, Y., Okada, M., et al. Thermochemical phosphorus recovery from sewage sludge and effectiveness of recovered phosphorus as fertilizer. 9th International Conference on: Combustion, Incineration/Pyrolysis, Emission and Climate change (i-CIPEC), Kyoto; 2016.

35. Kumar, M., Badruzzaman, M., Adham, S., and Oppenheimer, J. (2007). Beneficial phosphate recovery from reverse osmosis (RO) concentrate of an integrated membrane system using polymeric ligand exchanger (PLE). *Water Research* **41** (10): 2211–2219.
36. Lin, Y.M., Bassin, J.P., and van Loosdrecht, M.C.M. (2012). The contribution of exopolysaccharides induced struvites accumulation to ammonium adsorption in aerobic granular sludge. *Water Research* **46** (4): 986–992.
37. Solovchenko, A., Verschoor, A.M., Jablonowski, N.D., and Nedbal, L. (2016). Phosphorus from wastewater to crops: an alternative path involving microalgae. *Biotechnology Advances* **34** (5): 550–564.
38. Pfennig, A. and A. Léonard (2018). "Phos4You Fact Sheet I3 PULSE. Acid leaching of phosphorus from partially/fully dried sewage sludge: PULSE process." https://www.biorefine.eu/sites/default/files/publication-uploads/3_phos4you-pulse.pdf from https://www.biorefine.eu/sites/default/files/publication-uploads/3_phos4you-pulse.pdf
39. Mébarki, C. and M.-L. Daumer (2018). "6_Phos4You Struvia_Bioacidification. Biological phosphorus dissolution before P precipitation from sludge liquor." https://www.biorefine.eu/sites/default/files/publication-uploads/6_phos4you-struvia_bioacidification.pdffrom https://www.biorefine.eu/sites/default/files/publication-uploads/6_phos4you-struvia_bioacidification.pdf.
40. Schmelz, K.-G., et al. (2018). "Phos4You Fact Sheet I1 EuPhoRe.Thermochemical solution to recover phosphorus from sewage sludge: EuPhoRe®." https://www.biorefine.eu/sites/default/files/publication-uploads/1_phos4you-euphore.pdffrom https://www.biorefine.eu/sites/default/files/publication-uploads/1_phos4you-euphore.pdf.
41. Blöhse, D. and P. Herr (2018). "Phos4You Fact Sheet I2 Tetraphos. Acid extraction of phosphorus from sewage sludge incineration ash: REMONDIS TetraPhos®." https://www.biorefine.eu/sites/default/files/publication-uploads/2_phos4you-tetraphos.pdf
42. Escudero, A. and T. Kennedy (2018). "Phos4You Fact Sheet I4 microalgae. Microalgae to recover phosphorus from small-scale waste water treatment plants." https://www.nweurope.eu/media/3406/4_phos4you-microalgae.pdffrom https://www.nweurope.eu/media/3406/4_phos4you-microalgae.pdf.

3.3

Ammonia Stripping and Scrubbing for Mineral Nitrogen Recovery

Claudio Brienza, Ivona Sigurnjak, Evi Michels, and Erik Meers

Department of Green Chemistry and Technology, Faculty of Bioscience Engineering, Ghent University, Ghent, Belgium

3.3.1 Introduction

Unlike biological ammonia (NH_3) removal, NH_3 stripping and scrubbing from wastewater couples the reduction of NH_3 concentration in the treated effluent and its recovery. This results in the production of chemical building blocks (generally ammonium sulfate) suitable for the production of marketable biobased fertilizers or other chemical products. As explained in Chapter 3.1, the process consists of two steps: NH_3 stripping and NH_3 scrubbing.

The former (NH_3 stripping) is a gas–liquid mass-transfer process where NH_3 is stripped from the wastewater to a gas phase, usually by air or steam. Though similar to air stripping, steam stripping requires higher operational temperatures, making the process extremely energy-intensive [1]. When the stripping system is associated with anaerobic digestion (AD) installations for the processing of fermentation effluents, either biogas or combined heat and power (CHP) flue gas can be used as suitable strip gases, as an alternative to air or steam [2, 3]. During the stripping phase, the equilibrium between ammonium hydroxide ions (NH_4^+) and volatile aqueous NH_3 is influenced by pH, temperature, air-to-liquid ratio, air supply rate, and hydraulic loading rate. These parameters can be adjusted to shift the equilibrium toward NH_3 gas and increase the overall efficiency of NH_3 stripping [4]. The most common reagents used to adjust the pH are sodium hydroxide (NaOH), potassium

Biorefinery of Inorganics: Recovering Mineral Nutrients from Biomass and Organic Waste, First Edition.
Edited by Erik Meers, Gerard Velthof, Evi Michels and René Rietra.
© 2020 John Wiley & Sons Ltd. Published 2020 by John Wiley & Sons Ltd.

hydroxide (KOH), and calcium oxides (CaO and $Ca(OH)_2$). One downside of NH_3 stripping is represented by the formation of carbonate salts in the stripping tower, leading to scaling and fouling, which can be prevented by frequent cleaning [5].

In the second step (NH_3 scrubbing), NH_3 contained in the gas phase is stabilized by contact with an acid solution, which results in the formation of so-called air scrubber water (ASW). The type of ASW depends on the type of used acid. If NH_3 is absorbed onto sulfuric acid solution (H_2SO_4), ammonium sulfate ($(NH_4)_2SO_4$) will be formed, whereas the use of nitric acid (HNO_3) as scrubbing agent will result in ammonium nitrate (NH_4NO_3) solution [6]. The generated ammonium sulfate (AS) or ammonium nitrate (AN) contain nitrogen (N) entirely in mineral form, and as such represent an interesting alternative for the substitution of synthetic mineral N fertilizers.

This chapter is organized as follows: Section 3.3.2 deals with different configurations for the recovery of NH_3 as ASW from waste air and effluent streams, mainly focusing on livestock production and AD. Section 3.3.3 gives a brief overview of a variety of chemicals (organic acids, nitric acid, and gypsum) that have been investigated as alternatives to H_2SO_4 in order to capture NH_3 in scrubbing towers. Section 3.3.4 examines successful full-scale applications of NH_3 recovery from different waste origins. Finally, Section 3.3.5 looks at the quality characteristics of the recovered ammonia salts.

3.3.2 Ammonia Stripping and Scrubbing from Biobased Resources

Stripping and scrubbing is an established technology in different industrial realities and its implementation in wastewater treatment dates back to the 1970s [7]. Several studies have reported the application of NH_3 stripping combined with acid scrubbing for the treatment of different types of wastes, including dewatered sewage sludge [8], landfill leachate [9, 10], urea fertilizer plant wastes [11], condensates from sugar beet factory [12], and cellulose-acetate fiber wastewater [13]. Potential NH_3-rich waste streams are currently generated at animal stables, composting facilities, and biogas installations, making these suitable sites for N mining [14–16]. An overview of the possible routes for the production of ASW (AS or AN) is given in Figure 3.3.1. This section examines different studies conducted on NH_3 recovery from exhaust gases and wastewaters generated at livestock facilities and AD plants.

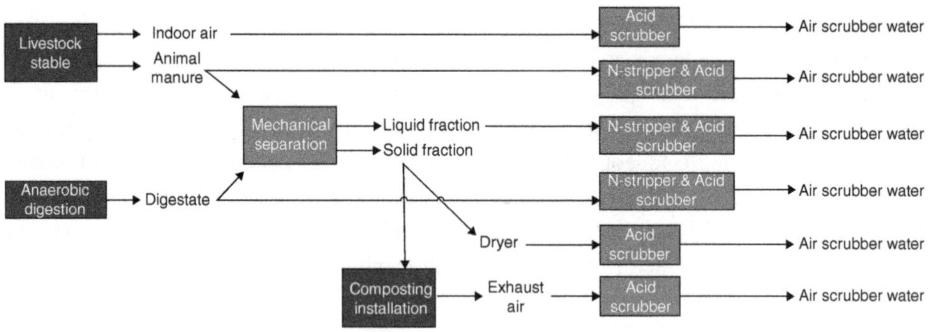

Figure 3.3.1 Production of air scrubber waters (AS or AN) from organic wastes via different stripping and scrubbing pathways.

3.3.2.1 Acid Scrubbing of Exhaust Air

Air scrubbers find their application at livestock facilities in the treatment of indoor air from stables as well as the exhaust gases emitted during the processing of manure (e.g. mechanical separation and drying process). Similarly, acid traps are operated at AD and composting installations to reduce NH_3 emissions to the atmosphere during the composting process or digestate handling. An air scrubber is a reactor (known as a packed air scrubber or trickling filter) filled with porous and inert inorganic material, characterized by a large specific surface area (1–2 $cm^2\ cm^{-3}$). Recirculating water is sprayed on top, and its pH can be adjusted below 4 to increase the mass transfer of NH_3 from the gas phase to the liquid solvent. Air can be discharged in the reactor either horizontally (cross-current) or upward (counter-current) [17]. Usually, H_2SO_4 is used to decrease the pH of the washing solution, producing AS, which can be used as a nitrogen-sulfur (N-S) fertilizer. In the extensive review by Van der Heyden et al. [16], the reported NH_3 recovery efficiencies at poultry and pig farms range from 40 to 100%. In order to improve the efficiency of NH_3 removal from exhaust air, the authors recommend careful maintenance and operation of these systems.

3.3.2.2 Stripping and Scrubbing from Manure

In 1995, one of the first studies on NH_3 removal from animal manure was conducted by Liao et al. [18]. These authors compared air stripping and aeration. Though cheap and simple, aeration implementation for NH_3 removal showed lower efficiency compared to air stripping. In both configurations, best results were achieved when high pH and temperature were tested. La et al. [19] have reported promising results with microwave radiation in alternatives to air stripping on pig slurry. At optimal conditions (pH 11, 5 minutes, 700 W), the researchers recovered 82.6% of the initial NH_3 contained in raw slurry in the form of AS. At pH 9.5, recovery efficiencies above 90% were recorded by Laureni et al. [20] from swine wastewater with low chemical oxygen demand (COD) content ($<10\ g\ l^{-1}$). The authors suggested that the low total organic carbon (TOC) contamination in the final product makes AS (6% N) a marketable fertilizing supplement. According to Zhang and Jahng [21], pH adjustment is achieved more effectively by using NaOH or KOH instead of $Ca(OH)_2$. In their study, the stripped manure was subsequently used as feedstock for AD. They attributed the inhibitory effects on methane production to the higher calcium toxicity following the addition of $Ca(OH)_2$.

3.3.2.3 Stripping and Scrubbing from Anaerobic Digestate

A considerable amount of literature has been published on the implementation of AD with stripping and scrubbing systems, mainly with the objective of circumventing NH_3 inhibition inside the digesters [22]. Different configurations have been assessed at bench scale: post-digestion NH_3 removal [23, 24], in situ NH_3 removal [25–27], side-stream NH_3 removal [28–30], and pre-digestion NH_3 removal [31]. In recent years, there has been an increasing amount of literature on AD effluent processing via stripping and scrubbing. Liu et al. [32] demonstrated that reducing the amount of alkali in order to adjust the pH is feasible by stripping CO_2 and reducing, in turn, the buffer capacity of the digestate. With this approach, 97% of NH_3 was recovered as AS. A recent study by Ukwuani and Tao [33] successfully reported the recovery of AS crystals from dairy manure, sludge, and food waste digestate through an innovative vacuum thermal stripping. By applying vacuum in the stripping tower, the boiling temperature of digestate decreased, allowing 95% of NH_3 to be recovered. In a different study, the researchers addressed different methods to produce granular AS by thermal stripping. High-purity crystals ($\geq$ 98%) were obtained when the neutralization step was performed with H_2SO_4 solution pre-saturated with AS and subsequently cooled [34].

3.3.2.4 Manure and Digestate Processing by Evaporation

An alternative way to recover NH_3 from wastewaters is represented by evaporation. The process results in condensed water and a concentrate with higher nutrient concentration. By adjusting process conditions (mainly pH), it is possible to shift the equilibrium of N ions toward NH_3, enhancing its transfer from the liquid to the gas phase, thus generating an NH_3-rich condensate (NH_4^+ water), whose fertilizing properties need to be further explored. The research to date has tended to focus on concentrating nutrients and producing clean water rather than recovering N as NH_4^+ water. This is usually achieved by prior acidification of slurry or digestate and subsequent evaporation [35–37]. One of the few examples of NH_4^+ water production occurs at the digestate processing facility of Warterleau New Energy (Ieper, Belgium). Digestate is first solid–liquid separated, and the liquid fraction (LF) undergoes a biological treatment to decrease the organic matter content. Subsequently, the COD-depleted substrate is sent to an evaporator to generate NH_4^+ water with 15% NH_4-N content.

3.3.3 Alternative Scrubbing Agents

3.3.3.1 Organic Acids

Currently, H_2SO_4 is the most common scrubbing agent due to a favorable low price. However, research is slowly focusing on alternative acids that can be used instead of inorganic acids such as H_2SO_4. Starmans and Melse [38] compared different organic and inorganic acids with respect to their efficiency, safety, and cost in order to assess the best option for the purification of exhaust gas in an acidic air scrubber. They identified two organic acids as replacements for H_2SO_4: citric acid and maleic acid. The former was best in terms of safety and efficiency, while the latter was ranked the second best option. The application of citric acid would result in the formation of ammonium citrate (AC) [38]. In a recent study by Jamaludin et al. [39], 90% of NH_3 from digestate was recovered either with citric acid or with acetic acid. Compared to AS, AC has a higher solubility in water and lacks S, which makes it a more suitable N fertilizer for S-abundant soils. Moreover, AC has the capacity to enhance the solubility of precipitated and absorbed phosphorous [39].

3.3.3.2 Nitric Acid

The utilization of HNO_3 as a washing agent in air scrubbers has been demonstrated to be effective at full scale. At the wastewater treatment plant (WWTP) of VEAS (Oslo, Norway), sludge is anaerobically treated (retention time of 20 days and mesophilic conditions) and slaked lime is added to it for conditioning and sanitization. The digested sludge is subjected to a membrane press adapted with a heated vacuum press in order to dewater it and achieve volume reduction, resulting in a solid (cake) and a liquid (filtrate) fraction. The cake is P-rich, as P precipitates as Ca-P by the addition of slaked lime. The cake is stockpiled and made available for farmers. The alkaline filtrate has pH 10, a condition under which most of the N is present as NH_3. The NH_3 is air stripped from the filtrate and pumped into an absorption tower, where 55% HNO_3 is applied to recover NH_3 as AN. Until 1998, H_2SO_4 was used as a scrubbing agent, but it was replaced by HNO_3 when farmers started to ask for a fertilizer that would not cause soil acidification. As the mixing of organic compounds and HNO_3 can cause explosions, Yara set a TOC limit in AN of 100 ppm [40].

3.3.3.3 Gypsum

Flue gas desulfurization (FGD) gypsum ($CaSO_4$) is the by-product of flue gas desulfurization processes. Coal-based electric-generating plants exploit this process to remove sulfur dioxide (SO_2) from exhaust gas [41]. The NH_3 stripping system installed at the AD plant BENAS (Germany) was developed and patented by GNS [42]. The system relies on the stripping of NH_3 by thermal treatment at reduced pressure from digestate to a gas phase. Subsequently, the gas phase, containing CO_2 and NH_3, is transferred to a scrubber, where NH_3 is absorbed to a liquid suspension containing FGD-gypsum. The excess gas not absorbed that contains CO_2 is recirculated into the circuit. Binding of NH_3 to $CaSO_4$ results in the production of AS and calcium carbonate ($CaCO_3$).

3.3.4 Industrial Cases of Stripping and Scrubbing

NH_3 stripping and scrubbing toward the production of N mineral fertilizers is already a reality. This section examines various case studies of N recovery from diversified sources.

3.3.4.1 Waste Air Cleaning Via Acid Scrubbing

As previously described, air scrubbers are used to process exhaust gases rich in NH_3. Van der Heyden [43] extensively reviewed the implementation of air scrubbers at animal housing systems, reporting numerous examples of air scrubbers at pig and poultry facilities. Similarly, composting installations generate considerable amounts of process gas rich in NH_3. The biogas plant of GMB BioEnergie (Lichtenvoorde, the Netherlands) treats about $40\,000\,t\,y^{-1}$ of organic wastes. The biogas effluent is solid/liquid separated and the LF is sent to a WWTP for further processing. The solid fraction (SF) of digestate is mixed with sludge transported from different WWTPs and composted in aerated tunnels for 25 days. The NH_3 gas generated during the composting process is captured in an absorption tower with H_2SO_4 and transformed into marketable AS fertilizer with 8% N content.

3.3.4.2 Raw Digestate Processing Via Stripping and Scrubbing and Recirculation of the N-Depleted Digestate

Sigurnjak et al. [44] describe two AD installations where NH_3 is recovered from raw digestate by means of stripping and scrubbing, and the stripping residue is recirculated to the bioreactors. The first plant is located in Vellezzo Bellini (Italy) and treats about 120 000 t y^{-1} of sewage sludge and food wastes at thermophilic regime. The installation is implemented with a side-stream NH_3 stripping unit operating at a temperature range of 60–80 °C, using the biogas generated as strip gas. Approximately 22% of the NH_4-N contained in the digestate is recovered as AS, which is applied to the surrounding agricultural fields as inorganic nutrient supplement. Both the N-depleted digestate and the biogas are recirculated back into the digester [44].

A second biogas plant, located in Ottersberg (Germany), processes about 100 000 t y^{-1} of chicken manure and agricultural feedstock at thermophilic temperatures. The NH_3 stripping unit consists of three stripping columns where NH_3 is transferred from the digestate to the gas phase at a temperature interval of 50–80 °C. FGD-gypsum is used to absorb NH_3, with a maximum recovery efficiency of 80%. The installation produces 25% AS and $CaCO_3$. The stripped digestate is solid/liquid separated and the LF is recirculated back to the digester [44].

3.3.4.3 Liquid Fraction Digestate Processing Via Stripping and Scrubbing

A recent study by Bolzonella et al. [45] reports the implementation of a double-column stripping and scrubbing system at a farm AD installation. The biogas plant (600 KW_{el}) was fed on cow manure, pig effluents, and energy crops. About 130 t d^{-1} of anaerobic digestate were first dewatered by means of a screw press, and the resulting LF was further treated in lamella settlers to remove suspended solids. The pH of the LF of digestate was adjusted over 9 by adding $Ca(OH)_2$, and the effluent entered into the first column, where hot air (60–70 °C) favored the transfer of NH_3 to the gas phase. In the second column, NH_3-rich air was neutralized with H_2SO_4 to form AS. The system allowed the recovery of 17% of the N contained in the digestate [45].

3.3.4.4 Liquid Fraction of Digestate Processing Via Membrane Separation and Stripping and Scrubbing

Preliminary work on full-scale NH_3 stripping and scrubbing from LF of digestate was undertaken by Ledda et al. [46]. The N-Free® process described by the authors was implemented as post-treatment of digested cow and swine manure, enabling the treatment of 50–100 m^3 day^{-1} of digestate. It included a multi-step solid/liquid separation prior to the NH_3 stripping and scrubbing unit, where biogas effluent was first separated by means of a screw press and the resulting permeate was further treated into a decanter centrifuge. The permeate liquid was finally processed via a membrane filtration system consisting of ultrafiltration and reverse osmosis (RO). The RO centrate entered into a cold stripping unit and pH was increased between 12 and 12.5 by adding lime. NH_3-rich air was washed with H_2SO_4, resulting in 22–31% AS solution. The authors obtained 17% and 33% N recovery from digested cow and swine manure, respectively [46].

3.3.5 Product Quality of Ammonium Sulfate and Ammonium Nitrate

There are a large number of published studies describing NH_3 stripping and scrubbing from manure and digestate, but far too little attention has been paid to the quality of the ammonium salt solutions recovered via this process.

3.3.5.1 Ammonium Sulfate

AS contains N entirely in mineral form as NH_4-N and is an important source of S. Depending on the amount of H_2SO_4 added during the process, the amount of N and S can vary, together with the pH and electrical conductivity (EC), as shown in Table 3.3.1. In two independent studies on lettuce, Sigurnjak investigated the chemical composition and agronomic performance of AS in both full-scale trials and pot experiments [44, 47]. Both studies demonstrated that AS and a conventional fertilizing regime with synthetic calcium ammonium nitrate (CAN) had comparable effects on crop dry weight (DW) and fresh weight (FW) yield, making AS an interesting alternative to synthetic fertilizers [44, 47]. Despite the low pH, AS did not induce soil acidification and EC values did not exceed the upper limit value of 1.8 dS m^{-1} [47]. Similarly, high EC and S content in soil treated with AS did not negatively affect lettuce yield.

In a maize field trial on loamy-sand soil, CAN and AS were applied on top of pig manure and no significant differences were registered in regard to DW and FW yield or N and S uptake. Post-harvest nitrate (NO_3-N) residue analyses performed in October–November 2015 did not show any significant difference between AS treatment and synthetic CAN application, indicating that the risk of NO_3-N leaching for AS is comparable to that of a conventional fertilizing regime. Furthermore, NO_3-N residues were below the maximum allowable level of 90 kg NO_3-N ha^{-1} in a 0–90 cm soil layer applied in force in Flanders [44].

In a 3-year field trial (2011–2014), Vaneeckhaute et al. [48] investigated the use of AS as a substitute for chemical fertilizer. The study was performed with energy maize crops cultivated on sandy-loam soil and included different treatment scenarios with the following substrates: N chemical fertilizer, AS, animal manure, digestate, and LF of digestate. The highest N uptake was recorded when mineral N reference was replaced by AS. Post-harvest NO_3-N residue analyses did not show any significant difference between the different treatments, indicating that the potential risk of NO_3-N leaching to ground and surface water for AS was not higher than that of the conventional fertilizing regime. In both cases, NO_3-N residues were below the Flemish environmental standards previously indicated [48].

Table 3.3.1 *Product characteristics of ammonium sulfate recovered via air scrubbing.*

DM (%)	14–33
OC (%)	0.03–0.04
pH	2.40–7.7
EC (mS cm^{-1})	157–262
NH_4-N (g kg^{-1})	30–86
NO_3-N (g kg^{-1})	0
N_{total} (g kg^{-1})	30–86
$N_{mineral}/N_{total}$ (%)	100
P (g kg^{-1})	0
K (g kg^{-1})	0
S (g kg^{-1})	30–114

DM, dry matter; OC, organic carbon; EC, electrical conductivity.
Source: Adapted from Sigurnjak et al. [44].

Table 3.3.2 Product characteristics of ammonium nitrate recovered via NH_3 stripping and scrubbing.

DM (%)	48
OC (%)	0.1
pH	6.92–7.85
EC ($mS\,cm^{-1}$)	332–342
NH_4-N ($g\,kg^{-1}$)	76–109
NO_3-N ($g\,kg^{-1}$)	56–89
N_{total} ($g\,kg^{-1}$)	132–198
$N_{mineral}/N_{total}$ (%)	100
P	0
K	0
S	0

DM, dry matter; OC, organic carbon; EC, electrical conductivity.
Source: Adapted from Sigurnjak et al. [44].

3.3.5.2 Ammonium Nitrate

So far, the production of AN from manure or digestate and its implementation in agriculture has been recorded only at a pig farm in Gistel (Belgium), where horse manure, pig manure, and food waste are processed via anaerobic treatment. The produced digestate was solid/liquid separated and the LF of digestate was processed in a stripping and scrubbing unit that employed DETRICON technology. LF of manure digestate was first conditioned with CaO or NaOH to adjust the pH to 7.5–9. NH_3 stripping was performed at a temperature interval of 42–50 °C, followed by acid trapping with HNO_3. Around 87.5% of the NH_3 fed to the system was recovered in the form of 48% AN [44].

AN contains N entirely in mineral form, as NH_4-N and NO_3-N. pH and EC can vary depending on the amount of HNO_3 used (Table 3.3.2). Compared to AS, the pH of AN is often higher, which may increase NH_3 volatilization on the one hand and decrease the risk of machinery corrosion on the other. Agronomic performances of AN have been evaluated in pot and field experiments. Crop yield was slightly higher in lettuce pot experiments when AN was applied in comparison with CAN fertilizer on both sandy loam and loamy sand soils. Despite the high EC of AN (332–342 $mS\,cm^{-1}$), no detrimental effects were registered on soil salinity. No differences in DW yield, FW yield, or N and S uptake were observed in a field experiment on maize where the combination of pig manure and AN was compared to a conventional fertilization treatment (pig manure and CAN application). In the same experiment, post-harvest NO_3-N residue of AN application was investigated and, as with AS, no significant difference was observed between AN treatment and synthetic CAN application [44].

3.3.6 Conclusion

In order to cover the food and feed demand, the consumption of fertilizers in the European Union (EU) is forecasted to grow in the coming years. Currently, mineral N fertilizers are synthesized via the Haber–Bosch process, which consumes about 5% of the globally produced natural gas, a non-renewable resource [49, 50]. The recovery of N is propelled by the impelling need to face the excess of nutrient loads in regions with intensive livestock husbandry, where leaching of nutrients to ground and surface waters leads to severe environmental issues [51]. Therefore, implementing novel, sustainable nutrient resources is of paramount importance for the transition toward a biobased and circular economy, while simultaneously reducing greenhouse gas emissions and dependency on non-renewable reserves. Organic waste streams (including manure and digestate) have been identified as a potential resource for the recovery of nutrients. Application to agricultural land of N from manure and digestate is limited by environmental legislation. Thus, it needs to be either stored, transported to regions with nutrient deficiency, or processed in order to meet discharge nutrient standards. On the other hand, N synthetic fertilizers are required to satisfy crop nutrient requirements. In this situation, several EU regions (Belgium, the Netherlands, Denmark, northern Germany, Brittany, Catalonia and Aragon, the Po Valley) are facing simultaneously an overabundance of N in organic wastes and the use of synthetic mineral fertilizers [52]. A large and growing body of literature has investigated different strategies to recover N from manure and digestate via NH_3 stripping and scrubbing, in order to dispose of it in an economically feasible and eco-friendly way. However, most studies have been conducted on the bench and pilot scale. Still, the implementation at large scale of NH_3 stripping and scrubbing needs to be consolidated. Furthermore, far too little attention has been paid to the quality of recovered fertilizer products, which are often still classified as wastes despite having intrinsic characteristics similar to those of synthetic mineral fertilizers.

References

1. Martin, E.J., Chawla, R.C., and Swartzbaugh, J.T. (2004). Hazardous waste site remediation technology selection. In: *Solid Waste: Assessment, Monitoring and Remediation*, Waste Management Series, vol. **4** (ed. I. Twardowska), 1019–1066. Amsterdam: Elsevier.
2. Serna-Maza, A., Heaven, S., and Banks, C.J. (2014). Ammonia removal in food waste anaerobic digestion using a side-stream stripping process. *Bioresource Technology* **152**: 307–315.
3. Bousek, J., Scroccaro, D., Sima, J. et al. (2016). Influence of the gas composition on the efficiency of ammonia stripping of biogas digestate. *Bioresource Technology* **203**: 259–266.
4. Gustin, S. and Marinsek-Logar, R. (2011). Effect of pH, temperature and air flow rate on the continuous ammonia stripping of the anaerobic digestion effluent. *Process Safety and Environment Protection* **89** (1): 61–66.
5. Zarebska, A., Nieto, D.R., Christensen, K.V. et al. (2015). Ammonium fertilizers production from manure: a critical review. *Critical Reviews in Environmental Science and Technology* **45** (14): 1469–1521.
6. Vaneeckhaute, C., Lebuf, V., Michels, E. et al. (2017). Nutrient recovery from digestate: systematic technology review and product classification. *Waste and Biomass Valorization* **8** (1): 21–40.
7. Adams, C.E. (1973). Removing nitrogen from waste water. *Environmental Science & Technology* **7** (8): 696–701.
8. Janus, H.M. and van der Roest, H.F. (1997). Don't reject the idea of treating reject water. *Water Science and Technology* **35** (10): 27–34.
9. Ferraz, F.M., Povinelli, J., and Vieira, E.M. (2013). Ammonia removal from landfill leachate by air stripping and absorption. *Environmental Technology* **34** (13–16): 2317–2326.
10. Raboni, M., Torretta, V., Viotti, P., and Urbini, G. (2013). Experimental plant for the physical-chemical treatment of groundwater polluted by municipal solid waste (MSW) leachate, with ammonia recovery. *Revista Ambiente & Agua* **8** (3): 22–32.

11. Minocha, V.K. and Rao, A.V.S.P. (1988). Ammonia removal and recovery from urea fertilizer plant waste. *Environmental Technology Letters* **9** (7): 655–664.
12. Benito GG, Cubero MG. Ammonia elimination from beet sugar factory condense steams by a stripping-reabsorption system. Zuckerindustrie 1996;**121**(9):721–726.
13. Saracco, G. and Genon, G. (1994). High-temperature ammonia stripping and recovery from process liquid wastes. *Journal of Hazardous Materials* **37** (1): 191–206.
14. Maurer, C. and Müller, J. (2012). Ammonia (NH_3) emissions during drying of untreated and dewatered biogas digestate in a hybrid waste-heat/solar dryer. *Engineering in Life Sciences* **12** (3): 321–326.
15. Witter, E. and Lopezreal, J. (1988). Nitrogen losses during the composting of sewage-sludge, and the effectiveness of clay soil, zeolite, and compost in adsorbing the volatilized ammonia. *Biological Wastes* **23** (4): 279–294.
16. Van der Heyden, C., Demeyer, P., and Volcke, E.I. (2015). Mitigating emissions from pig and poultry housing facilities through air scrubbers and biofilters: state-of-the-art and perspectives. *Biosystems Engineering* **134**: 74–93.
17. Melse, R.W. and Ogink, N.W.M. (2005). Air scrubbing techniques for ammonia and odor reduction at livestock operations: review of on-farm research in the Netherlands. *Transactions of ASAE* **48** (6): 2303–2313.
18. Liao, P.H., Chen, A., and Lo, K.V. (1995). Removal of nitrogen from swine manure wastewaters by ammonia stripping. *Bioresource Technology* **54** (1): 17–20.
19. La, J., Kim, T., Jang, J.K., and Chang, I.S. (2014). Ammonia nitrogen removal and recovery from swine wastewater by microwave radiation. *Environmental Engineering Research* **19** (4): 381–385.
20. Laureni, M., Palatsi, J., Llovera, M., and Bonmatí, A. (2013). Influence of pig slurry characteristics on ammonia stripping efficiencies and quality of the recovered ammonium-sulfate solution. *Journal of Chemical Technology and Biotechnology* **88** (9): 1654–1662.
21. Zhang, L. and Jahng, D. (2010). Enhanced anaerobic digestion of piggery wastewater by ammonia stripping: effects of alkali types. *Journal of Hazardous Materials* **182** (1–3): 536–543.
22. Walker, M., Iyer, K., Heaven, S., and Banks, C.J. (2011). Ammonia removal in anaerobic digestion by biogas stripping: an evaluation of process alternatives using a first order rate model based on experimental findings. *Chemical Engineering Journal* **178**: 138–145.
23. Nie, H., Jacobi, H.F., Strach, K. et al. (2015). Mono-fermentation of chicken manure: ammonia inhibition and recirculation of the digestate. *Bioresource Technology* **178**: 238–246.
24. Wu, S., Ni, P., Li, J. et al. (2016). Integrated approach to sustain biogas production in anaerobic digestion of chicken manure under recycled utilization of liquid digestate: dynamics of ammonium accumulation and mitigation control. *Bioresource Technology* **205**: 75–81.
25. Serna-Maza, A., Heaven, S., and Banks, C.J. (2017). In situ biogas stripping of ammonia from a digester using a gas mixing system. *Environmental Technology* **38** (24): 3216–3224.
26. De la Rubia, M.Á., Walker, M., Heaven, S. et al. (2010). Preliminary trials of in situ ammonia stripping from source segregated domestic food waste digestate using biogas: effect of temperature and flow rate. *Bioresource Technology* **101** (24): 9486–9492.
27. Yao, Y.Q., Yu, L., Ghogare, R. et al. (2017). Simultaneous ammonia stripping and anaerobic digestion for efficient thermophilic conversion of dairy manure at high solids concentration. *Energy* **141**: 179–188.
28. Pedizzi, C., Lema, J.M., and Carballa, M. (2017). Enhancing thermophilic co-digestion of nitrogen-rich substrates by air side-stream stripping. *Bioresource Technology* **241**: 397–405.
29. Serna-Maza, A., Heaven, S., and Banks, C.J. (2015). Biogas stripping of ammonia from fresh digestate from a food waste digester. *Bioresource Technology* **190**: 66–75.
30. Zhang, W., Heaven, S., and Banks, C.J. (2017). Continuous operation of thermophilic food waste digestion with side-stream ammonia stripping. *Bioresource Technology* **244** (1): 611–620.

31. Bonmatı, A. and Flotats, X. (2003). Air stripping of ammonia from pig slurry: characterisation and feasibility as a pre-or post-treatment to mesophilic anaerobic digestion. *Waste Management* **23** (3): 261–272.
32. Liu, L., Pang, C.L., Wu, S.B., and Dong, R.J. (2015). Optimization and evaluation of an air-recirculated stripping for ammonia removal from the anaerobic digestate of pig manure. *Process Safety and Environment Protection* **94**: 350–357.
33. Ukwuani, A.T. and Tao, W. (2016). Developing a vacuum thermal stripping–acid absorption process for ammonia recovery from anaerobic digester effluent. *Water Research* **106**: 108–115.
34. Tao, W.D. and Ukwuani, A.T. (2015). Coupling thermal stripping and acid absorption for ammonia recovery from dairy manure: ammonia volatilization kinetics and effects of temperature, pH and dissolved solids content. *Chemical Engineering Journal* **280**: 188–196.
35. Bonmati, A. and Flotats, X. (2003). Pig slurry concentration by vacuum evaporation: influence of previous mesophilic anaerobic digestion process. *Journal of the Air & Waste Management Association* **53** (1): 21–31.
36. Chiumenti, A., da Borso, F., Chiumenti, R. et al. (2013). Treatment of digestate from a co-digestion biogas plant by means of vacuum evaporation: tests for process optimization and environmental sustainability. *Waste Management* **33** (6): 1339–1344.
37. Li, X., Guo, J.B., Dong, R.J. et al. (2016). Properties of plant nutrient: comparison of two nutrient recovery techniques using liquid fraction of digestate from anaerobic digester treating pig manure. *Science of the Total Environment* **544**: 774–781.
38. Starmans DAJ, Melse RW. Alternatives for the use of sulphuric acid in air scrubbers [Dutch]. Livestock Research, Wageningen UR; 2011.
39. Jamaludin, Z., Rollings-Scattergood, S., Lutes, K., and Vaneeckhaute, C. (2018). Evaluation of sustainable scrubbing agents for ammonia recovery from anaerobic digestate. *Bioresource Technology* **270**: 596–602.
40. Evans T, ed. Recovering ammonium and struvite fertilisers from digested sludge dewatering liquors. Proceedings of the IWA Specialist Conference: Moving Forward – Wastewater Biosolids Sustainability; 2007.
41. Stout, W.L. and Priddy, W.E. (1996). Use of flue gas desulfurization (FGD) by-product gypsum on alfalfa. *Communications in Soil Science and Plant Analysis* **27** (9–10): 2419–2432.
42. GNS, inventor; GNS – Gesellschaft fuer nachhaltige stoffnutzung MBH, N.S.M.G., assignee. DE10354063A1 Production of nitrogen fertilizer from organic waste products, comprising heating waste product under reduced pressure to produce gas containing carbon dioxide and ammonia, and absorbing ammonia from gas. Germany; 2003.
43. Van der Heyden C. Operation and characterization of air scrubbers for the emission reduction of ammonia, hydrogen sulphide and greenhouse gases from animal housing systems. Ghent University; 2017.
44. Sigurnjak, I., Brienza, C., Snauwaert, E. et al. (2019). Production and performance of bio-based mineral fertilizers from agricultural waste using ammonia (stripping-) scrubbing technology. *Waste Management* **89**: 265–274.
45. Bolzonella, D., Fatone, F., Gottardo, M., and Frison, N. (2018). Nutrients recovery from anaerobic digestate of agro-waste: techno-economic assessment of full scale applications. *Journal of Environmental Management* **216**: 111–119.
46. Ledda, C., Schievano, A., Salati, S., and Adani, F. (2013). Nitrogen and water recovery from animal slurries by a new integrated ultrafiltration, reverse osmosis and cold stripping process: a case study. *Water Research* **47** (16): 6157–6166.
47. Sigurnjak, I., Michels, E., Crappe, S. et al. (2016). Utilization of derivatives from nutrient recovery processes as alternatives for fossil-based mineral fertilizers in commercial greenhouse production of *Lactuca sativa* L. *Scientia Horticulturae* **198**: 267–276.

48. Vaneeckhaute, C., Ghekiere, G., Michels, E. et al. (2014). Assessing nutrient use efficiency and environmental pressure of macronutrients in biobased mineral fertilizers: a review of recent advances and best practices at field scale. *Advances in Agronomy* **128**: 137–180.
49. Michalsky, R. and Pfromm, P.H. (2012). Thermodynamics of metal reactants for ammonia synthesis from steam, nitrogen and biomass at atmospheric pressure. *AIChE Journal* **58** (10): 3203–3213.
50. Michalsky, R. and Pfromm, P.H. (2011). Chromium as reactant for solar thermochemical synthesis of ammonia from steam, nitrogen, and biomass at atmospheric pressure. *Solar Energy* **85** (11): 2642–2654.
51. Vitousek, P.M., Naylor, R., Crews, T. et al. (2009). Agriculture. Nutrient imbalances in agricultural development. *Science* **324** (5934): 1519–1520.
52. Bernet, N. and Beline, F. (2009). Challenges and innovations on biological treatment of livestock effluents. *Bioresource Technology* **100** (22): 5431–5436.

Section IV

Inspiring Cases in Nutrient Recovery Processes

4.1

Struvite Recovery from Domestic Wastewater

Adrien Marchi[1], Sam Geerts[1], Bart Saerens[1], Marjoleine Weemaes[1], Lies De Clercq[2], and Erik Meers[3]

[1]*Aquafin n.v., Aartselaar, Belgium*

[2]*Department of Physical and Analytical Chemistry, Ghent University, Ghent, Belgium*

[3]*Department Green Chemistry and Technology, Faculty of Bioscience Engineering, Ghent University, Ghent, Belgium*

4.1.1 Introduction

Worldwide, large amounts of phosphorus (P) rock are consumed annually in intensive agriculture and industry. The production of mineral phosphate fertilizers requires a lot of energy, and the energy cost and mineral P fertilizer demand are increasing. Furthermore, global phosphorus reserves are gradually depleting, and 90% of phosphorus mines are situated in just five exporting countries (mainly in the United States, China, and Morocco), which can lead to geopolitical instability [1]. The European Commission has added phosphate rock to the Critical Raw Materials list [2], a short list of resources that are strategically important within Europe and are likely to become scarce in the near future. On the other hand, nutrients such as phosphorus are present in excess quantities in wastewater and sewage sludge. Studies have revealed that 15–20% of the annual quantity of mined phosphorus eventually ends up in human excreta, which makes the municipal sewage sector one of the most important hotspots for phosphorus depletion mitigation [3]. The most straightforward way to recycle phosphorus from wastewater treatment plants (WWTPs) is to valorize the sludge as fertilizer on agricultural land [4]. Because, in many countries (including Belgium), the application of municipal sludge on agricultural land is not allowed, other P recycling

Biorefinery of Inorganics: Recovering Mineral Nutrients from Biomass and Organic Waste, First Edition.
Edited by Erik Meers, Gerard Velthof, Evi Michels and René Rietra.
© 2020 John Wiley & Sons Ltd. Published 2020 by John Wiley & Sons Ltd.

techniques need to be developed. However, orthophosphate (PO_4^{3-}) concentrations are generally low throughout the wastewater treatment cycle. In this way, P can only significantly be recovered from the ashes of mono-incinerated sludge, from concentrated wastewater streams such as sludge water of dewatered digested sludge, or from digested sludge itself [5]. Elevated free orthophosphate concentrations are usually obtained after an anaerobic treatment of the sludge, such as digestion. It should be highlighted that enhanced biological phosphorus removal (EBPR) is a prerequisite for the direct implementation of this technique. EBPR is the biological uptake and removal of phosphorus by activated sludge systems in excess of the amount that is removed by normal completely aerobic activated sludge systems [6]. If P is chemically removed with iron or aluminum salts, as is the case for most WWTPs, the orthophosphates will remain bounded to these metals. Consequently, the presence of iron or aluminum phosphate will inhibit the precipitation of struvite [7].

The precipitation of orthophosphate as struvite (magnesium-ammonium-phosphate, $MgNH_4PO_4 \cdot 6H_2O$) is one of the most common P recovery strategies. In the past, struvite formation in WWTPs was reported as a problem of natural scaling, and only later as a way to recover P [8, 9]. Controlled struvite crystallization requires an excess of ammonium (NH_4^+), together with a sufficiently high pH (around 7.5) and a source of magnesium, often added as $MgCl_2$ or $Mg(OH)_2$ [10]. Once recovered, struvite can be used in agriculture as a slow-release fertilizer. A P recovery technique was tested at the WWTP of Leuven (120 000 inhabitant equivalent), where a full-scale demonstration plant of struvite recovery from digested sludge was used for the first time on sludge instead of water. The demonstration plant has been operated since April 2013, and intensive measurement campaigns have been carried out ever since.

4.1.2 Process Description

The raw municipal wastewater comes from combined sewers at a rate of approximately 40 000 m^3 d^{-1}. First, the water undergoes a pre-treatment by means of a coarse grid filtration followed by a sand and a fat trap. There is no primary sedimentation. The biological treatment takes place in anaerobic contact tanks, oxidation ditches, and settling tanks. Anaerobic contact tanks are necessary for enhanced biological phosphorus removal and have a retention time of about 3 hours. The oxidation ditches provide alternating aerated and non-aerated conditions. The duration of each is based on continuous online oxygen, ammonium, and nitrate monitoring. The hydraulic retention time in the ditches is about 15 hours. The sludge is then thickened and digested.

From the digested sludge, phosphorus is removed as struvite. The digested sludge first goes through a cutter. The precipitation process itself consists of (i) a CO_2 stripper tank, to increase the pH, and (ii) a crystallization reactor, in which $MgCl_2$ is dosed (Figures 4.1.1 and 4.1.2). Afterward, the sludge moves gravitationally to the harvester, which allows a partial separation of the crystals by means of a cyclone. The retained crystals can be either recirculated to the reactor or harvested. The $MgCl_2$ dose is set at a molar Mg : P ratio of 1.75 at the end of the trial period. The installation operates at pH 7.8 in the stripper and 7.5 in the reactor. In the reactor, additional NaOH can be added if necessary to correct the pH against the acidifying effect of the added $MgCl_2$ and the precipitated struvite. The struvite recovery technology was provided by NuReSys®. The installation in Leuven is operated by

Figure 4.1.1 Struvite installation with a stripper unit (right tank) and a crystallization unit (left tank), with sludge digester in the background.

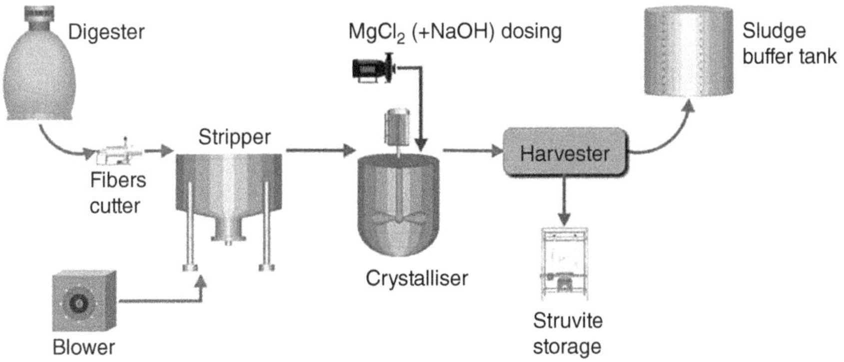

Figure 4.1.2 Schematic design of the full-scale installation for recovery of struvite from digested wastewater sludge at the Aquafin n.v. site in Leuven, Belgium.

Aquafin n.v. This type of struvite installation can remove PO_4^{3-} up to a lower limit between 20 and 40 mg PO_4-P l^{-1}. Consequently, this technology is not applicable for diluted streams.

4.1.3 Analyses and Tests

4.1.3.1 Mass Balance

Figure 4.1.3 shows a scheme of the sludge line. External sludge is added at three different locations in the line, drawn from other WWTPs without enhanced biological phosphorus removal. Therefore, the phosphorus in this sludge does not take the form of free orthophosphates and thus is not recoverable as struvite. When external sludge is added, there is a buffer to allow for a continuous operation of the sludge line.

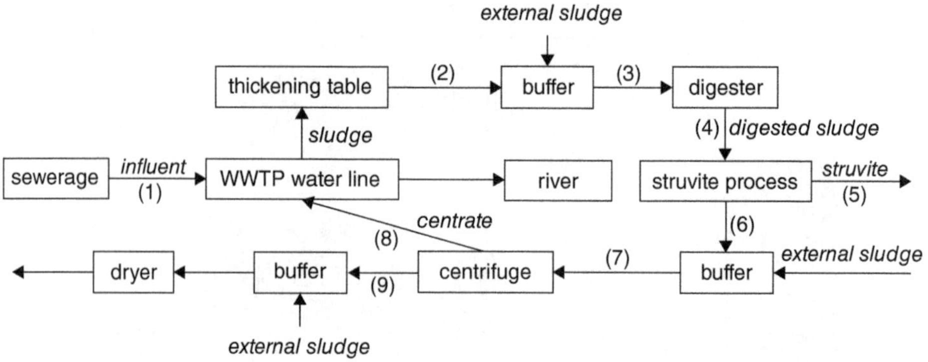

Figure 4.1.3 Scheme of the sludge line.

Table 4.1.1 Phosphorus mass balance of the sludge line; loads are expressed in kg P d^{-1} and as a percentage of the total plant load received from the sewers.

		Phosphorus load	
		(kg d^{-1})	(%)
(1)	Influent	200	100
(2)	Thickened sludge	160	80
(3)	Sludge to digester	260	130
(4)	Digested sludge	250	125
(5)	**Struvite produced**	**30**	**15**
(6)	Sludge after struvite process	220	110
(7)	Sludge to centrifuge	350	175
(8)	Centrate	15	7.5
(9)	Centrifuged sludge	320	160

Table 4.1.1 gives the mass balance of phosphorus in the sludge line. In reality, the mass balance undergoes important seasonal variations, due to changes in influent load (Leuven is a college town) and bacterial activity. Thus, the data are based on mass balances obtained during three intensive measurement campaigns in the spring, summer, and autumn, plus one additional campaign for dewaterability testing. The struvite recovery process allows for a conversion of 15% of the P load (30 kg P d^{-1}) received by the plant through the sewerage influent (200 kg P d^{-1}). In the digester, some P load is lost because of accumulation.

The impact of the struvite process on the N balance is minor. The withdrawn nitrogen in the struvite is 1–2% of the N load of the plant, but the decrease in the N load in the centrate is too small to be measurable.

4.1.3.2 Struvite Purity

A sample of the struvite produced (Figure 4.1.4) was analyzed by VITO (Flemish Institute for Technological Research) [11]. Several techniques were used:

- elemental analysis through combustion, quantitative measurement of N and H;
- inductively coupled plasma atomic emission spectroscopy (ICP-AES), quantitative measurement of Mg, P, and metals;

Figure 4.1.4 Struvite crystals recovered from wastewater.

- total organic carbon (TOC) analysis, quantitative measurement of C;
- Energy-dispersive X-ray fluorescence (ED-XRF), qualitative detection of all elements;
- X-ray powder diffraction (XRD), qualitative detection of crystalline materials;
- thermogravimetric analysis (TGA).

Table 4.1.2 summarizes the analytical mass composition of the dried struvite ($MgNH_4PO_4{\cdot}6H_2O$) sample produced by the installation. Some 97.6% of the mass in the analyzed sample is attributed to Mg, N, P, O, and H. The theoretical ratio of Mg : N : P in struvite is 1 : 1 : 1. Analysis of the struvite produced yields a ratio of 1.00 : 1.02 : 0.98. A low total carbon content of 0.5% indicates that during the process of struvite formation, nearly no pollution with organic matter occurs. The ICP-AES results show that the struvite is not contaminated with heavy metals, as the measured concentration for all these elements is low; see Table 4.1.3. The ED-XRF results show an elemental content of Al, Ca, Fe, Na, and Si, which indicates the presence of sand particles. No other contaminants are found. The observed XRD spectrum shows that the detected peaks correspond to mineral struvite. Only a small fraction of silicon dioxide (quartz) is noted, which confirms the detected Si measured by ED-XRF. No other products with a crystalline structure were formed during the process of struvite formation.

Table 4.1.2 Comparison between the theoretical mass fractions of struvite and the measured fractions.

	Mass fraction	
Component	Theoretical (%)	Measured (%)
Mg	9.91	9.2
N	5.71	5.4
P	12.62	11.5
O	65.19	65.2
H	6.57	6.3
C	0	0.5
Total	100	98.1

Table 4.1.3 Heavy metal content of the produced struvite as determined by ICP-AES.

Element	Concentration ($mg\,kg^{-1}$)
As	< 0.5
Cd	< 0.125
Cr	1.7
Cu	3.2
Pb	3.9
Ni	1.3
Zn	28

4.1.4 Operational Benefits

Compared to the implementation of phosphorus recovery on the water phase in WWTPs, implementation on the sludge phase offers advantages at full scale. These advantages are described here as "operational benefits," capable of compensating for the investment in a P recovery installation.

4.1.4.1 Enhanced Dewaterability

If dewatering of treated sludge could result in a higher dry solids concentration, this would lead to important energy savings when the sludge got transported, mono-incinerated, or dried, due to the diminished quantity of water available to transport or evaporate.

Controlled struvite formation in digested sludge can improve the sludge's dewaterability [12], which is however not related to the orthophosphate concentration.

Table 4.1.4 shows the results of a dewaterability test with a filter press across the four seasons. Statistically, there were no significant differences between the dewaterability of the sludge before it entered the struvite pilot and that after it came out (p-values with one-way ANOVA test all above 0.05).

The dewaterability in fall and winter (+3.3 %DS and +0.9 %DS, respectively) was better than that in spring and summer (−1.1 %DS and −1.3 %DS, respectively). The low dewaterability in spring and summer was due to improper polymer usage and NaOH dosing, respectively [13]. The $MgCl_2$ dose was higher in fall than in winter, explaining the more pronounced improvement of +3.3% of dry solids in the fall.

The results of the filter press tests were confirmed by the observation of dewatered sludge from the centrifuges at full scale. A moderate use of the caustic (NaOH) solution through

Table 4.1.4 Results of dewaterability tests in spring, summer, fall, and winter, expressed as percentage dry solids (%DS).

	Before struvite pilot	After struvite pilot	p-Value
Spring	21.9	20.8	0.23
Summer	30.4	29.1	0.12
Autumn	24.1	27.4	0.08
Winter	23.4	24.3	0.15

a sufficiently powerful aeration was required both for an improved dewaterability to limit the consumables costs [13].

The dewatering improvement is attributed mainly to the "bridging" effect of the magnesium divalent cations, whose concentration gets multiplied by 10 by the struvite recovery process due to an overdose of magnesium at a molar ratio of 1.75 Mg : P.

4.1.4.2 Enhanced Recovery Potential

Adding a polymer solution to liquid sludge in order to improve dewaterability means diluting the P content of the sludge. Therefore, the phosphate concentration of the liquid fraction (LF) of the sludge before dewatering is higher than the concentration of that same LF after dewatering to which the polymer solution has been added, together called the centrate.

For precipitation processes, it is known that higher initial concentrations of the reagents lead to higher precipitate yields, in this case of struvite. In practice, it is still unclear if the recovery potential (defined as 80% of PO_4^{3-} of the digested sludge that can be chemically formed, being 15% of the total P entering the WWTP) will be different when the recovery process is implemented on centrate. Calculations for an implementation on centrate lead to a similar phosphor removal. However, this residual concentration has never been tested, and a real implementation would be the only way to provide evidence for this. The influence of the additional secondary sludge from nearby stations (external sludge) on the mass balance is limited in this case, because it comes from other plants where EBPR is not implemented. Therefore, the P coming from the external sludge is not recoverable, since it is not present in the form of free orthophosphate. Consequently, only the influent from the sewers is set as a reference, and the external sludge is not taken into account for the recovery calculation.

4.1.4.3 Reduced Scaling

Natural struvite precipitation in digested sludge lines is known to cause operational problems such as pipe clogging and valve freezing, which requires regular and time-consuming pipe maintenance [14–16]. As shown in Table 4.1.5, the struvite recovery process strongly diminishes the orthophosphate concentration and somewhat diminishes the ammonium concentration in the outlet stream. In theory, this should reduce the speed of scaling downstream of the process. However, it is not yet possible to state that this will lead to significant economic savings due to reduced maintenance of pumps and centrifuges.

Table 4.1.5 *Observed concentrations in sludge before (in) and after (out) the struvite process; note that $MgCl_2$ is added during this process.*

	PO_4^{3-}		NH_4^+		Mg^{2+}		Ca^{2+}	
	In	Out	In	Out	In	Out	In	Out
Average concentration ($mg\,l^{-1}$)	179	29	1036	1018	5	52	43	68
Standard deviation ($mg\,l^{-1}$)	14	16	68	31	8	35	10	13
Samples d^{-1}	10	10	9	10	9	10	10	10

4.1.4.4 Reduced Phosphorus Content in the Sludge Pellets

Dried sludge can be valorized in the cement industry, which requires the phosphorus content to be as low as possible in order to guarantee good hardening properties of the cement. Therefore, a diminution of the P content would be considered an improvement [17]. From Table 4.1.1, it can be estimated that a P withdrawal with the struvite of $30\,kg\,P\,d^{-1}$ will induce a maximum reduction of 10% of the P content in the dewatered sludge ($320\,kg\,P\,d^{-1}$) and the dried sludge pellets. In the specific case of Leuven, where the same quantity of external dewatered sludge is added, the theoretical reduction of the P content in the dried sludge pellets would only be about 5%.

4.1.4.5 Reduced P and N Load in the Rejection Water

The withdrawn phosphorus ($15\,kg\,d^{-1}$) and nitrogen ($7\,kg\,d^{-1}$) from the sludge line provoke a reduction of the phosphate and ammonium concentration in the rejection water (centrate) that gets recycled from the centrifuge to the water line. This diminishes the P and N load in the water line, allowing a decrease in aeration (necessary for nitrification) and carbon source consumption (sometimes necessary for biological P removal or denitrification). However, the same benefit is also obtained when the struvite is recovered from the centrate water.

4.1.5 Economic Evaluation

The cost–benefit analysis of the precipitation process was based on a financial model that takes into account the most important revenues and expenditures of the struvite precipitation pilot. From the results, it can be seen that the required value of recovered struvite should be between €590 and €440 per ton at incoming phosphate concentrations between 150 and $450\,mg\,l^{-1}$ respective to be an economically viable recovery technique. However, the Monte Carlo runs show that there is a large scatter, both positive and negative, around these values (Figure 4.1.5), mostly due to the variation in improved dewaterability of the sludge. Given the current operational settings and results in Leuven, an average struvite subsidy (e.g. from the state) or selling price of €530/ton struvite recovered is required to guarantee a discounted payback time of 10 years. Though the current selling price of struvite is €50/ton, such a high price may be realistic someday – during the food crisis in 2008, international P rock prices were close to this value. Yet, this information alone is no solid basis for a business model. Without a price increase, or a subsidy, and expressed in terms of inhabitant equivalents, this would mean a cost of around €0.32/inhabitant equivalent on a yearly basis in order to have P recovered from digested sludge as struvite. The struvite recovery from digested sludge would, however, be more beneficial if the investment could be lowered by a higher production rate of recovery units, if the P price were higher, if the concentration of phosphate were higher (e.g. due to pre-treatment before digestion), or if there were more certainty about the improved dewaterability. The factors causing large deviations in necessary selling price are explained by Geerts et al. [18].

In the case of recovery of phosphorus from centrate rather than digestate, the profitability is very low due to the lower P concentrations ($<150\,mg\,P\,l^{-1}$). However, the uncertainty in this investment is less in comparison to the recovery from digested sludge. This is because the advantage of improved dewaterability is a highly sensitive parameter for the financial

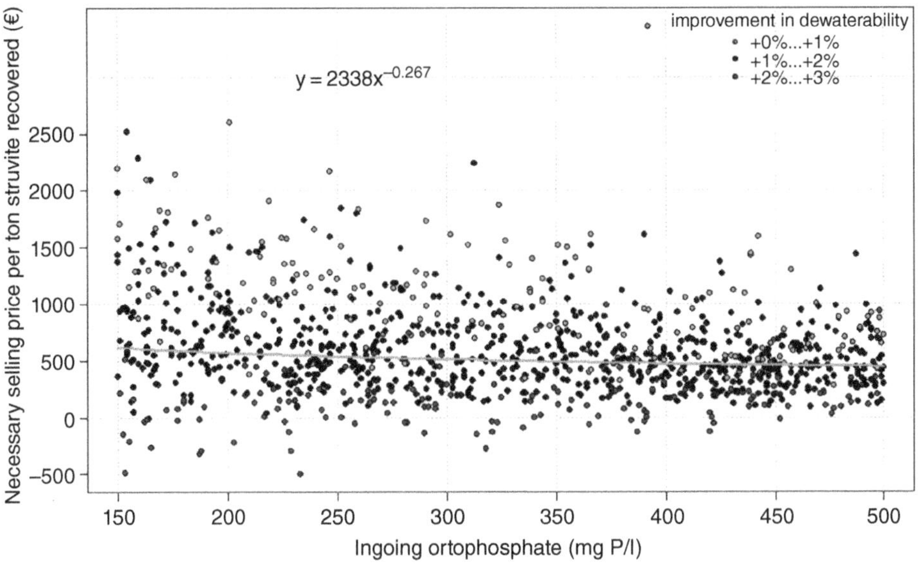

Figure 4.1.5 Selling price per ton of recovered struvite (€/ton) from digested sludge necessary for a discounted payback time of 10 years, in terms of the function of the varying incoming orthophosphate concentration (mg P l^{-1}). Dots are the result of the uncertainty analysis; full line is the result for the default inputs.

model for digested sludge but is not an influencing factor for centrate. Though struvite recovery from digested sludge has a higher risk, it is an investment with a potentially higher return in comparison to struvite recovery from centrate.

4.1.6 Future Challenges

4.1.6.1 In-Depth Quality Screening

Fertilizer quality tests with the struvite produced in Leuven have not yet been conducted. It is possible that the final product contains components with plant toxicological or pathogenic effects. In order to determine the presence of such micropollutants (e.g. pesticides, hormones, or pharmaceuticals) or pathogens (e.g. Salmonella), it is necessary to conduct plant assays and phytotoxicity tests.

4.1.6.2 Improved Crystal Separation

The greatest remaining challenge for the struvite recovery processes from digested sludge is the separation of the crystals. The high viscosity of the sludge and the small particle size of the crystals do not allow for their recover by simple sedimentation. Strategies are currently under investigation to increase the crystal size and therefore improve the recovery rate and make the washing step easier. To date, the harvester can only recover 25% of the chemically formed struvite, bringing the maximum recovery potential of 15% of effective phosphorus from the P total of the influent of the plant down to 4%.

4.1.7 Conclusion

During the first year of operation, the full-scale struvite recovery pilot achieved a chemical PO_4^{3-} removal efficiency of 80% with a Mg/P ratio of 1.75 and a pH of 7.5 in the reactor. To date, the process allows a maximum recovery potential of 15% of the total phosphorus load of the plant. The main obstacle to implementation of the technique in sludge is the difficult separation of the crystals. Optimization research into this process should be further explored to increase the crystal size and thereby improve the recovery rate. At present, the harvester can effectively recover 25% of the precipitated struvite, equivalent to an actual recovery of 4% of the phosphorus from the influent of the plant. The process decreases the P load in the centrate back to the water line by half, which normally accounts for 15–20% of the P load of the water line. It is reckoned that a dewaterability improvement of +1.5 %DS is achievable, given a sufficient magnesium dosage and the implementation of a powerful stripping to minimize the use of a caustic solution.

The struvite obtained is quite pure. Whether it is safe to use in agriculture should be evaluated by means of plant assays and phytotoxicity tests over an extended period.

To be economically viable, the required selling price of the obtained struvite with a payback time of 10 years is high (€530/ton) in comparison to the current price of struvite (€50/ton).

Furthermore, this full-scale experience should help determine the best way to recover P from municipal wastewater: from centrate or digestate, or at the end of sludge life from incinerated sludge ashes. This is currently under investigation within the wastewater sector and should be discussed based on technical performance evaluations, cost–benefit analyses, and comparative life-cycle analyses of primary and secondary P production.

References

1. Flanders Knowledge Center Water (Vlakwa) (2012) Nutriëntenplatform. Available from: https://www.vlakwa.be/ (accessed December 23, 2019).
2. European Commission (2014) 20 critical raw materials - major challenge for EU industry. Press Release, May 26.
3. Cordell, D., Drangert, J., and White, S. (2009). The story of phosphorus: global food security and food for thought. *Global Environmental Change* **19** (2): 292–305.
4. Linderholm, K., Tillman, A.-M., and Mattsson, J.E. (2012). Life cycle assessment of phosphorus alternatives for Swedish agriculture. *Resources, Conservation and Recycling* **66**: 27–39.
5. Cornel, P. and Schaum, C. (2009). Phosphorous recovery from wastewaters: needs, technologies and costs. *Water Science and Technology* **59** (6): 1069–1076.
6. Wentzel, M.C., Comeau, Y., Ekama, G.A. et al. (2008). Enhanced biological phosphorus removal. In: *Biological Wastewater Treatment – Principles, Modelling and Design* (eds. M. Henze, M.C.M. van Loosdrecht, G.A. Ekama and D. Brdjanovic), 155–220. London: IWA Publishing.
7. Parsons, S.A. and Smith, J.A. (2008). Phosphorus removal and recovery from municipal wastewaters. *Elements* **4**: 109–112.
8. Doyle, J.D. and Parsons, S.A. (2002). Struvite formation, control and recovery. *Water Research* **36**: 3925–3940.
9. Liu, Y., Kumar, S., Kwag, J.-H., and Ra, C. (2013). Magnesium ammonium phosphate formation, recovery and its application as valuable resources: a review. *Journal of Chemical Technology & Biotechnology* **88**: 181–189.

10. Hanhoun, M. (2011) Analysis and Modelling of Struvite Precipitation: Towards the Treatment of Industrial Wastewater Discharges. PhD Thesis: Université de Toulouse.
11. Vanhoof, C., Tirez, K. (2014) Analysemethode voor het bepalen van de zuiverheid van struviet. VITO eindrapport 2014/SCT/R/26.
12. Bergmans, B.J., Veltman, A.M., van Loosdrecht, M.C. et al. (2014). Struvite formation for enhanced dewaterability of digested wastewater sludge. *Environmental Technology* **35**: 549–555.
13. Marchi, A., Geerts, S., Weemaes, M. et al. (2015). Full-scale phosphorous recovery from digested waste water sludge in Belgium – Part I: technical achievements and challenges. *Water Science and Technology* **71** (4): 487–494.
14. Borgerding, J. (1972). Phosphate deposits in digestion systems. *Journal of the Water Pollution Control Federation* **44**: 813.
15. Munch, E. and Barr, K. (2001). Controlled struvite crystallisation for removing phosphorus from anaerobic digester sidestreams. *Water Research* **35** (1): 151.
16. Neethling, J. and Benisch, M. (2004). Struvite control through process and facility design as well as operation strategy. *Water Science and Technology* **49** (2): 191.
17. Rodriguez, N.H., Martinez-Ramirez, S., Blanco-Varela, T.M. et al. (2013). The effect of using thermally dried sewage sludge as an alternative fuel on Portland cement clinker production. *Journal of Cleaner Production* **52**: 94–102.
18. Geerts, S., Marchi, A., and Weemaes, M. (2015). Full-scale P-recovery from digested wastewater sludge in Belgium – Part II: economic opportunities and risks. *Water Science and Technology* **71** (4): 495–502.

4.2

Mineral Concentrates from Membrane Filtration

Paul Hoeksma and Fridtjof de Buisonjé
Livestock Research, Wageningen University & Research, Wageningen, The Netherlands

4.2.1 Introduction

Separation into a solid fraction (SF) and a liquid fraction (LF), subsequent polishing of the LF, and reverse osmosis (RO) as the final step is a treatment concept applied to livestock slurries [1, 2]. RO removes water from the LF. This process results in a concentrated nitrogen (N)-potassium (K) solution (RO-concentrate or "mineral concentrate"), in which most of the N is present as ammonium (NH_4^+). The water removed by RO has low concentrations of nutrients and contaminants and can be discharged into the sewer or into surface waters [3, 4]. Reduction of the volume by RO increases the desirability of transporting mineral concentrates from areas with a high livestock density to arable farming areas. The results of a monitoring program examining four full-scale production plants will be presented in this chapter.

4.2.2 Production of Mineral Concentrates

4.2.2.1 General Set-up

Figure 4.2.1 shows the general set-up of the production process for obtaining mineral concentrates from livestock slurry using RO.

- In the first step, pig or cattle slurry is mechanically separated into an SF and an LF. Examples of mechanical separation technologies applied are the decanter centrifuge,

Biorefinery of Inorganics: Recovering Mineral Nutrients from Biomass and Organic Waste, First Edition.
Edited by Erik Meers, Gerard Velthof, Evi Michels and René Rietra.
© 2020 John Wiley & Sons Ltd. Published 2020 by John Wiley & Sons Ltd.

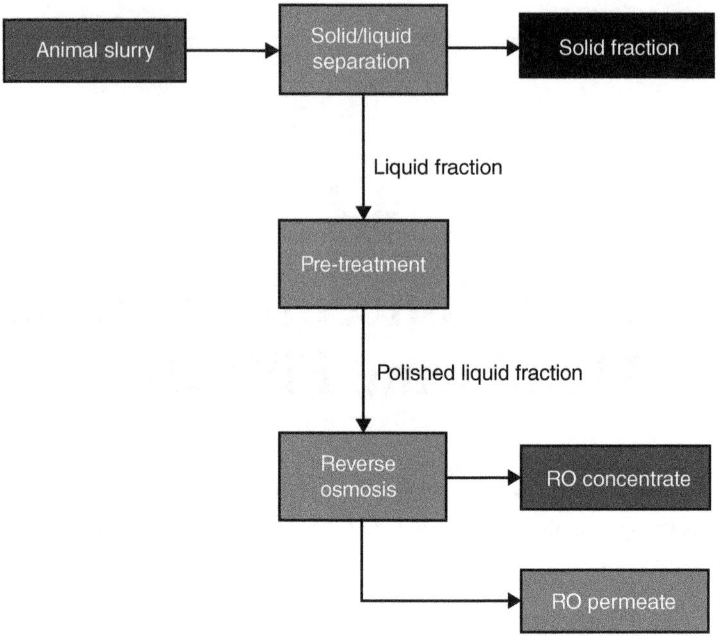

Figure 4.2.1 *General set-up of slurry treatment with RO.*

sieve belt press, and screw press filter. The LF is pre-treated to remove all solid, suspended, and colloidal particles. A polished liquid is required to avoid scaling and fouling of the RO membranes. Technologies used for pre-treatment of the LF are ultra-filtration (UF), microfiltration, and dissolved air flotation (DAF). Usually, coagulants and flocculants are added to support both the mechanical separation and the pre-treatment.

- The SF is rich in phosphorus (P) and organic matter (OM). This fraction can be used as a P fertilizer and a soil conditioner. It can also be dried and pelletized. Manure pellets are fit for storage and long-distance transport. The SF can also be used for biogas production.
- After pre-treatment, the LF passes an RO unit, ending up in two streams: an RO-permeate, which is water with a low nutrient concentration, and an RO-concentrate, which is a liquid with relatively high N and K concentrations. The RO-concentrate is often called "mineral concentrate."

4.2.2.2 Solid/Liquid Separation

A decanter centrifuge can be used for slurry separation based on a centrifugal force. Testing of a decanter centrifuge with cattle slurry by Hilhorst and Verloop [5] showed a separation efficiency for total mass of 34% and for P of 67%. The efficiency was defined as the mass of a compound in the SF compared to the mass in the raw slurry. In a review by Hjorth et al. [6], the mean separation efficiency for dry matter (DM) (from both pig and cattle slurry) by a decanter centrifuge was 61% ± 16%.

A sieve belt press can also be used for slurry separation. The liquid is first drained by gravity from solids in a dewatering section. The dewatered slurry is then compressed between two parallel belts, one of which is water-permeable; the filter cake is continuously removed as the belts rotate. The parallel belts pass between a number of pressure rollers, which squeeze the LF from the dewatered slurry. The liquid is then pressed through the water-permeable belt and drained off. In a review by Schröder et al. [7], the separation efficiency for P of a belt press was found to vary between 50 and 75% when coagulants and flocculants were added.

A screw press filter or press auger separator can also be used for pressurized filtration. The slurry is transported into a cylindrical screen and compressed by a slowly rotating screw. The liquid passes the screen and is collected. The separation efficiency for P of a screw press filter varies according to different sources between 14% (pig slurry) [8] and 38% (digested cattle slurry) [5]. The mean separation efficiency for DM (from both pig and cattle slurry) is $37 \pm 18\%$ [6].

The separation efficiency for DM and P can be significantly increased by the use of coagulants (e.g. $FeCl_3$, $Al_2(SO_4)_2$) and flocculants (e.g. polyacrylamide polymer). The separation efficiency for dissolved nutrients such as mineral nitrogen (NH_4^+) and potash (K_2O) is mostly less than 20% [8]. This means that more than 80% of the mineral nitrogen and potash remains in the LF. The use of coagulants and flocculants has little impact on the separation efficiencies for these dissolved nutrients [6].

4.2.2.3 Pre-treatment of the Liquid Fraction (Effluent from Mechanical Separation)

UF is a membrane filtration technique in which pressure (typically 2–10 bar) is the driving force for separation through a semi-permeable membrane (2 nm–0.1 μm). UF concentrates suspended and colloid solids and solutes of molecular weight greater than 1000 Da. The UF-permeate contains low-molecular-weight organic solutes and salts.

The LF can also be treated by DAF. LF in which air is dissolved is pumped in at the bottom of the DAF unit under release of small air bubbles. Organic particles are transported to the surface of the unit, where a layer of organic material is formed, which can be removed by scraping or filtration. Some heavier particles will sink to the bottom of the DAF unit, where the sediment can be removed periodically with a screw auger.

In most suspensions of organic material, colloidal particles will not aggregate because they are negatively charged and repel one another [6]. To enhance removal of these particles, coagulating and flocculating chemicals can be added to the liquid. These chemicals cause colloids and other suspended particles in liquids to aggregate. Often, aluminum or iron salts are used to neutralize negative charges (coagulation). The positively charged metal salts (multivalent cations) interact with the negatively charged organic particles, causing these particles to aggregate. A disadvantage of using coagulants and flocculants is that often chloride and sulfate ions are also added, and in some cases heavy metals. Therefore, polyacrylamide is often used instead of metal salts as coagulant. Adding polymers stimulates flocculation (increases the particle size after granulation from microfloc to suspended particles).

4.2.2.4 Reverse Osmosis

The principle of RO is based on the ability of RO-membranes to let water molecules pass while blocking salt ions. Water is pressed through the membrane, concentrating all of the organic molecules and minerals that are present in the influent. The pressure required for this process to take place depends on the electrical conductivity (EC) of the feed; it must exceed its osmotic pressure. During the process, the electrical conductivity of the concentrate increases; this is why an increasing pressure is required to reach the desired concentration. The maximal degree of concentration is limited by the osmotic pressure of the feed and the required driving force. Usually, RO-membranes, which were originally designed for desalination of sea water, are applied in livestock slurry treatment plants. The EC-value of pig slurry as produced at Dutch pig farms is typically 25–30 mS cm^{-1}, while that of influents (LFs) of RO is 30–35 mS cm^{-1}.

RO units as applied in livestock slurry treatment in the Netherlands are configured as multistage operations; in the first stage, the RO-concentrate is produced, and in subsequent stages (usually two), the permeate is produced stepwise. Full-scale RO units typically use 6–48 membranes, with a total membrane surface of 216–1728 m^2, a capacity of 2–17 m^3 h^{-1}, and a pressure of 40–70 bar [3]. RO installations as operated in the Netherlands today have the capacity to treat 15 000–200 000 tons of pig or cattle slurry per year.

RO-membranes need to be frequently cleaned to avoid fouling and scaling. Hydrochloric acid, sodium hydroxide, citric acid, and water are used in the cleaning process. During cleaning, the RO-membrane is out of production.

The permeate from RO still contains small amounts of ammonia nitrogen; this is the main reason it does not meet the quality standards for discharge into surface waters. Thus, discharge into the sewer is required. Alternatively, the remaining nitrogen can be removed using an ion exchanger.

4.2.2.4.1 Full-Scale Pilot Production Plants

A monitoring program of full-scale production plants (pilots) was carried out in the Netherlands from 2009 to 2011, designed to assess the chemical compositions of mineral concentrates and the agronomic and environmental effects of their use as a fertilizer. Data on the chemical compositions of the input materials, intermediates, and end products at the participating plants were collected by monthly sampling of the process flows. Samples were analyzed for primary nutrients, secondary nutrients, trace elements, and some heavy metals. The mineral concentrates were also analyzed for organic micropollutants. The monitoring covered a period of two years, 2009 and 2010, and included 10 production plants. The characteristics of eight plants are given in Table 4.2.1.

The monitoring was continued in 2011 at six plants (A, B, C, D, F, and G), after some technical modifications of the production processes. However, the processes at plants A and G were unstable due to technical problems and revision activities, so no representative data from these plants could be collected. Performance data from the remaining four plants (B, C, D, and F) will be presented in this chapter.

4.2.3 Mass Balance

Table 4.2.2 shows the average distribution of total mass, primary nutrients, and OM over the end-products of the four pig slurry (from sows and fatteners) treatment operations that participated in the monitoring program in 2011. All treatment plants had a set-up as shown in Figure 4.2.1, using a screw press filter or sieve belt press for solid/liquid separation and

Table 4.2.1 *Overview of technologies used and of the capacities of the pilot plants.*

Plant	Input	Solid/liquid separation	Pre-treatment liquid fraction	Final process step	Throughput (ton yr^{-1})
A	Pig slurry digestate	Decanting centrifuge	Ultra-filtration	Reverse osmosis	67 500
B	Fattening pig slurry	Belt press	Dissolved air flotation	Reverse osmosis + ion exchange	50 000
C	Fattening pig slurry	Belt press	Dissolved air flotation	Reverse osmosis	25 000
D	Sow slurry	Screw press	Dissolved air flotation	Reverse osmosis	10 000
E	Fattener/Sow slurry	Screw press	Dissolved air flotation	Reverse osmosis	5000
F	Fattening pig slurry	Belt press	Dissolved air flotation	Reverse osmosis	25 000
G	Fattening pig slurry	Belt press	Dissolved air flotation	Reverse osmosis	10 000
H	Cow slurry digestate	Decanting centrifuge	Ultra-filtration	Reverse osmosis	15 000

DAF for LF polishing. The input of raw slurry was set at 100. Note that the installations also used additives such as acids, coagulants, and flocculants in the process, causing the sum of the outputs of DM and OM to be higher than 100% for some installations.

The balance of total mass shows that almost half of the original slurry mass ended up as RO- permeate. This effluent does not have to be transported by truck from the farm but can be discharged into the sewer or into the surface water, which is of great economic benefit to Dutch farmers in highly concentrated livestock areas. The balance of total N shows that, on average, 44% of the N was recovered in the SF, 53% in the concentrate, and 2% in the permeate. The N balance suggests that, on average, 1% of the slurry N was lost during the treatment process. The largest part of both NH_4-N (70%) and K (78%) was recovered in the RO-concentrate, and most of the OM (94%) and P (96%) in the SF.

Figure 4.2.2 illustrates the average mass distribution of OM and primary nutrients in the end-products of the treatment plants, relative to the input.

4.2.4 Composition of Raw Slurry, Solid Fraction, and RO-Concentrate

Table 4.2.3 shows the average compositions of the feedstock (slurry from sows and fatteners) and the valuable end-products of four full-scale production plants of mineral concentrates in 2011.

4.2.4.1 Raw Slurry

The feedstock of the monitored production plants was mainly slurry from fattening pigs. One of the four plants operated at a single farm, mainly treating sow slurry. The other three

Table 4.2.2 *Relative mass distribution (and range) of total mass, DM, OM, total N, NH_4-N, total P, and K over the end-products of four slurry treatment plants [3].*[a]

	Total mass	Dry matter	Organic matter	Total N	NH_4-N	P	K
Input (raw slurry)	100	100	100	100	100	100	100
Solid fraction	18 (12–21)	86 (81–87)	94 (89–100)	44 (40–48)	29 (26–33)	96 (93–100)	18 (14–21)
RO-concentrate	36 (30–43)	21 (19–23)	12 (11–14)	53 (50–58)	70 (66–73)	4 (0–7)	78 (72–88)
RO-permeate	46 (36–56)	0 (0–0)	0 (0–0)	2 (0–3)	0 (0–0)	0 (0–0)	1 (0–2)
Output	100	106 (100–108)	106 (100–115)	99 (96–106)	99 (93–106)	100 (98–100)	97 (91–104)

[a] The addition of additives in the process such as acids, salts, and flocculants caused the output of DM and OM to be more than 100% in some cases.

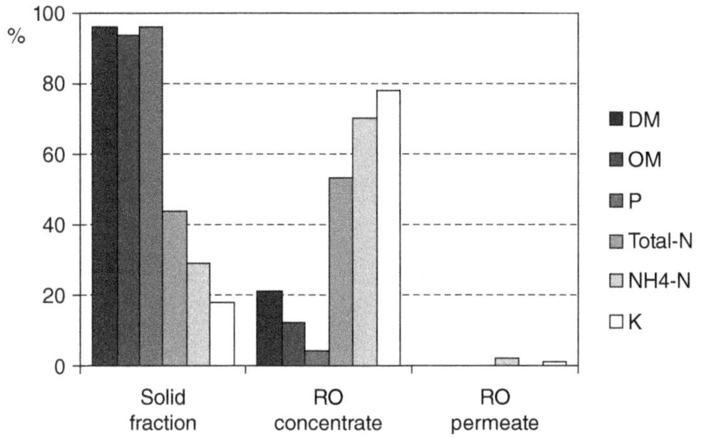

Figure 4.2.2 *Average mass distribution of DM, OM and nutrients in the end products of four slurry treatment plants, relative to the input.*

Table 4.2.3 *Average composition (and standard deviation) of raw pig slurry, SF, and RO-concentrate as measured at four full-scale production plants in the Netherlands in 2011 [3].*

		Raw pig slurry		Solid fraction		RO concentrate	
DM	$g\,kg^{-1}$	63.7	(17.5)	273	(29.4)	36.9	(9.18)
OM	$g\,kg^{-1}$	43.3	(13.9)	201	(31.0)	14.0	(3.97)
Total N	$g\,kg^{-1}$	5.53	(1.37)	12.5	(1.24)	8.15	(1.58)
NH_4-N	$g\,kg^{-1}$	3.91	(0.93)	5.71	(0.74)	7.51	(1.66)
P	$g\,kg^{-1}$	1.49	(0.32)	7.26	(0.32)	0.16	(0.11)
K	$g\,kg^{-1}$	3.79	(0.84)	3.38	(0.62)	8.02	(1.27)
Ca	$g\,kg^{-1}$	1.67	(0.31)	8.92	(1.67)	0.23	(0.10)
Cl	$g\,kg^{-1}$	1.41	(0.14)	0.93	(0.07)	2.93	(0.34)
Mg	$g\,kg^{-1}$	0.90	(0.18)	5.21	(1.21)	0.13	(0.19)
Na	$g\,kg^{-1}$	0.85	(0.09)	0.70	(0.08)	1.79	(0.15)
S	$mg\,kg^{-1}$	642	(186)	3414	(1080)	1787	(2843)
B	$mg\,kg^{-1}$	5.53	(0.82)	24.0	(3.80)	2.91	(0.45)
Fe	$mg\,kg^{-1}$	171	(28)	2825	(3676)	25.3	(28.1)
Mn	$mg\,kg^{-1}$	33.2	(13.0)	203	(29.0)	2.92	(1.70)
Mo	$mg\,kg^{-1}$	0.19	(0.13)	1.71	(0.80)	0.04	(0.03)
Cd	$mg\,kg^{-1}$	0.01	(0.01)	0.01	(0.01)	0.01	(0.01)
Co	$mg\,kg^{-1}$	0.03	(0.02)	0.43	(0.52)	0.07	(0.02)
Cr	$mg\,kg^{-1}$	0.34	(0.11)	2.22	(0.91)	0.26	(0.24)
Cu	$mg\,kg^{-1}$	23.5	(1.33)	142	(57.1)	1.52	(1.94)
Ni	$mg\,kg^{-1}$	0.60	(0.07)	2.69	(0.70)	0.58	(0.04)
Zn	$mg\,kg^{-1}$	103	(33.2)	488	(175)	9.30	(12.8)

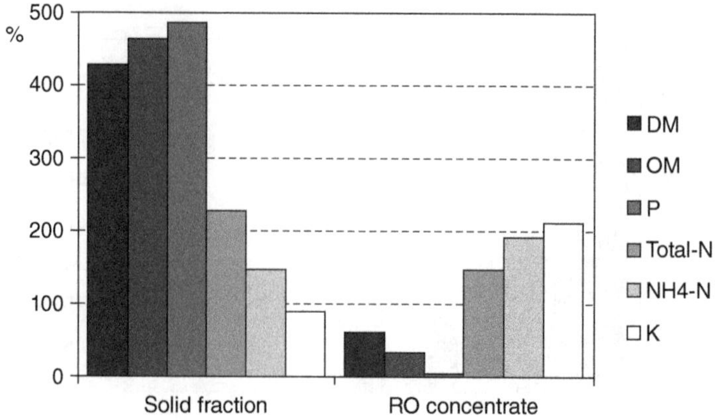

Figure 4.2.3 *Composition of the SF and RO-concentrate from pig slurry treatment relative to the raw slurry.*

plants treated mixed pig slurry, from both sows and fatteners, collected from farms in the region. Raw slurry from sows contains low concentrations of OM and nutrients compared to slurry from fatteners, which is expressed by a large standard deviation of these components in the mixture. The raw slurry that was used for treatment during the monitoring period was representative of the slurry produced at Dutch pig farms.

4.2.4.2 Solid Fraction

The SF is high in DM, OM, and phosphorus compared to the raw slurry, as shown in Figure 4.2.3, which makes it highly qualified as an organic P fertilizer and soil conditioner. This is the product that should be transported, after pasteurization (for export), drying, or granulation if necessary, from manure-surplus areas to areas with fertilizer demand.

4.2.4.3 RO-Concentrate

4.2.4.3.1 Nutrients and Minerals

RO-concentrate can be characterized as a mineral N—K fertilizer containing very small amounts of OM (partly volatile fatty acids) and P. It also contains secondary nutrients and trace elements, as well as organic and inorganic microcontaminants – components that contribute to the DM content, which is substantial. Most of the nitrogen (>90%) in the RO-concentrate is present as ammonia, which is one of the properties that distinguishes RO- concentrate from liquid slurry fractions obtained by simple separation technologies. The high NH_4-N to total-N ratio also guarantees that the mineral concentrate has comparable N fertilizer efficiency to the chemical fertilizer, if applied with low-emission technology. The RO-concentrate has higher concentrations of total-N, NH_4-N, and K compared to raw pig slurry (Figure 4.2.3).

4.2.4.3.2 Secondary Nutrients and Trace Elements

RO-concentrate also contains the secondary nutrients Ca, Mg, and S and a large number of trace elements or micronutrients (Table 4.2.2). Concentrations of Ca and Mg are too low to attribute an agronomic value to them [9]. The level of Na amounts to roughly 20–25% of that of K. The use of RO-concentrate as a N or K fertilizer is associated with a non-negligible gift of Na (20–40 $kg\,ha^{-1}$). Na has fertilizer value for some arable crops (e.g. sugar beet). S, present in various concentrations, is a valuable component of RO-concentrate. Over-application of RO-concentrate can result in so-called "burning" (salt damage) of sensible crops [10] and, especially on clay soils, an increased risk of grass tetany (hypomagnesemy) in cattle, caused by overdosing of K [11].

RO-concentrate also contains small amounts of B, Fe, Mn, and Mo, which are considered micronutrients. On the use of RO-concentrate as an N or K fertilizer, minor amounts of these elements are provided; gifts are normally limited to a few tens of grams per ha and are of little value for crop and soil [9].

4.2.4.3.3 Inorganic Microcontaminants

RO-concentrate contains low concentrations of the heavy metals Cd, Co, Cr, Cu, Ni, and Zn. The presence of Cu and Zn can be traced back to pig feed, the other elements possibly to wearing of the production units. Concentrations of heavy metals in the RO-concentrate do not exceed environmental limits [12].

4.2.4.3.4 Organic Microcontaminants

RO-concentrate was also examined for dioxins, non-ortho-PCB, mono-ortho-PCB, PAK, organic-chlorine pesticides, and mineral oil. Detection limits were not exceeded in nearly any samples; only for octa-chloro-dibenzo-p-dioxin (OCDD) was a value just above the detection limit found. Concentrations of organic microcontaminants in the mineral concentrate do not exceed environmental limits [12].

4.2.4.3.5 Volatile Fatty Acids

Part of the OM in RO-concentrate is volatile fatty acids (VFAs). Concentrations of VFA (C2-C5) are comparable to those in raw pig slurry; in Dutch pig slurry, this is roughly 5–7 $g\,kg^{-1}$. VFAs are products of the anaerobic microbial degradation processes of larger organic structures (proteins) occurring during storage. They are easily degradable components and are an important carbon source in denitrification processes in the soil [13].

4.2.5 Quality Requirements

In 2014, quality requirements for mineral concentrates were introduced, such that they have an effective agronomic function, are harmless to the environment, and are distinguishable from other animal manures. An NH_4-N/total-N ratio ≥90% is required, to ensure that the nitrogen in mineral concentrate is mostly present as mineral nitrogen. Furthermore, total-N/P (as P_2O_5) ≥ 15 and EC ≥ 50 mS cm^{-1} are also required, in order to distinguish mineral concentrates from other animal slurries and LFs after simple separation. The EC value of a slurry can quickly be checked using a handheld instrument.

The average total-N content of a mineral concentrate lies within the range of the total-N content of animal slurries as produced in Dutch livestock farming. The average P content of mineral concentrate is lower than that of slurries. A low P concentration differentiates mineral concentrate from most livestock slurries, but after slurry dilution comparable P concentrations can be obtained. The total-N/P ratio of mineral concentrate is much higher than that of other slurries and therefore is a distinguishing parameter.

4.2.6 Conclusion

The production and use of mineral concentrates supports the ambition to recycle nutrients from animal manure as much as possible and to reduce the nitrogen surplus at a national level. This will be especially the case if the content of valuable components can be raised to levels that make the product an attractive alternative to chemical fertilizers and valuable enough to be worth transporting over long distances.

The production of mineral concentrates from animal slurry involves rather complex technology, comes with high (investment and operational) costs, is labor-intensive, and is therefore best executed at a regional level in large-scale production units. Today, the production cost of mineral concentrates is between 10 and 14 euros per ton of raw slurry, to which the costs of storage and marketing of the end products (SF and mineral concentrate) must be added. The production of mineral concentrates is only financially viable in regions with high costs for manure disposal (in areas with intensive livestock production) and a high demand for mineral concentrates.

References

1. Thörneby, L., Persson, K., and Trägårdh, G. (1999). Treatment of liquid effluents from dairy cattle and pigs using reverse osmosis. *Journal of Agricultural Engineering Research* **73** (2): 159–170.
2. Masse, L., Masse, D., and Pellerin, Y. (2007). The use of membranes for the treatment of manure: a critical literature review. *Biosystems Engineering* **98** (4): 371–380.
3. Hoeksma, P., de Buisonjé, F.E., Ehlert, P.A.I. et al. (2011). *Mineralenconcentraten uit dierlijke mest: monitoring in het kader van de pilot mineralenconcentraten*. Lelystad: Wageningen UR Livestock Research.
4. Hoeksma P., de Buisonjé F.E., Aarnink A.A. (2012) Full-Scale Production of Mineral Concentrates from Pig Slurry Using Reverse Osmosis. Ninth International Livestock Environment Symposium, Valencia Spain.
5. Hilhorst G., Verloop K. (2010) Scheiden van rundveemest met decanter van GEA Westfalia Separator: testresultaten van scheiden met vergiste en onvergiste rundveemest. Wageningen UR Livestock Research. Report No.: 0169-3689.
6. Hjorth, M., Christensen, K.V., Christensen, M.L. et al. (2010). Solid-liquid separation of animal slurry in theory and practice. A review. *Agronomy for Sustainable Development* **30** (1): 153–180.
7. Schröder J.J., de Buisonjé F.E., Kasper G.J., et al. (2009) Mestscheiding: relaties tussen techniek, kosten, milieu en landbouwkundige waarde. Wageningen: Plant Research International.

8. Lemmens, B., Ceulemans, C., Elslander, H. et al. (2007). *Beste Beschikbare Technieken (BBT) voor mestverwerking*. Gent: Academia Press.
9. Ehlert, P.A.I. and Hoeksma, P. (2011). *Landbouwkundige en milieukundige perspectieven van mineralenconcentraten: deskstudie in het kader van de Pilot Mineralenconcentraten*. Wageningen: Alterra Wageningen UR.
10. Velthof, G.L. (2011). *Kunstmestvervangers onderzocht: tussentijds rapport van het onderzoek in het kader van de pilot mineralenconcentraten*. Wageningen: Alterra Wageningen UR.
11. Buning S. (2015) Mineralenconcentraat Kunstmestvervanger nog in ontwikkeling. Melkvee (No. 10, okt. 2015), pp. 21–25.
12. Ehlert P.A.I., Hoeksma P., Velthof G.L. (2009) Anorganische en organische microverontreinigingen in mineralenconcentraten: resultaten van de eerste verkenningen. Animal Sciences Group. Report No.: 1570-8616.
13. Velthof G., van Beek C., Brouwer F., et al. (2004) Denitrificatie in de zone tussen bouwvoor en het bovenste grondwater in zandgronden. Alterra. Report No.: 1566-7197.

4.3

Pyrolysis of Agro-Digestate: Nutrient Distribution

Evert Leijenhorst
BTG Biomass Technology Group B.V., Enschede, The Netherlands

4.3.1 Introduction

4.3.1.1 Background

Co-digestion of animal manures for biogas production has been widely implemented in recent years. Alongside biogas, digestate is obtained as a side-product from the digestion process. Digestate is a mixture of partly decayed organic matter and inorganic matter, and is usually a mixture of liquid and solid components. Recovery and reuse of the nutrients present in the digestate is highly desirable.

For further processing, it is often desirable to separate the liquid (thin) and solid (thick) fraction [1]. The thick digestate fraction can be converted by pyrolysis into solid (biochar) and liquid (pyrolysis oil) products in order to steer the product nutrient composition. The influence of the pyrolysis process conditions on the product distribution, and especially the fate of the nutrients, has been experimentally investigated. Figure 4.3.1 presents a schematic representation of this route.

4.3.1.2 The Pyrolysis Process

Pyrolysis is the thermal decomposition of an organic material in the absence of oxygen. Where chemically complex feedstocks such as biomass or organic waste streams are pyrolyzed, a complex product slate is generated. Typically, the products are separated based on their state of aggregation. The solid product is referred to as char or biochar, the liquid product as pyrolysis oil or bio-oil, and the gas phase as pyrolysis gas [2].

For the production of biochar, which can be used as a soil improver, a slow heating rate is applied. At low temperatures (<300 °C), the process is referred to as torrefaction, which

Biorefinery of Inorganics: Recovering Mineral Nutrients from Biomass and Organic Waste, First Edition.
Edited by Erik Meers, Gerard Velthof, Evi Michels and René Rietra.
© 2020 John Wiley & Sons Ltd. Published 2020 by John Wiley & Sons Ltd.

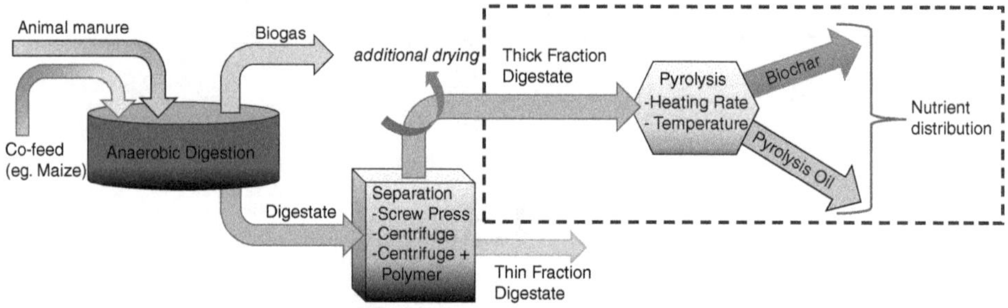

Figure 4.3.1 *Schematic representation of the investigated nutrient recovery route.*

generally aims to maximize the energetic recovery. Higher temperatures (>400 °C) are used in the carbonization process, typically to produce char with a high fixed carbon (the amount of carbon that is not volatilized upon heating) content [3].

To maximize the pyrolysis oil yield, a fast heating rate is required, and temperatures of 500 °C are typically applied. Currently, fast pyrolysis processes aim at producing pyrolysis oil for energy purposes [4]. The properties of the char by-product of fast pyrolysis are reported to be significantly affected by the heating rate [5]. Therefore, the influence of the heating rate on the nutrient distribution will also be investigated. However, it should be noted that most commercial fast pyrolysis processes combust the char to produce the energy required in the pyrolysis process; thus, the nutrients are recovered in the ash by-product rather than as char. The majority of the nitrogen present in the biomass is lost to the air in the form of N_2 when the char is combusted.

4.3.1.3 Pyrolysis of Agro-Digestate

The pyrolysis of agro-digestate for biochar or pyrolysis oil production is relatively unexplored. Though anaerobic co-digestion of manure is implemented commercially and significant digestate volumes are available, attempts to pyrolyze the thick fraction are scarce. The most likely reason is the high ash and moisture content, which limits the energetic efficiency. Justification for a digestate pyrolysis process lies primarily in the recovery of nutrients. In one of the few publications on pyrolysis of agro-digestate, Schouten et al. [6] investigate the carbon and nitrogen dynamics in soil applied with biochar produced by fast pyrolysis at 500 °C. Untreated cattle manure and digestate from anaerobic digestion are used as reference materials. All three samples are applied to loess and sandy soil. The loss of soil C from biological activity is highest for the untreated manure (39 and 32% for the loess and sandy soil, respectively), followed by the digestate (31 and 18%), and lowest for the biochar (15 and 7%). The carbon in biochar is clearly more resistant to biological conversion. Emissions of N_2O range from 0.6% for the biochar to 4.0% for the manure. ^{15}N isotope labeling shows that manure N is mostly mineralized, contributing 50% to soil inorganic N. The anaerobic digestate is the only by-product, increasing the mineral N pool while reducing emissions of N_2O compared with manure. For the biochar, 18.3% of soil mineral N is derived from the biochar, without constraining mineralization of the native soil N. Details on recovery products and nutrients in the pyrolysis process are unfortunately not reported.

4.3.2 Investigation

To investigate the distribution of nutrients between the solid and liquid products of the pyrolysis process, 18 pyrolysis runs were performed. Variables included the digestate separation technique, the pyrolysis temperature, and the heating rate in the pyrolysis reactor.

4.3.2.1 Materials and Methods

Three thick digestate fractions were obtained from the anaerobic co-digestion of pig and cow manure, maize, and organic waste streams, each using a different separation method. Details on the feedstocks used are presented in Table 4.3.1.

The digestate samples were pyrolyzed in an electrically heated screw reactor at a feed capacity of ~250 g h^{-1}. The screw reactor could be used in slow pyrolysis mode. In addition,

Table 4.3.1 *Digestate feedstocks and their properties.*

Description	Press	Polymer	Centrifuge
Manure type	Pig	Cow	Cow
Digestate separation technique	Screw press	Centrifuge with polymer flocculent	Centrifuge separation without flocculent
Total ash (dry, 550 °C) (wt%)	18	60	60
Moisture content (wet) (wt%)	6	4	3
Elemental analysis (dry)			
Carbon (g kg^{-1})	383	175	184
Hydrogen (g kg^{-1})	51.6	26.5	27.0
Nitrogen (g kg^{-1})	30.9	25.5	15.8
Phosphorus (g kg^{-1})	10.1	16.1	15.7
Potassium (g kg^{-1})	16.1	7.56	5.54
Sulfur (g kg^{-1})	7.24	4.38	3.70
Alkali and alkali earth metals (dry)			
Na (g kg^{-1})	10.7	3.99	2.97
K (g kg^{-1})	16.1	7.56	5.54
Ca (g kg^{-1})	13.3	41.7	42.2
Mg (g kg^{-1})	4.55	7.17	11.3
Other metals (dry)			
Fe (g kg^{-1})	7.12	12.6	11.6
Al (g kg^{-1})	0.94	5.52	5.23
Cd (mg kg^{-1})	0.19	0.59	0.35
Co (mg kg^{-1})	1.02	2.84	2.36
Cr (mg kg^{-1})	16.9	17.8	10.6
Cu (mg kg^{-1})	42.0	51.0	20.6
Mn (mg kg^{-1})	132	319	271
Ni (mg kg^{-1})	8.57	15.3	11.8
Pb (mg kg^{-1})	8.20	10.3	11.8
Zn (mg kg^{-1})	177	215	136

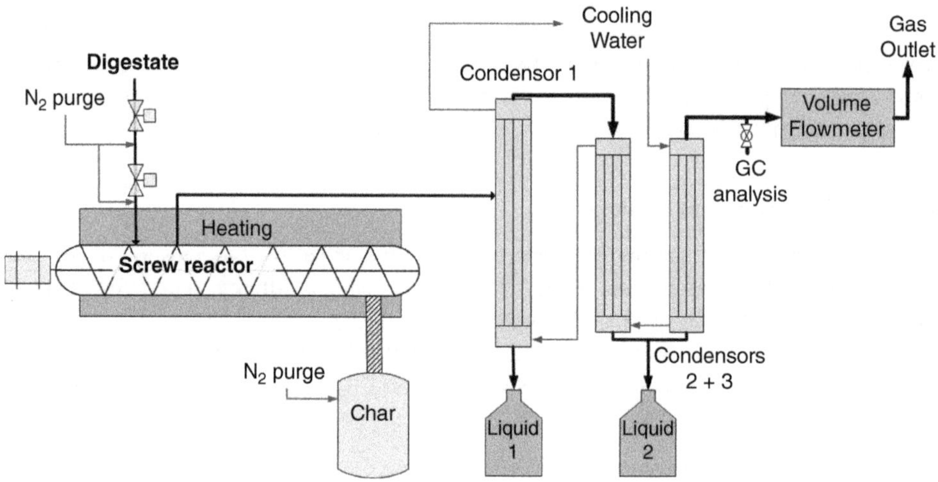

Figure 4.3.2 Schematic representation of the pyrolysis process.

a special heating system was installed in order to obtain the heating rate required for fast pyrolysis. After construction, the set-up was commissioned to compare the product yield with fast pyrolysis systems. With a pyrolysis oil yield of 65% for clean wood biomass, the set-up seemed to give results representative for fast pyrolysis.

A schematic representation of the experimental setup is shown in Figure 4.3.2. The biomass is fed to the screw reactor through a sluice system. Nitrogen is used as purge gas in the sluice system to prevent backflow of gases produced in the reactor. The biochar product is transported through the screw reactor to the char collection vessel. The gases formed during pyrolysis are transported to the condensing system on the right side of the figure. Three condensers are placed in series, to recover all condensable components. After each experiment, the liquid product from collection vessels 1 and 2 were combined to give the total pyrolysis oil product. The non-condensable gas components (mainly CO, CO_2, and CH_4) were analyzed by a gas chromatograph (GC), and the total gas volume measured by a flowmeter.

4.3.2.2 Product Analysis and Evaluation

After each experiment, the char and pyrolysis oil products were collected and weighed to determine the product yields. The gas volume and composition were measured as well to determine the overall mass balance closure

Because the total ash content of the feedstocks is quite high and deviates significantly between them (Table 4.3.1), the results are compared on "dry-ash-free" basis. This gives a better representation of the actual product yields than the "as-received" mass balance. Figure 4.3.3 shows a schematic representation of the differences between the two mass balances. On the left of the figure, the as-received balance is shown. Here, the entire feedstock is considered, including moisture and ash. The products are based on their state of matter. On the right, the "dry-ash-free" balance is shown. Here, the moisture and ash present in the feedstock are not considered. With the dry-ash-free balance, only the organic part

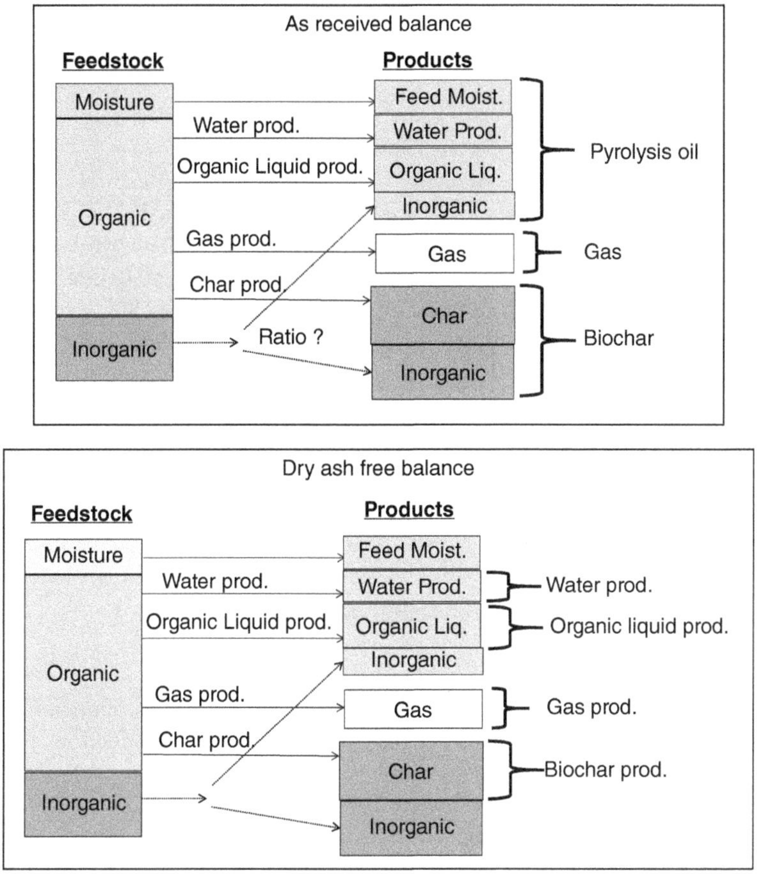

Figure 4.3.3 *Schematic overview of the two mass balance types and products from the process.*

of the feedstock and the products derived from the organic part are considered. Though moisture and ash from the feedstock can react in the process, this approach gives a fairer representation of the actual process performance.

The average mass balance closure for the dry-ash-free balance was 89% ± 8%, while that for the as-received balance was 95% ± 3%.

In the as-received balance (Figure 4.3.3), the term "ratio" represents the major research goal: to determine the ratio between inorganic elements and the liquid phase and solid phase during the pyrolysis process. The distribution of each element is calculated as the total amount in the product divided by the total amount in the feedstock, according to Eq. (4.3.1):

$$X_{i,j} = \frac{\phi_j \cdot C_{i,j}}{\phi_{BM} \cdot C_{i,BM}} 100\% \tag{4.3.1}$$

where $X_{i,j}$ represents the recovery of element *i* in product *j*, as a percentage of the amount present in the biomass feedstock; $\emptyset_j$ represents the mass flow of product *j* (pyrolysis oil, gas, biochar) in kg s^{-1}; $C_{i,j}$ represents the concentration of element *i* in product *j* (in mg/kg);$\emptyset_{BM}$

is the mass flow of the biomass feedstock into the process in kg s^{-1}; and $C_{i,BM}$ is the concentration of element i in the biomass feedstock in mg kg^{-1}.

4.3.3 Results and Discussion

This section presents the influence of the pyrolysis temperature on the product and nutrient distribution using a fast heating rate, followed by the influence of the heating rate. The latter provides insight into the differences in application of fast versus slow pyrolysis.

Some of the results obtained in this work show overall recovery rates significantly higher than 100%, which is theoretically not possible unless material is added or lost in the processing step. Most of these deviations were the result of the inhomogeneous nature of the digestate. When calculating the ash content based on the elemental analysis of the feedstocks,[1] the total ash content was only 8, 12, and 12% for the press-, polymer-, and centrifuge-separated digestate, respectively. The remaining part was believed to be sand. Unfortunately, the silicon content could not be measured to validate this. In addition, the sand content seemed to vary between the individual experiments, so no correction for the sand content could be made.

4.3.3.1 Fast Pyrolysis: Influence of Temperature

4.3.3.1.1 Product Distribution

Figure 4.3.4a–d presents the product distribution as a function of temperature for the fast pyrolysis of the three digestate streams. The dotted lines are added as a visual aid to show the general trend. The graphs show relatively similar product distributions for the three feedstocks. Several points deviate from the expected trend. A decreasing char yield would be expected with increasing temperature (Figure 4.3.4a), caused by increased thermal cracking of the organic material. For the centrifuge-separated stream at 400 °C, a lower-than-expected char yield is observed. Increased gas production (Figure 4.3.4d) with increasing temperature is a direct result of the increased thermal cracking. In Figure 4.3.4d, the polymer- and centrifuge-separated streams give a relatively low gas yield at 500 °C. The organic liquid yield (Figure 4.3.4b) usually shows an optimum around 500 °C for woody biomass materials. With the digestate streams, this optimum is found at a lower temperature of 400 °C, likely as a result of the high concentrations of inorganic elements increasing the pyrolysis reaction rates, thereby lowering the optimum pyrolysis temperature. The water production is relatively constant for all temperatures, as can be seen in Figure 4.3.4c. Only for the lowest temperature, with the centrifuge- and polymer-separated digestate streams, are water production rates below 10% obtained.

4.3.3.1.2 Nutrient Recovery

The recovery of nutrients in the products of the pyrolysis process is of primary interest. The most important fertilizer nutrients (N, P, K) are presented separately. For clarity, it has been

[1] The total ash content is corrected for the oxygen content. For example, copper will be measured in the analysis as elemental copper, while in the total ash content, CuO is likely measured.

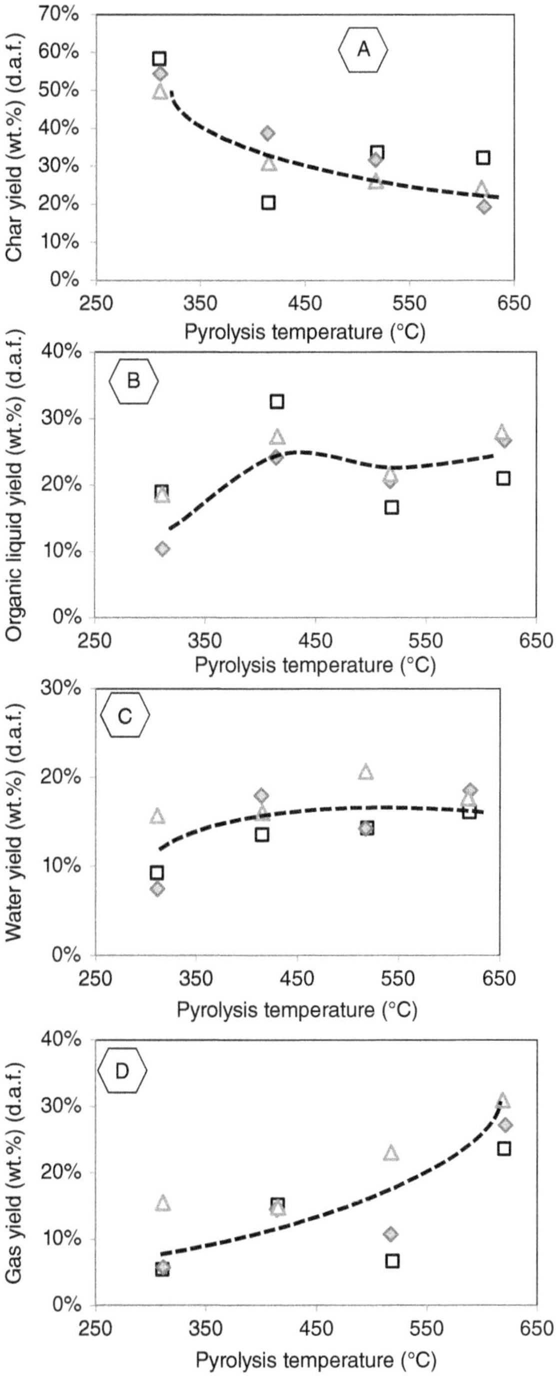

Figure 4.3.4 *Product distribution in weight percentage of the dry-ash-free mass balance as a function of pyrolysis temperature with fast heating rate for three digestate samples: (a) char yield, (b) organic liquid yield, (c) water yield, (d) gas yield.*

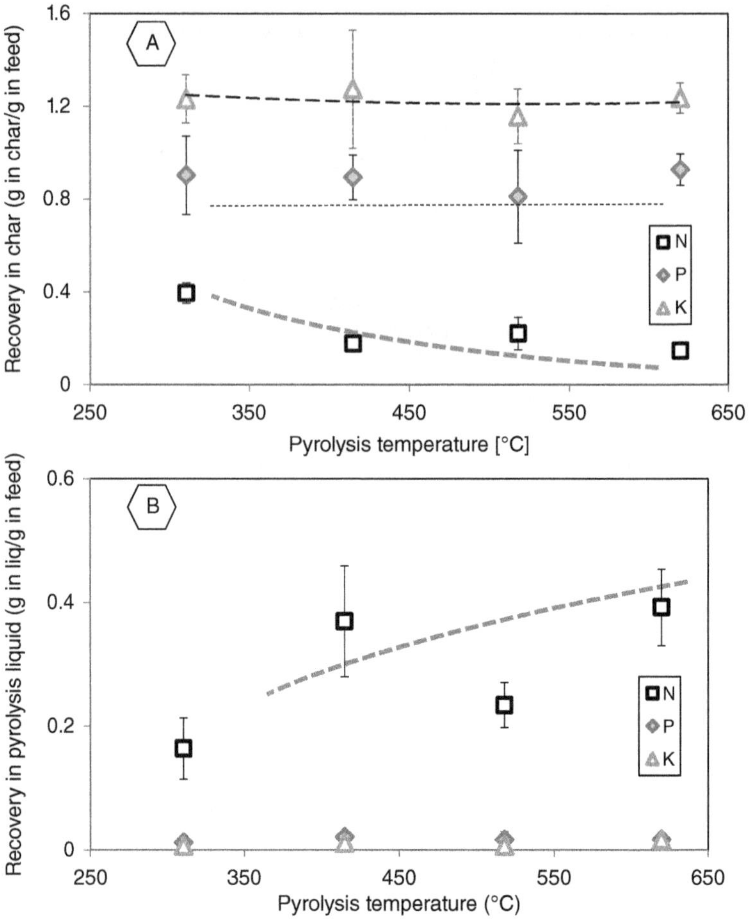

Figure 4.3.5 *NPK recovery calculated according to Eq. (4.3.1) in (a) biochar and (b) pyrolysis liquid as a function of temperature for fast pyrolysis.*

chosen to group some of the other elements in the data evaluation. In the alkali and alkali earth metal (AAEM) group, calcium was the dominant element, while for the metal group, iron was the largest contributor in all three feedstocks.

For the three different digestate streams, the distribution of the elements was found to be quite similar in the measured range. Therefore, the averaged values for the three streams were used in the evaluation of the results. Figure 4.3.5 shows the averaged recovery of the major fertilizer components N, P, and K in (a) the char and (b) the total pyrolysis liquid. The error bars indicate the standard deviation found between the three feedstocks.

The distribution of nitrogen between the products clearly depended on the pyrolysis temperature. The recovery of nitrogen in the char decreased from approximately 40% at 300 °C to 15% at 600 °C. Nitrogen recovery in the pyrolysis oil increased with increasing temperature. Nitrogen losses from the biochar with increasing pyrolysis temperature are often encountered; see, for example, Chan and Xu [7]. Part of the nitrogen is not recovered in the process; this is transferred to the gas phase (mainly N_2).

Table 4.3.2 *Carbon, sulfur, AAEM, and metals recovery, calculated according to Eq. (4.3.1), in char and pyrolysis oil as a function of temperature for fast pyrolysis.*

Recovery (wt%)		311 °C	415 °C	518 °C	620 °C
Carbon	Char	105 ± 24%	50 ± 6%	60 ± 10%	52 ± 8%
	Pyrolysis oil	21 ± 4%	44 ± 6%	31 ± 3%	46 ± 10%
Sulfur	Char	67 ± 12%	68 ± 15%	55 ± 17%	72 ± 22%
	Pyrolysis oil	10 ± 11%	25 ± 15%	16 ± 20%	40 ± 38%
AAEM	Char	83 ± 12%	83 ± 15%	82 ± 15%	88 ± 6%
	Pyrolysis oil	1 ± 1%	3 ± 1%	1 ± 0%	5 ± 5%
Metals	Char	117 ± 40%	213 ± 136%	221 ± 174%	129 ± 29%
	Pyrolysis oil	2 ± 2%	4 ± 1%	3 ± 2%	3[a] ±2%

[a]One very high iron measurement (10.9 wt% Fe in pyrolysis oil derived from the centrifuge-separated digestate) is removed from these data, because the value is likely a measurement error.

Both potassium and phosphor were (almost) fully retained in the char product. For potassium, the recovery was around 120%, and for phosphorus, around 90%. The deviations from 100% were caused by the inhomogeneous nature of the feedstock. The limited recovery of potassium and phosphorus in the liquid confirms that both are almost fully retained in the solid phase, regardless of the applied temperature. Azuara et al. [8] investigated phosphorus recycling from pig manure by fast pyrolysis of the thick fraction. For pyrolysis temperatures of 400, 500, and 600 °C, the yield of phosphorus from the thick fraction was 92, 96, and 97% respectively, confirming phosphorus can be retained in the biochar almost completely.

The recovery of the other element (groups) as a function of temperature for fast pyrolysis is presented in Table 4.3.2. The values represent average recovery rates for the three digestate streams, with the standard deviation indicating variation between the different feeds.

From Table 4.3.2, it can be seen that the carbon recovery in the char decreased with increasing temperature, while the carbon recovery in the pyrolysis oil increased with increasing temperature. The total recovery of carbon decreased because the loss of carbon to the gas phase increased. For both carbon and sulfur, the data at 518 °C deviate from the trend. Because similar C and S concentrations were measured, this is a direct result of the deviating product distribution found at 518 °C; see Figure 4.3.4.

The sulfur recovery in the char was quite stable at around two-thirds of the sulfur in the feed, while a slight increase with increasing temperature was found for the pyrolysis oil. It must be noted that the standard deviation of the measurements at 600 °C was very high, caused by one high sulfur concentration in the organic liquid produced from the polymer-separated digestate. For this particular product, the sulfur concentration was double that of the other samples from the same feedstock.

The transfer of AAEMs and metals to the liquid product was very limited, as was expected based on the low vapor pressures of the elements at the applied temperatures. The total AAEM recovery was only slightly below 100%, with relatively modest error values. For the metal group, recovery rates (far) above 100% were found. These are primarily caused by increased iron recoveries. Iron can potentially be derived from the stainless-steel material from which the pyrolysis set-up is constructed; the same holds for nickel and chrome. However, the iron recovery seemed to depend more on the feedstock than on the process conditions. The average iron recoveries were 86 ± 14%, 137 ± 24%, and 397 ± 190% for the press-, polymer-, and centrifuge-separated digestates, respectively. These numbers indicate that the deviation should be attributed to the inhomogeneous nature of the feed rather than to leaching of iron from the reactor material.

4.3.3.1.3 Product Composition

The composition of biochar in general depends heavily on the composition of the starting material, and a wide range of compositions can be found in the literature [7]. It is important to point out that the total content of many nutrients does not necessarily reflect the availability of these nutrients to plants. Organically bound nitrogen, for example, is often "fixed" in the char, while mineral nitrogen (ammonium and nitrate groups) is freely available. According to Bagreev et al. [9], nitrogen elements initially present as amine functionalities are gradually transformed into pyridine-like compounds at higher temperatures, reducing the nitrogen availability in the char.

The C/N ratio of biochar is sometimes used as an indicator for the availability of nitrogen to plants. Generally, a C/N ratio of 20 is used as a critical limit, above which nitrogen immobilization may occur [7]. Though detailed investigations are required to determine the actual plant availability, the C/N ratio does give some information. In Figure 4.3.6a–c, the C/N ratios of the biochar, organic liquid, and aqueous liquid are presented for the three digestate streams.

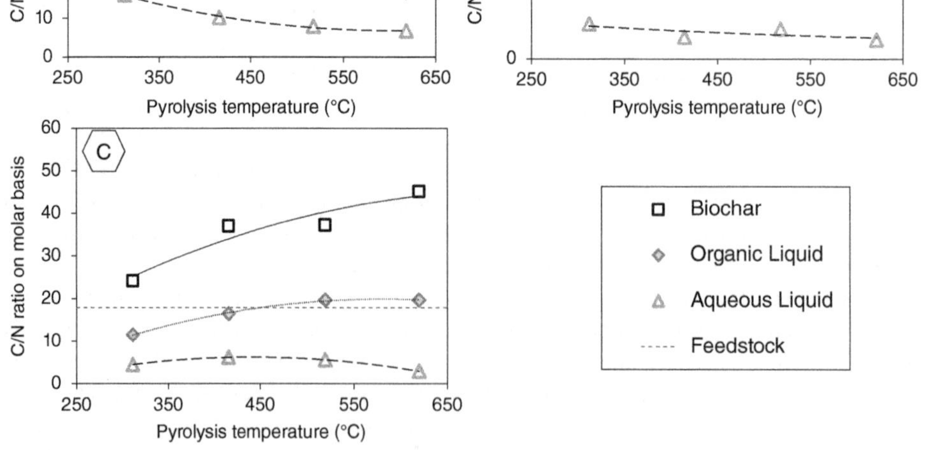

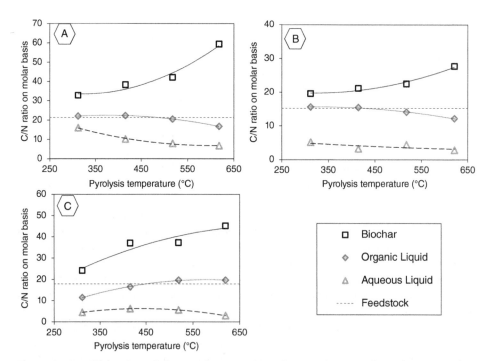

Figure 4.3.6 *C/N ratios of the pyrolysis products from (a) press-, (b) polymer-, and (c) centrifuge-separated digestate.*

The trend for each of the digestate streams is similar, while the absolute C/N ratios are primarily a result of the initial feedstock value. All the biochar samples produced here had C/N ratios above 20. In order to determine the actual fertilizer value of the biochar products, detailed nutrient availability tests with crops should be performed; see, for example, Lehmann et al. [10]

4.3.3.2 Influence of Heating Rate

4.3.3.2.1 Product Distribution

In addition to the final temperature, the influence of the heating rate was investigated as well. Figure 4.3.7 shows the char and liquid production obtained for fast and slow heating at 400 and 500 °C using the three digestate streams.

For both the press- and the polymer-separated digestate, the char production behavior is according to expectations. A slower heating rate resulted in a higher char yield, both at 400 and at 500 °C. The relative difference is quite modest, however, at between 6 and 13%. For the centrifuge-separated digestate, deviating behavior was found. With an averaged mass balance closure of only 82% for the centrifuge-separated data, the uncertainty in the centrifuge results are quite high. This uncertainty is likely caused by the heterogeneous nature of the feedstock.

4.3.3.2.2 Nutrient Recovery

The influence of the heating rate on the nutrient recovery of NPK and C between the char and the pyrolysis liquid was also investigated. Figure 4.3.8 shows the averaged results, where the standard deviation between the three feedstocks is again used to plot the error bars. Based on these results, it can be concluded that the heating rate had a very limited effect on the recovery of phosphorus and potassium. For both nitrogen and carbon, a slow heating rate at a final temperature of 400 °C yielded the highest recovery rates to the char; however, the difference was only minor. Furthermore, it was observed that the heating rate had a very limited influence on the recovery of sulfur, AAEM, and metal elements (data not shown).

4.3.4 Conclusion

Three digestate streams obtained from the anaerobic co-digestion of animal manure, maize, and organic waste streams were converted by pyrolysis. The influence of heating rate and pyrolysis temperature on the distribution of products and nutrients was found to be relatively low. Virtually all of the phosphorus and potassium available in the digestate ended up in the char product. Nitrogen was distributed almost evenly between the char and the liquid product, where higher pyrolysis temperatures led to lower nitrogen recovery in the char and increased recovery in the pyrolysis oil. Part of the nitrogen was also lost to the gas phase (mainly N_2). C/N ratios found in the biochar ranged from 20 to 60 C/N. These values suggest that part of the nitrogen might not be available to the plants. However, to determine the actual fertilizer value of the char, further research with crops test is needed. If the products

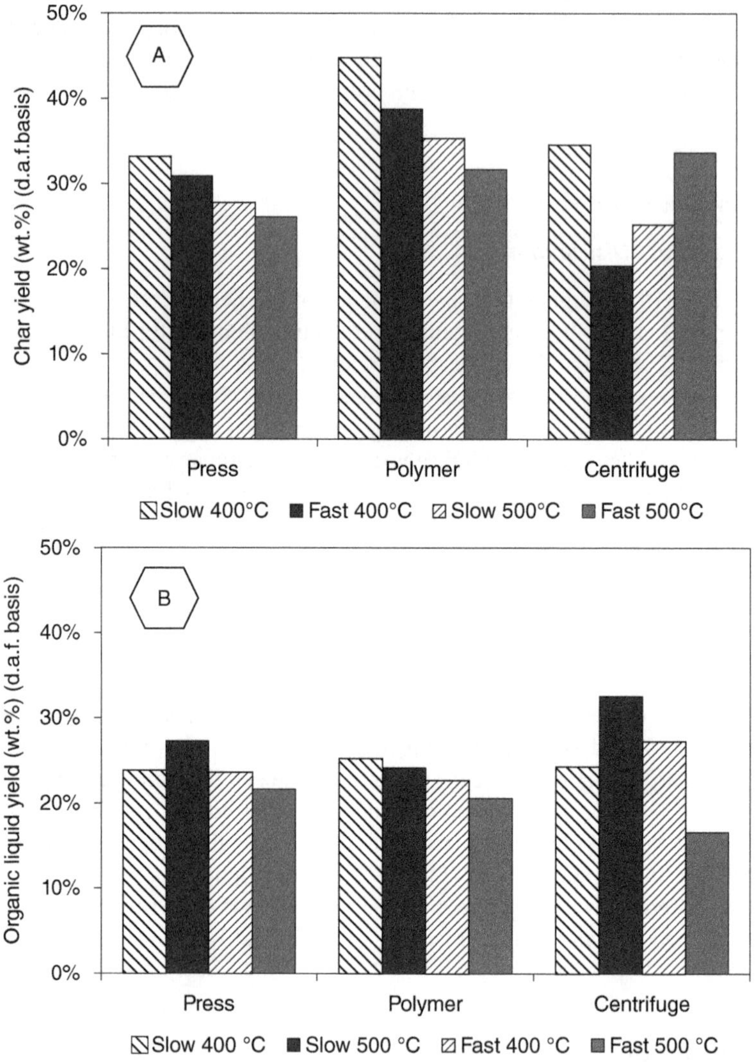

Figure 4.3.7 *(a) Char and (b) organic liquid production on dry ash free basis as function of heating rate for three digestate samples.*

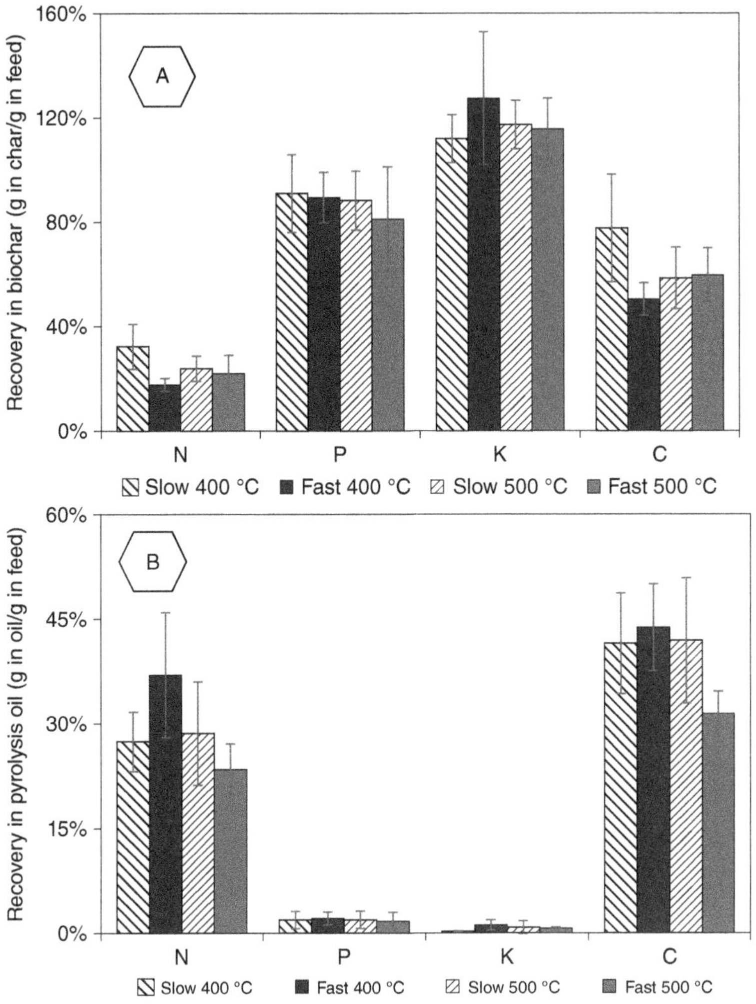

Figure 4.3.8 *NPK-C recovery in (a) char and (b) pyrolysis oil as a function of heating rate.*

from the fast pyrolysis show no or limited added value compared to the slow pyrolysis products, slow pyrolysis is probably more economic for this application. If a carbon-free solid product is desired, use of the mineral ash product after the char combustion applied in most fast pyrolysis processes might be considered. However, all nitrogen (and all carbon) is lost when the char is combusted.

Acknowledgment

This work is partly financed by the European Union within the KP-7 project INEMAD, grant agreement no. 289712. This financing is greatly appreciated. Special thanks go out to Reinhart van Poucke at Ghent University for his contributions to the analysis work.

References

1. Rehl, T. and Müller, J. (2011). Life cycle assessment of biogas digestate processing technologies. *Resources, Conservation and Recycling* **56** (1): 92–104.
2. Venderbosch, R.H. and Prins, W. (2010). Fast pyrolysis technology development. *Biofuels, Bioproducts and Biorefining* **4** (2): 178–208.
3. Ronsse, F., Nachenius, R.W., and Prins, W. (2015). Carbonization of biomass. In: *Recent Advances in Thermochemical Conversion of Biomass* (eds. A.K. Pandey, T. Bhaskar, M. Stocker and R.K. Sukumaran), 293–324. Amsterdam: Elsevier.
4. Guda, V.K., Steele, P., Penmetsa, V.K., and Li, Q. (2015). Fast pyrolysis of biomass: recent advances in fast pyrolysis technology. In: *Recent Advances in Thermochemical Conversion of Biomass* (eds. A.K. Pandey, T. Bhaskar, M. Stocker and R.K. Sukumaran), 177–211. Amsterdam: Elsevier.
5. Kong, Z., Liaw, S.B., Gao, X. et al. (2014). Leaching characteristics of inherent inorganic nutrients in biochars from the slow and fast pyrolysis of mallee biomass. *Fuel* **128**: 433–441.
6. Schouten, S., van Groenigen, J.W., Oenema, O., and Cayuela, M.L. (2012). Bioenergy from cattle manure? Implications of anaerobic digestion and subsequent pyrolysis for carbon and nitrogen dynamics in soil. *GCB Bioenergy* **4** (6): 751–760.
7. Chan, K.Y. and Xu, Z.H. (2009). Biochar: nutrient properties and their enhancement. In: *Biochar for Environmental Management Science and Technology* (eds. J. Lehmann and S. Joseph). London: Earthscan.
8. Azuara, M., Kersten, S.R.A., and Kootstra, A.M.J. (2013). Recycling phosphorus by fast pyrolysis of pig manure: concentration and extraction of phosphorus combined with formation of value-added pyrolysis products. *Biomass and Bioenergy* **49**: 171–180.
9. Bagreev, A., Bandosz, T.J., and Locke, D.C. (2001). Pore structure and surface chemistry of adsorbents obtained by pyrolysis of sewage sludge-derived fertilizer. *Carbon* **39** (13): 1971–1979.
10. Lehmann, J., Pereira da Silva, J. Jr., Steiner, C. et al. (2003). Nutrient availability and leaching in an archaeological anthrosol and a ferralsol of the central amazon: fertilizer, manure and charcoal amendments. *Plant and Soil* **249** (2): 343–357.

4.4

Agronomic Effectivity of Hydrated Poultry Litter Ash

Phillip Ehlert

Wageningen University & Research, Wageningen Environmental Research, Wageningen, The Netherlands

4.4.1 Introduction

BMC Moerdijk produces green energy and hydrated poultry litter ash (HPLA), a PK fertilizer, by incineration of approximately 430 000 tons of poultry manure per year. The green energy produced amounts to 285 000 MWh, which serves the energy requirements of more than 70 000 households. After incineration, 60 kton of ash remains, which is converted to HPLA. Overall, about a third of the total quantity of poultry litter produced each year in the Netherlands is incinerated (http://www.bmcmoerdijk.nl/en/home.htm).

4.4.2 Energy Production Process

BMC Moerdijk incinerates poultry litter produced by 500 poultry farms united in the cooperative DEP (http://www.cooperatiedep.nl). The solid poultry manure comes from broilers (55%), laying hens (20%), turkeys, broiler parents, and laying hen parents (25%). The power plant BMC Moerdijk produces electricity 24 hours a day, 7 days a week. The fuel poultry litter is delivered daily by truck. A strict protocol for the delivery is followed. After delivery, each truck is sanitized.

The process of BMC Moerdijk consists of five main systems (Figure 4.4.1):

1. fuel handling and transport system;
2. steam generation (boiler) and its ancillary systems;

Biorefinery of Inorganics: Recovering Mineral Nutrients from Biomass and Organic Waste, First Edition.
Edited by Erik Meers, Gerard Velthof, Evi Michels and René Rietra.
© 2020 John Wiley & Sons Ltd. Published 2020 by John Wiley & Sons Ltd.

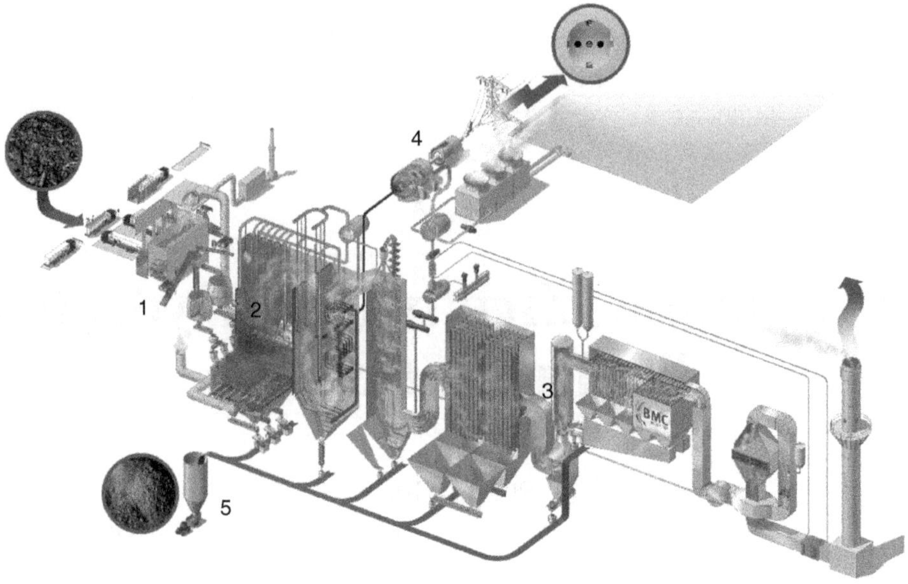

Figure 4.4.1 Overview of the production process of BMC Moerdijk: (1) fuel bunker; (2) bubbling fluidized bed combuster or BFBC boiler; (3) flue gas cleaning installation, including an electrostatic precipitator (ESP); (4) steam turbine and electricity generator; and (5) cooling unit and ash collectors.

3. flue gas cleaning;
4. power generation (turbine) and its ancillary systems; and
5. production of hydrated poultry litter ash.

The fuel handling and transport system starts at the gate, with each truck passing through a sampling station where the poultry litter is weighed, sampled, and unloaded into four pits. The litter is then collected by an automated crane system and discharged into the main storage bunker. In the bunker, the poultry litter is constantly mixed by the automated crane system in order to obtain a homogeneous mixture. From there, it is transported by conveyor belts, sieved in a star screen separator to remove particles larger than 10 cm, and passed through a magnet separator.

The poultry litter is then transported to the boiler, where it is fed to a fluidized bed. The bed, consisting of a bubbling mass of sand and ash, is kept at constant temperature of 750 °C. There, the poultry litter is gasified in a reducing atmosphere produced by the use of secondary air (two stages) and recirculation air. The burner operates with secondary air in two stages and an intermittent stage of recirculation of air. In a second phase with these additional levels of secondary air and recirculation air, complete combustion of the fuel takes place at a temperature around 1000 °C. In this phase, an oxidizing atmosphere is used, which guarantees complete and uniform combustion and low emissions in order to comply with the waste incineration directive. From here, flue gas and fly ash leave the combustion chamber and travel through several boiler passes to the flue gas cleaning system. In the boiler, fuel ash is collected in each of the passes. In the first pass, the ash is collected below the fluidized bed; in the second and third, fly ash is collected at the bottom and removed by screws. All ashes are then transported to the storage silos. Flue gas cleaning is composed of three subsystems:

1. an electrostatic precipitator to remove ash particles;
2. a semi-dry absorption system with baghouse filter, which removes acid components (HCl; SO_2) through sorption on calcium hydroxide; the remaining ash coming from the electrostatic precipitator is also removed; and
3. a selective catalytic reduction, De-NOx, which reduces NOx.

The steam generation results in electricity for approximately 70 000 households. The ashes collected during the different phases of the boiler passes is secondary raw material for the production of HPLA. The production consists in the treatment of the homogenized ashes with water up to about 10%. The main reason for water addition is to prevent the formation of dust. Next, HPLA is stored in bunkers. Within the production line, there is no contact between fuel handling systems, storage of poultry litter, and systems for the storage and transport of HPLA.

4.4.3 Composition of HPLA

The average composition of poultry litter from egg-laying hens and broiler chicken is given in Table 4.4.1. The main minerals of poultry litter are calcium (Ca), potassium (K), and phosphorus (P). Calcium originates from poultry litter and the adsorbent used at fly gas cleaning, calcium hydroxide, which is converted to calcium carbonate during the cleaning process.

The main nutrients of HPLA from the combustion of poultry litter are given in Table 4.4.2. The main constituents are P and K. HPLA has a neutralizing value (NV) due to the presence of carbonates and hydrated burned lime from flue gas cleaning. In addition, other minerals are present in high (Ca) or lower (Mg, S) percentages. Table 4.4.2 gives an average composition for HPLA.

The mineral composition of HPLA indicates value as a PK fertilizer. This can be seen when HPLA is analyzed following the standard analytical method used for conformity analyses of PK fertilizers under the European Fertilizer Regulation 2003/2003. A large part of mineral phosphorus is soluble in citrate, but none is soluble in water. A large part of mineral potassium is water-soluble. HPLA has an acid-neutralizing value of 27% (as CaO), caused by calcium carbonate and hydrated burned lime. HPLA is a fine material that passes through sieves (wet sieving) of 3.15 mm (99.8%), 1 mm (87.8%), and 0.50 mm (70.7%). A new facultative European regulation on fertilizing products will replace the Fertilizer Regulation 2003/2003, by July 16, 2022.

4.4.4 Agronomic Effectivity of HPLA

A pot experiment was carried out with so-called Mitscherlich pots (height 22 cm, Ø 20 cm, 0.0314 m^2, 5.2 l). Soil (5 kg dry) was homogeneously mixed with the fertilizers and slightly compacted on top of the nylon mesh to provide a good soil structure. Soil moisture content was conditioned at 60% (on weight basis) of the liquid limit of the soil. Next, an unfertilized

Table 4.4.1 Average composition of poultry litter (data from BMC Moerdijk; n = 47).

Parameter	Unit	Average	Standard deviation
Moisture	wt%	57.8	2.5
Ash	wt% DM	16.0	2.3
OM	wt% DM	84.0	2.3
N	wt%	2.6	0.2
P_2O_5	wt%	2.1	0.4
K_2O	wt%	2.0	0.3
CaO	wt%	2.9	0.7
MgO	wt%	0.8	0.1
Na_2O	wt%	0.4	0.1
SO_3	wt%	0.9	0.1
B	$mg\,kg^{-1}$ DM	35.4	4.2
Co	$mg\,kg^{-1}$ DM	<5.0	*
Cu	$mg\,kg^{-1}$ DM	97.9	12.3
Fe	$mg\,kg^{-1}$ DM	1038	239
Mn	$mg\,kg^{-1}$ DM	509	70
Mo	$mg\,kg^{-1}$ DM	3.9	0.7
Zn	$mg\,kg^{-1}$ DM	500	80.9
As	$mg\,kg^{-1}$ DM	<3.0	*
Cd	$mg\,kg^{-1}$ DM	0.18	0.0
Cr	$mg\,kg^{-1}$ DM	10.4	1.0
Hg	$mg\,kg^{-1}$ DM	<0.05	*
Ni	$mg\,kg^{-1}$ DM	6.7	1.8
Pb	$mg\,kg^{-1}$ DM	<5.0	*
C	wt% DM	40.4	0.8
H	wt% DM	5.86	0.5
Cl	wt% DM	0.54	0.03
F	wt% DM	<0.01	*
Al	$mg\,kg^{-1}$ DM	741	287
Se	$mg\,kg^{-1}$ DM	<5.0	*
Si	$mg\,kg^{-1}$ DM	1610	719

soil layer of 1 kg (dry) was placed on top, also conditioned at 60% (w) of the liquid limit of the soil. After the pot was filled with soil, a plastic pipe was placed in the middle. A quartz sand layer of 2 cm (0.5 kg dry sand) with a moisture content of 60% (w) of the liquid limit was placed on the soil surface to prevent evaporation and salt crust formation. The total weight of the pot was recorded. During the pot experiment, demineralized water was given by means of this pipe to prevent destruction of the soil structure. Pots were weighted daily. Loss of water due to evapotranspiration was daily corrected by adding demineralized water to compensate loss of weight, taking into account the weight of the growing crop. The position of the pots was rotated within their replications during the daily watering.

The agronomic effectivity of HPLA was tested in pot and incubation experiments [1–5]. Sandy soils with a low phosphorus and potassium status and low pH were used for all these experiments [6, 7]. Rating of the nutrient and pH status of the sandy soil was according to the fertilizer recommendation scheme of the Netherlands for arable land [6, 7]. Nutrients

Table 4.4.2 Average composition of processed hydrated poultry litter ash.

Parameter	Unit	Average	Standard deviation[a]
Dry matter	wt%	90.3	3.6
Ash	wt% DM	96.2	10.6
OM	wt% DM	3.6	10.6
N[a]	wt%	0.03	0.02
P_2O_5 total[a]	wt%	12.1	1.5
P_2O_5 soluble in 2% citric acid[a]	wt%	8.2	1.1
P_2O_5 soluble in neutral ammonium citrate[a]	wt%	5.1	1.1
P_2O_5 soluble in water[a]	wt%	0.1	0.0
K_2O total[a]	wt%	13.0	1.9
K_2O soluble in water[a]	wt%	10.1	2.1
CaO	wt%	21.4	2.4
MgO	wt%	5.5	0.7
Na_2O	wt%	2.4	0.3
SO_3	wt%	6.5	0.9
Acid-neutralizing value[a]	kg CaO/100 kg	27	*
B	mg kg^{-1} DM	160	18.7
Co	mg kg^{-1} DM	<5	*
Cu	mg kg^{-1} DM	421	55.0
Fe	mg kg^{-1} DM	5369	1018
Mn	mg kg^{-1} DM	2158	267
Mo	mg kg^{-1} DM	17.4	4.3
Zn	mg kg^{-1} DM	2292	585
As	mg kg^{-1} DM	<3.0	*
Cd	mg kg^{-1} DM	0.98	0.2
Cr	mg kg^{-1} DM	17.8	10.9
Hg	mg kg^{-1} DM	<0.05	*
Ni	mg kg^{-1} DM	25.4	6.6
Pb	mg kg^{-1} DM	7.0	3.7
C	wt% DM	1.2	0.4
H	wt% DM	0.2	0.1
Cl	wt% DM	2.3	0.5
F	wt% DM	<0.01	*
Al	mg kg^{-1} DM	7306	2136
Se	mg kg^{-1} DM	6.9	1.3
Si	mg kg^{-1} DM	8.483	3450

[a]n = 47.
Source: Data from BMC Moerdijk; n = 47 or 48.

were tested according to the *ceteris paribus* principle; that is, only the tested nutrient was varied – all other nutrients were applied to the same optimal application rate.

The agronomic effectivity of phosphorus and potassium was tested on crops with a short growing season. For this purpose, green bean (*Phaseolus vulgaris* L.) was used. Yields of beans and bean leaves were determined, as well as their chemical compositions. The agronomic effectivity was also tested on crops with a longer growing season. For this, rye grass (*Lolium perenne* L.) was used. During the growing season, five or six cuts were harvested and the yield of each and its chemical composition were measured.

The effectivity of the acid-neutralizing value was tested in an incubation experiment. Soil was amended with HPLA to application rates of phosphorus or potassium fitted to good agricultural practices, avoiding P and K application rates higher than recommended. Soil samples were stored in audiothene plastic bags at 15 °C and sampled at given time intervals for pH analysis.

The effectivity of HPLA was compared to reference fertilizers. For phosphorus, triple superphosphate and dicalcium phosphate were the reference fertilizers. The effectivity of potassium of HPLA was compared with that of potassium sulfate. The effectivity of acid-neutralizing value was compared with calcium carbonate (chalk) and calcium hydroxide (hydrated burned lime). All reference fertilizers met the requirements of the European regulations for fertilizers [8].

The effectivity of HPLA was derived from the apparent recovery efficiency (ARE) and fertilizer replacement values. The ARE of applied phosphorus or potassium (mg P or K taken up by the crop per pot) with reference fertilizer (triple superphosphate, dicalcium phosphate, or potassium sulfate) or HPLA was calculated according to Dobermann [9]:

$$\mathrm{ARE} = 100 * (\mathrm{U_p}-\mathrm{U_0})/\mathrm{F_p} \qquad (4.4.1)$$

where ARE is the apparent recovery efficiency of phosphorus or potassium as percentage (%); Up is the uptake of phosphorus or potassium of fertilizer treatment (mg P or K pot^{-1}); U0 is the uptake of phosphorus or potassium of unfertilized (control) treatment, no P or K fertilization (mg P or K pot^{-1}); and Fp is the application rate of fertilizer treatment (mg P or K pot^{-1}). ARE depends on the congruence between plant demand for P or K and the release of P or K from fertilizer. The phosphorus or potassium fertilizer replacement value (PFRV) of HPLA can be calculated if the differences in ARE between the HPLA and reference fertilizers are statistically significantly different, according to:

$$\mathrm{PFRV} = 100 * \mathrm{ARE}_{\text{Hydrated poulty litter ash}}/\mathrm{ARE}_{\text{Reference fertiliser}} \qquad (4.4.2)$$

where $\mathrm{ARE}_{\text{Hydrated poultry litter ash}}$ is the apparent recovery efficiency of phosphorus or potassium of HPLA (%) and $\mathrm{ARE}_{\text{Reference fertilizer}}$ is the apparent recovery efficiency of phosphorus or potassium of reference fertilizer dicalcium phosphate or triple superphosphate or potassium sulfate (%). By this definition, PFRV of a given reference fertilizer is 100%.

The effectivity of the acid-neutralizing value of HPLA followed a similar approach in its calculation. The effectivity was derived from the change in soil pH with 0.01 M $CaCl_2$ [10].

4.4.5 Phosphorus

Green bean had significant yields of dry matter (not shown) and phosphorus uptake only at the highest phosphorus application rate (Figure 4.4.2).

Triple superphosphate significantly increased the phosphorus uptake of leaves compared to other treatments (Figure 4.4.2). The phosphorus uptake of leaves of the dicalcium treatment was lower, but not significantly different from treatment with triple superphosphate.

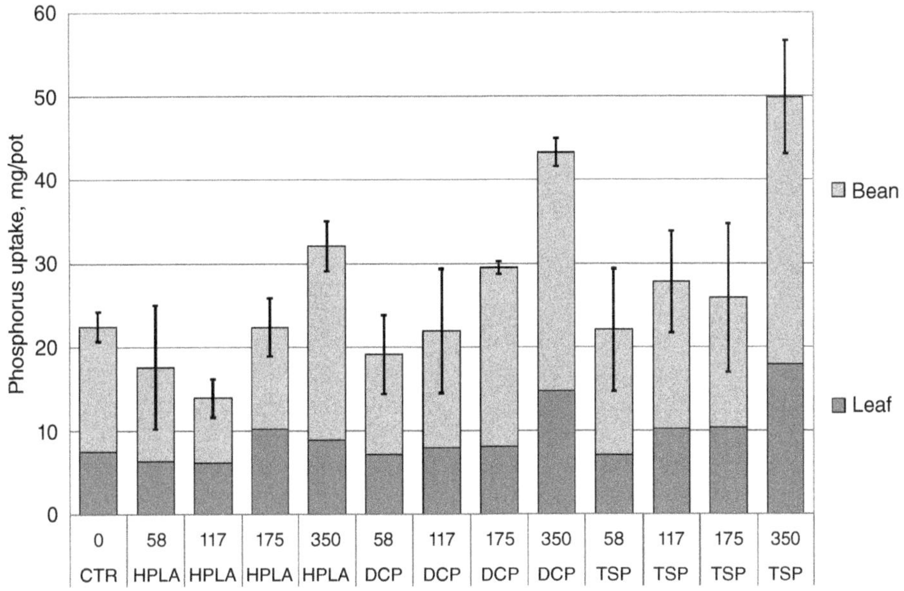

Figure 4.4.2 *Phosphorus uptake of leaves and beans of green bean (*Phaseolus vulgaris *L.) and their totals and standard errors of means of total phosphorus uptake in mg P/pot for control (CTR), hydrated poultry litter ash (HPLA), dicalcium phosphate (DCP), and triple superphosphate (TSP) and for application rates of 0, 58, 117, 175, and 350 kg P_2O_5 ha^{-1}.*

Table 4.4.3 *Apparent recovery efficiency (ARE) and phosphorus fertilizer replacement value (PFRV) in per cent for dicalcium phosphate, triple superphosphate, and hydrated poultry litter ash based on total phosphorus uptake by green bean (*Phaseolus vulgaris *L.) at the highest application rate of 350 kg P_2O_5 ha^{-1}. Different letters indicate a difference for p values >0.05 and <0.10.*

Parameter	Dicalcium phosphate	Triple superphosphate	Hydrated poultry litter ash
Apparent recovery efficiency, %	2.1(a)	2.7(a)	1.0(b)
Phosphorus fertilizer replacement value, %, dicalcium phosphate as reference	100(a)	134(a)	49(b)
Phosphorus fertilizer replacement value, %, triple superphosphate as reference	83(a)	100(a)	37(b)

Differences in phosphorus uptake between control, HPLA, and dicalcium phosphate were not significantly different.

Table 4.4.3 summarizes the average results for ARE and phosphorus fertilizer replacement values with dicalcium phosphate or triple superphosphate as reference fertilizer.

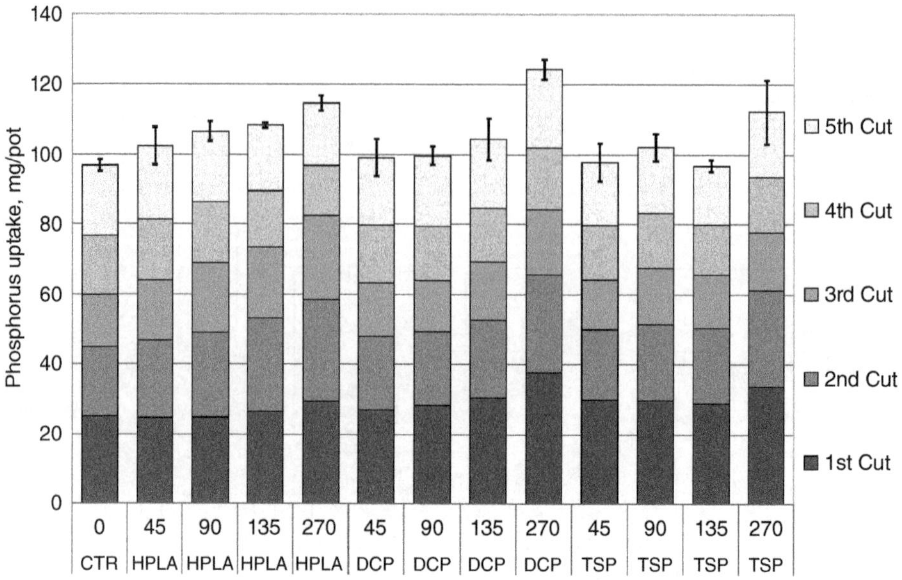

Figure 4.4.3 *Total uptake of phosphorus uptake per fertilizer treatment of rye grass (*Lolium perenne *L.) per cut for application rates 0, 45, 90, 135, and 270 kg P_2O_5 ha^{-1} with their standard errors of means for total phosphorus uptake.*

Values for ARE were low and ranged on average from 1.0 to 2.7% (Table 4.4.3). The ARE of HPLA tended to be lower than that of reference fertilizers. The PFRV for poultry litter ash with dicalcium phosphate as reference fertilizer was 46%, while that for litter ash with triple super phosphate as reference fertilizer was 37%. Dry matter yields of rye grass hardly benefited from phosphorus fertilization (not shown). Overall, fertilizer treatments did not differ from the control treatment. Phosphorus uptake per fertilizer treatment per application rate is given in Figure 4.4.3. HPLA had comparable uptakes to the reference fertilizers dicalcium phosphate and triple superphosphate.

Overall, total phosphorus uptake from HPLA was not different to that from reference fertilizers. HPLA and dicalcium phosphate had significantly higher phosphorus uptake compared to the control treatment. Triple superphosphate had a higher phosphorus uptake, but the difference was not significant.

ARE was 4.4% for dicalcium phosphate, 2.6% for triple superphosphate, and 7.0% for HPLA. There was no statistically significant effect of fertilizers or phosphorus application rates on ARE. The results point to a similar efficacy of phosphorus of HPLA and of dicalcium phosphate and triple superphosphate.

4.4.6 Potassium

The K uptake by green bean responded significantly on potassium fertilization. Increasing potassium application increased potassium uptake (Figure 4.4.4). The ARE of HPLA per application rate was equal to the ARE of potassium sulfate. AREs decreased with increasing

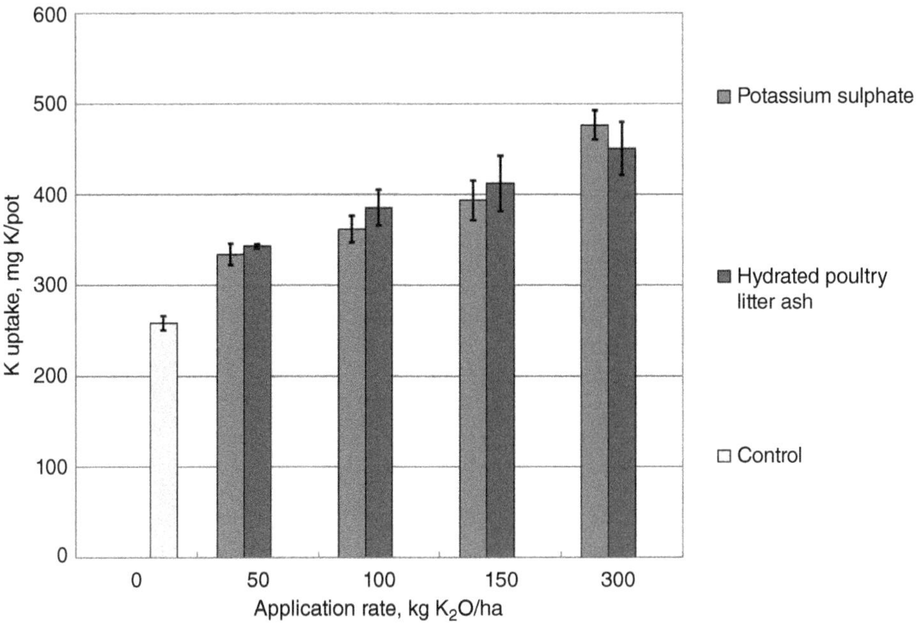

Figure 4.4.4 *Total uptake of potassium (K uptake) per fertilizer treatment of green bean* (Phaseolus vulgaris *L.) for application rates 0, 50, 100, 150, and 300 kg K_2O ha^{-1} with their standard errors of means.*

Table 4.4.4 *Apparent recovery efficiency (ARE) of potassium sulfate and hydrated poultry litter ash per application rate. Different letters given after the values as suffix points indicate significant differences between treatments ($\alpha = 0.05$).*

Application rate, kg K_2O ha^{-1}	Potassium sulfate	Hydrated poultry litter ash
50	58a	65a
100	40b	49b
150	35bc	39bc
300	28c	25c

Tested using *Phaseolus vulgaris* L.

application rate, clearly indicating a less efficient use of potassium at higher rates. Potassium fertilizer replacement values (KFRVs) of HPLA were derived from ARE. Table 4.4.4 shows that there are no differences between HPLA and the reference mineral fertilizer potassium sulfate. Thus, the KFRV is 100%.

4.4.7 Rye Grass

The test crop rye grass responded on potassium fertilization (Figure 4.4.5). The higher the application rate, the higher the yield of dry matter. During the growing season and on

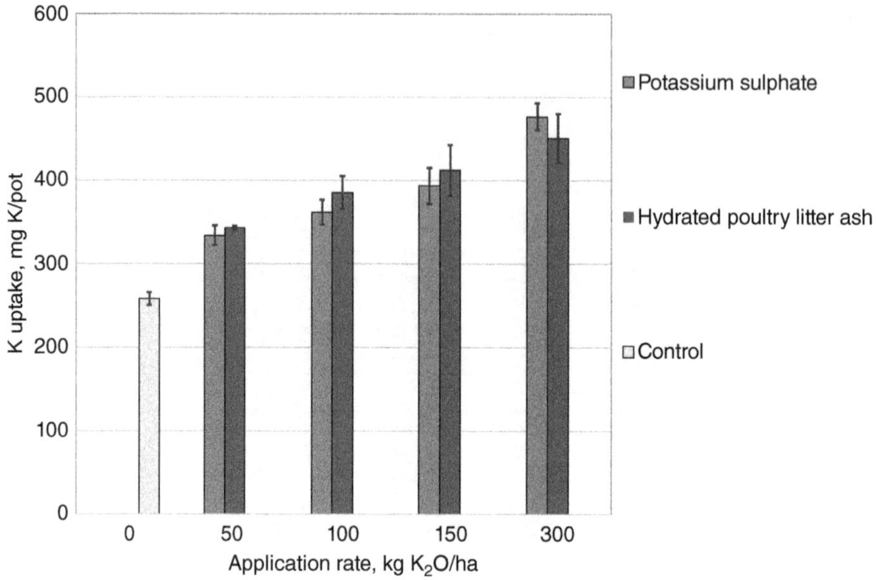

Figure 4.4.5 *Total uptake of potassium (K uptake) per fertilizer treatment of rye grass (*Lolium perenne *L.) per cut for application rates 0, 100, 200, 300, and 600 kg $K_2O\,ha^{-1}$ with their standard errors of means for total potassium uptake.*

Table 4.4.5 *Apparent recovery efficiency (ARE) of potassium sulfate and hydrated poultry litter ash per application rate, expressed as per cent. Different letters given after the values indicate differences between treatments ($\alpha = 0.05$).*

Application rate, kg $K_2O\,ha^{-1}$	Potassium sulfate	Hydrated poultry litter ash
100	97a	100a
200	99a	97a
300	96a	95a
600	92b	90b

Tested using *Lolium perenne* L.

subsequent clipping of the grass, differences between low and high application rates became larger. The K uptake of the grass increased with increasing application rate but decreased on clipping (Figure 4.4.5).

ARE and PFRV values are shown in Tables 4.4.4 and 4.4.5.

ARE of HPLA per application rate was equal to that of potassium sulfate. AREs decreased with increasing application rates, clearly indicating a less efficient use of potassium at higher application rates. This difference was significant for the highest application rate.

KFRVs of HPLA were derived from ARE. As the AREs for HPLA and potassium sulfate are similar, the KFRV of HPLA is 100%.

4.4.8 Acid-Neutralizing Value

HPLA contains some calcium hydroxide and calcium carbonate, and thus has an acid-neutralizing value. However, the use of HPLA within good agricultural practices is determined by P and K, and this use means that the actual application rate of acid-neutralizing value is modest. Higher application rates of acid-neutralizing value, required to repair a too-low pH, would lead to excessive applications of P and K, which does not fit with good agricultural practices. HPLA was tested as a soil amendment to maintain soil pH. For this purpose, relatively modest application rates of acid-neutralizing value are needed [6]. Figure 4.4.6 summarizes the results of the development of soil pH over 0–60 weeks of incubation at 15 °C. Due to the low rates, the increase of soil pH was low.

4.4.9 Efficacy

A higher rate of HPLA led to a greater increase in pH.

The average relative liming effects of HPLA compared with chalk and hydrated burned lime were 44 and 47%, respectively (Table 4.4.6). This is due to the relatively coarse particles of HPLA compared with the reference liming materials. The effect of grinding HPLA on the relative liming effect was tested [4]. If HPLA was ground to 100% < 0.50 mm and 90% < 0.15 mm, the liming replacement values equaled 100%.

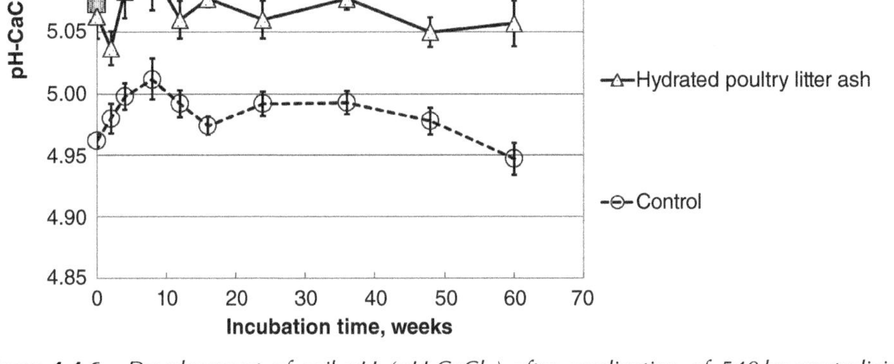

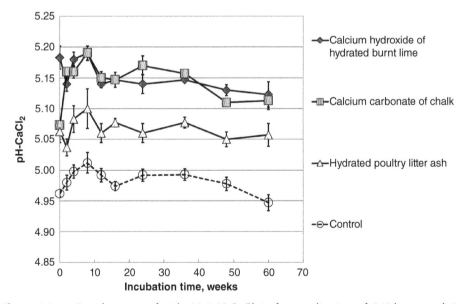

Figure 4.4.6 *Development of soil pH (pH-CaCl$_2$) after application of 540 kg neutralizing value ha^{-1} with hydrated burned lime, chalk, and hydrated poultry litter ash compared with the control treatment, plus their standard errors of means per time step.*

Table 4.4.6 Liming replacement values of hydrated poultry litter ash for an application rate of 540 kg NV ha⁻ᵃ, expressed as per cent, compared with the reference liming products chalk and hydrated burned lime. Different letters given after the values indicate significant differences between them[a] ($\alpha = 0.05$).

Reference liming product	Acid-neutralizing value, kg ha^{-1}	Chalk	Hydrated burned lime	Hydrated poultry litter ash
Chalk	540	100a	94a	44b
Hydrated burned lime	540	109a	100a	47b

[a] Results of T-test: If values have different letters, then they are significantly different (P-value 0.05). If they have the same letter, they are not significantly different.

4.4.10 Conclusion

Incineration of poultry manure prevents emissions from fossil fuel combustion, resulting in a reduced environmental impact in the impact category "climate change" [11, 12]. Electricity production from manure outperforms land spreading of manure in the impact categories "terrestrial acidification," "particulate matter formation," "marine eutrophication," and "photochemical oxidant formation" [11, 12]. The ash is recovered as a PK fertilizer, which is odorless, dry, and sterile, and has a lower mass and volume than poultry manure, making it more suitable for export to regions with a high P demand [11].

HPLA contains the same contaminants as poultry manure. Incineration does not lead to the formation of organic contaminants. As organic matter (OM) is incinerated, the concentrations of heavy metals are higher than those in the feedstock manure, but their ratios remain unchanged. Thus, the use of HPLA as a PK fertilizer will lead to similar loads of heavy metals as caused by animal manure. Animal manure contains low concentrations of heavy metals due to the low concentrations in animal feed and fodder. As copper (Cu) and zinc (Zn) are added to animal feed, animal manure has elevated levels of these elements, and thus so too does HPLA. Use of HPLA as P and K fertilizers will lead to similar Cu and Zn loads as with animal manure applied to good agricultural standards. Cadmium loads with HPLA are much lower compared with regular mineral phosphorus fertilizers (~60 mg Cd kg^{-1} P_2O_5) as the Cd/P_2O_5 ratio is 7 mg Cd kg^{-1} P_2O_5.

A life cycle assessment (LCA) study showed that though nitrogen and OM are lost, it is still environmentally more beneficial to process poultry litter in the BMC plant by means of thermal conversion than to use other poultry manure processing routes [13]. Other routes examined in the study were co-digestion at a digestion plant in Germany, which generated electricity and marketed the digestate as a fertilizer substitute; co-firing in a wood-fired biomass power plant; direct application of raw manure in the Netherlands and Germany; and composting with or without granulation for use as fertilizer or for mushroom-growing [13]. If thermal conversion on poultry farms can be combined with effective use of most of the heat produced, this route scores better than the conversion on plant scale of BMC Moerdijk [13].

The loss of nitrogen and OM by incineration is not in line with resource conservation, but these aspects require optimization of an (environmental) cost–benefit analysis given that the Netherlands is confronted with a surplus of manure. Loss of nitrogen can be compensated by synthetic nitrogen production. The energy required for this step would equal approximately 20% of the energy BMC Moerdijk produces from incineration of poultry litter (BMC Moerdijk, pers. comm.). OM is a renewable resource, and energy production is based on the use of renewable resources. Loss of nitrogen has to be balanced against other uses of nitrogen from other renewable resources. Recovery of mineral nitrogen from poultry manure before incineration can increase the sustainability of the incineration process. This requires an additional cost–benefit analysis.

Though nitrogen and OM are lost, the valuable components for fertilization and maintenance of soil fertility can be obtained from other sources. Phosphate and potassium are finite resources and therefore important from the perspective of circular economy and resource conservation.

The results of the experiments with green beans show that in the short term the fertilizer replacement value of HPLA is about half that of regular mineral phosphate fertilizers. In the longer term, the value of HPLA equals that of these mineral phosphate fertilizers.

Potassium of HPLA has an equal effectivity to potassium sulfate in both the short and the longer term.

The liming replacement value of HPLA is about half that of regular liming materials but can be increased by grinding to finer particles.

This technical study shows that the phosphorus and potassium of HPLA have agricultural value as a PK fertilizer. The acid-neutralizing value adds to the agronomic value. The main agronomic function of the use of HPLA is in maintenance of soil fertility.

References

1. Ehlert, P.A.I. and Nelemans, J.A. (2015). *Efficacy of Phosphorus of Hydrated Poultry Litter Ash; Phosphorus Use Efficiency of Rye Grass*. Wageningen: Alterra Wageningen UR.
2. Ehlert, P.A.I. and Nelemans, J.A. (2015). *Efficacy of Potassium of Hydrated Poultry Litter Ash; Potassium Use Efficiency of Green Bean*. Wageningen: Alterra Wageningen UR.
3. Ehlert, P.A.I. and Nelemans, J.A. (2015). *Efficacy of Potassium of Hydrated Poultry Litter Ash; Potassium Use Efficiency of Ryegrass*. Wageningen: Alterra Wageningen UR.
4. Ehlert, P.A.I. and Nelemans, J.A. (2015). *Efficacy of Hydrated Poultry Litter Ash as Liming Material*. Wageningen: Alterra Wageningen UR.
5. Ehlert, P.A.I. and Nelemans, J.A. (2015). *Efficacy of Phosphorus of Hydrated Poultry Litter Ash; Phosphorus Use Efficiency of Green Bean*. Wageningen: Alterra Wageningen UR.
6. De Haan J., van Geel W. (2014) Handboek Bodem en Bemesting. Available from: http://www.handboekbodemenbemesting.nl/nl/handboekbodemenbemesting.htm (accessed December 23, 2019).
7. Commissie Bemesting Grasland en Voedergewassen. (2017) Bemestingsadvies (laatste wijziging: augustus 2017). Available from: https://www.bemestingsadvies.nl/nl/bemestingsadvies.htm (accessed December 23, 2019).
8. European Comission. (2003) Regulation No 2003/2003 of the European Parliament and of the Council of 13 October 2003 Relating to Fertilisers. Availabel from: https://eur-lex.europa.eu/legal-content/EN/TXT/?uri=celex:32003R2003 (accessed December 23, 2019).
9. Dobermann, A. (2007). Nutrient use efficiency – measurement and management. In: *Fertiliser Best Management Practices General Principles, Strategy for their Adoption and Voluntary Initiatives vs Regulations* (eds. A. Krauss, K. Isherwood and P. Heffer), 1–28. Paris: International Fertiliser Industry Association.
10. Houba, V.J.G., Temminghoff, E.J.M., and Gaikhorst, G.A. (2000). Soil analysis procedures using 0.01 M calcium chloride as extraction reagent. *Communications in Soil Science and Plant Analysis* **31** (9–10): 1299–1396.

11. Billen, P., Costa, J., Van der Aa, L. et al. (2014). Electricity from poultry manure: a cleaner alternative to direct land application. *Journal of Cleaner Production* **96**: 467–475.
12. Williams, A.G., Leinonen, I., and Kyriazakis, I. (2015). Environmental benefits of using Turkey litter as a fuel instead of a fertiliser. *Journal of Cleaner Production* **113**.
13. De Graaff L., Odegard I., Nusselder S. (2017) In Dutch: LCA thermische conversie pluimveemest BMC Moerdijk. Delft: CE Delft. Report No.: 17.2H94.01. Available from: http://www.ce.nl/publicatie/lca_thermische_conversie_pluimveemest_bmc_moerdijk/1933 (accessed December 23, 2019).

4.5

Bioregenerative Nutrient Recovery from Human Urine: Closing the Loop in Turning Waste into Wealth

Jayanta Kumar Biswas[1,2], Sukanta Rana[2], and Erik Meers[3]

[1]*Department of Ecological Studies, University of Kalyani, Nadia, West Bengal, India*

[2]*International Centre for Ecological Engineering, University of Kalyani, Kalyani, West Bengal, India*

[3]*Department of Applied Analytical and Physical Chemistry, Faculty of Bioscience Engineering, Ghent University, Ghent, Belgium*

4.5.1 Introduction

In recent years, there has been growing interest in promoting innovative approaches to recovering resources from wastewater, including human urine (HU) [1–3]. HU is a concentrated biogenic solution full of nutrients and ions, and hence should be considered not a waste but a valuable resource. However, due to lack of knowledge and superstition, it has traditionally been treated as a foul material – a waste demanding to be discarded and disposed. To usher in a paradigm shift toward a recycling and regenerative society, source separation of HU coupled with resource recovery could be seen as a potential solution to the prevailing "linear flow" sanitation systems [4]. Ecological sanitation sees HU as a nutrient rich renewable resource, and its management should follow an integrated and holistic approach that dovetails water resources, nutrient flow, and sanitation, synthesizing the concepts of life-cycle thinking, systems theory, circular economy, industrial ecology, and biomimicry [5]. Ecological sanitation has been advocated as a safe, ecofriendly, and sustainable approach to promoting closed-loop flows of nutrients and resources from

---

Biorefinery of Inorganics: Recovering Mineral Nutrients from Biomass and Organic Waste, First Edition.
Edited by Erik Meers, Gerard Velthof, Evi Michels and René Rietra.
© 2020 John Wiley & Sons Ltd. Published 2020 by John Wiley & Sons Ltd.

sanitation to agri(aqua)culture. It seeks to challenge the conception and dialectics of two human constructs, "resources" and "wastes," by asserting that HU is indeed a resource of a natural cycle that circulates biological nutrients and can feed another production system (the agri/aquaculture system). This new paradigm adopts an ecosystems approach aimed at moving from a linear to a circular flow of resources in order to create a closed-loop fertility (nutrient) cycle. It is designed to mimic ecosystems where human waste is a resource or feed for other organisms. The bottom line is that excreta from water brings major benefit to water quality, while excreta applied in water or on to land can provide huge service to food security. To mitigate the reliance on chemical fertilizers, and to build a sustainable society based on sound resource recycling and low-carbon practices, it is necessary to build a bridge between agriculture/aquaculture and sanitation: a holistic strategy for waste management, agricultural production, and food security. The objective of this chapter is to present a state-of-the-art review of nutrient recovery and recycling from HU in bioregenerative production practices, along with associated environmental constraints and concerns.

4.5.2 Composition and Fertilizer Potential

HU is a major source of the nutrients present in domestic wastewater, contributing approximately 80–90%, 50–65%, and 50–80% of the N, P, and K, respectively, despite only making up 0.4–1.0% of the volume [6–11]. It contains essential nutrients needed for plant growth, making it an appealing fertilizer alternative (Table 4.5.1; [16–19]). In developing countries, it could be a low-cost, easily available, and safe alternative to fertilizer [20, 21]. It releases plant nutrients faster than sources such as livestock/avian excreta, green manure and compost [22]. It has an N : P : K ratio of 18 : 2 : 5, which may change to 15 : 1 : 3 when mixed with flushing water [23]. It is composed of eight main ionic species: Na^+, K^+, NH_4^+, Ca^{2+}, Cl^-, SO_4^{2-}, PO_4^{3-}, and HCO_3^- (Table 4.5.1), with nitrogen dominating (90%), mainly in the forms of ammonical-N and ammonium bicarbonate [12]. Amino acids, glucose, vitamins, and hormones are excreted via urine [24]. Due to differences in food habits among people inhabiting different regions of the world, fresh HU contains 80–90% N, 50–60% P, and 50–80% K [12, 23–26]. With a per capita per annum urine excretion rate of 500 l [27], each individual annually contributes about 4.6 kg of P and 1.1 kg of K, along with some growth-promoting substances [28]. According to Jaatinen [29], each person produces approximately 1.2 l of urine per day, containing 11.5 gN/p/d and 1.2 gP/p/d. In fresh HU, about 85% of nitrogen remains as urea-N and about 5% as ammonia-N [13]. Urea mineralizing bacteria transform urea-N into ammonium-N, which is further converted by nitrifying bacteria into nitrate (NO_3^-); both NH_4^+ and NO_3^- are readily bioavailable nitrogen (N) forms for phytoplanktons [27].

4.5.3 State of the Art of Regenerative Practices

4.5.3.1 HU in Agriculture

The possibility of using HU in agriculture as an alternative fertilizer source is not a novel concept; since ancient times, it has regularly been used in some way in some parts of the world to recharge soil with nutrients [27, 30–32]. Due to a dearth of organic fertilizers and the detrimental environmental impact of chemical fertilizers, HU is again drawing attention

Table 4.5.1 *Physicochemical profile of human urine (HU) of diverse sources and dilutions.*

Parameter	Household [12]	School [12]	Workplace [13]	Workplace [14]	Household [15]	Workplace [13]	Fresh urine [13]	Fresh urine [14]
Dilution	0.33	0.33	0.26		0.75	1.0	1.0	1.0
pH	9.0	8.9	9.0	9.0	9.1	9.1	6.2	5.6
N_{tot} ($g\,m^{-3}$)	1795	2610	1793	—	3631	9200	8830	
$NH_4^+ + NH_3$ –N ($g\,m^{-3}$)	1691	2499	1720	4347	3576	8100	463	438
$NO_3 + NO_2$ –N ($g\,m^{-3}$)	0.06	0.07	—	—	<0.1	0	—	
P_{tot} ($g\,m^{-3}$)	210	200	76	154	313	540	800–2000	388
COD ($g\,m^{-3}$)	—	—	1650	6000	—	10 000	—	7660
K ($g\,m^{-3}$)	875	1150	770	3284	1000	2200	2737	1870
S ($g\,m^{-3}$)	225	175	98	273	331	505	1315	
Na ($g\,m^{-3}$)	982	938	837	1495	1210	2600	3450	3240
Cl ($g\,m^{-3}$)	2500	2235	1400	2112	1768	3800	4970	6620
Ca ($g\,m^{-3}$)	15.75	13.34	28	—	18	0	233	
Mg ($g\,m^{-3}$)	1.63	1.5	1.0	—	11.1	0	119	
Mn ($g\,m^{-3}$)	0	0	—	—	0.037	—	0.019	
B ($g\,m^{-3}$)	0.435	0.44	—	—	—	—	0.97	

as a fertilizer substitute to be used in supplementing agricultural productivity in developing countries, particularly in Asia and Africa [30, 33–35]. It has been used to fertilize edible plants (Table 4.5.2) with no considerable changes in taste and equivalent yields compared to artificial fertilizers [36, 37, 42]. One of the advantages of the reuse of HU in agriculture is the virtual absence or insignificant load of the hazardous chemical compounds and heavy metal(loid)s (e.g. Cd, As, Pb, Cr, Hg, Ni, and V) that may occur in high concentrations in chemical phosphate fertilizers [24, 49].

The use of HU as a source of plant nutrients has already been attempted in several countries for a variety of crop types (Table 4.5.2). The results show that crops cultivated with HU register an appreciable yield, with a nutritional value and taste comparable to crops grown in normal soil. Given proper sanitization, HU does not pose much hygienic threat or leave any distinct flavor in food products [24, 38, 42]. Another avenue is to treat HU with the magnesium (Mg^{2+}) leading to synthesis of struvite ($NH_4MgPO_4{\cdot}6H_2O$) [40], a slow-release phosphatic fertilizer [7, 14, 43].

Since HU contains valuable plant nutrients, research has been conducted to assess the comparative efficacy of urine relative to commercial fertilizers using diverse types of vegetables and cereals such as wheat [27], maize [50, 51], spinach [52], cabbage [37, 52, 53], cucumber [36], carrot [31], beetroot [31], pumpkin [38], tomato [54], red beet [37], cauliflower [55], broadleaf mustard [55], potato [55], and radish [55]. While Heinonen-Tanski and Wijk-Sijbesma [6] applied fresh HU as fertilizer in the cultivation of vegetables, stored urine was applied at 145 kg N ha^{-1} for wheat cultivation in the cold climate of Sweden, with a yield of 5650 kg ha^{-1} [42]. An NGO (ACTS, Bangalore) has obtained appreciable growth of banana, chilli, pappya, and brinjal using 10% solution of HU [41].

Application of urine has resulted in plant growth characteristics of field-grown cabbage (*Brassica oleracea*) similar to those treated with mineral fertilizers [52], and even better plant height and leaf length in maize (*Zea mays*) [50]. Likewise, tests on tomatoes (*Solanum lycopersicum*) [38, 40], red beet (*Beta vulgaris*) [37], pumpkin (*Cucurbita maxima*) [54], maize (*Zea mays*) [31, 43], and okra crops (*Abelmoschus esculentus*) [22] showed similar results, with comparable plant biomass and yield to those found with chemical fertilizers, and greater than is found with an unfertilized control. Pradhan et al. [52] reported lower pest attack on urine-fertilized cabbage than on controls (cabbage grown without urine or mineral fertilization), with no deterioration in flavor or taste. Studies show that the annual amount of HU produced by one person is equivalent to the fertilizer value of 250 kg of cereal; that is, the annual consumptive need of an individual [24]. Agri/horticultural crops fertilized with HU show no considerable changes in taste and quality, but offer equipotence in terms of yield compared to artificial fertilizers [22, 36–38, 42, 49].

4.5.3.2 HU in Aquaculture

Use of urine in aquaculture, such as in hydroponic cultivation of plants or growing zoo- and phytoplankton, has been attempted as a potential source of renewable nutrient resource (Table 4.5.2). HU can be used in culture of microalgae, zoo- and phytoplankton, tomatoes [17, 40, 44], water spinach [19], and fish and prawns [20]. In most cases, cultivation in diluted urine leads to similar or higher yields compared to various nutrient solutions and sources (cow urine, vermi-compost, etc.), depending upon the rate of application. Rana

Table 4.5.2 *Human urine (HU) tested as a fertilizer in agri(aqua)cultural studies.*

Conditions	Applications	Results	References
		Agriculture	
Fresh urine	Cultivation of high N-demand vegetables – cabbage, barley, and cucumber	High yield, quick maturation, palatability improvement, and bag damage reduction	Heinonen-Tanski and Wijk-Sijbesma [6]
Outdoor cultivation with 233 $kgN\,ha^{-1}$ HU and 34 $kgN\,ha^{-1}$ mineral fertilizer	Cucumber	Yield with HU similar to or slightly better than yield with mineral fertilizer	Heinonen-Tanski et al. [36]
Outdoor cultivation with HU (133 $kgN\,ha^{-1}$) only and urine plus wood ash	Red beet	Urine-plus-wood-ash and urine-only fertilizer produced 1720 and 556 $kg\,ha^{-1}$ more biomass than chemical fertilizer	Pradhan et al. [37]
Greenhouse trial with HU (136 $kgN\,ha^{-1}$)	Tomato	Equally effective to chemical fertilizer in terms of yield, and 4.2 times more yield compared to unfertilized counterparts	Pradhan et al. [38]
Fresh urine	Leek vegetable	Excellent growth	Båth [39]
Fresh urine and urine feces blend	Maize	Cob growth comparatively good in urine	Vinneras et al. [23]
Fresh urine	Tomato cultivation in green house	1.3 kg tomato from eight plants in 114 d	Adamsson [40]
Cultivation with 10% fresh HU (10 times dilution)	Banana, chili, brinjal, and pappya	Average weight of a single banana as much as 500 g	ACTS Ministry, Bangalore, India (UNESCO and IHP[41])
Stored urine as per 145 kg of N ha^{-1}	Wheat cultivation	Yield 5650 $kg\,ha^{-1}$	Tidaker et al. [42]
Greenhouse pot experiment with HU-derived struvite (0.17$gN\,kg^{-1}$)	Maize and ryegrass	Struvite induced similar to or significantly higher biomass production as mineral fertilizer	Antonini et al. [43]

Table 4.5.2 *(continued)*

Conditions	Applications	Results	References
		Aquaculture	
Hydroponic culture 2% solution of fresh HU, 0.5, 1, and 2% of fresh urine	Culture of microalgae (*Scenedesmus acuminatus*), zooplankton (*Daphnia magna*), and tomato	Excellent growth within 3 d 100% mortality in 2% 50% mortality in 1% No mortality in 0.5% Good growth, reproduction, and appearance of first neonate after 8 d	Adamsson [40]
	Culture of zooplankton		
Culture with 0.01% solution of fresh urine	Culture of zooplankton (*Moina micrura*)	Population size as much as 1256 $10\,l^{-1}$ after 3 mo; early maturity of neonates; reproduction of neonates 4 d earlier	Golder et al. [17]
Culture in small holding tanks with HU compared to cattle manure	Indian carp and freshwater prawn	Yield highest in cattle manure CM (621.5 g $tank^{-1}$), followed by mixed treatments under iso-nitrogenous (428 g $tank^{-1}$) and iso-phosphorus (333 g $tank^{-1}$) conditions, aerated HU (321 g $tank^{-1}$), and HU (319 g $tank^{-1}$)	Rana et al. [21]
Culture in 5000 l tanks with 0.02% urine	Phytoplankton	Primary production highest with stored urine	Jana et al. [44]

Supernatants of stored HU after precipitation of struvite	Culture of *Daphnia magna*	50% mortality after 24 h	Ganrot et al. [1]
Hydroponic cultivation with 2, 3.3, 5, and 10% urine	Water spinach (*Ipomoea aquatica*)	2% urine produced comparable growth characteristics to nutrient solution	Yang et al. [19]
Culture with nitrified stored HU	*Spirulina platensis*	Excellent biomass increase	Feng et al. [45]
Batch cultivation in photobioreactor, bubble column with HU	*Spirulina platensis*	Significant biomass increase	Yang et al. [46]
On microtiter plate in photobioreactor with HU	*Chlorella sorokiniana*	Successful yield	Zhang et al. [47], Tuantet et al. [48]

et al. [21] demonstrated that naturally nutrient-rich HU is a resource that can be reclaimed and recycled as a sanitized multinutrient liquid fertilizer and conditioner in carp and prawn aquaculture as an alternative to cattle manure or chemical fertilizer or as a blend with cattle manure. HU has been effectively used as a liquid fertilizer in plankton culture (Table 4.5.2). Golder et al. [17] made an attempt to assess the nutrient potential of HU for the culture of zooplankton (*Daphnia magna*). Not only can HU be used for struvite synthesis, but its supernatant, which is free from phosphorus, can be used for zooplankton production [1].

Recently, use of HU has been attempted (Table 4.5.2) in the recovery of nutrients in micro-algaculture [56], though large-scale cultivation is not yet economically feasible [47]. Adamsson [40] observed the algal biomass of *Scenedesmus acuminatus* was initially (0–3 days) higher in HU compared to artificial culture medium, but after 10 days it was higher in the latter. Using nitrified stored urine, Feng et al. [45] observed excellent biomass increase of blue green algae (*Spirulina platensis*). After harvesting, microalgal biomass can be utilized as a potential feedstock for the biobased production of chemicals and biofuels as a new energy source in the production of methane in anaerobic digestion [57]. It can also be used in biodiesel production by extracting algal lipids [58, 59], and as a nutritional supplement, natural pigment, or aquaculture feed [60]. Different factors that influence the HU-based algal cultivation system include dilution factor, light intensity, cultivation strategies (batch vs. continuous, open pond vs. closed photobioreactor, etc.), climate conditions, infrastructure and logistics, supplementary supply of carbon dioxide, algal harvesting, and biomass post-processing [58, 61–63]. Tuantet et al. [47] cultivated *Chlorella sorokiniana* in microtiter plates of HU as well as synthetic urine. The non-diluted HU contained an abundance of required elements for microalgal cultivation, namely N, P, K, Mg, and Ca. While *Scenedesmus* sp. [40] and *Spirulina* sp. [45] were found to be incapable of growing in concentrated urine, *C. sorokiniana* was reported to grow well in microtiter plates with non-diluted fresh urine, with a highest growth rate of $0.104\,h^{-1}$ [47]. Ammonia is considered as a growth inhibitor, reducing photosynthetic activity by 50% at 0.675 mM of NH_3 and inflicting toxic effects on microalgae at higher concentrations [64]. Source-separated, hydrolyzed HU diluted to 3.5% has been compared to other bioenergy feedstocks for algal cultivation [65], but hydrolyzed HU needs excessive dilution, making it more water-intensive compared to wastewater-based algaculture. HU contains a high concentration of urea, which could hydrolyze rapidly, especially under aerobic conditions, to form ammonium. Very high ammonia concentrations resulted from the hydrolysis of urea. The pH increase during hydrolysis of urea is also high, causing the production of ammonia. Both can cause algal growth inhibition in test cultures, especially those with higher urine concentrations [48]. The immediate occurrence of urea hydrolysis and self-precipitation inherent in concentrated urine are obstacles for sustained microalgal growth, resulting in little nutrient recovery as microalgal biomass. Nitrification of ammonium for stabilization can overcome this obstacle, but it requires significant pH adjustments and maintenance of elevated levels of dissolved oxygen [45].

4.5.4 Cautions, Concerns, and Constraints

The prevention of ammonia loss during storage and soil application of HU is condicio sine qua non for efficient fertilizer use [12, 66]. The application technique has to be adjusted to the high ammonia content of HU [67]. The rate of ammonia volatilization can be minimized when the nutrients present in the HU are recovered as struvite – a slow-release fertilizer [6, 66]. Soil ureases rapidly hydrolyze urea, leading to ammonia emissions when urine is applied into the soil [66, 68–70]. In order to prevent this, the use of urease inhibitors is recommended [66]. The most promising urease inhibitor in various laboratory and field studies is N-(n-butyl) thiophosphoric triamide (nBPT), a non-toxic, stable, and inexpensive structural analog of urea that is effective in low concentrations [66, 69, 70]. Granular urea amended with nBPT has reduced losses under laboratory conditions from 11 to 1.9% of urea-N applied, and no decline in the efficiency of nBPT has been detected following repeated applications to the same soil over a span of 3 years [71]. Injection or incorporation of urea is very effective in decreasing ammonia emission.

Though HU has been successfully compared with commercial chemical fertilizers [31, 36–38, 53], it contains some salts [12, 31, 52], pathogens [53], and pharmaceuticals [11, 67] that may pose some hygenic threat. Excessive use of HU in agricultural fields as a fertilizer may lead to increased soil salinity and accumulation of sodium (Na) and nitrogen (N) in the soil [72]. The soil concentration of built-up Na may eventually exceed plants' demand and cause deleterious effects. Excess Na inhibits plant growth, which can be explained by the fact that, subject to salt stress conditions, excess Na may limit root growth, disperse soil particles as a result of breakdown of soil aggregates, alter water uptake capacity in the root zone, and causes ion-specific toxicities and interference among competitive nutrients [73, 74]. Mnkeni et al. [52] reported depressed growth of spinach and cabbage in a single booster dose of HU as a result of increased soil salinity.

On the other hand, excess nitrogen can adversely affect the amount of sugar and vitamins in vegetables and build up in plant tissues, which may have significant health and taste implications for consumers [31, 75]. Inappropriate urine application has been reported to inhibit growth of cabbage, spinach, and carrot, accompanied by increased soil electrical conductivity (EC) [31, 52]. The time and frequency of HU application in agriculture need to be adequately addressed for sustainable agricultural production. Jönsson et al. [27] proposed that HU be applied to a crop field once, prior to or at the time of sowing/planting. Furthermore, Kirchmann and Pettersson [12] reported increased N loss from soil under intensive use of HU. Single dosing of urine in undiluted form (100%) before seeding is not recommended, and frequent application might be the better option given nitrogen leaching, which may pose a problem in case of N-excess accompanied by sufficient precipitation. Nishihara et al. [76] showed retarded growth of spinach due to the combined effects of its sensitivity to acidic ambience and high P demand, which could not be met by HU [36, 38]. In agriculture, this deficiency can be overcome by supplementing with ash [36] or compost [77].

To get maximum fertilizer effect, raw HU should be applied into the soil afresh [23]. This can be done by applying it in 1–4 cm-deep furrows, which are then covered after application, or by applying water after HU input, thereby washing the nutrients into the soil. When applying urine, spraying of leaves should be avoided as this can cause foliar burning due to high concentration of salts if it is left to dry on them. The whole root of the plant should not be thoroughly soaked with raw undiluted urine, as this might be toxic or even lethal, especially for small plants. Instead, the urine should be applied either prior to sowing or implanting or at such a distance from the plants that the nutrients are within reach of the root but not soaked. It is effective to apply HU in the evening when evaporation is lower [6].

HU may contain microorganisms, pharmaceuticals, and hormones. Despite its immense potential as a fertilizer, some developed nations have imposed certain restrictions on its use in aquaculture and agriculture due to apprehension over disease transmission. In healthy individuals, urine is sterile in the bladder [78], and only a few pathogens are normally excreted with it [6, 36]. But urine from unhealthy humans may contain pathogenic organisms [79]. Furthermore, fecal contamination can result in high counts of fecal indicator organisms [78]. When fertilizing with HU, transmission of infectious diseases by pathogenic microorganisms is the major concern [10]. The microbes and parasites that may be emitted in HU consist of pathogens (*Schistosoma haematobium*, *Salmonella typhi*, *Salmonella paratyphi*, *Mycobacterium tuberculosis*, and *Leptospira interrogans*), enteric microorganisms, viruses (e.g. Cytomegalovirus), protozoa (e.g. Microsporidia), and helminth and parasitic eggs (e.g. *S. haematobium*) [6, 10, 78]. It is wise to address their fate and removal during urine storage and reuse of HU in order to make it environmentally safe. Therefore, separating urine from other wastewaters can help reduce the microbial load. The recommended large-scale sanitation treatment for HU is storage in closed tanks [7]. It is well known that many microorganisms die off during the hygienization (simple storage) of HU. Long-term (>3 months) storage of urine may not be necessary in tropical developing countries, even though storage for 6 months is recommended in the Nordic countries [27], because the heat (>20 °C), UV radiation, and increase in pH to 9 caused by urea hydrolysis may be beneficial for the inactivation of pathogenic microorganisms [36, 78]. Based on some studies and risk-assessment guidelines [80], storage of collected urine at 20 °C for 6 months is recommended for a satisfactory reduction in load of pathogenic bacteria [81], some viral agents [82], and *Cryptosporidium parvum* [83]. In urine, an ammonia concentration of 40 mM, which can be found during storage, has been recommended as an inactivation threshold for pathogenic microorganisms [10]. One main issue with the sterilization of urine before use in a hydroponic system is the required storage time of 4–5 weeks. The conversion of urea to ammonia and its volatilization requires a period of 4–5 weeks and results in an increase of pH to ≥9, required for sterilization of HU [53]. Though storage has been recommended for using HU as fertilizer in agriculture, it has not been demonstrated as totally safe and pathogen-free. Arnbjerg-Nielsen et al. [84] reported that 12 months' storage of human excreta was not sufficient to stop the risk of transmission of infectious disease. Though most are killed in stored urine, certain pathogens like the bacteria *Clostridium* spp. and the ova of the intestinal round worm *Ascaris suum* can survive with high population size after several months [78]. Streptococci are also found in stored urine at high concentration [81]. The survival of eggs and miracidia of helminth (*Schstosoma* sp.) is reduced in HU under the influence of high salt concentration and pH [6].

The heavy metal content in urine is generally low compared to commercially available fertilizers and manures [12, 15], remaining in a ppb range that may vary depending on the storage time and possible metal precipitation [10, 21]. Human pharmaceuticals and hormones (60–70%) are excreted [85–87]. One study reported that 80% of the natural estrogens and 67% of the synthetic hormones (17*a*-ethinyl estradiol) and endocrine-disrupting substances are excreted via HU, but their impact on the environment has not been clearly ascertained [88, 89]. Certain antibiotics (ampicillin, streptomycin, amoxycillin, etc.) and their metabolites, if present in fresh or stored urine of medicated persons, may kill biogeochemical cycling bacteria, resulting in significant nutrient subtraction from the nutrient cycle, leading to a decline in ecosystem productivity. Pharmaceuticals, including antibiotics, may be accumulated in the soil [90] and adversely affect plant growth and soil microbial and enzymatic activities [91]. HU fertilization may transfer carbamazepine into the roots and aerial parts of ryegrass [67].

The main obstacles to using fresh urine in aquaculture are the high ammonia concentration and elevated pH, which may adversely affect the aquaculture organisms [20, 92, 93]. Aeration in urine-fed treatment was thought to reduce the toxicity of ammonia at high pH by its rapid transformation into nitrate via the oxygen-dependent nitrification process [20, 94]. Another benefit of the aeration in such a treatment would be the reduction of anaerobic pathogens [20]. The toxicity of urine at higher concentrations can pose risks to zooplanktons survival, so dilution is a must before use [17]. Stored urine is recommended for use as it reduces nitrite toxicity [95]. pH plays a very significant role, with pH levels of 8.5–8.8 offering the optimum nitrification rate [95]. In the case of stored urine with a pH value exceeding 9, 90% of nitrogen exists as dissolved ammonia (NH_3) and dissolved ammonium ion (NH_4^+); at pH 9.25, the concentration of NH_4^+ and NH_3 is in the ratio 1 : 1 [96]. It has also been demonstrated that, when applied directly to arable soil, the high pH ($\geq$9) of stored urine inhibits the conversion of nitrite into nitrate by nitrifying microbes, causing nitrite accumulation in the soil [13, 97]. Application of excessive quantum of stored urine in aquaculture may disrupt the buffering capacity of the aquatic system [95].

In future, better treatment of HU will comprise a combination of treatment processes, removal of phosphate and ammonium by struvite precipitation, and a biological process to eliminate organic pollutants and nitrogen [7]. One innovative approach may use microbial fuel cells (MFCs) to recover nitrogen while simultaneously producing electricity [98, 99]. Above all, the prejudice-based "urine blindness" in modern societies is a major stumbling block for its popularization [50, 100]. Cultural customs, ethoses, taboos, and attitudes have imposed boundary conditions for the exploitation of the nutrient and energy potential of HU. Nevertheless, the scope for closing the loop of the enormous nutrient flow from HU to agri/aquaculture cannot be denied. Therefore, safe and rational use of HU in biological production may be a solution for a large selection of poverty-ridden farmers in developing countries who are looking for a low-cost eco-friendly fertilizer in place of expensive and energy-intensive chemical ones.

4.5.5 Conclusion

HU is not a waste but a precious renewable resource, a potential organic fertilizer. The ecological sanitation approaches discussed in this chapter demonstrate a "win–win" closed-loop strategy and cradle-to-cradle (C2C) approach for turning waste into wealth, trash into treasure, effluent into affluence by recovering and recycling resources and nutrients from urine back into bioregenerative practices like agri(aqua)culture rather than pouring them down the drain, where they may diffuse into water bodies and cause downstream pollution. Such recycling in agri(aqua)culture would fulfill the twin objectives of guaranteeing the conservation of environmental equilibria so as to promote sustained productivity and resource sustainability without dissipation of non-renewable materials or energy and ensuring income generation and livelihood security (i.e. profit to people, economic sustainability). As an available resource, HU can play pivotal role in the uplifting of rural economies and promotiuon of sustainable development. The overlooked hidden nutrient potential of liquid humanure is neither a flight of fancy nor a mysterious myth but a realizable reality that may be adopted as a promising, cost-effective, and eco-friendly ecotechnological nutrient management option in agri(aqua)culture practices as an alternative to chemical fertilizers. However, in order to contain and eliminate constraints and risks, continued research efforts are needed to implement integrated technological pathways that ensure simultaneous source separation, nutrient recovery, pathogen inactivation, and reduction of pharmaceuticals for the use of bioactive substances and nutraceuticals present in HU.

References

1. Ganrot, Z., Dave, G., and Nilsson, E. (2007). Recovery of N and P from human urine by freezing, struvite precipitation and adsorption to zeolite and active carbon. *Bioresource Technology* **98**: 3112–3121.
2. Wilsenach, J.A., Schuurbiers, C.A.H., and Van Loosdrecht, M.C.M. (2007). Phosphate and potassium recovery from source separated urine through struvite precipitation. *Water Research* **41**: 458–466.
3. Ganesapillai, M., Simha, P., and Gugalia, A. (2014). Recovering urea from human urine by bio-sorption onto microwave activated carbonized coconut shells: equilibrium, kinetics, optimization and field studies. *Journal of Environmental Chemical Engineering* **2**: 46–55.
4. Ganesapillai, M., Simha, P., and Zabaniotou, A. (2015). Closed-loop fertility cycle: realizing sustainability in sanitation and agricultural production through the design and implementation of nutrient recovery systems for human urine. *Sustainable Production and Consumption* **4**: 36–46.
5. Langergraber, G. and Muellegger, E. (2005). Ecological sanitation – a way to solve global sanitation problems? *Environment International* **31**: 433–444.
6. Heinonen-Tanski, H. and Wijk-Sijbesma, C.V. (2005). Review: human excreta for plant production. *Bioresource Technology* **96**: 403–411.
7. Maurer, M., Pronk, W., and Larsen, T.A. (2006). Review: treatment process for source separated urine. *Water Research* **40**: 3151–3166.
8. Pronk, W., Palmquist, H., Biebow, M. et al. (2006). Nanofiltration for the separation of pharmaceuticals from nutrients in source separated urine. *Water Research* **40**: 1405–1412.
9. Vinneras, B., Palmquist, H., Balmer, P. et al. (2006). Characteristics of household wastewater and biodegradable waste – a proposal for new Swedish norms. *Urban Water* **3**: 3–11.
10. Vinneras, B., Nordina, A., Niwagaba, C. et al. (2008). Inactivation of bacteria and viruses in human urine depending on temperature and dilution rate. *Water Research* **42** (15): 4067–4074.
11. Winker, M., Tettenborn, F., Faika, D. et al. (2008). Comparison of analytical and theoretical pharmaceutical concentrations in human urine in Germany. *Water Research* **42**: 3633–3640.
12. Kirchmann, H. and Petterson, S. (1994). Human urine – chemical composition and fertilizer use efficiency. *Fertilizer Research* **40** (2): 149–154.
13. Udert, K.M., Larsen, T.A., and Gujer, W. (2006). Fate of major compounds in source separated urine. *Water Science and Technology* **54**: 413–420.
14. Ronteltap, M., Maurer, M., Hausherr, R. et al. (2010). Struvite precipitation from urine - influencing factors on particle size. *Water Research* **44**: 2038–2046.
15. Jonsson, H., Stenstrom, T.A., Svensson, J. et al. (1997). Source separated urine – nutrient and heavy metal content water saving and faecal contamination. *Water Science and Technology* **35**: 145–152.
16. Simons, J., Clemens, J. (2003). The use of separated human urine as mineral fertilizer. Proceedings of the 2nd International Symposium "Ecosan – Closing the Loop," April 7–11, Lubeck, Germany. pp. 595–600.
17. Golder, D., Rana, S., Sarkar, Paria, D. et al. (2007). Human urine is an excellent liquid waste for the culture of fish food organism, *Moina micrura*. *Ecological Engineering* **30**: 326–332.

18. Paruch, A.M. (2012). Preservation of nutrients during long-term storage of source-separated yellowwater. *Water Science and Technology* **66**: 804–809.
19. Yang, L., Giannis, A., Chang, V.W.-C. et al. (2015). Application of hydroponic systems for the treatment of source-separated human urine. *Ecological Engineering* **81**: 182–191.
20. Jana, B.B., Bag, S.K., and Rana, S. (2012). Comparative evaluation of the fertilizer value of human urine, cow manure and their mix for the production of carp fingerlings in small holding tanks. *Aquaculture International* **20**: 735–749.
21. Rana, S., Biswas, J.K., Rinklebe, J. et al. (2017). Harnessing fertilizer potential of human urine in a mesocosm system: a novel test case for linking the loop between sanitation and aquaculture. *Environmental Geochemistry and Health* **39** (6): 1545–1561.
22. Akpan-Idiok, A.U., Udo, I.A., and Braide, E.I. (2012). The use of human urine as an organic fertilizer in the production of okra (*Abelmoschus esculentus*) in south eastern Nigeria. *Resources, Conservation and Recycling* **62**: 14–20.
23. Vinneras, B., Jonsson, H., Salomon, E., et al. (2003) Tentative guidelines for agricultural use of urine and faeces. Proceedings of the 2nd International Symposium "Ecosan – Closing the Loop," April 7–11, Lubeck, Germany. pp. 101–108.
24. Karak, T. and Bhattacharyya, P. (2011). Human urine as a source of alternative natural fertilizer in agriculture: a flight of fancy or an achievable reality. *Resources, Conservation and Recycling* **55**: 400–408.
25. Schouw, N.L., Danteravanich, S., Mosbaek, H. et al. (2002). Composition of human excreta- a case study from Southern Thailand. *Science of the Total Environment* **286**: 155–166.
26. Jonsson, H., Baky, A., Jeppsoon, U., Hellstrom, D., Karman, E. (2005). Composition of urine, faeces, graywater and biowaste for utilization in the URWARE model. Urban Water Report of the MISTRA Programme, Report 2005: 6, Chalmers University of Technology, Gothenburg, Sweden.
27. Jonsson, H., Stintzing, R., Vinneras, B., Salomon, E. (2004). Guidelines on the use of urine and faeces in crop production. Report 2004-2, Ecosanres, Stockholm Environment Institute, Stockholm, Sweden.
28. Ek, M., Bergstrom, R., Bjurhem, J.-E. et al. (2006). Concentration of nutrients from urine and reject water from anaerobically digested sludge. *Water Science and Technology* **54** (11–12): 437–444.
29. Jaatinen, S. (2016). *Characterization and Potential Use of Source-Separated Urine*, vol. **1391**. Tampere: Tampere University of Technology.
30. Jensen, P.K.M., Phuc, P.D., Knudsen, L.G. et al. (2008). Hygiene versus fertiliser: the use of human excreta in agriculture – Vietnamese example. *International Journal of Hygiene and Environmental Health* **211**: 432–439.
31. Mnkeni, P.N.S., Kutu, F.R., Muchaonyerwa, P. et al. (2008). Evaluation of human urine as a source of nutrients for selected vegetables and maize under tunnel house conditions in the Eastern Cape, South Africa. *Waste Management & Research* **26** (2): 132–139.
32. Tang, F.H.M. and Maggi, F. (2016). Breakdown, uptake and losses of human urine chemical compounds in barley (*Hordeum vulgare*) and soybean (*Glycine max*) agricultural plots: effectiveness of human urine use in agriculture. *Nutrient Cycling in Agroecosystems* **104** (2): 221–245.
33. Mang, H., Jurga, I.P., and Xu, Z. (2007). Experience in improving fertilizer value of compost by enriching with urine. *International Journal of Environmental Technology and Management* **7**: 464–471.
34. Fatunbi, A.O. (2009). Suitability of human urine enriched compost as horticultural growing medium. *World Applied Sciences Journal* **6** (5): 637–643.
35. Winker, M., Vinneras, B., Muskolus, A. et al. (2009). Fertilizer products from new sanitation systems: their potential values and risks. *Bioresource Technology* **100**: 4090–4096.
36. Heinonen-Tanski, H., Sjoblom, A., Fabritius, H. et al. (2007). Pure human urine is a good fertilizer for cucumbers. *Bioresource Technology* **98**: 214–217.

37. Pradhan, S.K., Holopainen, J.K., Weisell, J. et al. (2010). Human urine and wood ash as plant nutrients for red beet (*Beta vulgaris*) cultivation: impacts on yield quality. *Journal of Agricultural and Food Chemistry* **58** (3): 2034–2039.
38. Pradhan, S.K., Holopainen, J.K., and Heinonen-Tanski, H. (2009). Stored human urine supplemented with wood ash as fertilizer in tomato (*Solanum lycopersicum*) cultivation and its impacts on fruit yield and quality. *Journal of Agricultural and Food Chemistry* **57** (16): 7612–7617.
39. Båth, B. (2003) Field trials using human urine as fertilizer to leeks [Swedish]. Department of Ecology and Plant Production Science, Swedish University of Agricultural Sciences, Uppsala.
40. Adamsson, M. (2000). Potential use of human urine by greenhouse culturing of microalgae (*Scenedesmus acuminatus*), zooplankton (*Daphnia magna*) and totatoes (*Lycopersicon*). *Ecological Engineering* **16**: 243–254.
41. United Nation Educational, Scientific and Cultural Organization, International Hydrological Programme and Deutsche Gesellschaft fur Technische Zusammenarbeit for Capacity Building (2006). *Capacity Building for Ecological Sanitation: Concepts for Ecologically Sustainable Sanitation in Formal and Continuing Education*, 156. Paris: UNICEF/IHP and GTZ.
42. Tidaker, P., Mattsson, B., and Jonsson, H. (2007). Environmental impact of wheat production using human urine and mineral fertilizers – a scenario study. *Journal of Cleaner Production* **15**: 52–62.
43. Antonini, S., Nguyen, P.T., Arnold, U. et al. (2012). Solar thermal evaporation of human urine for nitrogen and phosphorus recovery in Vietnam. *Science of the Total Environment* **414**: 592–599.
44. Jana, B.B., Rana, S., and Bag, S.K. (2012). Use of human urine in phytoplankton production as a tool for ecological sanitation. *Water Science and Technology* **65**: 1350–1356.
45. Feng, D., Wu, Z., and Xu, S. (2008). Nitrification of human urine for its stabilization and nutrient recycling. *Bioresource Technology* **99**: 6299–6304.
46. Yang, C., Liu, H., Li, M. et al. (2008). Treating urine by Spirulina platensis. *Acta Astronaut* **63**: 1049–1054.
47. Tuantet, K., Janssen, M., Temmink, H. et al. (2014). Microalgae growth on concentrated human urine. *Journal of Applied Phycology* **26**: 287–297.
48. Zhang, S., Lim, C.Y., Chen, C.-L. et al. (2014). Urban nutrient recovery from fresh human urine through cultivation of *Chlorella sorokiniana*. *Journal of Environmental Management* **145**: 129–136.
49. Hellstrom, D., Johansson, E., and Grennberg, K. (1999). Storage of human urine: acidification as a method to inhibit decomposition of urea. *Ecological Engineering* **12**: 253–269.
50. Guzha, E., Nhapi, I., and Rockstrom, J. (2005). An assessment of the effect of human faeces and urine on maize production and water productivity. *Physics and Chemistry of the Earth* **30** (11–16): 840–845.
51. Kassa, K., Meinzinger, F., Zewdie, W. (2010) Experience from the use of human urine in Arba Minch, Ethiopia. EcoSan Club Sustainable Sanitation Pratice, Issue 3.
52. Mnkeni, P.N.S., Austin, A., Kutu, F.R. (2005). Preliminary studies on the evaluation of human urine as a source of nutrients for vegetables in the Eastern Cape Province, South Africa. In: Ecological Sanitation: A Sustainable Integrated Solution. Proceedings of the 3rd International Ecological Sanitation Conference, Durban, South Africa, May 23–26. pp 418–426.
53. Pradhan, S.K., Nerg, A., Sjoblom, A. et al. (2007). Use of human urine fertilizer in cultivation of cabbage (*Brassica oleracea*)-impacts on chemical, microbial, and flavor quality. *Journal of Agricultural and Food Chemistry* **55**: 8657–8663.
54. Pradhan, S.K., Pitkanen, S., and Heinonen-Tanski, H. (2009). Fertilizer value of urine in pumpkin (*Cucurbita maxima* L.) cultivation. *Agricultural and Food Science* **19** (1): 57–67.
55. Pradhan, S.K., Piya, R.C., and Heinomen-Tanski, H. (2011). Eco-sanitation and its benefits: an experimental demonstration program to raise awareness in central Nepal. *Environment, Development and Sustainability* **13** (3): 507–518.
56. Wijffels, R.H., Barbosa, M.J., and Eppink, M.H.M. (2010). Microalgae for the production of bulk chemicals and biofuels. *Biofuels, Bioproducts and Biorefining* **4** (3): 287–295.

57. Sialve, B., Bernet, N., and Bernard, O. (2009). Anaerobic digestion of microalgae as a necessary step to make microalgal biodiesel sustainable. *Biotechnology Advances* **27**: 409–416.
58. Chisti, Y. (2007). Biodiesel from microalgae. *Biotechnology Advances* **25**: 294–306.
59. Griffiths, M.J. and Harrison, S.T.L. (2009). Lipid productivity as a key characteristic for choosing algal species for biodiesel production. *Journal of Applied Phycology* **21**: 493–507.
60. Vandamme, D., Foubert, I., and Muylaert, K. (2013). Flocculation as a low-cost method for harvesting microalgae for bulk biomass production. *Trends in Biotechnology* **31**: 233–239.
61. Lam, M.K. and Lee, K.T. (2012). Microalgae biofuels: a critical review of issues, problems and the way forward. *Biotechnology Advances* **30**: 673–690.
62. Cho, S., Lee, N., Park, S. et al. (2013). Microalgae cultivation for bioenergy production using wastewaters from a municipal WWTP as nutritional sources. *Bioresource Technology* **131**: 515–520.
63. Rawat, I., Bhola, V., Kumar, R.R. et al. (2013). Improving the feasibility of producing biofuels from microalgae using wastewater. *Environmental Technology* **34**: 1765–1775.
64. Konig, A., Pearson, H., and Silva, S.A. (1987). Ammonia toxicity to algal growth in waste stabilization ponds. *Water Science and Technology* **19**: 115–122.
65. Clarens, A.F., Resurreccion, E.P., White, M.A. et al. (2010). Environmental life cycle comparison of algae to other bioenergy feedstocks. *Environmental Science & Technology* **44**: 1813–1819.
66. Watson, C.J., Akhonzada, N.A., Hamilton, J.T.G. et al. (2008). Rate and mode of application of the urease inhibitor N-(n-butyl) thiophosphoric triamide on ammonia volatilization from surface-applied urea. *Soil Use and Management* **24**: 246–253.
67. Winker, M., Clemens, J., Reich, M. et al. (2010). Ryegrass uptake of carbamazepine and ibuprofen applied by urine fertilization. *Science of the Total Environment* **408**: 1902–1908.
68. Krajewska, B. (2009). Ureases I. Functional, catalytic and kinetic properties. *Journal of Molecular Catalysis B: Enzymatic* **59**: 9–21.
69. Saggar, S., Singh, J., Giltrap, D.L. et al. (2013). Quantification of reductions in ammonia emissions from fertiliser urea and animal urine in grazed pastures with urease inhibitors for agriculture inventory: New Zealand as a case study. *Science of the Total Environment* **465**: 136–146.
70. Singh, J., Kunhikrishnan, A., Bolan, N.S. et al. (2013). Impact of urease inhibitor on ammonia and nitrous oxide emissions from temperate pasture soil cores receiving urea fertilizer and cattle urine. *Science of the Total Environment* **465**: 56–63.
71. Watson, C.J., Poland, P., and Allen, M.B.D. (1998). The efficacy of repeated applications of the urease inhibitor N-(n-butyl) thiophosphoric triamide for improving the efficiency of urea fertilizer utilization on temperate grassland. *Grass and Forage Science* **53**: 137–145.
72. Dagerskog, L., Bonzi, M. (2010) Opening mind and closing loops productive sanitation initiatives in Burkina Faso and Niger. EcoSan Club Sustainable Sanitation Practice, Issue 3.
73. Dasgan, Y.H., Aktas, H., Abak, K. et al. (2002). Determination of screening techniques to salinity tolerance in tomatoes and investigation of genotype response. *Plant Science* **163**: 695–703.
74. Asano, T., Burton, F.L., Leverenz, H.L. et al. (2007). *Water Reuse: Issues Technologies and Applications*. New York: McGraw-Hill.
75. Brady, N.C. and Weil, R.R. (1996). *The Nature and Properties of the Soils*, 14e. New York: Prentice Hall.
76. Nishihara, E., Inoue, M., Kondo, K. et al. (2001). Spinach yield and nutritional quality affected by controlled soil water matric head. *Agricultural Water Management* **51**: 217–229.
77. Hijikata, N., Yamauchi, N., Yabui K., et al. (2011). Characterization of several agricultural wastes as a matrix of composting toilet – from fecal degradation to reuse as a soil conditioner. Proceedings of 8th IWA International Symposium on Waste Management Problems in Agro-Industries. pp. 317–324.
78. Höglund, C., Stenstrom, T.A., and Asbolt, N. (2002). Microbial risk assessment of source separated urine used in agriculture. *Waste Management & Research* **20**: 150–161.
79. AdeOluwa, O.O. and Cofie, O. (2012). Urine as an alternative fertilizer in agriculture: effects in amaranths (*Amaranthus caudatus*) production. *Renewable Agriculture and Food Systems* **27** (4): 287–294.

80. World Health Organization (2006). *Guidelines for the Safe Use of Wastewater, Excreta and Grey Water: Excreta and Grey Water Use in Agriculture*, vol. **4**. Geneva: WHO.
81. Höglund, C., Stenstrom, T.A., Jonsson, H. et al. (1998). Evaluation of faecal contamination and microbial die-off in urine-separating systems. *Water Science and Technology* **38**: 17–25.
82. Höglund, C., Ashbolt, N., Stenstrom, T.A. et al. (2002). Viral persistence in source separated human urine. *Advances in Environmental Research* **6**: 265–275.
83. Höglund, C.E. and Stenström, T.A.B. (1999). Survival of *Cryptosporidium parvum* oocysts in source separated human urine. *Canadian Journal of Microbiology* **45** (9): 740–746.
84. Arnbjerg-Nielsen, K., Kjolholt, J., Stuer-Lauridsen, F., et al. (2003) Risk assessment of local handling of human faeces with focus on pathogens and pharmaceuticals. Proceedings of the 2nd International Symposium "Ecosan – Closing the Loop," April 7–11, Lubeck, Germany.
85. Khetan, S.K. and Collins, T.J. (2007). Human pharmaceuticals in the aquatic environment: a challenge to green chemistry. *Chemical Reviews* **107**: 2319–2364.
86. Lienert, J., Burki, T., and Escher, B. (2007). Reducing micropullants with source control (human urine). *Water Science and Technology* **34** (3–4): 87–94.
87. Kogan, M.I., Naboka, Y.L., Ibishev, K.S. et al. (2015). Human urine is not sterile – shift of paradigm. *Urologia Internationalis* **94** (4): 445–452.
88. Christiansen, L.B., Winther-Nielsen, M., and Helweg, C. (2002). *Feminization of Fish – The Effect of Estrogenic Compounds and their Fate in Sewage Treatment Plants and Nature*. Environmental Project No. 729. Copenhagen: Danish Environmental Protection Agency.
89. Bhandari, R.K., Deem, S.L., Holliday, D.K. et al. (2015). Effects of the environmental estrogenic contaminants bisphenol A and 17a-ethinyl estradiol on sexual development and adult behaviors in aquatic wildlife species. *General and Comparative Endocrinology* **214**: 195–219.
90. Kinney, C.A., Furlong, E.T., Werner, S.L. et al. (2006). Presence and distribution of wastewater-derived pharmaceuticals in soil irrigated with reclaimed water. *Environmental Toxicology and Chemistry* **25**: 317–326.
91. Liu, F., Ying, G.-G., Tao, R. et al. (2009). Effects of six selected antibiotics on plant growth and soil microbial and enzymatic activities. *Environmental Pollution* **157**: 1636–1642.
92. Abbas, H.H. (2006). Acute toxicity of ammonia to common carp fingerlings (*Cyprinus carpio*) at different pH levels. *Pakistan Journal of Biological Sciences* **9** (12): 2215–2221.
93. Biswas, J.K., Sarkar, D., Chakraborty, P. et al. (2006). Density dependent ambient ammonium as the key factor for optimization of stocking density of common carp in small holding tanks. *Aquaculture* **261**: 952–959.
94. Masic, A., Santos, A.T.L., Etter, B. et al. (2015). Estimation of nitrite in source-separated nitrified urine with UV spectrophotometry. *Water Research* **85**: 244–254.
95. Lucas, J.S. and Southgate, P.C. (eds.) (2003). *Aquaculture: Farming Aquatic Animals and Plants*. Oxford: Blackwell.
96. Lee, J.D. (1996). *Concise Inorganic Chemistry*, 5e. Oxford: Blackwell Science.
97. Burns, L.C., Stevens, R.J., Smith, R.V. et al. (1995). The occurrences and possible sources of nitrite in a grazed fertilized grassland soil. *Soil Biology and Biochemistry* **27**: 47–59.
98. Kuntke, P., Śmiech, K.M., Bruning, H. et al. (2012). Ammonium recovery and energy production from urine by a microbial fuel cell. *Water Research* **46**: 2627–2636.
99. Santoro, C., Ieropoulos, I., Greenman, J. et al. (2013). Power generation and contaminant removal in single chamber microbial fuel cells (SCMFCs) treating human urine. *International Journal of Hydrogen Energy* **38**: 11543–11551.
100. Drangert, J.O. (1998). Urine blindness and the use of nutrients from human excreta in urban agriculture. *Geochemical Journal* **45**: 201–208.

4.6

Pilot-Scale Investigations on Phosphorus Recovery from Municipal Wastewater

Marie-Edith Ploteau[1], Daniel Klein[1], Johan te Marvelde[2], Luc Sijstermans[3], Anders Nättorp[4], Marie-Line Daumer[5], Hervé Paillard[6], Cédric Mébarki[6], Ania Escudero[7]Ole Pahl[7], Karl-Georg Schmelz[8], and Frank Zepke[9]

[1]*Lippeverband, Essen, Germany*
[2]*HVC Groep, Alkmaar, The Netherlands*
[3]*Slibverwerking Noord-Brabant, Moerdijk, The Netherlands*
[4]*School of Life Sciences FHNW, Muttenz, Switzerland*
[5]*Institut national de recherche en sciences et technologies pour l'environnement et l'agriculture, Rennes, France*
[6]*Véolia Environnement, Aubervilliers, France*
[7]*Glasgow Caledonian University, Glasgow, Scotland*
[8]*Emschergenossenschaft, Essen, Germany*
[9]*EuPhoRe GmbH, Telgte, Germany*

4.6.1 Introduction

Consequential to the development towards a higher proportion of the population being connected to urban wastewater treatment plants in the European Member States (36% in Romania to 99% in Austria in 2013), the quantity of the residual of wastewater treatment, sewage sludge, to be disposed increases. EUROSTAT published at the end of 2016 the amount of sewage sludge produced from urban wastewater in the European Member states. According to the most recent data available (2012–2014), almost ten million tons of sewage sludge (in dry substance) are produced in EU-28 and Switzerland [1].

Biorefinery of Inorganics: Recovering Mineral Nutrients from Biomass and Organic Waste, First Edition.
Edited by Erik Meers, Gerard Velthof, Evi Michels and René Rietra.
© 2020 John Wiley & Sons Ltd. Published 2020 by John Wiley & Sons Ltd.

Sewage sludge is rich in mineral nutrients (such as phosphorus and nitrogen) and organic carbon. To make use of the value of this sludge on average 50% of it is spread in agriculture. However, the high concentrations of pollutants such as heavy metals and pharmaceutical residues in sewage sludge have led countries to seek for alternative pathways to sludge disposal, rather than agricultural use. Whereas futher 12% of sludge is recycled through landscaping measures and road underground, almost 40% is incinerated or continues to be disposed in landfill. High disparities between EU Member States are observed: in some countries such as Portugal, Ireland, the United Kingdom, and France, more than 75% of sludge is spread in agriculture or used for landscaping measures, while more than 60% of the sludge is incinerated in the Netherlands, Switzerland, Belgium, and Germany.

As a result of the implementation of the Urban Waste Water Treatment Directive [2], the more stringent tertiary wastewater treatment is being implemented in more and more catchments throughout Europe. The positive effect is the reduction of the nutrient pollution (nitrogen and phosphorus) and protection of aquatic systems, by avoiding eutrophication. The downside is that continuously more nutrients, especially phosphorus, end up in the growing quantity of sludge that's increasingly banned for land use.

According to the United States Geological Survey [3], the exploitable reserves of phosphate rock (68 billion tons) are sufficient for a few hundred years by current worldwide mine production (261 million tons per year reported in 2016) and "there are no imminent shortages of phosphate rock." Nevertheless, the awareness regarding phosphorus resources has been raised in Europe. As phosphorus is essential for all living organisms on earth, as there is no recycling of the imported phosphorus, the EU has added phosphate rock to its list of 20 critical raw materials in 2014 [4]. In addition, Europe depends on foreign reserves that are mainly owned by states and located in politically unstable regions of the world (Morocco, China, Algeria, Syria, South Africa, and Russia). Another aspect of encouraging alternatives to mineral phosphorus is the depletion of cleaner sources of phosphate rock. Phosphate rock deposits can be characterized mainly with cadmium, but also uranium.

Based on those considerations, the phosphorus flows concentrated in waste water treatment plants (WWTP) constitute an interesting secondary phosphorus raw material for fertilizer production. Considering a phosphorus content in sewage sludge ranging between 25 and 40 $g \cdot kg^{-1}$ of dry solids, recovery of the phosphorus contained in sewage sludge not used for agriculture or landscape would allow substitution of 16–26% of the import of mineral phosphorus fertilizers (amounting to 1.0 million tons according to EUROSTAT) in EU-28 plus Switzerland. Some countries (e.g. the Netherlands) would even produce a surplus of secondary raw phosphorus compared to their fertilizer needs [5].

4.6.2 European and National Incentives to Act on Market Drivers

Market drivers will determine the breakthrough of the secondary raw phosphorus. Therefore it is required that its price and its quality enable the acceptance of the farmers. The Environmental legislation will be decisive to boost the use of organic and waste-based fertilizers.

For this reason, in 2019, the European Commission published the revised Fertilizer Regulation, taking into account recovered nutrient products as part of the Circular Economy Package [6].

The national context varies widely between EU Member States. Where a ban of sewage sludge spreading applies in e.g. Switzerland or is foreseen as in Germany, technological phosphorus recovery is fostered through legislation. In Switzerland, phosphorus recovery is mandatory before 2026 [7], in Germany, it will be before 2029 for bigger WWTPs with population equivalents higher than 100 000 and before 2032 for those higher than 50 000 [8].

Where a phosphorus surplus occurs, such as in the Netherlands, Belgium, and part of France, because of competition with animal manure, the technological recovery is driven by the need to export the phosphorus contained in sludge from the production's location. This can be accompanied by legislative adjustments such as in the Netherlands, where regulatory requirements for recovered phosphates have been added in 2016 in the Dutch Fertilizer Regulation [9].

In other regions of Europe where the management of growing amount of sewage sludge through a more stringent implementation of the wastewater directive is on the agenda, the opportunity for the simultaneous recovery of nutrients is of concern. In Scotland, decentralized phosphorus recovery from effluent could help overcome eutrophication issues. In this sense, it supports the Scottish government's drive towards Hydro Nation and circular economy policies. In Ireland, phosphorus recovery fits especially with the National Wastewater Sludge Management Plan revised in 2016 [10].

4.6.3 Pilot Investigations

The European INTERREG VB project Phos4You has implemented pilot-scale investigations involving seven EU Member States in North-West Europe, following various strategies:

* leaching solutions to recover phosphorus from sewage sludge ashes;
* thermal solutions to recover phosphorus from sewage sludge; and
* struvite solutions with biological acidification to recover P from sewage sludge liquor.

Selected processes are described in the following sections.

4.6.3.1 Acid Leaching Solutions to Recover Phosphorus from Sewage Sludge Ashes

In this investigation, experiences in acid ash leaching from Switzerland and the Netherlands/Belgium have been transferred to a German context. The German waterboards Emschergenossenschaft and Lippeverband (EG/LV), the Dutch mono-incinerators Slibverwerking Noord-Brabant (SNB) and HVC Groep (HVC), and the Swiss School of Life Sciences (FHNW HLS) have demonstrated together the feasibility of ash leaching. The demonstration was based on recovery trials, assessment of results with regard to feasibility, costs, and other factors.

EG/LV operates over 50 waste water treatment plants (WWTPs) of various sizes. Sewage sludge is usually digested on-site and subsequently transported to central facilities for further treatment. This includes two sewage sludge mono-incinerations operated by EG and a partner company.

In a first step toward valorization of the phosphorus, EG/LV seeks to evaluate P-recovery technologies. EG/LV mainly focuses on technologies that are able to recover phosphorus from sewage sludge ashes through (acid) leaching. The tested process was selected by a public (EU-wide) procurement process. The service to be procured included "pre-industrial

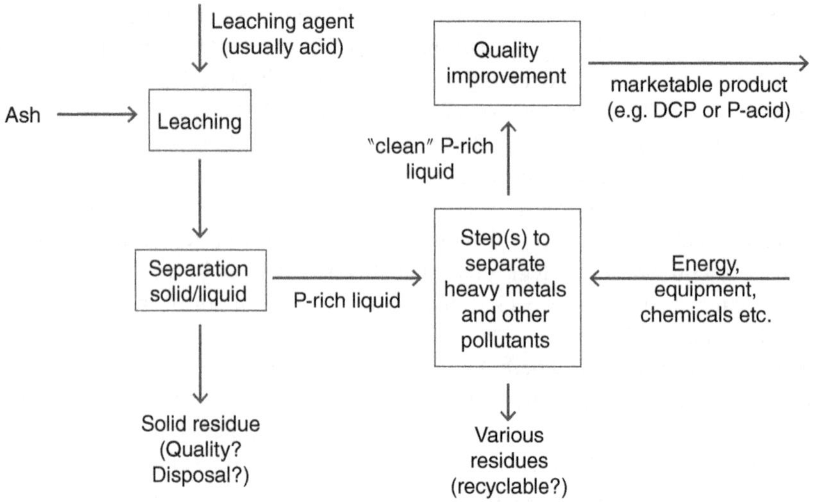

Figure 4.6.1 *A general overview of the process. Source: Lippeverband.*

scale" trials, including all process steps. Scale and extent of trials should be sufficient to determine the relevant process parameters, the needed utilities, and costs in general. This approach is aimed at high-quality products that already have a market, such as phosphoric acid and defined phosphoric fertilizer. A general overview of leaching processes is given in Figure 4.6.1.

Based on the results of the trials, a comprehensive business plan has been developed by EG/LV. In this, the question is addressed on how the assessed technology can be implemented at a regional scale.

4.6.3.2 Pilot Demonstration of Thermal Solutions to Recover Phosphorus from Sewage Sludge: The EuPhoRe® Process

The main interest of the EuPhoRe process is to enable the recovery of phosphate simultaneously with the double thermal treatment of sewage sludge. The ashes generated are characterized by a high plant availability of the phosphorus and low heavy metal content. This end-product can be directly used as a phosphorus fertilizer.

The EuPhoRe process consists of two steps (anoxic and oxic) occurring in a rotary kiln: first, a pyrolysis/dry carbonation, directly followed by incineration. Via the thermal treatment, an essential amount of heavy metals evaporate and get separated in the flue gas cleaning. Organic pollutants are fully destroyed in the thermal process.

The pilot plant is designed for energy and material recovery of sewage sludge. Depending on the input quality, additives can be used to further improve the quality of the end-product. The sewage sludge is loaded into a rotary kiln via an airlock system. Here it is predried by using hot flue gas from the two EuPhoRe® steps in a counter-current manner.

There are two main process steps within the kiln, which effectively separated it into two zones (Figures 4.6.2 and 4.6.3). In the middle part of the rotary kiln after predrying,

Figure 4.6.2 EuPhoRe® pilot plant in operation. Source: EuPhoRe GmbH.

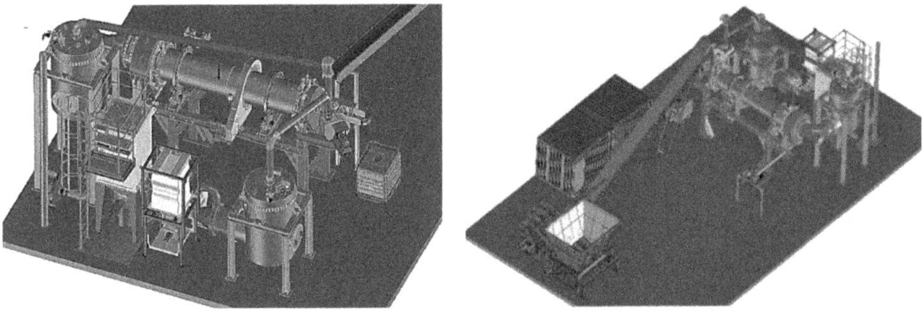

Figure 4.6.3 3D-Simulation of EuPhoRe® demonstration plant. Source: LUKSON AG.

pyrolysis occurred due to the oxygen-deprived atmosphere and heat. Volatile substances evaporate and are mixed into the gas due to the heat. In the end, part of the kiln remaining oxygen in the flue gas out of the burning chamber is consumed by oxidizing the carbon from the pyrolysis coke.

The mixture of flue gas, pyrolysis gas and vapour form the so-called process gas that is extracted out of the rotary kiln at the sludge feeding side. It is split and sent to two burning chambers. A part is combusted to generate flue gas for the rotary kiln. The hot flue gas is charged via the outlet housing where the ash is discharged due to the counter-current system. The balance of the process gas gets fully incinerated in a burning chamber with downstream flue gas cleaning. This flue gas is treated via a fine particle filter (fabric filter) and adsorbent injections to comply with air emission regulations.

The ash exits the kiln via a cooling unit into a bag or a container, for transport.

Specification of the demonstration plant:

Input material: sewage sludge, 25–30% dry matter (DM)
Input mass: approx. 100 $kg{\cdot}h^{-1}$
Output: phosphate minerals (12–20% P_2O_5)
Output mass: approx. 10–15 $kg{\cdot}h^{-1}$
Heat value (wet): approx. 1180 $kJ{\cdot}kg^{-1}$
Combustion heat performance
Location: outside, with dodger

4.6.3.3 Demonstration of struvite solution with biological acidification to increase the P recovery from sewage sludge

The mineral fertilizer production as struvite or calcium phosphate is currently one of the most popular ways to recycle P from wastewater as a valuable product. Several technologies are available on the market as discussed in Chapter 3.2. Most of them can only recover 10–20% of the WWTP influent P concentration.

The aim of the technology innovation proposed in the frame of the Phos4You project (cooperation between Véolia and IRSTEA (incorporated in INRAE, FR) is to raise the overall P recycling rate by increasing the dissolve P flow entering the crystallization process.

The addition of an external easily bio-degradable co-product (grease, food waste, bio-wastes from the F&B industry ...) is needed to boost production of VFA for acidification of the sludge. By adding sugar rich co-substrate to EBPR WWTP, the process is providing suitable conditions to phosphorus accumulations bacteria (PAO) for P release but also to lactic bacteria for decreasing the pH down to 4–5 by in situ production of lactic acid. Up to 75% of the P in sludge can be dissolved by this biological acidification process at the lab scale and confirmed at pilot scale. Moreover, sugar rich co-substrate is also promoting activity of the iron reducing bacteria. This is highly valuable mechanism for the WWTPs where P is removed with the iron salts. A significant part of the P precipitated with a Fe salt can also be solubilized thanks to this bo-acidification process as proved by the iron dissolution that can reach up to 80% of the total iron concentration in sludge (11).

Its implementation in a WWTP is described in figure 4.6.4. This promising technology was scaled up at a small and large pilot scale in Lille and Tergnier (France), and combined with the Struvia crystallization (using magnesium or calcium chemistry) process in a WWTP during the Phos4You project.

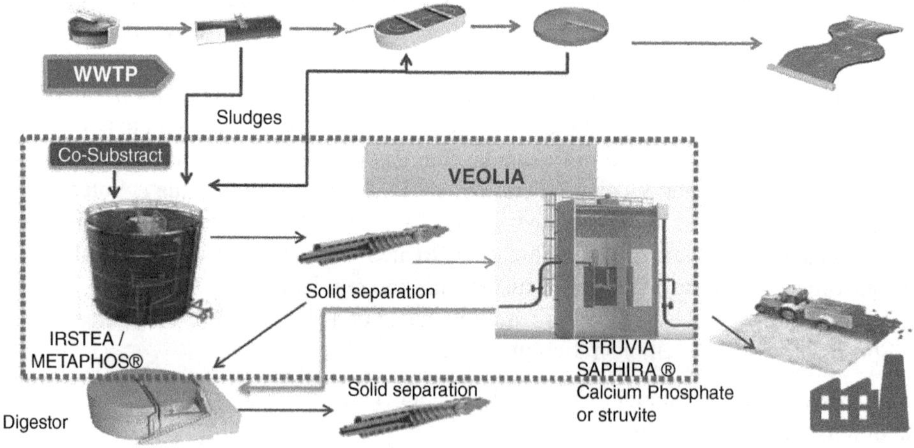

Figure 4.6.4 *Implementation of a bio-acidification stage on liquid sludge in a municipal waste water treatment plant to boost P recovery. Source: Véolia.*

4.6.3.4 Innovative Technical Solutions to Recover P from Small-Scale WWTPs: Downscaling Struvite Precipitation for Rural Areas

Limited supplies of naturally occurring phosphorus (P) resources and their decreasing purity, coupled with legislative pressure on water-service companies to remove P from both their effluent and waste streams, has focused considerable attention on recycling P recovered from domestic wastewater in existing P markets [12].

Up to now, P-recovery technologies producing struvite or calcium phosphate have been developed on the sludge treatment line of large WWTPs above 100 000 PE. "Downscaling" a technology designed for >100 000 to 10–1000 PE for rural areas faces several challenges:

- more than 50% of the P is fixed in the sludge of these small wastewater treatment systems [13], then the concentration of P-PO_4 to recover at their outlet is low (about 10–15 $mg \cdot l^{-1}$) compared to a high concentration of P-PO_4 (more than 100 mgP-PO_4/l) in sludge digestate of large WWTP;
- remote accessibility, hence maintenance, robustness, and operation constraints;
- high variability in both flows and concentrations (seasonal and diurnal effect); and
- final product outcome (quantity, quality, etc.) and devolution on rural markets.

As phosphorus becomes increasingly scarce, it would seem likely that irrigation specifically for the purpose of phosphorus recovery ("Fert-irrigation") in rural areas will increase. This may be far more challenging than it initially appears due to quality, temporal, and spatial constraints [14].

Similarly, urine has a much higher phosphate concentration than a conventional septic tank's wastewater outlet, enabling simpler (and less expensive) processes for the precipitation of phosphates. An average of 95% of influent phosphate was removed regardless of ammonium or potassium struvite precipitation [15]. However, these approaches require further investment in segregating networks, issues with product quality and safety records which tend to make them unrealistic.

In regard to a small sewage treatment (<1000 PE), low-velocity horizontal and vertical filters with apatite as P adsorbent have been available for several years. Notwithstanding these filters are using imported P rocks, the P removal is in the range of 80% in the first period of treatment and decreases with time; there is also a low saturation rate (0.8 g $P \cdot kg^{-1}$ of apatite). Furthermore, these filters require a high footprint (0.8–2 m^2 per PE) and are significantly impacted by the loading rate (HLR) and temperature variations [16].

The Phos4You project worked on reengineering the STRUVIA struvite precipitation technology to address the following main challenges:

- higher compactness than current apatite filters;
- low or nearly no operation along 1 year; robust design which means no chemical adjustment or daily delivery – extraction of product, low automation;
- innovative final products and P forms, together with a high recovery rate (>75%) even on low concentrated streams; and
- versatility of system, suitable for various streams (septic to low aerobic design such as aerated ponds, etc.).

The technology innovation proposed under Phos4You looks forward to sustainable and efficient solutions using the framework of the STRUVIA® and FILTRAFLO® technologies, as shown on Figure 4.6.5. Changing the chemistry- and efficiency-centric approaches and reshaping the new design around a high-capacity adsorbent will allow implementation of such units with low maintenance needs. Developing new media and carriers, enabling phosphorous adsorption with no chemical requirements until exhaustion, will enable standalone operation, despite any seasonal or diurnal impact. A significant part (>75% removal as P-PO_4) will be adsorbed either from small-scale septic tanks or low-loaded

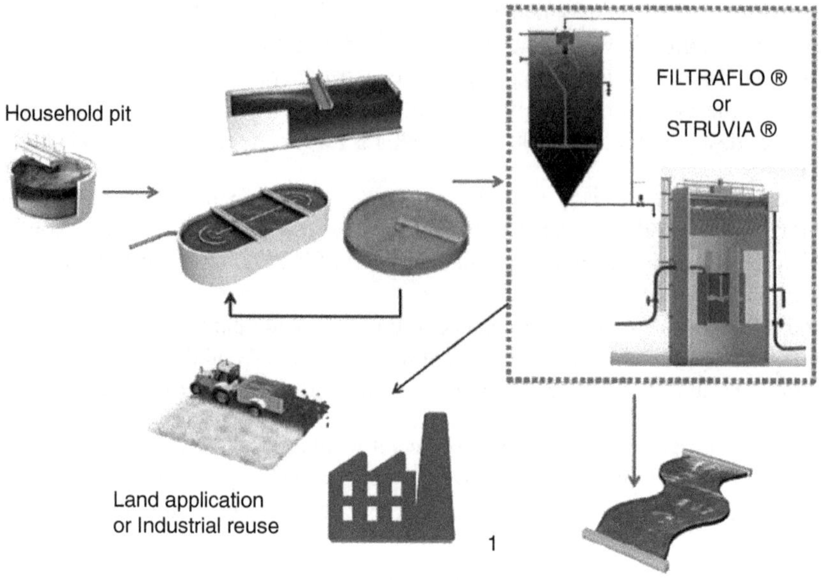

Figure 4.6.5 Implementation of a STRUVIA® or FILTRAFLO® downstream small-scale municipal WWTP to recover dissolved P. Source: Véolia.

aerobic systems (aerated lagoons, ponds, etc.), giving small communities the possibility to enrich-concentrate the dissolved phosphorous on specific media. These systems will be versatile enough to allow the use of "directly consumable" media for agricultural purposes such as fertilizers or "intermediate-feedstock" media that are desorbed in centralized P-recovery plants and reused afterwards (reusable media).

Giving the opportunity to rural communities to choose the most applicable recovery method depending on their spatial and temporal constraints is the core of the project developed here, together with a circular-economy aspect (use of second-hand or by-products-based material).

4.6.3.5 Algal-Based Solutions to Recover Phosphorus from Small-Scale WWTPs: A Promising Approach for Remote, Rural, and Island Areas

Scotland is 97% rural, with 70% of the land mass defined as "remote rural" (areas with a population of less than 3000 people, and with a drive time of over 30 minutes to a settlement of 10 000 or more). Thus any nutrient recovery occurring in these areas requires technologies that are small-scale, low-maintenance, and can dove-tail into existing infrastructure and recovery mechanisms. Microalgae have proven efficient in recovering nutrients from different effluents such as municipal waste [17, 18], agricultural waste [19], and industrial wastewater [20], thus avoiding nutrient runoff and eutrophication. The algae can then be used for different purposes: as feedstock for producing biogas, as third-generation biofuel, for commercial uses, and for recycling on farms as feed, compost, or fertilizer [21].

Cultivation of extremophilic microorganisms has gained interest in recent years due to the ability of these microorganisms to accumulate and produce valuable compounds such as metabolites, enzymes, and surfactants [22]. *Chlamydomonas acidophila* has been shown to be a promising agent for the removal of nutrients from different wastewaters, as it has high phosphorus and nitrogen uptake rates (up to 90%) whilst operating at low temperatures and low light requirements, which is well suited to Scotland. Furthermore, the growth of *C. acidophila* prevents ammonia stripping in the treatment process. It can grow in media containing up to 1000 mg NH_4-N·l-1 [23].

In recent years, there has been increasing scientific and regulatory attention toward micropollutants such as pharmaceuticals in the aquatic environment. They are generally originated in urban environments and typically found at very low concentrations in the environment. They are unlikely to affect human health but could cause chronic exposure damage to aquatic organisms. Some are included as watchlist priority substances in the Water Framework Directive (WFD). The selected microalga (*C. acidophila*) was unaffected in studies at Glasgow Caledonian University (GCU) by concentrations of pharmaceuticals exceeding those reported in the literature and therefore could be successfully used for nutrient removal in wastewaters containing these compounds. Additionally, this organism appears to increase the removal of the antibiotics erythromycin and clarithromycin, two of the three antibiotics added in 2015 to the WFD watchlist of priority substances [24]. This unicellular green microalga is a dominant phytoplankton species in acidic mining lakes, so it is also very resistant to classic pollutants. Additionally, it accumulates high concentrations of lutein, a well-known antioxidant, which has been discovered to be useful for the treatment of oxidative diseases such as macular degeneration [25].

GCU, one of the research partners within Phos4You, aimed at demonstrating the potential of microalgae to recover phosphorus from small-scale WWTPs. A demonstration plant is constructed in real conditions for a typical location of 200 PE, with a typical daily flow of 30 m^3 and an aim of removing 50–75% of phosphorous. The large-scale production and the harvesting of algae are the two major challenges to implementing a microalgae system in order to produce a valuable bioproduct [26]. Most of the current algal bioreactors rely on suspended cultures. These culture systems require a large amount of energy to harvest algal cells, via centrifugation, coagulation, and filtration, and to eliminate water downstream – estimated to be up to 30% of the costs [27–29]. Centrifugation is rapid but energy-intensive; coagulation may be a pre-treatment to centrifugation to improve recovery of cells but coagulant may influence the quality of the final product, and the high cost of membrane replacement and pumping restricts the application of filtration in large-scale operations [21, 28, 30]. Therefore, new methods for microalgal biomass recovery are required. Growing microalgae on surfaces is an attractive option and are showing promising results [31]. When microalgae are cultivated as biofilms, it is much easier to separate the biomass from the medium and the costs of scaling the system up can be significantly reduced [30, 32, 33].

Building on past experience, GCU demonstrated that a fixed-bed reactor design is suitable for *C. acidophila* cultivation, as similar nutrient recovery rates were observed in cultures with suspended cells and immobilized microalgae. Of the substrates evaluated, untreated wool appears to be most suitable for microalgae cell attachment. However, the adoption of a biofilm system of culture provides specific design challenges relating to the construction of a unique fixed-bed reactor to meet the demands of operating in the Scottish environment. These challenges include optimizing effluent flow through the system and maximizing the algal nutrient recovery rate. These together with evaluating the effectiveness of the system in terms of phosphorus recovery, gauging the potential usage of the end-products, as well as quantifying the energy demand, are addressed to identify the suitability of using microalgae to recover phosphate from small-scale WWTPs in remote, rural, and island areas.

References

1. EUROSTAT. Sewage sludge production and disposal from urban wastewater (in dry substance (d.s)). 2017. Available from: http://ec.europa.eu/eurostat/web/products-datasets/-/ten00030 (accessed December 23, 2019).
2. European Comission. Council Directive 91/271/EEC of 21 May 1991 Concerning Urban Waste-Water Treatment (UWWTD). 1991.
3. Survey USG (2017). *Mineral Commodity Summaries 2017*. Reston, VA: US Geological Survey.
4. European Comission. Report on Critical Raw Materials for the EU. 2014.
5. EUROSTAT. Phosphorus Fertilizer Consumption by Agriculture. EU-27, NO and CH, 2000–2012.
6. European Comission. Regulation (EU) 2019/1009 of the European Parliament and of the Council of 5 June 2019 laying down rules on the making available on the market of EU fertilising products and amending Regulations (EC) No 1069/2009 and (EC) No 1107/2009 and repealing Regulation (EC) No 2003/2003. Official Journal of the European Union. 2019;L 170/1.
7. Verordnung über die Vermeidung und die Entsorgung von Abfällen. December 4, 2015. Swiss Federal Council.
8. Verordnung zur Neuordnung der Klärschlammverwertung. September 27, 2017. German Federal Government.
9. Meststoffenwet. Dutch parliament. June 13, 2019.
10. National Wastewater Sludge Management Plan. Irish Water; 2016.
11. Braak, E., Auby, S., Piveteau, S., Guilayn, F., and Daumer, M.L. (2016). Phosphorus recycling potential assessment by a biological test applied to wastewater sludge. *Environmental Technology* **37** (11): 1398–1407.
12. Gaterll, M.R., Gay, R., Wilson, R. et al. (2011). An economic and environmental evaluation of the opportunities for substituting phosphorus recovered from wastewater treatment works in existing UK fertiliser markets. *Environmental Technology* **21**: 1067–1084.
13. Centre Européen d'Etude des Polyphosphates, Scope Newsletter n°63, January 2006.
14. Shilton, A.N., Powell, N., and Guieysse, B. (2012). Plant based phosphorus recovery from wastewater via algae and macrophytes. *Current Opinion in Biotechnology* **23** (6): 884–889.
15. Wilsenach, J.A., Schuurbiers, C.A.H., and van Loosdrecht, M.C.M. (2007). Phosphate and potassium recovery from source separated urine through struvite precipitation. *Water Research* **41** (2): 458–466.
16. Harouiya, N., Rue, S.M., Prost-Boucle, S. et al. (2011). Phosphorus removal by apatite in horizontal flow constructed wetlands for small communities: pilot and full-scale evidence. *Water Science and Technology* **63** (8): 1629–1637.
17. Li, Y.C., Chen, Y.F., Chen, P. et al. (2011). Characterization of a microalga *Chlorella* sp well adapted to highly concentrated municipal wastewater for nutrient removal and biodiesel production. *Bioresource Technology* **102** (8): 5138–5144.
18. Martinez, M.E., Sanchez, S., Jimenez, J.M. et al. (2000). Nitrogen and phosphorus removal from urban wastewater by the microalga *Scenedesmus obliquus*. *Bioresource Technology* **73** (3): 263–272.
19. Mulbry, W., Kondrad, S., Pizarro, C., and Kebede-Westhead, E. (2008). Treatment of dairy manure effluent using freshwater algae: algal productivity and recovery of manure nutrients using pilot-scale algal turf scrubbers. *Bioresource Technology* **99** (17): 8137–8142.
20. Valderrama, L.T., Del Campo, C.M., Rodriguez, C.M. et al. (2002). Treatment of recalcitrant wastewater from ethanol and citric acid production using the microalga *Chlorella vulgaris* and the macrophyte *Lemna minuscula*. *Water Research* **36** (17): 4185–4192.

21. Brennan, L. and Owende, P. (2010). Biofuels from microalgae-a review of technologies for production, processing, and extractions of biofuels and co-products. *Renewable and Sustainable Energy Reviews* **14** (2): 557–577.
22. Schiraldi, C. and De Rosa, M. (2002). The production of biocatalysts and biomolecules from extremophiles. *Trends in Biotechnology* **20** (12): 515–521.
23. Escudero, A., Blanco, F., Lacalle, A., and Pinto, M. (2014). Ammonium removal from anaerobically treated effluent by *Chlamydomonas acidophila. Bioresource Technology* **153**: 62–68.
24. European Comission. Directive 2008/105/EC, annex X. European Parliament and Council; 2008.
25. Cuaresma, M., Casal, C., Forjan, E., and Vilchez, C. (2011). Productivity and selective accumulation of carotenoids of the novel extremophile microalga *Chlamydomonas acidophila* grown with different carbon sources in batch systems. *Journal of Industrial Microbiology & Biotechnology* **38** (1): 167–177.
26. Christenson, L. and Sims, R. (2011). Production and harvesting of microalgae for wastewater treatment, biofuels, and bioproducts. *Biotechnology Advances* **29** (6): 686–702.
27. Gross, M., Henry, W., Michael, C., and Wen, Z.Y. (2013). Development of a rotating algal biofilm growth system for attached microalgae growth with in situ biomass harvest. *Bioresource Technology* **150**: 195–201.
28. Grima, E.M., Belarbi, E.H., Fernandez, F.G.A. et al. (2003). Recovery of microalgal biomass and metabolites: process options and economics. *Biotechnology Advances* **20** (7–8): 491–515.
29. Uduman, N., Qi, Y., Danquah, M.K. et al. (2010). Dewatering of microalgal cultures: a major bottleneck to algae-based fuels. *Journal of Renewable and Sustainable Energy* **2** (1).
30. Zhuang, L.L., Hu, H.Y., Wu, Y.H. et al. (2014). A novel suspended-solid phase photobioreactor to improve biomass production and separation of microalgae. *Bioresource Technology* **153**: 399–402.
31. Katarzyna, L., Sai, G., and Singh, O.A. (2015). Non-enclosure methods for non-suspended microalgae cultivation: literature review and research needs. *Renewable and Sustainable Energy Reviews* **42**: 1418–1427.
32. Lin, Y.H., Leu, J.Y., Lan, C.R. et al. (2003). Kinetics of inorganic carbon utilization by microalgal biofilm in a flat plate photoreactor. *Chemosphere* **53** (7): 779–787.
33. Schnurr, P.J. and Allen, D.G. (2015). Factors affecting algae biofilm growth and lipid production: a review. *Renewable and Sustainable Energy Reviews* **52**: 418–429.

Section V

Agricultural and Environmental Performance of Biobased Fertilizer Substitutes: Overview of Field Assessments

5.1

Fertilizer Replacement Value: Linking Organic Residues to Mineral Fertilizers

René Schils[1,2], Jaap Schröder[2], and Gerard Velthof[1]

[1] Wageningen Environmental Research, Wageningen University & Research, Wageningen, The Netherlands

[2] Wageningen Plant Research, Wageningen University & Research, Wageningen, The Netherlands

5.1.1 Introduction

Organic residues of animal origin are an important nutrient source for crop production. In the European Union, the annual nitrogen (N) input to agricultural soils is 25 490 Gg [1], or approximately 138 kg N ha^{-1}. Mineral fertilizer is the dominant source, representing 45% of the input, while applied and grazing manure represent 15 and 12%, respectively. Crop residues (15%), biological fixation (4%), and deposition (8%) complete the budget. For phosphorus (P), the data show similar large contributions from animal origin. The European phosphorus balance shows a total annual input to crops of 2967 Gg P [2], which equates to approximately 23 kg P ha^{-1}, comprising animal manure (53%), fertilizer (36%), sewage sludge (6%), compost (4%), and deposition (1%).

It is evident that the valuable nutrients from animal manures and other organic residues need to be applied to soils in an effective and efficient way, serving both agronomic efficiency and environmental quality. Due to the presence of organic nutrients, manures are more difficult to manage than mineral fertilizers. Nevertheless, it is essential that these "wastes" are reused to increase the nutrient use efficiency of our systems of food, feed, and other biomass production [3].

Biorefinery of Inorganics: Recovering Mineral Nutrients from Biomass and Organic Waste, First Edition.
Edited by Erik Meers, Gerard Velthof, Evi Michels and René Rietra.
© 2020 John Wiley & Sons Ltd. Published 2020 by John Wiley & Sons Ltd.

In areas where livestock is present, both mineral fertilizers and manures are applied to a single crop. The contrasting characteristics of these nutrient sources present a permanent challenge for farmers. While mineral fertilizers have a constant composition of inorganic components, manures are heterogeneous and also contain different organic fractions. Therefore, it is recognized that the fertilizing values of all nutrient sources need to be merged into a single useful figure. The increasing diversification in available organic residues makes the need for a correct assessment of the fertilizing value of these products more urgent. Next to straightforward traditional products such as slurry and farmyard manure, there is an increasing supply of other organic residues from animal and other origins. Manure processing, alone or in combination with feedstock from crops or industrial residues, will lead to an even wider range of products.

The objective of this chapter is to outline the theoretical concept of the fertilizer replacement value (FRV), and how it is derived in science and applied in farming practices. First, we explain the nutrient pathways from application of organic residues to crop uptake, identifying the different routes to nutrient losses. Then, we review the concept of FRVs and discuss methods of obtaining estimates, including potential pitfalls. We conclude with some examples of how FRVs are applied in fertilizer plans, comprising mineral fertilizers and manures.

The concept of FRVs may be applied to all nutrients, but here we focus on N and P. Both are important from an agronomic and environmental point of view. Furthermore, they are contrasting nutrients in the way they appear in manures and in their behavior in soil, water, crop, and air compartments. The behaviors of many other nutrients can be compared with either N or P.

5.1.2 Nutrient Pathways from Land Application to Crop Uptake

Manures are a mixture of feces and urine and are rest products of feed metabolized by animals. The feed digestibility and its nutrient contents are reflected in the excretion products. In view of its fate following application, N is present as inorganic N (ammonium, urea, and uric acid), easily decomposable organic N (proteins and amino acids), and more resistant organic matter (lignocellulosic fiber complex) [4]. In contrast to N, P in feed is of both organic and added inorganic origin, resulting in a wide variation in manure composition at the moment of excretion. The inorganic P content of manure can range from approximately 50 to 90% [5]. In a study by Sharpley and Moyer [6], the proportion of inorganic P increased in the order: dairy manures (63%) < poultry manure (84%) < swine slurry (91%).

Manure composition at the moment of land application depends not only on animal type and diet composition, but also on housing system, storage time and method, and further processing (Table 5.1.1). The N and P contents of slurries generally increase in the order cattle < pigs < poultry [7], but the ranges are large. Diet composition, in particular protein and P content, is a dominant factor in manure composition [12, 17]. The housing system affects manure composition in various ways. It is evident that there is a large contrast between solid manures from straw-based systems and slurries from cubicle systems [7]. Furthermore, floor type affects ammonia losses and thus the resulting N content. Further down the manure chain, storage method and time affect the ammonia losses and the mineralization of organic components [18]. Taking into account all these factors, the nutrient content

Table 5.1.1 *Non-exhaustive list illustrating the variation in composition of manures and other organic residues applied to agricultural land.*

Source	Animal	Manure type	n	DM ($g\,kg^{-1}$)	OM[b] ($g\,kg^{-1}$)	total N ($g\,kg^{-1}$)	NH4-N ($g\,kg^{-1}$)	NO3-N ($g\,kg^{-1}$)	Org. N ($g\,kg^{-1}$)	Inorg. N (%)	P ($g\,kg^{-1}$)	N/P
Menzi[a] [7]	Cattle	Slurry	27	67	57	4	2.3		1.7	58	0.61	6.6
	Pig	Slurry	27	52	38	4.8	3.4		1.4	71	0.92	5.2
	Poultry	Slurry	27	170	122	11.2	5.3		5.9	47	3.75	3.0
	Cattle	Farmyard manure[c]	27	207	165	5.2	1.4		3.8	27	1.05	5.0
	Pig	Farmyard manure[c]	27	238	161	6.8	2.2		4.6	32	2.35	2.9
	Poultry	Farmyard manure[c]	27	455	347	22.5	6.2		16.3	28	7.28	3.1
Nicholson et al. [8]	Layer	Deep pit	52	356	168	21	8		13	38	7	3.0
		Belt-scraped	27	286	147	17	5		12	29	5	3.4
		Perchery	5	397	209	22	3		19	14	5	4.4
		Free-range	4	579	242	34	5		29	15	9	3.8
	Broiler	Litter	21	642	385	33	6		27	18	11	3.0
	Turkey	Litter	10	523	314	27	7		20	26	10	2.7
	Duck	Litter	1	167	94	27	7		20	26	10	2.7
Fangueiro [9]	Pig	Untreated slurry	1	82.5	36.9	4.2	1.3		2.9	31	1.1	3.8
		Liquid fraction – centrifugation	1	4.9	1.9	1.4	1.2		0.2	86	0.02	70.0
		Solid fraction – centrifugation	1	251.7	118.1	10.4	1.6		8.8	15	5.7	1.8
		Liquid fraction – sedimentation	1	5.7	2.2	1.4	1.3		0.1	93	0.04	35.0
		Solid fraction – sedimentation	1	83.5	41.0	4.3	1.1		3.2	26	1.8	2.4
Schils et al. [10]	Dairy cattle	Nitric acid-treated slurry	12	106		7.6	1.9	3.4	2.3	70		
Velthof et al. [11]	Pig	Liquid mineral concentrate	16	36.9	14	8.15	7.51		0.64	92	0.16	50.9

Table 5.1.1 *Non-exhaustive list illustrating the variation in composition of manures and other organic residues applied to agricultural land.*

Source	Animal	Manure type	n	DM ($g\,kg^{-1}$)	OM[b] ($g\,kg^{-1}$)	total N ($g\,kg^{-1}$)	NH4-N ($g\,kg^{-1}$)	NO3-N ($g\,kg^{-1}$)	Org. N ($g\,kg^{-1}$)	Inorg. N (%)	P ($g\,kg^{-1}$)	N/P
Sorensen et al. [12]	Dairy cattle	Slurry; diet of early cut-grass silage and concentrates	3	740	65.3	5.32	3.52		1.8	66		
		Slurry; diet of hay without concentrates	3	790	65.7	2.92	1.4		1.52	48		
Möller et al. [13]	Dairy cattle	Undigested slurry	19	113	79.0	3.8646	1.7		2.2	43.4	0.62	6.2
		Digested slurry	19	92	58.7	3.9284	2.1		1.9	52.9	0.61	6.5
		Digested slurry and cereal straw	14	183	164.0	2.4156	0.2		2.2	9.5	0.27	8.8
		Digested slurry and pea straw	5	172	153.1	3.1992	0.2		3.0	5.6	0.41	7.8
		Digested slurry and cover crop	12	176	129.4	5.7376	1.5		4.3	25.7	0.7	8.2
		Digested slurry and grass/clover	10	203	168.5	4.7502	0.8		3.9	17.7	0.79	6.0
Barrena [14]		Industrial compost	25	550	286	14.3					5.5	2.6
		Home compost	27	215	123	4.7					1.72	2.8
D'Hose et al. [15]		Farm compost	4	504	131	3.9					1.35	2.9
Stutter [16]		Domestic sewage sludge, cambi process	1	280	553	11.088					5.82	1.9
		Anaerobic digestate slurry and abattoir waste	1	50	631	2.17					0.74	3.0
		Food waste compost	1	480	443	9.168					1.49	6.2
		Seaweed	1	200	633	1.38					0.56	2.5
		Green compost	1	600	657	8.1					1.62	5.0

[a] Based on a survey among 27 countries.
[b] Some sources presented carbon; converted to organic matter assuming 58% carbon in organic matter.
[c] Unit ($g\,l^{-1}$).

of slurries may vary from approximately 1.2 to 18 g N kg^{-1} and from 0.2 to 15 kg P_2O_5 kg^{-1}. Solid manures show variations from 2.0 to 58 g N l^{-1} and from 1.0 to 39 g P_2O_5 l^{-1}, respectively [7].

Manure processing adds further variation to manure composition. Mechanical separation results in a solid fraction (SF) and a liquid fraction (LF), with a higher proportion of the dry matter (DM) and organic matter contained in the SF [19]. Recently, more advanced technologies have been developed like ultrafiltration and reverse osmosis that further concentrate inorganic nutrients in the LF [11]. Almost complete water removal can be obtained by forced drying [20]. Anaerobic digestion breaks down organic components, leading to increased inorganic components compared to the original manure [21]. To increase the energy output of anaerobic digestion, other organic feedstocks are often co-digested. These may be specially grown annual or perennial "energy" crops or residues from food-processing industries. The amount of nutrients in the incoming feedstock also increases the nutrient content of the resulting digestate. Another biological treatment is composting, in which easily decomposable organic matter is broken down and inorganic components are immobilized [22]. Organic residues from non-animal origin, like household organic wastes and biomass from nature conservation areas, may also find their way to agricultural land through digestion or composting without manure. Finally, the use of additives can significantly alter manure composition. Dilution with water, often through addition of residual cleaning water, leads to reduced nutrient concentrations. Acids may be used to reduce slurry pH, and if this is carried out with nitric acid, the slurry will also contain nitrate [10]. Clay minerals and other chemical-physical additives may adsorb ammonium and prevent ammonia losses during storage [23].

5.1.2.1 Nitrogen

Inorganic N in applied manure is predominantly present as ammonium and is therefore vulnerable to ammonia losses (Figure 5.1.1). The main factors affecting ammonia losses are weather, slurry composition, and application method [24–26]. Ammonia losses are promoted by high wind speeds, high temperatures, and lack of rainfall. With respect to manure composition, ammonia losses are higher for manures with a relatively high DM content, high pH, and high total ammonia nitrogen (TAN). These effects of weather and manure composition significantly interact with application method. The dominant aspects of application technique that determine the ammonia loss are the area and duration of the existence of an exchange surface between manure and air. There are a wide variety of application methods, with surface spreading at one end of the spectrum and direct injection at the other. In between are methods like surface spreading followed by incorporation or irrigation, band spreading, trailing hose, trailing shoe, and shallow injection. In a summary of Dutch measurements using cattle and pig slurries, emission factors (% of TAN) varied from on average 2% (range 1–3%) for deep injection on arable land to on average 74% (range 28–100%) for surface spreading on grassland [24]. In another review, Sommer and Hutchings [27] found surface-application emission factors of 8–10% for pig slurries, 9–67% for cattle slurries, and 37–65% for farmyard manures. Irrigation or incorporation of farmyard manures reduced the emission to around 19%. Webb et al. [26] estimated mean abatement efficiencies as simple means and weighted means, taking account of the number of experiments

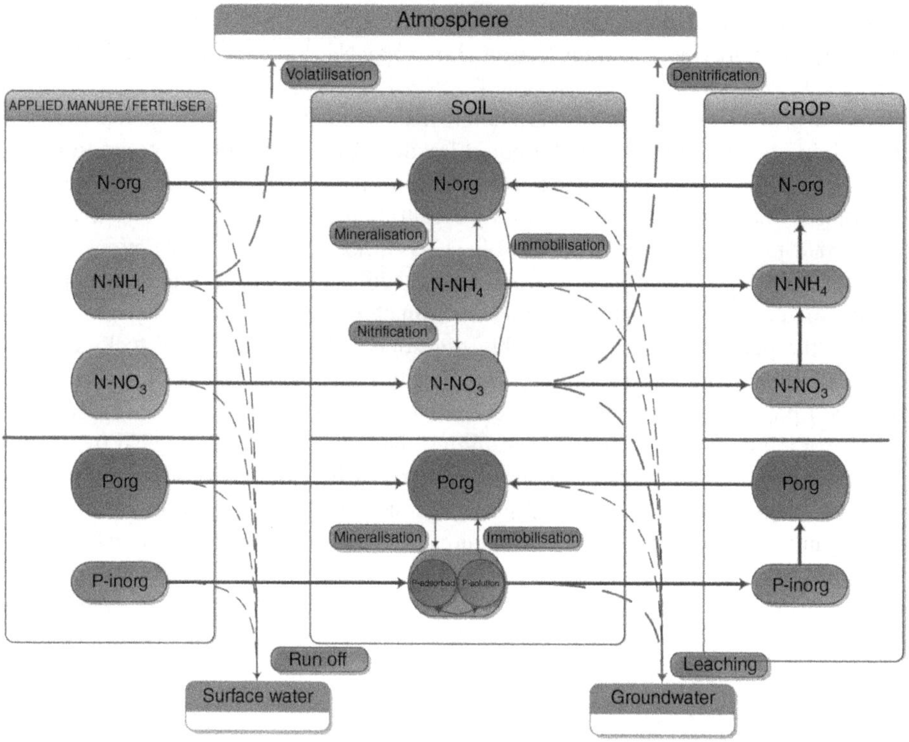

Figure 5.1.1 *Simplified representation of nitrogen and phosphorus pathways from land application of manures and fertilizers.*

reported in each paper. They concluded that abatement, compared to surface spreading, is greater from the use of trailing shoe (65%) and open-slot injection (70–80%) than from trailing hose (35%).

Ammonia that is not lost during and after application is relatively immobile in soils as it is adsorbed to clay particles and organic matter. Ammonium is however easily nitrified into nitrate. Crops take up most of their N during the vegetative growth period. In agricultural systems, most N is taken up in inorganic form. Generally, plants adapted to soils that are acid or which have a low redox potential preferably take up ammonium, whereas plants adapted to calcareous, high-pH soils prefer nitrates [28].

Nitrate is susceptible to denitrification through several pathways, leading to either nitrous oxide or dinitrogen losses [29]. The pathways and total losses depend on temperature, soil pH, and the availability of nitrate, degradable carbon, and oxygen. High quantities of nitrous oxide losses are observed in low-oxygen environments with sufficient nitrate and degradable carbon. Application techniques developed to reduce ammonia emissions generally promote nitrous oxide losses [24], and most likely also dinitrogen losses. Nitrous oxide emissions from manures increase with increasing contents of N and easily degradable carbon, such as in pig slurries compared to cattle slurries [30].

Following manure application, organic N components are turned over at different rates, depending on the breakdown characteristics of decomposable compounds. During the initial

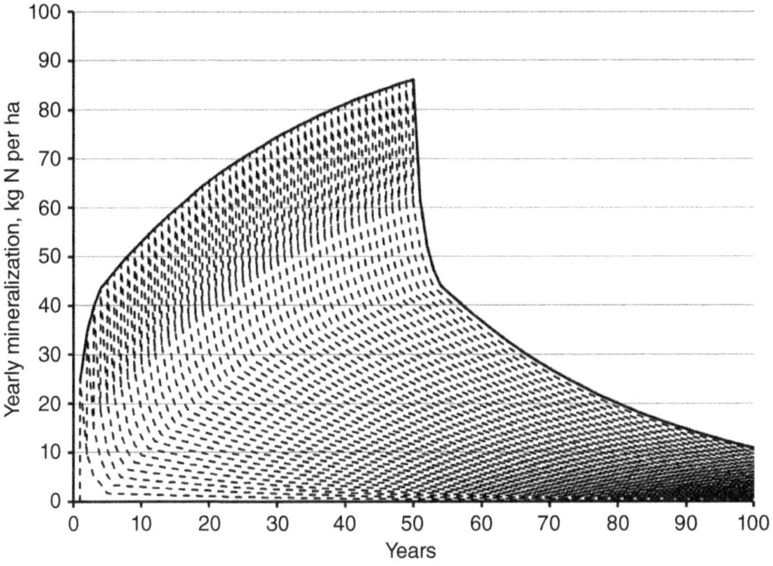

Figure 5.1.2 *Simulation of cumulative mineralization due to 50 yearly repeated manure applications. Source: Reprinted from Schröder et al. [31].*

days or weeks after application, immobilization is normally greater than mineralization. Part of the immobilized N is again mineralized soon after, whereas part of it is stabilized. The net mineralization is affected by the breakdown characteristics of the organic material, soil temperature, and C/N ratio. Nitrogen release from organic residues therefore varies greatly. This has important consequences for assessing the fertilizing value of manures, as illustrated by Schröder [3]. In the first year after its application, only a fraction of the organic input mineralizes (Figure 5.1.2). When a similar amount of organic material is applied in the subsequent year, the total mineralization equals the sum of a fraction of the total organic input in the second year and a fraction of the organic pool left over from the first. Eventually, an equilibrium will be established between the annual organic inputs and the annual mineralization.

Surface runoff may result in losses of organic and inorganic manure compounds following application to sloping land or on areas near watercourses, especially on bare land [32].

5.1.2.2 Phosphorus

Though applied manure generally contains large proportions of inorganic P, not all of it is readily available for crop uptake [6]. Several methods of sequential extraction, using for example water, bicarbonate, hydroxide, and acid, have been developed to characterize the P availability [5, 6]. Following application, the different compounds are included in the soil P complex (Figure 5.1.1). The different soil P fractions exist in a complex equilibrium as transformations between the forms occur continuously, from very stable to sparingly available to plant-available pools such as labile P and solution P in soils [33]. Phosphorus in soils is often categorized in three pools: solution, active, and fixed. The solution pool is usually in the orthophosphate form, which can be taken up by plants, but organic P may occur as well. The active or labile pool contains inorganic P, which is adsorbed to calcium, iron, or aluminum particles in the soil, and organic P, which is easily mineralized. The fixed or stable pool of P contains insoluble inorganic P compounds and organic compounds that are resistant to mineralization. Manure application affects the relative sizes of the different pools, directly through the different P compounds or indirectly through changes in soil pH.

Crops take up dissolved phosphates from the soil solution, which is then replenished from labile pools. Leaching of P to groundwater may occur, depending on adsorption capacity of the soil [34, 35]. Following manure application, P may be directly lost to surface water through runoff, with a high risk on sloping land, near watercourses, and on bare land [36].

5.1.3 Fertilizer Replacement Value

The importance of organic residues for crop production and soil fertility has long been recognized. In the nineteenth century, Von Liebig stated that "the fertility of a soil cannot remain unimpaired, unless we replace in it all those substances of which it has been … deprived" [37]. That was the function of manures and organic residues, such as animal bones, urine, guano, and "night soil" (human excreta). Following the successful large-scale introduction of mineral fertilizers, the need for manures and other organic residues declined. Though their appreciation as a crop nutrient never disappeared completely [38], mineral fertilizers became the benchmark. Gradually, the fertilizing value of manures was expressed relative to that of a standard mineral fertilizer. For example, Herron and Erhart [39] defined "nitrogen replacement value" as the amount of N from a commercial fertilizer required to produce an equivalent yield of grain. The more general expression of "fertilizer replacement value" was possibly coined for manures by Wilkinson [40] in a review paper on the value of animal manures. Against a background of rising oil prices, and thus rising fertilizer prices, the author pleaded for increased efforts to better use plant nutrients in animal manures. However, his paper did not present a formal definition. That seems to be the case in general; the fertilizing value is labeled with many different names with different underlying definitions. Besides FRV, examples of other terms used are manure nutrient efficiency, nutrient equivalents, mineral fertilizer equivalents, efficiency index, working coefficient, and fertilization effectiveness [41–43]. Adding to the confusion is the use and misuse of a wide range of underlying calculation procedures. Here, we propose a general definition of FRV, including essential specified elements that need to accompany any presentation of FRV's (Table 5.1.2):

> *The fertilizer replacement value (FRV) of an organic manure – given crop type, soil type, application time, and application method – specifies how much standard mineral fertilizer – given formulation, application time, and application method – is needed for a similar crop response measured over a given period. The FRV is expressed as kilograms of mineral fertilizer-nutrient per 100 kg manure-nutrient.*

Table 5.1.2 *Non-exhaustive list illustrating the variation in nitrogen fertilizer replacement values (NFRVs) of manures and other organic residues applied to agricultural land.*

Source	Manure type	Experiment	Test crop	Soil type	Test parameter	Period	Application method	Application time	Reference fertilizer	Treatments	NFRV (%)
Schröder et al. [44]	Liquid mineral concentrates (pig)	Field	Maize	Sand	N uptake whole crop	Growing season	Injection	April/May	CAN	Response curve, control plus three levels	78
			Potato	Sand	N uptake tuber	Growing season	Injection		CAN	Response curve, control plus three levels	83
			Potato	Loam	N uptake tuber	Growing season	Injection		CAN	Response curve, control plus three levels	76
Klop et al. [45]	Liquid mineral concentrates (pig)	Pot, controlled environment	Grass	Sand	N uptake whole crop	26 d	Injection		CAN	Control plus one level	96
							Surface application				62
	Slurry (pig)		Grass	Sand	N uptake, whole crop	26 d	Injection				79
							Surface application				41
Schils and Kok [46]	Slurry (cattle)	Field	Grass	Sand	N uptake, whole crop	Growing season	Injection	March, May, and June/July	CAN	Response curve, control plus two levels	56
							Surface application		CAN	Response curve, control plus two levels	38
Schröder et al. [47]	Slurry (cattle)	Field	Grass	Sand	N uptake, whole crop	Growing season	Injection	April, May, and June	CAN	Control plus one level	63

Table 5.1.2 *Non-exhaustive list illustrating the variation in nitrogen fertilizer replacement values (NFRVs) of manures and other organic residues applied to agricultural land.*

Source	Manure type	Experiment	Test crop	Soil type	Test parameter	Period	Application method	Application time	Reference fertilizer	Treatments	NFRV (%)
	Digested slurry (cattle)						Injection	April, May, and June			54
	Farm Yard Manure (cattle)						Surface application	February			34
						Four consecutive growing seasons	Injection		CAN	Control plus one level	71
							Injection				74
							Surface application				37
D'Hose et al. [15]	Farm compost	Field	Potato	Sand	DM yield tuber	Growing season	Surface application	Before sowing	AN	Control plus one level, fertilizer at more levels	17
			Fodder beet	Sand	DM yield beet						29
			Forage maize	Sand	DM yield, whole crop						−8
			Brussels sprouts	Sand	DM yield, sprouts						1
Sorensen et al. [12]	Slurry (cattle); diet of early cut grass silage and concentrates	Open-ended PVC cylinders placed in a field	Spring barley	Sand	N uptake, grain and straw		Injection	Before sowing	AN	Control plus one level, fertilizer at more levels	67
	Slurry (cattle); diet of hay without concentrates										57

Motavalli et al. [48]	Cattle		Corn	Sand	Nutrient uptake, whole crop	Growing season	Injection	Before sowing	AN/DAP	Response curve, control plus three levels	35
				Loam							47
				Loam							13
Lalor et al. [49]	Slurry (cattle)	Field	Grass		Nutrient uptake, whole crop	One cut	Surface application	March	CAN	Control plus one level, fertilizer at more levels	30
						Growing season					32
							Trailing shoe				40
											38
					DM yield, whole crop	One cut	Surface application		CAN		21
						Growing season					32
							Trailing shoe				30
											38
Delin et al. [50]	Manure (Layer)	Field	Spring barley	Clay	Grain yield	Growing season	Soil incorporation	Early spring	AN	Control plus one level, fertilizer at more levels	33
								At sowing			34
	Dry manure (Broiler)							Early spring			33
								At sowing			42
Cavalli et al. [51]	Dairy cattle slurry	Field	Silage maize	Loam	Nitrogen uptake, whole crop	Growing season	Soil incorporation	Before sowing	AS	Control plus one level	21
	Digested slurry–maize mix										46
	Liquid fraction of digested slurry										43
	Solid fraction of digested slurry										14

5.1.3.1 Crop Response

The most essential part of the FRV definition is the crop response. Often, this is crop nutrient uptake, crop DM yield, crop fresh yield, or marketable fresh yield. Van Der Meer et al. [43], for instance, found that FRVs based on N uptake were consistently higher than those based on DM yield, following deep injection of cattle slurry to grassland. Crop responses are expressed relative to unfertilized controls, using indices as apparent nutrient recoveries and apparent nutrient efficiencies (see assessment methods, 5.1.4.3).

Instead of a crop response, there are also studies that use the soil nutrient status as a response parameter, especially for phosphate [42]. Instead of crop yield, a soil P test measurement with and without fertilizer or manure is used to calculate the FRV. These assessments are, however, often carried out in pots without a crop.

5.1.3.2 Response Period

FRVs are generally presented on an annual basis, but some studies have measured during shorter periods, such as one or two grass cuts. Schröder et al. [31] stress the importance of understanding the difference between short-term (one growing season) and long-term FRVs. The mineralization of organic nutrients in manures generally extends over a longer period than just 1 year. Manures therefore have a residual effect (Figure 5.1.2). FRVs established over a period of one growing season underestimate the actual FRV. The higher the C/N ratio in manures, the greater the difference between short- and long-term FRV.

5.1.4 Reference Mineral Fertilizer

The properties of the reference mineral fertilizer need to be known. These may vary between regions, depending on the most commonly used standard fertilizer (e.g. calcium ammonium nitrate or urea).

5.1.4.1 Crop and Soil Type

It is evident that crop types need to accompany any presentation of FRVs due to differences in length of growing season and nutrient uptake patterns. Soil type is relevant as nutrient losses and net mineralization rates differ between soils with different textures and organic matter contents [1, 52].

5.1.4.2 Application Time and Method

The application time and method of the manure under study and of the reference fertilizer need to be known. Application techniques affect ammonia losses and consequently also FRVs. Another important aspect of application technique is the horizontal placement, especially for crops with a wide row distance [53]. Application times in relation to the specific crop growing season affect nutrient losses from the soil and therefore FRVs. The timing of fertilizer and manure application present a particular problem when comparing crop responses. The timing of mineral fertilizers generally follows good agricultural practice and is applied around sowing for arable crops or is determined by other indicators, such as temperature sums on grassland. In farming practice, the time window for manure application is smaller due to other factors such as the carrying capacity of soils. Therefore, there is an inherent interaction between application time and nutrient source (manure vs. fertilizer).

5.1.4.3 Assessment Method

Ideally, the FRV is calculated as the ratio between the apparent nutrient efficiencies or recoveries of the organic residue and those of the reference mineral fertilizer. The apparent nutrient recovery equals the amount of additional nutrient crop uptake on treated plots compared to that of an untreated control, expressed per kilogram of applied nutrient. The apparent nutrient efficiency is similar but uses the amount of DM or fresh marketable crop yield as the response variable.

- apparent nutrient recovery ($kg\,kg^{-1}$) = (nutrient uptake manure- or fertilizer-treated plot – nutrient uptake untreated control)/nutrient application rate
- apparent nutrient efficiency (kg crop yield kg $nutrient^{-1}$) = (crop yield manure- or fertilizer-treated plot – crop yield untreated control)/nutrient application rate
- FRV ($kg\ 100\,kg^{-1}$) =
  - apparent nutrient $recovery_{manure}$ * 100/apparent nutrient $recovery_{fertilizer}$, or
  - apparent nutrient $efficiency_{manure}$ * 100/apparent nutrient $efficiency_{fertilizer}$

The recoveries and efficiencies can be determined in field experiments in which the effects of one or several manure and fertilizer application rates, including an untreated control, are compared. The mineral fertilizer is applied according to best practice; that is, at the start of the growing season. Several levels of application rate are preferred because the response curve to nutrients is generally not linear. In an ideal set-up, both reference fertilizer and manure are applied at different rates. However, many studies use only one rate of manure, which is compared with a set of different fertilizer application rates. There is some risk to this approach as assessment of the fertilizing value needs to be done at suboptimal nutrient supply. The objective of this type of experiment is to assess the FRV of a single nutrient in manure. Therefore, all other macro- and micronutrients should be supplied adequately. Even in such a seemingly ideal set-up there are still some remaining pitfalls. Organic residues contain organic matter that may have other beneficial properties than the nutrient studied. Examples are improving water retaining capacity [54] and changing disease pressure [55]. Furthermore, the application method may negatively affect crop yields, through either soil compaction in tire tracks, smothering and scorching of leaves, or desiccation along application slits. These negative effects can be measured on plots where manure application is combined with ample mineral fertilizer, making sure that none of the nutrients are limiting. Using this approach, Prins and Snijders [56] observed approximately 10% yield reductions in the first cut following deep cattle slurry injection on grassland. An alternative approach to account for these negative effects is to include an additional control plot where the manure application equipment passes over the plot but without applying any manure.

Similar experimental set-ups can also be carried out in pot experiments in laboratory facilities, or under controlled field conditions. These types of experiments are suitable for comparing the relative FRVs of many organic residues simultaneously or in combination with different test crops or soil types, under pre-defined controlled circumstances [57]. However, the translation of results to practical field conditions requires a cautious approach, preferably with some identical treatments carried out in field trials to establish the absolute FRV.

If either field or laboratory experiments are not (yet) feasible, a chemical characterization of the manure in question can give an initial assessment of the FRV. By analyzing the total inorganic and organic nutrient content, a first estimate is possible. Additional incubation experiments to quantify the decomposition rate of organic constituents will allow for an improved estimate. Combining this with information on intended application time and method, crop type, and soil type will specify and improve the assessment. Models can assist in calculating the nutrient supply from inorganic and organic manure components [58].

5.1.5 Fertilizer Replacement Values in Fertilizer Plans

Fertilizer plans or nutrient management plans aim to match nutrient supply from fertilizers, manures, and other sources to nutrient demand from crops. A correct match between supply and demand increases nutrient use efficiency, improves crop yield and quality, reduces costs, and protects the environment. Basically, there are four steps in nutrient planning: (i) determine crop demand; (ii) determine soil supply; (iii) calculate nutrient supply from manures; and (iv) calculate additional fertilizer requirement. The third step, in which nutrient supply from manures is quantified, requires adequate information on the FRV. Farmers should therefore have access to tables with FRVs for all combinations of manure types, crops, soils, application times, and methods. Experimental results generally present an incomplete picture, not representing all combinations of manures, crops, and soils. Therefore, researchers and extension specialists will need to assess and translate the empirical results into a consistent picture, relevant for the specific country. This type of work is normally not peer-reviewed and published, or even worse, it may be poorly documented. Table 5.1.3 presents some FRVs from official fertilizer recommendations in the Netherlands and part of the United Kingdom. In the Netherlands, FRV is presented in relation to manure source, application time and method, and crop type, while the UK recommendations also distinguish between soil types. For some manures in the Dutch recommendations, nitrogen FRVs are presented separately for inorganic and organic fractions. Preferably, farmers should possess manure analyses with information on inorganic and organic N and should not have to resort to default values for nutrient contents. Even when comparing FRVs from only two countries, there are already quite remarkable differences. Webb et al. [62] compared N replacement values used in the European Union and concluded that none of the Member States presented a specific reference to, or definition of, manure N efficiency in their Action Program or Code of Good Agricultural Practice. While there were differences in estimates of the proportions of N available for crop uptake in any given type of manure, there was reasonable agreement in the ranking of manures, with N availability decreasing in the order pig slurry > cattle slurry > poultry manure > farmyard manure. They did not find any systematic variation in the reported manure N efficiency related to climate.

5.1.6 Conclusion

Organic residues from animal origin are an important nutrient source for crop production. While mineral fertilizers have a consistent composition of inorganic components, manures are heterogeneous and also contain different organic fractions. The increasing diversification in available organic residues necessitates a correct assessment of the fertilizing value of these products. The fertilizing value of manures is labeled with many different names with different underlying definitions. We propose the following general definition:

The fertilizer replacement value (FRV) of an organic residue – given crop type, soil type, application time, and application method – specifies how much standard mineral fertilizer – given formulation, application time, and application method – is needed for a similar crop response measured over a given period. The FRV is expressed as kilograms mineral fertilizer-nutrient per 100 kg manure-nutrient.

The FRV is calculated as the ratio between the apparent nutrient efficiencies or recoveries of the organic residue and the reference mineral fertilizer. The recoveries and efficiencies can be determined in field or laboratory experiments in which the effects of one or several manure and fertilizer application rates, including an untreated control, are compared. Alternatively, chemical characterization in combination with incubation studies or modeling exercises can be used to assess the FRV. We propose that FRVs should be established in the following order of preference:

1. field experiment with appropriate response curves for both reference fertilizer and manure;
2. field experiment with appropriate response curves for reference fertilizer and one moderate level of manure;
3. controlled experiment in field or laboratory with appropriate response curves for both reference fertilizer and manure;
4. controlled experiment in field or laboratory with response curves for reference fertilizer and one moderate level of manure;
5. chemical characterization of nutrient contents and fractions, in combination with incubation study or modeling; and
6. chemical characterization of nutrient contents and fractions.

FRVs are used in fertilizer plans to calculate the nutrient supply from manures. As experimental data are generally only partly available, an additional assessment is necessary by researchers and extension specialists to present all relevant combinations of manure types, crops, soils, application times, and methods.

Table 5.1.3 *Example of annual fertilizer replacement values (FRVs) of manures in the Netherlands and United Kingdom [59–61].*

Manure	Application method	Application time	Soil	Crop	NFRV_inorg	NFRV_org	NFRV_tot	PFRV
The Netherlands								
Cattle slurry	Shallow injection	Spring	All	Grass	76	24	49[a]	100
Cattle slurry	Trailing shoe	Spring	All	Grass	64	24	44[a]	100
Pig slurry	Shallow injection	Spring	All	Grass	76	24	58[a]	100
Pig slurry	Trailing shoe	Spring	All	Grass	64	24	50[a]	100
Cattle and pig farmyard manure	Surface application	Spring	All	Grass			20	80
Cattle and pig farmyard manure	Surface application	Fall	All	Grass			10	80
Poultry manure	Surface application	Spring	All	Grass			35	80
Poultry manure	Surface application	Fall	All	Grass			20	80
Cattle slurry	Injection	Spring	All	Arable crops	95	20	57[a]	60
Cattle slurry	Incorporation after surface application	Spring	All	Arable crops	80	20	49[a]	60
Pig slurry	Injection	Spring	All	Arable crops	95	60	83[a]	100
Pig slurry	Incorporation after surface application	Spring	All	Arable crops	80	60	73[a]	100
Cattle farmyard manure	Incorporation after surface application	Spring	All	Arable crops	80	15	26[a]	60
Poultry dry manure	Incorporation after surface application	Spring	All	Arable crops	80	60	62[a]	70
Poultry litter manure	Incorporation after surface application	Spring	All	Arable crops	80	60	63[a]	70
Broiler manure	Incorporation after surface application	Spring	All	Arable crops	80	60	65[a]	70
Compost (Champost)	Incorporation after surface application	Spring	All	Arable crops	80	35	37[a]	80

England, Wales, Northern Ireland							
Farmyard manure	Surface application	Fall	Sandy shallow	All crops		5	60
Farmyard manure	Surface application	Fall	Medium/heavy	All crops		10	60
Farmyard manure	Surface application	Winter	Sandy shallow	All crops		10	60
Farmyard manure	Surface application	Winter	Medium/heavy	All crops		10	60
Farmyard manure	Surface application	Spring	All	All crops		10	60
Farmyard manure	Surface application	Summer	All	Grass		10	60
Fresh farmyard manure	Incorporation after surface application	Fall	Sandy shallow	All crops		5	60
Fresh farmyard manure	Incorporation after surface application	Fall	Medium/heavy	All crops		10	60
Fresh farmyard manure	Incorporation after surface application	Winter	Sandy shallow	All crops		10	60
Fresh farmyard manure	Incorporation after surface application	Winter	Medium/heavy	All crops		10	60
Fresh farmyard manure	Incorporation after surface application	Spring	All	All crops		10	60
Old farmyard manure	Incorporation after surface application	Fall	Sandy shallow	All crops		5	60
Old farmyard manure	Incorporation after surface application	Fall	Medium/heavy	All crops		10	60
Old farmyard manure	Incorporation after surface application	Winter	Sandy shallow	All crops		10	60
Old farmyard manure	Incorporation after surface application	Winter	Medium/heavy	All crops		10	60

(*Continued*)

Table 5.1.3 *(continued)*

Manure	Application method	Application time	Soil	Crop	NFRV_inorg	NFRV_org	NFRV_tot	PFRV
Old farmyard manure	Incorporation after surface application	Spring	All	All crops			15	60
Cattle slurry 2% DM	Surface application	Fall	Sandy shallow	Arable crops, except winter oilseed			5	50
Cattle slurry 2% DM	Surface application	Fall	Medium/heavy	Arable crops, except winter oilseed			30	50
Cattle slurry 2% DM	Surface application	Winter	Sandy shallow	Arable crops, except winter oilseed			30	50
Cattle slurry 2% DM	Surface application	Winter	Medium/heavy	Arable crops, except winter oilseed			30	50
Cattle slurry 2% DM	Surface application	Spring	All	Arable crops, except winter oilseed			45	50
Cattle slurry 2% DM	Surface application	Fall	Sandy shallow	Grassland and winter oilseed rape			10	50
Cattle slurry 2% DM	Surface application	Fall	Medium/heavy	Grassland and winter oilseed rape			35	50

Cattle slurry 2% DM	Surface application	Winter	Sandy shallow	Grassland and winter oilseed rape	30	50
Cattle slurry 2% DM	Surface application	Winter	Medium/heavy	Grassland and winter oilseed rape	30	50
Cattle slurry 2% DM	Surface application	Spring	All	Grassland and winter oilseed rape	45	50
Cattle slurry 2% DM	Surface application	Summer	All	Grassland	35	50
Cattle slurry 10% DM	Surface application	Fall	Sandy shallow	Arable crops, except winter oilseed	5	50
Cattle slurry 10% DM	Surface application	Fall	Medium/heavy	Arable crops, except winter oilseed	20	50
Cattle slurry 10% DM	Surface application	Winter	Sandy shallow	Arable crops, except winter oilseed	20	50
Cattle slurry 10% DM	Surface application	Winter	Medium/heavy	Arable crops, except winter oilseed	20	50
Cattle slurry 10% DM	Surface application	Spring	All	Arable crops, except winter oilseed	25	50

(*Continued*)

Table 5.1.3 *(continued)*

Manure	Application method	Application time	Soil	Crop	NFRV_inorg	NFRV_org	NFRV_tot	PFRV
Cattle slurry 10% DM	Surface application	Fall	Sandy shallow	Grassland and winter oilseed rape			10	50
Cattle slurry 10% DM	Surface application	Fall	Medium/heavy	Grassland and winter oilseed rape			25	50
Cattle slurry 10% DM	Surface application	Winter	Sandy shallow	Grassland and winter oilseed rape			20	50
Cattle slurry 10% DM	Surface application	Winter	Medium/heavy	Grassland and winter oilseed rape			20	50
Cattle slurry 10% DM	Surface application	Spring	All	Grassland and winter oilseed rape			25	50
Cattle slurry 10% DM	Surface application	Summer	All	Grassland			20	50
Cattle slurry 10% DM	Shallow injection	Fall	Sandy shallow	Arable crops, except winter oilseed			5	50
Cattle slurry 10% DM	Shallow injection	Fall	Medium/heavy	Arable crops, except winter oilseed			20	50
Cattle slurry 10% DM	Shallow injection	Winter	Sandy shallow	Arable crops, except winter oilseed			25	50

Cattle slurry 10% DM	Shallow injection	Winter	Medium/heavy	Arable crops, except winter oilseed	25	50
Cattle slurry 10% DM	Shallow injection	Spring	All	Arable crops, except winter oilseed	35	50
Cattle slurry 10% DM	Shallow injection	Fall	Sandy shallow	Grassland and winter oilseed rape	10	50
Cattle slurry 10% DM	Shallow injection	Fall	Medium/heavy	Grassland and winter oilseed rape	25	50
Cattle slurry 10% DM	Shallow injection	Winter	Sandy shallow	Grassland and winter oilseed rape	25	50
Cattle slurry 10% DM	Shallow injection	Winter	Medium/heavy	Grassland and winter oilseed rape	25	50
Cattle slurry 10% DM	Shallow injection	Spring	All	Grassland and winter oilseed rape	35	50
Cattle slurry 10% DM	Shallow injection	Summer	All	Grassland	30	50

[a]Calculated from separate NFRVs for inorganic and organic fractions, assuming a default ratio for the different manure types.

References

1. Sutton, M.A., Howard, C.M., Erisman, J.W. et al. (2011). *The European Nitrogen Assessment: Sources, Effects and Policy Perspectives*. Cambridge: Cambridge University Press.
2. Withers, P.A., van Dijk, K., Neset, T.-S. et al. (2015). Stewardship to tackle global phosphorus inefficiency: the case of Europe. *Ambio* **44** (2): 193–206.
3. Schröder, J. (2005). Revisiting the agronomic benefits of manure: a correct assessment and exploitation of its fertilizer value spares the environment. *Bioresource Technology* **96** (2): 253–261.
4. Van Faassen, H. and Van Dijk, H. (1987). *Manure as a Source of Nitrogen and Phosphorus in Soils*. New York: Springer.
5. Dou, Z., Toth, J., Galligan, D. et al. (2000). Laboratory procedures for characterizing manure phosphorus. *Journal of Environmental Quality* **29** (2): 508–514.
6. Sharpley, A. and Moyer, B. (2000). Phosphorus forms in manure and compost and their release during simulated rainfall. *Journal of Environmental Quality* **29** (5): 1462–1469.
7. Menzi, H. (2002). Manure management in Europe. In: *10th FAO Ramiran Conference on Recycling of Organic Residues in Agriculture* (eds. J. Venglovsky and G. Greserova), 93–102. Kosice: University of Veterinary Medicine.
8. Nicholson, F., Chambers, B., and Smith, K. (1996). Nutrient composition of poultry manures in England and Wales. *Bioresource Technology* **58** (3): 279–284.
9. Fangueiro, D., Lopes, C., Surgy, S., and Vasconcelos, E. (2012). Effect of the pig slurry separation techniques on the characteristics and potential availability of N to plants in the resulting liquid and solid fractions. *Biosystems Engineering* **113** (2): 187–194.
10. Schils, R., Van Der Meer, H., Wouters, A. et al. (1999). Nitrogen utilization from diluted and undiluted nitric acid treated cattle slurry following surface application to grassland. *Nutrient Cycling in Agroecosystems* **53** (3): 269–280.
11. Velthof, G.L., Hoeksma, P., Schröder, J.J. et al. (eds.) (2012). *Agronomic Potential of Mineral Concentrate from Processed Manure as Fertiliser*. York: International Fertiliser Society.
12. Sørensen, P., Weisbjerg, M.R., and Lund, P. (2003). Dietary effects on the composition and plant utilization of nitrogen in dairy cattle manure. *Journal of Agricultural Science* **141** (1): 79–91.
13. Möller, K., Stinner, W., Deuker, A., and Leithold, G. (2008). Effects of different manuring systems with and without biogas digestion on nitrogen cycle and crop yield in mixed organic dairy farming systems. *Nutrient Cycling in Agroecosystems* **82** (3): 209–232.
14. Barrena, R., Font, X., Gabarrell, X., and Sánchez, A. (2014). Home composting versus industrial composting: influence of composting system on compost quality with focus on compost stability. *Waste Management* **34** (7): 1109–1116.
15. D'Hose, T., Cougnon, M., De Vliegher, A. et al. (2012). Farm compost application: effects on crop performance. *Compost Science & Utilization* **20** (1): 49–56.
16. Stutter, M.I. (2015). The composition, leaching, and sorption behavior of some alternative sources of phosphorus for soils. *Ambio* **44** (2): 207–216.
17. Maguire, R., Dou, Z., Sims, J. et al. (2005). Dietary strategies for reduced phosphorus excretion and improved water quality. *Journal of Environmental Quality* **34** (6): 2093–2103.
18. Webb, J., Menzi, H., Pain, B.F. et al. (2005). Managing ammonia emissions from livestock production in Europe. *Environmental Pollution* **135** (3): 399–406.
19. Hjorth, M., Christensen, K.V., Christensen, M.L., and Sommer, S.G. (2010). Solid–liquid separation of animal slurry in theory and practice: a review. *Agronomy for Sustainable Development* **30** (1): 153–180.
20. Ghaly, A. and Alhattab, M. (2013). Drying poultry manure for pollution potential reduction and production of organic fertilizer. *American Journal of Environmental Sciences* **9** (2): 88.
21. Sommer, S.G., Christensen, M.L., Schmidt, T., and Jensen, L.S. (2013). *Animal Manure Recycling: Treatment and Management*. Chichester: Wiley.

22. Barker, A.V., Rechcigl, J., and Mac Kinnon, H. (1997). Composition and uses of compost. In: *Agricultural Uses of By-Products and Wastes* (eds. J.E. Rechcigl and H.C. MacKinnon), 140–162. Washington, DC: ACS.
23. Lefcourt, A. and Meisinger, J. (2001). Effect of adding alum or zeolite to dairy slurry on ammonia volatilization and chemical composition. *Journal of Dairy Science* **84** (8): 1814–1821.
24. Huijsmans, J. and Schils, R. (eds.) (2009). *Ammonia and Nitrous Oxide Emissions Following Field Application of Manure: State of the Art Measurements in The Netherlands*. York: International Fertiliser Society.
25. Misselbrook, T., Nicholson, F., and Chambers, B. (2005). Predicting ammonia losses following the application of livestock manure to land. *Bioresource Technology* **96** (2): 159–168.
26. Webb, J., Pain, B., Bittman, S., and Morgan, J. (2010). The impacts of manure application methods on emissions of ammonia, nitrous oxide and on crop response – a review. *Agriculture, Ecosystems & Environment* **137** (1): 39–46.
27. Sommer, S.G. and Hutchings, N. (2001). Ammonia emission from field applied manure and its reduction – invited paper. *European Journal of Agronomy* **15** (1): 1–15.
28. Marschner, H. and Marschner, P. (2012). *Marschner's Mineral Nutrition of Higher Plants*. New York: Academic Press.
29. Butterbach-Bahl, K., Baggs, E.M., Dannenmann, M. et al. (2013). Nitrous oxide emissions from soils: how well do we understand the processes and their controls? *Philosophical Transactions of the Royal Society B: Biological Sciences* **368** (1621): 20130122.
30. Velthof, G.L., Kuikman, P.J., and Oenema, O. (2003). Nitrous oxide emission from animal manures applied to soil under controlled conditions. *Biology and Fertility of Soils* **37** (4): 221–230.
31. Schröder J, Bechini L, Bittman S, et al., eds. Residual N effects from livestock manure inputs to soils. RAMIRAN International Conference 2013: Recycling Agricultural, Municipal and Industrial Residues in Agriculture Network (RAMIRAN).
32. Smith, K., Jackson, D., and Pepper, T. (2001). Nutrient losses by surface run-off following the application of organic manures to arable land. 1. Nitrogen. *Environmental Pollution* **112** (1): 41–51.
33. Datta, A., Shrestha, S., Ferdous, Z., and Win, C.C. (2015). Strategies for enhancing phosphorus efficiency in crop production systems. In: *Nutrient Use Efficiency: From Basics to Advances* (eds. A. Rakshit, H.B. Singh and A. Sen), 59–71. New York: Springer.
34. Breeuwsma, A., Reijerink, J., and Schoumans, O. (1995). Impact of manure on accumulation and leaching of phosphate in areas of intensive livestock farming. In: *Animal Waste and the Land–Water Interface* (ed. K.F. Steele), 239–249. Boca Raton, FL: CRC Press.
35. Schoumans, O.F. and Chardon, W.J. (2015). Phosphate saturation degree and accumulation of phosphate in various soil types in The Netherlands. *Geoderma* **237–238**: 325–335.
36. Smith, K., Jackson, D., and Withers, P. (2001). Nutrient losses by surface run-off following the application of organic manures to arable land. 2. Phosphorus. *Environmental Pollution* **112** (1): 53–60.
37. Brock, W.H. (2002). *Justus von Liebig: The Chemical Gatekeeper*. Cambridge: Cambridge University Press.
38. Wadman, W., Sluijsmans, C., and Cremer, L.D.L.L. (1987). *Value of Animal Manures: Changes in Perception*. New York: Springer.
39. Herron, G.M. and Erhart, A.B. (1965). Value of manure on an irrigated calcareous soil 1. *Soil Science Society of America Journal* **29** (3): 278–281.
40. Wilkinson, S. (1979). Plant nutrient and economic values of animal manures. *Journal of Animal Science* **48** (1): 121–133.
41. Webb, J., Sorensen, P., Velthof, G.L. et al. (2011). *Study on Variation of Manure N Efficiency throughout Europe*. Didcot: AEA Technology.
42. Oenema, O., Chardon, W.J., Ehlert, P. et al. (eds.) (2012). *Phosphorus Fertilisers from By-Products and Wastes 2012*. Peterborough: International Fertiliser Society.

43. Van Der Meer, H., Thompson, R., Snijders, P., and Geurink, J. (1987). Utilization of nitrogen from injected and surface-spread cattle slurry applied to grassland. In: *Animal Manure on Grassland and Fodder Crops Fertilizer or Waste?* 47–71. New York: Springer.
44. Schroder, J.J., De Visser, W., Assinck, F.B.T. et al. (2014). Nitrogen fertilizer replacement value of the liquid fraction of separated livestock slurries applied to potatoes and silage maize. *Communications in Soil Science and Plant Analysis* **45** (1): 73–85.
45. Klop, G., Velthof, G., and Van Groenigen, J. (2012). Application technique affects the potential of mineral concentrates from livestock manure to replace inorganic nitrogen fertilizer. *Soil Use and Management* **28** (4): 468–477.
46. Schils, R. and Kok, I. (2003). Effects of cattle slurry manure management on grass yield. *NJAS – Wageningen Journal of Life Sciences* **51** (1–2): 41–65.
47. Schroder, J.J., Uenk, D., and Hilhorst, G.J. (2007). Long-term nitrogen fertilizer replacement value of cattle manures applied to cut grassland. *Plant and Soil* **299** (1–2): 83–99.
48. Motavalli, P., Kelling, K., and Converse, J. (1989). First-year nutrient availability from injected dairy manure. *Journal of Environmental Quality* **18** (2): 180–185.
49. Lalor, S.T.J., Schroder, J.J., Lantinga, E.A. et al. (2011). Nitrogen fertilizer replacement value of cattle slurry in grassland as affected by method and timing of application. *Journal of Environmental Quality* **40** (2): 362–373.
50. Delin, S. (2011). Fertilizer value of nitrogen in hen and broiler manure after application to spring barley using different application timing. *Soil Use and Management* **27** (4): 415–426.
51. Cavalli, D., Cabassi, G., Borrelli, L. et al. (2016). Nitrogen fertilizer replacement value of undigested liquid cattle manure and digestates. *European Journal of Agronomy* **73**: 34–41.
52. Sharpley, A., Daniel, T., Sims, J., and Pote, D. (1996). Determining environmentally sound soil phosphorus levels. *Journal of Soil and Water Conservation* **51** (2): 160–166.
53. Schröder, J., Vermeulen, G., van der Schoot, J. et al. (2015). Maize yields benefit from injected manure positioned in bands. *European Journal of Agronomy* **64**: 29–36.
54. Haynes, R. and Naidu, R. (1998). Influence of lime, fertilizer and manure applications on soil organic matter content and soil physical conditions: a review. *Nutrient Cycling in Agroecosystems* **51** (2): 123–137.
55. Bailey, K. and Lazarovits, G. (2003). Suppressing soil-borne diseases with residue management and organic amendments. *Soil and Tillage Research* **72** (2): 169–180.
56. Prins, W.H. and Snijders, P.J. (1987). *Negative Effects of Animal Manure on Grassland Due to Surface Spreading and Injection*. New York: Springer.
57. Antil, R., Janssen, B., and Lantinga, E. (2009). Laboratory and greenhouse assessment of plant availability of organic N in animal manure. *Nutrient Cycling in Agroecosystems* **85** (1): 95–106.
58. Nicholson, F., Bhogal, A., Chadwick, D. et al. (2013). An enhanced software tool to support better use of manure nutrients: MANNER-NPK. *Soil Use and Management* **29** (4): 473–484.
59. Hoeks P. Adviesbasis bemesting grasland en voedergewassen. Praktijkonderzoek Veehouderij. Available from: www.bemestingsadvies.nl (accessed December 23, 2019).
60. Van Dijk W, Van Geel W. Adviesbasis voor de bemesting van akkerbouw-en vollegrondsgroentegewassen. Available from: http://www.kennisakker.nl/kenniscentrum/handleidingen/adviesbasis-voor-de-bemesting-van-akkerbouwgewassen-secundaire-hoofdelem (accessed December 23, 2019).
61. Department for Environment, Food & Rural Affairs. Fertiliser Manual RB209. Available from: https://www.gov.uk/government/publications/fertiliser-manual-rb209--2 (accessed December 23, 2019).
62. Webb, J., Sorensen, P., Velthof, G. et al. (2013). An assessment of the variation of manure nitrogen efficiency throughout Europe and an appraisal of means to increase manure-N efficiency. In: *Advances in Agronomy*, vol. **119** (ed. D.L. Sparks), 371–442. Amsterdam: Elsevier.

5.2

Anaerobic Digestion and Renewable Fertilizers: Case Studies in Northern Italy

Fabrizio Adani[1], Giuliana D'Imporzano[1], Fulvia Tambone[1], Carlo Riva[1], Gabriele Boccasile[2], and Valentina Orzi[1]

[1]*Gruppo Ricicla Lab., Università degli Studi di Milano, DISAA, Milan, Italy*

[2]*Regione Lombardia DG-Agricoltura, Milan, Italy*

5.2.1 Introduction

Lombardy Region is one of the largest and most productive agricultural areas in Italy and in Europe; it has a total surface area of 23 861 km^2 and a Utilized Agricultural Area (UAA) equal to 986 853 ha. Livestock breeding is one of its main activities, and Lombardy is home to 4 500 000 pigs, 1 700 000 cows, and 18 000 000 poultry (LSU of 1.35 Mg Ha^{-1}), all of which generate a surplus of nutrients (nitrogen and phosphorus) due to the concentration of breeding activities on limited land.

Lombardy and EU meat and dairy production are dependent on protein feed imports (e.g. Europe imports over 20 million Mg of soymeal from outside the EU countries, and Lombardy alone almost 1 million). The net input of feed from outside the region into Lombardy entails the import of 60 000 Mg of N per year (elaborated data, source: EUROSTAT, ISTAT). As a consequence, Lombardy produces manure and slurry equal to 38 million m^3 (data of 2013), causing 140 000 Mg of N to be spread on the fields in the region. In addition, intensive agriculture uses N and P from chemical fertilizers to add nutrients to the soil (96 396 Mg N a^{-1}) so that total N loading on soil is about 270–290 kg ha^{-1}.

Biorefinery of Inorganics: Recovering Mineral Nutrients from Biomass and Organic Waste, First Edition.
Edited by Erik Meers, Gerard Velthof, Evi Michels and René Rietra.
© 2020 John Wiley & Sons Ltd. Published 2020 by John Wiley & Sons Ltd.

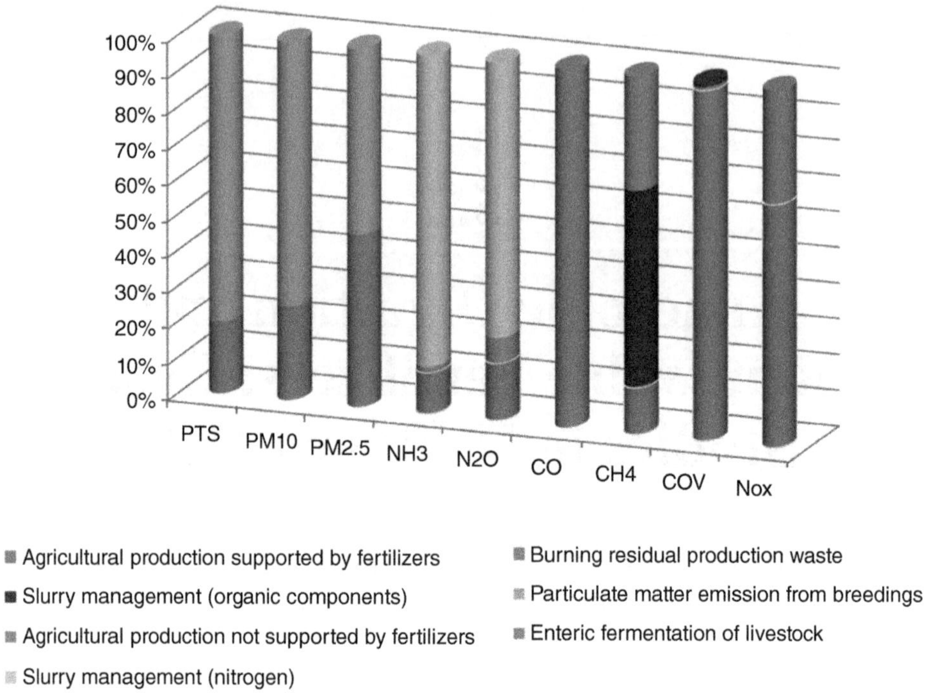

Figure 5.2.1 *Percentage contribution of agriculture activities to different impact categories. PTS, total suspended particles; PM10, particulate of 10 μm; PM 2.5, particulate of 2.5 μm; NH3, ammonia-N; N2O, ammonia dioxide; CO, carbon monoxide; CH4, methane; COV, volatile organic carbon; Nox, nitrogen oxides.*

Agricultural activities have affected pollutant production, particularly around slurry management (Figure 5.2.1). An EU Directive aimed at reducing the impact of the use of slurries was introduced in 1991 [1].

Regions appoint Nitrate Vulnerable Zones (NVZs) under the coordination of the Ministry of Environment. Currently, NVZs cover approximately 56% of the agricultural land in Lombardy, accounting for the whole territory of 419 municipalities and partially covering 216 more. The Italian Regulation concerns both NVZs and Nitrate Not-Vulnerable Zones (non-NVZs): the main difference between them is the amount of N from manure that is allowed to be spread, which is limited to 170 kg ha^{-1} in NVZs (or 250 kg ha^{-1} for farms applying for derogation) and 340 kg ha^{-1} in non-NVZs.

Efforts have been made in Lombardy Region in the last 10 years to improve the efficiency of N contained in slurry (NUE) by reducing chemical N input (thus fostering the use of manure in place of mineral fertilizers) and hence the total input of N in the agricultural system.

5.2.2 Anaerobic Digestion as a Tool to Correctly Manage Animal Slurries

Recently, a tangible opportunity to improve the sustainable management of nutrient wastes from animal slurries has been the diffusion in many EU intensive-farming areas

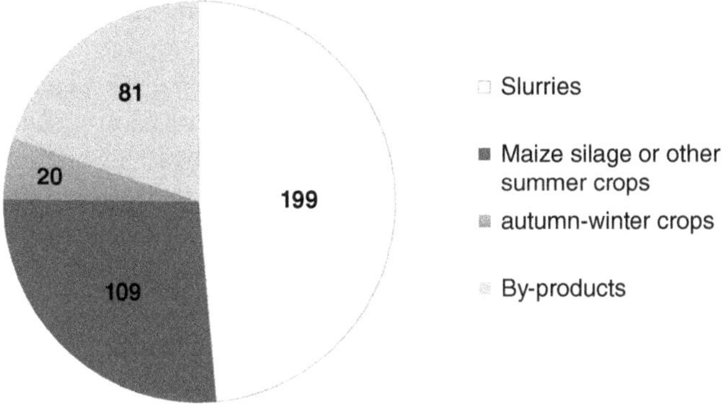

Figure 5.2.2 Feedstock for a biogas plant in Lombardy. Total amount as Mg × 10^6. Source: Project Ecobiogas, Lombardy Region, 2013, unpublished.

of anaerobic digestion (AD) technology. AD efficiently mineralizes nutrients present in the liquid digestion phase, so that digestate can be useful recycled as fertilizer [2], increasing NUE.

In the past 6 years Lombardy Region has developed a wide AD chain based on the livestock sector, including more than 400 AD biogas plants that allow the sustainable management of slurry and nutrients. In Lombardy, unlike what has happened in other countries (e.g. Germany) [3], dedicated energy crops represent only a small proportion of total AD feedstock, with the greater part of the biomass being slurry and by-products (Figure 5.2.2).

A recent project, financed by Lombardy Region (Ecobiogas Project – Lombardy Region – N. 1713, 2013 – unpublished), aimed to measure the potential impact of slurry management by AD and subsequent digestate use as a substitute for mineral fertilizers, and has indicated how effectively this approach reduced impacts on *human health*, *ecosystems*, and *resources* (Figure 5.2.3). The study was performed by Life Cycle Assessment (LCA),

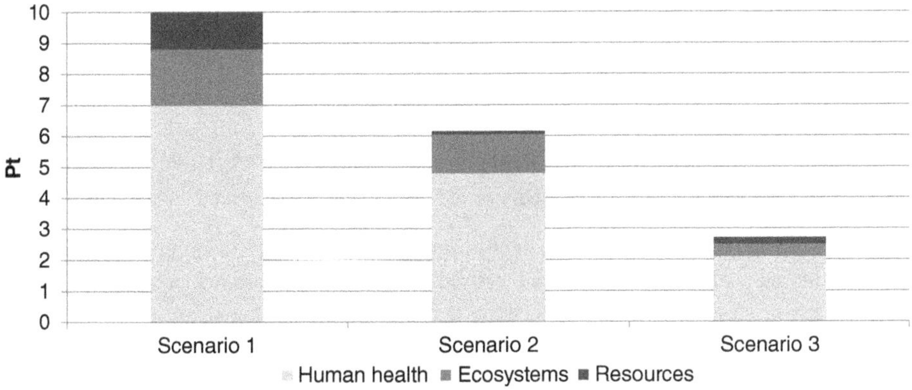

Figure 5.2.3 Impacts on human health, ecosystem, and resources of three different scenarios: 1. Agricultural standard management; 2. Good practices in slurry management; 3. Biogas and best practices in digestate management. Source: Project Ecobiogas, Lombardy Region, 2013, unpublished.

using Simapro software and according to the calculation method Recipe V1.06 (2008) Endpoint H/A Europe, starting from data collected on the territory (Lombardy Region). The functional unit was the hectare of land used for agricultural production. Each scenario considered looked at a "typical farm" based on average Lombardy livestock production (pigs and cattle), using production from the field to grow livestock and exploiting slurry (in accordance with the Nitrate Directive) to grow maize and winter crops. Scenario 1 described a farm in which slurry produced during livestock breeding was stored in an open tank for 180 days and distribution of slurry in the field was performed by surface spreading at the beginning of the season (at pre-sowing). Crop production was supported by the use of synthetic fertilizers (urea) due to low NUE in slurry and according to the nitrogen budget. In Scenario 2, slurry was separated into liquid fractions (LFs) and solid fractions (SFs) and stored in an open tank for 180 days. Slurry distribution in the field was performed by injection (LF) and incorporation in soil (SF) without any addition of synthetic fertilizers. In Scenario 3, digestate was separated into LFs and SFs and biogas was recovered. The LF was stored for 180 days in a closed tank and distributed in substitution for chemical fertilizers by injection to increase the N efficiency, at pre-sowing and as a topdressing. The SFs were incorporated into the soil at pre-sowing. Emission coefficients for ammonia, nitrous oxide, methane, and volatile organic carbon (VOC) were collected from the literature [4–8] and from measured data (unpublished). Emission coefficients for agricultural operations such as fuel, transport, chemical fertilizer production, and so on were collected from Ecoinvent v2.

The results obtained outlined a significant decrease in environmental impacts of agriculture due to the inclusion of AD and related practices (Figure 5.2.3). The introduction of virtuous practices in slurry management significantly reduced damage to human health in Scenario 2, but it was the introduction of AD and the use of digestate in substitution for chemical fertilizers that provided the most drastically decreased impact due to animal slurry management.

The overview of the Lombardy case, with its intensive agriculture and management practices for slurry and nitrogen fertilizer, shows a very interesting picture: agriculture has a large margin of improvement to become more environmentally friendly. Biogas, if properly used, will allow farmers to significantly improve the impact of the agricultural system through the efficient use of nitrogen in digestate.

5.2.3 Chemical and Physical Modification of Organic Matter and Nutrients during Anaerobic Digestion

AD can be considered as an ad hoc biotechnology for the recovery of organic matter (OM) and nutrients as a consequence of the complex biotransformations occurring during the biological process [9]. The digestate is characterized by higher concentrations than biomass of readily available nitrogen (i.e. ammonium dissolved into the liquid phase). This allows us to consider the digestate as a potential nitrogen ready-effect fertilizer, the efficiency of which does not differ from that of the traditional chemical fertilizers, such as urea and ammonium sulfate [10]. Digestate with high ammonia-N content is very close to chemical N-fertilizer (similar efficiency of N-uptake in plants), showing similar N-efficiency use by plants [9, 10].

During AD, a part of the OM contained in the ingestate is transformed by microorganism into biogas, essentially composed of methane and carbon dioxide; this fact determines a loss of OM ranging from 20 to 95% [11], depending on the characteristics of the biomass, the retention time, and other conditions. As a consequence, the content of volatile solids (VS) and total organic carbon (TOC) decreases and the relative concentration of more recalcitrant organic molecules increases [12–15]. In various studies [12, 15, 16] carried out on samples of digestates collected in AD plants located in Lombardy Region, spectroscopic (^{13}C CPMAS NMR), chemical (fiber analysis), and biological substrate oxygen uptake rate test (SOUR test) and anaerobic biogas potential test (ABP test) approaches indicate that digestates are characterized by a high degree of biological stability as a consequence of AD treatment. Experimental data relative to fiber analysis [12] show that AD caused a significant reduction (close to or more than 60% with respect to initial mass) of cell-soluble compounds, cellulose, and hemicellulose, with a relative increase in recalcitrant molecules (10%) such as lignin and non-hydrolyzable lipids. These results are confirmed by ^{13}C CPMAS NMR analysis: if ingestate is characterized by signals (chemical shift) typical of polysaccharides, the spectra of digestates show a relative increase in long-chain aliphatic carbon, characteristic of more recalcitrant organic compounds [12]. These molecules can be considered as humus precursors [17], fundamental to soil OM turnover and to maintaining soil fertility.

Studies on the biological stability of digestates performed by respirometric technique (SOUR-test), as oxygen consumption in 20 hours (OD_{20}), clearly show that the cumulative oxygen uptake rate decreases during AD, while biological stability increases [12], becoming similar to that of a mature compost [18]. Table 5.2.1 shows the more important variations occurring during AD, such as those revealed by studies performed on Lombardy biogas plants. OM degradation leads, also, to a reduction in the phytotoxicity of

Table 5.2.1 *Biological stability and CPMAS ^{13}C NMR tools for the study of anaerobic digestion OM evolution.*

Parameter	Ingestate	Digestate	Compost	References
		Biological stability		
OD_{20} (g O_2 kg^{-1} TS 20 h^{-1})	235 ± 62	30 ± 20	25.49	Tambone et al. [16] and Scaglia et al. [18]
ABP (NL kg^{-1} VS)	568 ± 142	284 ± 92	222 ± 54	Tambone et al. [16]
SOUR (mg O_2 g^{-1} SV h^{-1})	70.2 ± 4.8	11.4 ± 2.1	4.4 ± 1.3	Provenzano et al. [15]
		CPMAS ^{13}CNMR		
Total aliphatic-C (Average; n = 8)	21.8 ± 4	35.6 ± 8.3	12.42 ± 1.66	Tambone et al. [16, 19]
O–CH_3 or N-alkyl O-alkyl-C di-O-alkyl-C (Average; n = 8)	63.6 ± 4	46.8 ± 10.5	71.47 ± 4.37	
Aromatic-C phenol or phenyl ether-C	6.7 ± 1.6	7.9 ± 1.9	11.14 ± 1.54	
Carboxyl-C keto-C	8 ± 0	9.5 ± 2.4	4.98 ± 1.39	

organic matrices, according to previous findings [13, 20], though contrasting results have also been reported [21, 22]. Recent studies have shown the presence of bioactive substances in digestate [11, 23], such as phytohormones, which are capable of promoting growth [24].

5.2.4 From Digestate to Renewable Fertilizers

5.2.4.1 N-Fertilizer from the LF of Digestate

Thanks to biological and chemical processes, AD might be considered a bioprocess capable of increasing the availability of nutrients contained in biomasses, so that digestate can be used as a "mineral fertilizer" in substitution for non-renewable ones [10]. This is due to the fact that organic-N is mineralized during AD, producing $N\text{-}NH_4^+$, which is relatively concentrated, enhancing the fertilization properties of digestate [25]. Simple solid–liquid (S/L) separation permitted the production of an ammonia-N-enriched LF, providing N-fertilizers that can substitute for synthetic ones. Data from Lombardy Region indicate that, on average, the AD of pig or cow slurry mixed with energy crops and other organic wastes – having an average dry matter (DM) content of 12% weight/weight (w/w) and a total N content of 2% DM – allows the production of an LF with a DM content of 6% (w/w) and a total N content of 7.8% DM, 55% of which is ammonia (average of 15 full-scale plants) [19]. The successive centrifugation of the LF allows for the concentration of ammonia-N, which becomes 84% of the total N content (DM 1.22% w/w, total N 2.17% DM, $N\text{-}NH_4^+$ 1.83% DM, $N\text{-}NH_4^+$/NT 84.3%) [19]. This LF could be used in total substitution of synthetic N-fertilizers for crops requiring high N input for their growth (e.g. corn).

Lombardy Region in the last 5 years strongly supported full field experiments to show how effectively AD in conjunction with S/L separation allows for reductions in the use of N-fertilizers. A field study conducted in the years 2012–2013 on an experimental field of 5.5 Ha located near the AD plant indicated that digestate (above all, the LF) could substitute completely all synthetic N-fertilizers, resulting in similar or higher crop production (Figure 5.2.4) [10].

The ability of digestate to substitute mineral fertilizers is important if one takes into consideration that of a total agricultural area of 986,853 Ha, about 56% is NVZ and so has a limitation in using animal-N. The impossibility of using animal-N represents a problem as it was calculated that about 175.000 kg of animal-N is available each year and that 100.000 kg of N from mineral fertilizer must be added to this amount, creating environmental problems. Therefore, AD represents an opportunity for transforming organic-N into readily available N to be used in substitution of mineral-N.

As consequence of these results, our research group proposed in 2010 a simplified scheme for the use of digestate in substitution for N-fertilizers, to be adopted by the Italian government (Table 5.2.2). In 2015, the Italian government released a draft proposal that included, substantially, these suggestions. The aim of this proposal was to push toward the reuse of N and other nutrients through AD, reducing the use of synthetic fertilizers and overcoming the limitations of the Nitrate Directive [1], which by now represents an obstacle to the circular economy (i.e. nutrient recovery).

Therefore, AD and successive S/L separation can be considered as an industrial biotechnology capable of transforming animal slurries and organic wastes into fertilizers for use in total substitution for synthetic fertilizers.

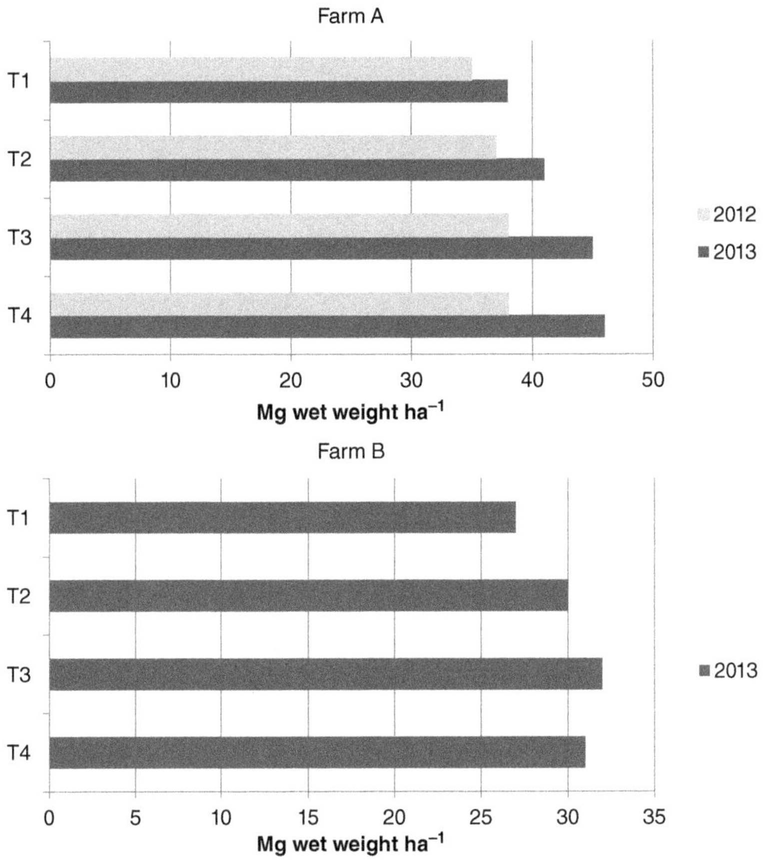

Figure 5.2.4 *Corn production obtained at two different farms using liquid fraction (LF) of digestate in comparison with urea. T1, control (0 N); T2, superficial liquid digestate; T3, urea; T4, injected liquid digestate. Source: [10], revised.*

Table 5.2.2 *Guidelines for digestate liquid fraction (LF) use in complete substitution of N-fertilizers, as suggested by the Gruppo Ricicla – DiSAA and proposed by the Italian government.*

Parameter	Obligation/suggestion
a. Nitrogen	N-NH_4^+/TKN > 70%.
b. N/P/K content	To be declared on w/w and DM basis
c. Meso and oligoelements (Ca^{2+}, Mg^{2+}, etc.)	To be declared on w.w and DM basis
d. Agronomic use	Distribution by spreading plus immediate incorporation or by direct injection
e. Biological stability	To ensure reduced odor emissions, OM must be highly stable (i.e. low oxygen demand in 20 h respirometric test: $OD_{20} < 100$ g O_2 kg^{-1}DM 20 h^{-1})
f. Microbiological quality	*Salmonella* spp.: absent *Escherichia coli* < 1000 MPN

In many cases, the N produced on a farm was greater than the total N requirement of crops, so that alternative N allocations needed to be reduced in order to avoid the impacts deriving from an improper use of digestate/slurry. The introduction of AD technology in the treatment of animal slurries allows investment in research and technology in solving the overdose of N on farms. Thanks to this approach, new technologies have been developed at full scale in Lombardy Region and are currently under operation. These technologies couple AD with the removal of N (i.e. the transformation of organic-N into mineral-N) and ease successive N recovery from digestates by applying N-stripping technologies [26, 27].

As previously described for digestate use in agriculture, AD has become a key factor in the success of these technologies, in three ways: (i) government benefits for the production of renewable energy allow farmers investing in these technologies to solve Nitrate Directive problems and propose digestate as a fertilizer in substitution of mineral fertilizers; (ii) organic-N transformation into mineral-N by AD eases successive N recovery from digestates by applying stripping technologies; and (iii) combined heat and power (CHP) from biogas provides heat to be used in successive N recovery from digestates, lowering costs.

In particular, Lombardy Region approached N reduction problems by adopting chemical-physical and direct ammonia-stripping technologies. The chemical-physical approach consists of sequential physical and chemical treatments: centrifugation, ultra-centrifugation, ultra-filtration (UF), reverse osmosis (RO), low-temperature ammonia stripping, and zeolite refining (Figure 5.2.5) [26]. These sequential technologies allow the accumulation of organic and mineral fractions in a series of concentrated streams, leading to the purification of a consistent part of the aqueous fraction by removal of N (Table 5.2.3). In particular, this approach allows the removal of 70% of ammonia, producing ammonia sulfate (N = 8% TS), and the reduction of 60–80% of the total digestate volume, producing clean water to be discharge directly in shallow water bodies. Total costs are as low as 6 € m^{-3} of digestate (the avoided costs for transportation and distribution and N fertilizers sold need to be subtracted: 1–1.5 € m^{-3}).

The second technology consists in N-stripping directly from digestates after S/L separation. This approach allows the removal of 70% of N-NH_4^+, producing concentrated ammonium sulfate solution (7–8% DM as N) by using thermal power from the AD plant, at an estimated cost of 2.3–2.8 € m^{-3} of digestate, including payback and service

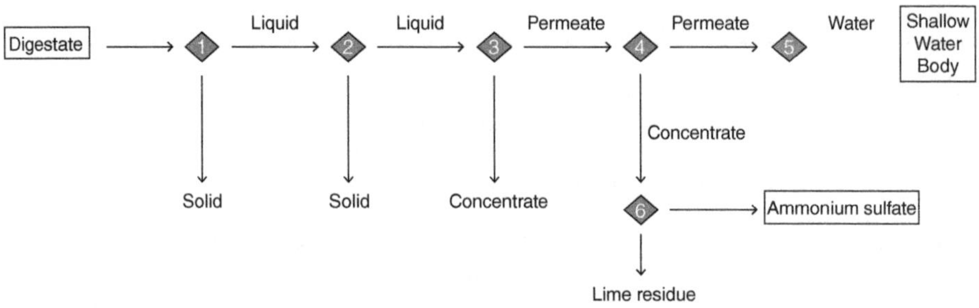

Figure 5.2.5 *Simplified scheme of a plant. 1, screw press separation; 2, centrifugation; 3, ultra-filtration (UF); 4, reverse osmosis (RO); 5, zeolite refining; 6, cold ammonia stripping.*

Table 5.2.3 *Total nitrogen and ammonia contents in fractions obtained after anaerobic digestion (AD) and successive chemical-physical treatment and N stripping [27].*

Process	Stream	TKN (mg kg^{-1})	NH_4^+ (mg kg^{-1})
Chemical-physical technology			
Digestate cattle manure		2185 ± 50	1225 ± 50
Centrifuge	LF	2160 ± 260	1172 ± 57
	SF	2670 ± 151	767 ± 146
Decanter	Liquid after decanter	1756 ± 273	1062 ± 88
	Concentrate after decanter	5415 ± 410	1922 ± 260
Ultra-filtration	Permeate of UF	987 ± 311	697 ± 120
	Retentate of UF	4023 ± 702	833 ± 98
Reverse osmosis	Permeate of RO	52 ± 15	72 ± 15
	Retentate of RO	4137 ± 6	3920 ± 412
Low-temperature stripping	Ammonium sulfate	39.8 ± 1.2	39 800 ± 1200
	Retentate of RO after stripping	507 ± 6	372 ± 19
Water after zeolites		7.4 ± 1.4	7.4 ± 1.5
Hot stripping technology			
	Slurry	2600 ± 100	1700 ± 0
Anaerobic digestion			
	Digested slurry	2200 ± 400	1700 ± 300
S/L separation			
	SF	7200 ± 1800	2300 ± 500
	LF	1800 ± 200	1500 ± 100
Hot stripping			
	LF after stripping	700 ± 100	400 ± 100

5.2.4.2 Organic Fertilizer from the SF of Digestate

The simple mechanical S/L separation of digestate, allowing recovery of a SF and a LF with different characteristics and fertilizing properties, is helping in digestate management [28]. These products can be used in agriculture directly [12, 16, 29] without any further treatment, thereby conserving their fertilizing value. In particular, the SF can be used as an organic fertilizer or amendment without composting because of the high biological stability of the OM contained within it [30]. Data reported from several experiments [11, 13, 15, 19, 31] show how composting of the SF of digestate did not produce significant benefits in terms of increase of stability/maturity and fertilizer properties. This is because AD itself performs a strong modification of the OM contained in the ingestate, as previously discussed [12]; in addition, the successive solid/liquid fractionation allows us to concentrate about 80–85% of the total P and part of the N and K in the SF. All these facts suggest using the SF of digestate to produce an NP-organic fertilizer.

Recently, Lombardy Region, with our research group and a number of small and medium-sized enterprises (SME), promoted the procedure to officially recognize the SF of digestate as a marketable NP-organic fertilizer. In doing so, a set of procedures and

Table 5.2.4 Characteristics of the two proposed fertilizers obtained from the solid fraction (SF) of digestate.[a]

	pH	DM	VS	TOC	TKN	$N\text{-}NH_4^+$	C/N
	% w/w				% DM		
DSF[b]	8.46 ± 0.08	86.44 ± 0.96	85.33 ± 1.25	41.56 ± 0.72	2.19 ± 0.2	0.06 ± 0	18.97
DSF + ashes	8.69 ± 0.04	90 ± 0.1	85.25 ± 1.58	34.37 ± 0.83	1.76 ± 0.1	0.05 ± 0	19.53
	P_2O_5	K_2O	Cd	Cu	Ni	Pb	Zn
	$g\,kg^{-1}$				$mg\,kg^{-1}$		
DSF	39.2 ± 0.1	40.7 ± 0.4	0.35 ± 0.02	54.86 ± 0.02	6.15 ± 1.37	2.58 ± 0.44	209 ± 16
DSF + ashes	20.5 ± 0.0	26.4 ± 0.9	0.18 ± 0.01	44.41 ± 0.15	5.53 ± 0.97	3.72 ± 0.21	178 ± 17

[a]Project PSR124 – Dom. Mis.124, Lombardy Region, unpublished.
[b]Dried solid fraction.

criteria have been assessed to prove the feasibility and utility of using the SF of digestate as a fertilizer, taking into consideration both fertilizer value and environmental impact. In particular, two products have been proposed: the dry solid fraction (DSF) of digestate and the dry fraction of the SF mixed with virgin wood ashes (DSF + ashes), the latter with the aim of enriching the organic fertilizer with mineral elements useful for plant nutrition (e.g. potassium), as well as of reusing ashes. The characteristics of these two NP organic fertilizers are reported in Table 5.2.4 and the proposal is presented in Table 5.2.5. Data indicate the high OM content and significant N, P, and K contents. In addition, the low heavy metals concentration (Table 5.2.2) and the absence of human pathogens (*Escherichia coli* and *Salmonella* spp.) provide further guarantees for the use of these organic fertilizers in high-value agriculture. Agronomic trials performed on corn with DSF and DSF + ashes, in comparison to mineral fertilizers, did not indicate any differences in terms of yield ($8.21 \pm 0.41\,Mg\,ha^{-1}$ for mineral fertilizer, $8.40 \pm 0.59\,Mg\,ha^{-1}$ for DSF, $8.27 \pm 1.10\,Mg\,ha^{-1}$ for DSF + ashes).

5.2.5 Environmental Safety and Health Protection Using Digestate

AD represents a useful option for reducing the impact of animal slurry management and use, by lowering emissions of methane and greenhouse gases. Correct use of the slurry after AD and restitution of the material to the soil allow significant decreases in methane emissions, contributing to the reduction of the total emission of greenhouse gases. AD, promoting the biostabilization of the OM, reduces odor production. A recent investigation into odor reduction in Lombardy Region plants (10 full-scale plants) indicated that AD allowed a reduction of 99%. Due to the complex biological processes occurring, AD determines high reductions in the "putrescibility" of the organic fraction and so in the availability of nutrients for the microorganisms responsible for producing odors. An important index for estimating the quality of a material and its environmental impact, when distributed to agricultural land, is biological stability. The measure of biological stability under standardized conditions was found to be an optimal parameter for the indirect estimation of odors production. The measure of biological stability as oxygen consumption in the 20-hour respirometry test

Table 5.2.5 *Limits proposed for the NP organic fertilizers DSF and DSF + ashes.*

NP organic fertilizer	Description	Preparation/origin	Minimal content in nutrients (w/w basis)		Other nutrients (optional; w/w basis)
			Total	For each nutrient	
DSF	Dried solid fraction of digestate	Product obtained from the SF of digestate from livestock manure or agro-industrial and subsequent drying	3.5% N + P_2O_5	1.5% N as TKN 2% P_2O_5 TOC (% FM): 30 Dry matter (% FM): 75	K_2O declarable only if >1%
DSF + ashes	Dried solid fraction of digestate mixed with heavy ash from the combustion of virgin wood	Product obtained by mixing the SF of digestate, obtained by mechanical separation of the residue of anaerobic digestion (AD), mixed with heavy ash from the combustion of virgin wood	3.5% N + P_2O_5	1.5% N as TKN 2% P_2O_5 TOC (% FM): 30 Dry matter (% FM): 75	K_2O declarable only if >2%

Table 5.2.6 *Values of odor impact (odor units, OU) and corresponding biological stability measured from untreated and treated slurries/wastes.*

	Odor impact ($OU\ m^{-2}\ h^{-1}$)	Biological stability ($g\ O_2\ kg^{-1}\ TS\ 20\ h^{-1}$)	Note
Input	119 446	316	
Digestate	13 314	107	
Input	42 773	130	This work
Digestate	5317	62	
Input	76 017	258	
Digestate	40 213	83	
Input	99 106 ± 149 173	212 ± 146	Average of 15 full-scale plants [32]
Digestate	1106 ± 771	37.7 ± 36.2	

Table 5.2.7 *Odor emission from soil treated with digestate and slurry (corn at 4–5 leaves).*

Treatment	Odor units ($OU\ m^{-2}\ h^{-1}$)	VOC ($\mu g\ m^{-2}$ soil h^{-1})
Control	1202	2.15
Superficial LF of digestate	1563	7.7
Urea	862	3.37
Injected LF of digestate	1563	2.13
Superficial slurry	6474	3.9

Source: [10] and Progetto Nero, Lombardy Region, readapted.

(OD_{20}) (a test routinely used) has been proposed as an index to predict odor impacts during digestate distribution to agricultural land (Table 5.2.6) [32, 33].

Digestate has been reported to have a higher biological stability than compost [12, 32], so that it can be used in full-sized fields without problems. In this way, Lombardy Region is also promoting AD in order to prevent the nuisance caused to the population by slurry use in agriculture (Table 5.2.7).

Obtaining biological stability also determined the partial hygienization of the digestate because of the competition of OM-degrading microorganisms with pathogens [32]. In addition, $N\text{-}NH_4^+$ produced during the AD process because of organic-N mineralization and the increase in pH determined a strong reduction of both pathogens and indicators of pathogens [32].

Monitoring of 10 full-scale plants in Lombardy Region for 2 years indicated that mesophilic AD was able to reduce pathogens and indicators of pathogens significantly: *Enterobacteriaceae* from Log 3.83 ± 1.25 to Log 1.26 ± 1.58; fecal coliform from Log 4.26 ± 1.95 to Log 1.39 ± 1.58; *E. coli* from Log 3.94 ± 1.38 to Log 1.25 ± Log 1.46; and *Clostridium perfringens* from Log 4.8 ± Log 0.88 to Log 3.71 ± log 0.9 (all data as $CFU\ g^{-1}$ w/w) [32]. More interesting was the fact that pathogen reduction was strongly correlated not only to pH and toxic ammonia content, as well documented in the literature, but also to the biological stability data. This indicated that the biological process (i.e. substrate degradation) plays an important role in pathogen content reduction. Getting pH higher than 8.3 and toxic ammonia concentration higher than 2.9 g kg^{-1} allowed the *C. perfringens* content to be reduced by Log 3.

5.2.6 Conclusion

AD is a suitable biotechnology for transforming slurry/organic wastes into useful fertilizers. SFs of digestate can be proposed as both organic amendments and organic fertilizers, while LFs can substitute synthetic N-fertilizers. By doing so, AD allows the proposal of renewable fertilizers, promoting a circular economy.

References

1. Euroepan Union. (1991). Council Directive 91/676/EEC of 12 December 1991 Concerning the Protection of Waters Against Pollution Caused By Nitrates from Agricultural Sources. Official Journal of the European Communities No. L 375/1.
2. Schievano, A., Scaglia, B., D'Imporzano, G. et al. (2009). Prediction of biogas potentials using quick laboratory analyses: upgrading previous models for application to heterogeneous organic matrices. *Bioresource Technology* **10** (23): 5777–5782.
3. Britz, W. and Delzeit, R. (2013). The impact of German biogas production on European and global agricultural markets, land use and the environment. *Policy Energy* **62**: 1268–1275.
4. Dinuccio, E., Berg, W., and Balsari, P. (2008). Gaseous emissions from the storage of untreated slurries and the fractions obtained after mechanical separation. *Atmospheric Environment* **42**: 2448–2459.
5. Rodhe, L., Pell, M., and Yamulki, S. (2006). Nitrous oxide, methane and ammonia emissions following slurry spreading on grassland. *Soil Use and Management* **22**: 229–237.
6. Amon, B., Kryvoruchko, V., Amon, T., and Zechmeister-Boltenstern, S. (2006). Methane, nitrous oxide and ammonia emissions during storage and after application of dairy cattle slurry and influence of slurry treatment. *Agriculture, Ecosystems and Environment* **112**: 153–162.
7. Erisman, J.W., Grinsven, H.V., Grizzetti, B. et al. (2011). The European nitrogen problem in a global perspective. In: *The European Nitrogen Assessment Sources, Effects and Policy Perspectives* (eds. M.A. Sutton, C.M. Howard, J.W. Erisman, et al.), 1–6. Cambridge: Cambridge University Press.
8. Carozzi, M., Ferrara, R.M., Rana, G., and Acutis, M. (2013). Evaluation of mitigation strategies to reduce ammonia losses from slurry fertilisation on arable lands. *Science of the Total Environment* **449**: 126–133.
9. Søresen, P. and Möller, M. (2009). Fate of nitrogen in pig and cattle slurries applied to the soil-crop. In: *Anaerobic Digestion: Opportunities for Agriculture and Environment* (eds. F. Adani, A. Schievano and G. Boccasile), 19–26. Milan: DiProVe, University of Milan.
10. Riva, C., Orzi, V., Carozzi, M. et al. (2016). Short-term experiments in using digestate products as substitutes for mineral (N) fertilizer: agronomic performance, odours, and ammonia emission impacts. *Science of the Total Environment* **547**: 206–214.
11. Möller, K. and Müller, T. (2012). Effects of anaerobic digestion on digestate nutrient availability and crop growth: a review. *Engineering in Life Sciences* **12** (3): 242–257.
12. Tambone, F., Genevini, P., D'Imporzano, G., and Adani, F. (2009). Assessing amendment properties of digestate by studying the organic matter composition and the degree of biological stability during the anaerobic digestion of the organic fraction of MSW. *Bioresource Technology* **100**: 3140–3142.
13. Marcato, C.E., Mohtar, R., Revel, J.C. et al. (2009). Impact of anaerobic digestion on organic matter quality in pig slurry. *International Biodeterioration and Biodegradation* **63**: 260–266.

14. Menardo, S., Balsari, P., Dinuccio, E., and Gioielli, F. (2011). Thermal pre-treatment of solid fraction from mechanically separated raw and digested slurry to increase methane yield. *Bioresource Technology* **102**: 2026–2032.
15. Provenzano, M.R., Malerba, A.D., Pezzolla, D., and Gigliotti, G. (2014). Chemical and spectroscopic characterization of organic matter during the anaerobic digestion and successive composting of pig slurry. *Waste Management* **34**: 653–660.
16. Tambone, F., Scaglia, B., D'Imporzano, G. et al. (2010). Assessing amendment and fertilizing properties of digestates from anaerobic digestion through a comparative study with digested sludge and compost. *Chemosphere* **81**: 577–583.
17. Lorenz, K., Lal, R., Preston, C.M., and Nierop, K.G.J. (2007). Strengthening the soil organic carbon pool by increasing contributions from recalcitrant aliphatic bio(macro)molecules. *Geoderma* **142**: 1–10.
18. Scaglia, B., Erriquens, F.G., Gigliotti, G. et al. (2007). Precision determination for the specific oxygen uptake rate (SOUR) method used for biological stability evaluation of compost and biostabilized products. *Bioresource Technology* **98**: 706–713.
19. Tambone, F., Terruzzi, L., Scaglia, B., and Adani, F. (2015). Composting of the solid fraction of digestate derived from pig slurry: biological processes and compost properties. *Waste Management* **35**: 55–61.
20. Gell, K., van Groenigen, J.W., and Cayuela, M.L. (2011). Residues of bioenergy production chains as soil amendments: immediate and temporal phytotoxicity. *Journal of Hazardous Materials* **186**: 2017–2025.
21. Poggi-Varaldo, H.M., Trejo-Espino, J., Fernandez-Villagomez, G. et al. (1999). Quality of anaerobic compost from paper mill and municipal solid wastes for soil amendment. *Water Science and Technology* **40**: 179–186.
22. Abdullahi, Y.A., Akunna, J.C., White, N.A., and Hallett, P.D. (2008). Investigating the effects of anaerobic and aerobic post treatment on quality and stability of organic fraction of municipal solid waste as soil amendment. *Bioresource Technology* **99**: 8631–8636.
23. Scaglia, B., Pognani, M., and Adani, F. (2015). Evaluation of hormone-like activity of the dissolved organic matter fraction (DOM) of compost and digestate. *Science of the Total Environment* **514**: 314–321.
24. Sensel, K., Wragge, V. (2008). Pflanzenbauliche Verwertung von Gärrückständen aus Biogasanlagen unter besonderer Berücksichtigung des Inputsubstrats Energiepflanzen. Available from: http://www.nachwachsenderohstoffe.de (accessed December 23, 2019).
25. Birkmose, T. and Pedersen, R.P. (2009). Contribution of biogas plants to nutrient management planning. In: *Anaerobic Digestion: Opportunities for Agriculture and Environment* (eds. F. Adani, A. Schievano and G. Boccasile), 19–26. Milan: DiProVe, University of Milan.
26. Ledda, C., Schievano, A., Salati, S., and Adani, F. (2013). Nitrogen and water recovery from animal slurries by a new integrated ultrafiltration, reverse osmosis and cold stripping process: a case study. *Water Research* **47** (16): 6157–6166.
27. Adani, F., D'Imporzano, G., Schievano, A. et al. (2011). Suistanable managment of nitrogen and nutrient. *Biocycle* **52** (10): 54–57.
28. Hjorth, M., Christensen, K.V., Christensen, M.L., and Sommer, S.G. (2010). Solid–liquid separation of animal slurry in theory and practice. A review. *Agronomy for Sustainable Development* **30**: 153–180.
29. Iacovidou, E., Vlachopoulou, M., Mallapaty, S. et al. (2013). Anaerobic digestion in municipal solid waste management: part of an integrated, holistic and sustainable solution. *Waste Management* **33**: 1035–1036.
30. Liedl, B.E. and Shafflelf, J.M. (2006). Fertilizer potential of liquid and solid effluent from thermophilic anaerobic digestion of poultry waste. *Water Science and Technology* **53** (8): 69–79.

31. Meissl, K. and Smidt, E. (2007). High quality composts by means of cocomposting of residues from anaerobic digestion. *Compost Science & Utilization* **15**: 78–83.
32. Orzi, V., Scaglia, B., Lonati, S. et al. (2015). The role of biological processes in reducing both odor impact and pathogen content during mesophilic anaerobic digestion. *Science of the Total Environment* **526**: 116–126.
33. Orzi, V., Riva, C., Scaglia, B. et al. (2018). Anaerobic digestion coupled with digestate injection reduced odour emissions from soil during manure distribution. *Science of the Total Environment* **621**: 168–176.

5.3

Nutrients and Plant Hormones in Anaerobic Digestates: Characterization and Land Application

Shubiao Wu and Renjie Dong

Key Laboratory of Clean Utilization Technology for Renewable Energy in Ministry of Agriculture, College of Engineering, China Agricultural University, Beijing, China

5.3.1 Introduction

Anaerobic digestion (AD) provides a versatile technology platform for reducing organic agricultural waste and energy utilization. This process not only controls environmental pollution but also simultaneously produces bioenergy. As a result, AD has been attracting considerable attention worldwide. Furthermore, in recent years, there has been a rapid increase in the number of biogas plants using AD as their core technology. This has led to enormous environmental and economic benefits to the communities around them. For example, in China, biogas plants processed approximately 42% of the 3.8 billion tons of waste generated from livestock [1]. The biogas thus generated can then be further purified or upgraded to cooking gas, which can be used by farmers in surrounding communities or converted to electricity. However, during the AD process, a large amount of digestates are also produced. The adequate management or disposal of these digestates must be addressed in order to avoid any constraint on the sustainable development of AD systems [2].

Digestate, which is typically a mixture of liquid and solid material, remains a by-product of the AD process. Due to its high content of organic matter (OM) and nutrients, the

---

Biorefinery of Inorganics: Recovering Mineral Nutrients from Biomass and Organic Waste, First Edition.
Edited by Erik Meers, Gerard Velthof, Evi Michels and René Rietra.
© 2020 John Wiley & Sons Ltd. Published 2020 by John Wiley & Sons Ltd.

utilization of digestate as a fertilizer is beneficial both economically, through improved crop production, and environmentally, by reducing the quantities of artificial fertilizers with large carbon footprints. In addition to its high fertilizer value, AD improves the hygienization and disposal safety of the digested organic waste. Earlier studies have shown that pathogens including *Escherichia coli* O157:H7, Paratuberculosis, Salmonella, and Cryptosporidium are effectively inactivated during thermophilic (above 40 °C) and mesophilic AD, within a retention time of 20–30 days at digester temperatures of 35–40 °C.

Previous work has demonstrated the recovery of important macromolecules such as mineral trace elements and plant hormones from the AD process [3]. The nutrient content in the digestate is generally dependent on the nutrients present in the digested feedstock; however, the digestate mineral fertilizer value is primarily determined by its nitrogen availability after AD. Moreover, the nutrient characteristics of digestate vary and are specific to each digester and its operation parameters. Therefore, in order to effectively plan a fertilizer application based on the nutrient requirements of a given crop, it is necessary to quantitatively characterize the nutrients within different digestates. Data on the fertilizer value of digestates is useful in order to maximize farmers' profits through application of the proper amounts of nutrients and to minimize the unintended negative environmental impacts that result from overuse of fertilizer.

The benefits of soil application of digestates include improvement in seedling growth, crop yield, and fruit and vegetable quality, as well as an increase in insect and disease resistance [4]. Moreover, the physical, chemical, and biological properties of soil are also affected. These effects on agriculture have been reported in numerous papers. For this chapter, research and experiments pertaining to the application of digestates are analyzed with respect to the number and varieties of crops using pie-chart analysis; the results are shown in Figure 5.3.1. They show that there is a wide scope for soil application of digestates, including varieties of cereal, grass, vegetables, and fruit, indicating the significance of the application of digestates in agriculture.

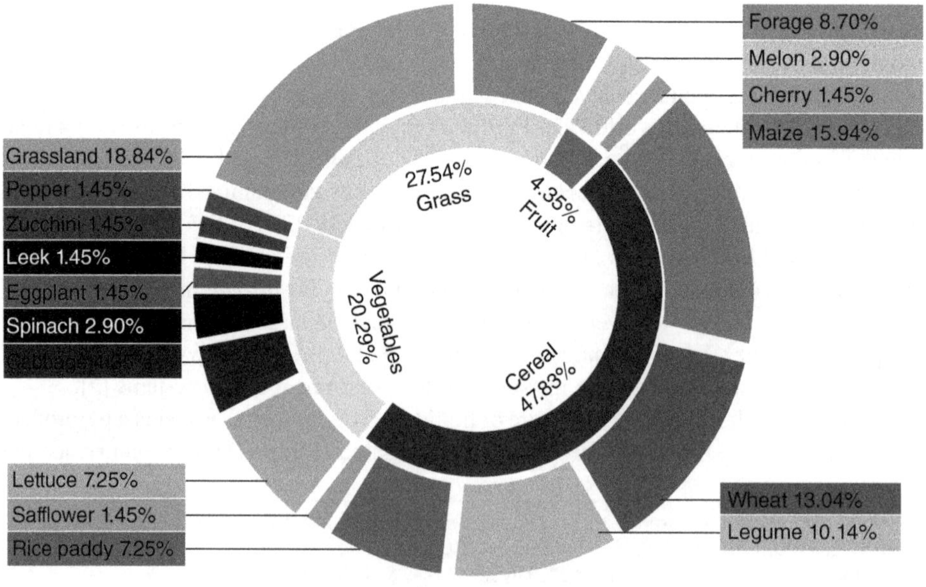

Figure 5.3.1 *Reports on the application of digestates summarized with respect to the number and varieties of crops. Source: Web of Science (2006–2015).*

From a policy perspective, identification of the nutrients and fertilizer values of a digestate is important to ensuring conformity of farms to the fertilizer use regulations. Rules regarding the maximum applicable amounts of nutrients in the digestate differ between countries, as well as between states and regions. For example, nitrogen application from the use of digestate from animal manure in the European Union (EU) must not exceed 170 kg N ha^{-1} in designated Nitrate Vulnerable Zones (NVZs) [5], but in China there is no similarly published regulation.

Prior to application, it is common practice to store a digestate for a period of time, typically 6–9 months in the EU and approximately 4–6 months in China; however, the nutrients and properties of the digestate may change during storage. For instance, ammonium, the form in which most of the nitrogen stored in digestate exists, has been reported to be volatized into ammonia gas and released during storage. In addition, methane is emitted due to the residual decomposition of OM in the digestate. Given that both ammonia and methane are harmful gases, it is frequently emphasized that digestate should be stored in closed vessels in order to control emissions. Conversely, ammonium loss from stored digestate decreases its nitrogen fertilizer value and results in loss of economic value. Therefore, mitigation of ammonia losses from digestate is beneficial not only in reducing pollution but also in presevering nutrient value. In this chapter, studies assessing changes in nutrient and plant hormone availability in digestate are presented and the use of digestate for stimulating seedling germination and interactions with soil is reported.

5.3.2 Nutrient Characterization in Anaerobic Digested Slurry

After solid–liquid (S/L) separation, digestate can be found in two forms, differing in nutrient compositions and in physical and chemical properties. The liquid fraction (LF) of digestate is rich in available nutrients that can be easily absorbed by plants; hence, it is possible to directly apply it as fertilizer. The solid fraction (SF) is mainly composed of decomposed raw materials with high dry matter (DM) content, which can be directly used as compost or soil amendment [6]. The potential for digestate to be used as a fertilizer and soil amendment, and the risk it poses to the environment and human health, are largely a result of its physicochemical and biological characteristics. The characteristics of digestate are determined mainly by the characteristics of the original fermentation substrate.

The main sources of feedstock for AD in agriculture are livestock manures and energy crops/silages. Of these, livestock manure is the main raw material used for agriculture-related biogas projects in most countries. Even though energy crops and silage are still the main raw materials used in Germany, the Renewable Energy Law of Germany has been modified twice to encourage new biogas projects to increase the use of livestock manure and reduce the use of energy crops. Among livestock manures, chicken, pig, and cattle manure are found in the largest amounts. Hence, in this chapter, the physicochemical and biological characteristics of digestates (especially those derived from AD of three types of livestock manure) are reviewed.

5.3.2.1 N, P, and K Contents

The contents of N, P, and K in digestates are key indicators for evaluating the efficiency of fertilizers. The N, P, and K contents in digestates obtained from different biogas plants that treated three types of animal manures or their mixtures are summarized in Table 5.3.1. The

Table 5.3.1 *Characteristics of digestates obtained from three animal manures treated in different biogas plants.*

Type	Site	pH	TN	TKN	NH_4-N	TP	TK	Reference
CH	Virginia, USA		2473		1787	214	2100	[7]
	Beijing, China	8.40	5308		5202	308		[8]
	Guizhou, China	8.30	2860		2465	80	2130	[9]
CA	Jiangsu, China	7.66	674.4		554.2	187.4	880.4	[10]
	Lincoln, USA		2313		893	1119	1303	[11]
	Braunschweig, Germany	7.80		2280	1510	370		[12]
PM	Guizhou, China	8.10	1660		580	24	1800	[9]
	Hainan, China		1124.8		869	129.8	1003	[13]
	Barcelona, Spain	7.10			201.83	188.94	250.9	[14]
	Iowa, USA	8.10		2660	1500	780	1120	[15]
	Istanbul, Turkey	7.95		1580	1318	790		[16]
PM + KW	Sanwa, Japan	7.50	1770		1510	432		[17]
CA + CH	Zaria, Nigeria	7	2360			2370		[18]
CA + PO	Braunschweig, Germany	7.8		2460	1650	260		[12]

CH, chicken manure; CA, cattle manure; PM, pig manure; KW, kitchen waste; PO, potato starch; TN, total nitrogen; TP, total phosphorus; TK, total potassium. Unit: $mg\,l^{-1}$.

concentrations of N, P, and K in the LF of digestates, based on 191 data points collected from the literature, are shown in Figure 5.3.2.

N in digestates was generated mainly from degradation of protein in raw materials and mineralization of organic N during AD [19]. Data in Table 5.3.1 indicate that the majority of total N in digestates (at least 50%, on average) is in the mineral form, as ammonium. NH_4^+-N can be absorbed directly by the plants and plays an important role in their growth. The content of total N in digestates obtained from AD of three types of livestock manures was highest in chicken manure, followed by cattle manure, and then pig manure, with average concentrations of 2892.91 ± 1337.56 ($n = 13$), 1114.71 ± 660.43 ($n = 34$), and 1102.75 ± 640.98 ($n = 13$) $mg\,l^{-1}$, respectively (Figure 5.3.2). The content of total N in digestate from chicken manure was different than that of the other manures, and this difference was statistically significant (one-way analysis of variance [ANOVA]) ($p < 0.05$). The digestive tracts of poultry are shorter than those of mammals; hence, the digestate from chicken manure had a higher content of total N than did that obtained from mammals. Moreover, the absorption rate of forage by poultry was quite low, resulting in N-rich manure.

P is one of the indispensable nutrients required for plant growth. The total P content in digestates from cattle manure was slightly higher than that from chicken and pig manure, with average concentrations of 352.19 ± 211.86 ($n = 15$), 244.04 ± 125.99 ($n = 18$), and 226.40 ± 308.50 ($n = 34$) $mg\,l^{-1}$, respectively. These differences were not statistically

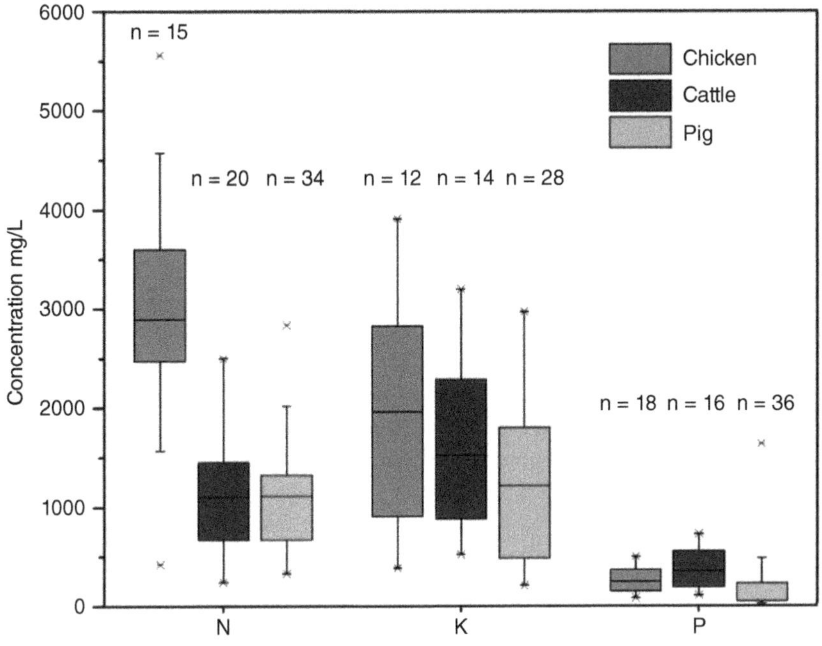

Figure 5.3.2 Macronutrients in liquid digestates obtained from three animal manures. The line within the box denotes the median. The box ranges from the lower to the upper quartile. Whiskers extend up to the minimal and maximal values within 1.5 interquartile ranges of the box. Dots indicate outliers.

significant ($p > 0.05$). The content of P does not significantly decrease during AD, being present mainly in the SF of digestates as a result of precipitation, while the LF contains only the soluble form. The soluble form of P, which can be utilized by plants, accounts for 24.0–48.4% of total P in the LF of digestates [20].

The average concentrations of K in chicken manure, pig manure, and cattle manure were 1960.62 ± 1153.804 (n = 12), 1519.95 ± 954.05 (n = 9), and 1216.16 ± 822.68 (n = 26) mg l^{-1}, respectively. Total K contents in different digestates varied according to the fermentation feedstock and were in the range of 250.9–2130 mg l^{-1} (Table 5.3.1), but the differences between the three types were not statistically significant ($p < 0.05$).

The contents of N, P, and K in beef cattle and dairy cattle manure are different, and these distinctions are dependent on their feed. The forage for dairy cattle (e.g. barley) should have balanced amounts of protein, energy, and fiber, whereas beef cattle are mainly fed on maize and hay. Thus, because of the differences in species and forage, the nutrient content in digestates from dairy cattle manure is slightly higher than that from beef cattle manure, which further causes increased nutrient contents in subsequent digestates [21]. Previous studies have shown that the nutrient contents in the LF of digestates from pig manure are higher than those from beef cattle manure, but data in Figure 5.3.2 indicate that the characteristics of the LF of digestates from pig and cattle manure were similar. This difference can be attributed to the data resource of cattle manure in Figure 5.3.2, which was gathered from beef cattle and dairy cattle. The LFs of digestates from chicken manure, pig manure, and cattle manure had high contents of N and K but low contents of P. This would result in an incomplete fertilizer characteristic , and an additional chemical fertilizer would need to be used.

5.3.2.2 Bioactive Substances

In addition to the macronutrients N, P, and K, anaerobic digestates also contain certain bioactive substances, which are supplementary nutrients that promote crop production. The bioactive substances that have been detected in digestates are phytohormones, including gibberellic acid, indoleacetic acid, and abscisic acid, nucleic acids, monosaccharides, free amino acids (lysine, tryptophan, methionine, and histidine), vitamins, unsaturated fatty acids, proline, linoleic acid, and fulvic acid [4]. These substances can be easily absorbed by plants, and can promote expansion of the root system, plant growth, plant resistance to biotic and abiotic stress, and tolerance to certain plant pathogens, as well as delay senescence. Gibberellins have been reported to stimulate seed germination and plant growth, and vitamins can promote disease resistance [22].

The concentrations of bioactive substances in anaerobic digestates are summarized in Table 5.3.2. Digestates have higher contents of indoleacetic acid than the original feedstock [23]. The reason for this is likely microbial synthesis during the digestion process. The auxin-like properties of digestates have been principally attributed to the presence of auxin generated from AD of aromatic amino acids [25]. The bioactive substances in digestates are still poorly understood. Hence, further studies to detect specific bioactive substances and determine their mechanism of action in promotion of plant growth are required.

Table 5.3.2 *Bioactive substances in digestates.*

Feedstock	GA	IAA	ABA	TAA	Reference
Chicken manure[a]	44.83 ± 1.68	36.84 ± 0.23	13.23 ± 9.15		[23]
Pig manure[a]	16.37 ± 2.16	21.17 ± 2.02	35.59 ± 3.42		[23]
Cattle manure[a]	38.53 ± 1.40	17.38 ± 2.31	23.53 ± 2.27		[23]
Chicken manure[b]		27 900		4940	[24]
Maize silage[a]	26.17	11.13	11.91		Unpublished data
Chicken manure[a]	11.21	12.03	12.37		Unpublished data
Chicken manure[a]	15.68	13.88	12.76		Unpublished data
Chicken manure[a]	32.29	24.74	18.34		Unpublished data
Pig manure [a]	9.39	9.95	21.25		Unpublished data
Pig manure[a]	38.30	23.85	6.91		Unpublished data
Cattle manure[a]	15.68	13.88	12.76		Unpublished data
Cattle manure[a]	35.27	14.92	5.23		Unpublished data
Cattle manure[a]	11.59	9.85	13.45		Unpublished data

GA, gibberellic acid; IAA, indoleacetic acid; ABA, abscisic acid; TAA, total amino acid. Unit: $mg\,l^{-1}$.
[a]Collected directly from the liquid outlet of the digester.
[b]Concentrated liquid digestate.

5.3.3 Use of Digestates as Fertilizers for Plant Growth

Digestates contain high levels of N, P, and K in the bioavailable form; hence, they have the potential to replace mineral fertilizers. Currently, the most popular method for investigating a digestate and its potential as a fertilizer is to evaluate its agrochemical characteristics and effect on plant growth. The crop yield and quality after application of various digestates have been compared with other well-studied fertilizers such as animal manure and mineral fertilizers [26, 27]. Fifteen organic fertilizers were evaluated with respect to mineral–fertilizer equivalents, based on the yield and average values of N content of different crops. The digestates from co-fermentation and AD of plant biomass had the fifth and sixth highest fertilizer value, respectively, behind organic wastes such as urine, poultry slurry, meat/blood/bone meal, and dried poultry excrements. An assessment of the agronomic characteristics of 12 digestates (from anaerobic co-digestion of agro-industrial residues) has shown that the ammonia nitrogen (NH_4^+-N) contents of digestates are high; hence, they have the potential to be used as fertilizers [2].

Moreover, enhancement of seedling growth, crop yield, fruit and vegetable quality, and improvement of insect resistance and disease resistance with the application of liquid digestates have also been discussed [4]. Reports from Web of Science, relating to the application of digestates, have been summarized with respect to the number and varieties of crops; the results are shown in Figure 5.3.1. Crop yields are equal or improved on application of digestate compared to application of the corresponding mineral fertilizer. A field experiment where the LF of digestates, undigested slurry, inorganic N, and NPK compound fertilizer were applied in a grassland showed equal or enhanced yield when applied with digestate, but relatively low impact on the environment [28]. Moreover, the application of digestates led to an increase in root length density ($1.10\,cm\,cm^{-3}$) of wheat (*Triticum aestivum*) compared to untreated control ($0.58\,cm\,cm^{-3}$) [29]. The effect of digestate application on the quality of kohlrabi has been previously shown, with the digestate-treated kohlrabi having a higher content of ascorbic acid (100.8%) and decreased nitrate content (33.6%) compared to urea treatment (100%) [30].

The application of digestates as a fertilizer also affects the physical (bulk density, saturated hydraulic conductivity, and moisture retention capacity), chemical (nutrients and heavy metals), and biological (microbial community) properties of soils. OM is considered a key indicator of the status of soil amendment (especially for agricultural soils) [31]. Nkoa [32] reviewed the organic value of digestates with 38.6–75.4% DM and suggested that digestates have the potential to be used as organic amendment. Tambone et al. [33] proposed that soil amendment properties could be sequentially ranked as follows: compost $\cong$ digestate $>$ digested sludge $\gg$ ingestate. Anaerobic digested pig manure was determined to be the most appropriate soil fertilizer among sewage sludge, hen manure, fruit and vegetable waste, piggery slurry, and grain [34].

The benefits of the application of anaerobic digestates to land in terms of crop production and improvement of soil properties have been well documented in the literature [35]. However, the risk of chemical contamination, pathogenicity, and odor impact during the process of application of anaerobic digestates have also been reported, raising issues around their effect on environment and human health [36]. These drawbacks can be attributed to improper methods of application. Moreover, N losses owing to NH3 volatilization also occur during the digestate storage process. A better understanding of digestate properties is essential to determining appropriate methods for digestate application and storage.

5.3.4 Effect of Digestate on Seed Germination

Plants are mainly propagated by seed, and seed germination is a critical stage in the life cycle of crop plants, as it often controls population dynamics [37].

Due to its high nutrient content, the anaerobic digestate is an excellent source of nutrients for the germinating seed, particularly when these nutrients are sufficiently imbibed. The anaerobic digestate typically contains a higher proportion of ammoniacal nitrogen than the undigested material. Therefore, when used in the crop field prior to sowing, AD can provide nitrogen to plants in a readily available form, similar to that of inorganic fertilizers [38]. In addition, most of the phosphorus and potassium detained in the digestate are converted from the organic state to the chemical state, thus becoming readily available to the seedlings [39]. Trace elements such as magnesium, nickel, copper, and zinc are also readily replenished from the digestate to the soil [2].

While the digestate is known to contain substantial amounts of nutrients, studies on the actual effects of digestate application on stimulation of seed germination have yielded conflicting results. Negre et al. [40] reported that digestate did not promote the germination of cress, white mustard, or radish. Likewise, Gulyás et al. [41] documented toxicity occurring at high application of AD that limited root and plant growth, suggesting that the high concentration of ammonium in the digestate was the underlying cause of the observed phenomena. In additional studies, Abdullahi et al. [42] reported that the digestate could be phytotoxic to germinating seeds, and that phytotoxicity decreased in parallel with decreasing amounts of readily biodegradable organics in the waste. The decomposition of easily biodegradable organics in the digestate consumed oxygen, and thus impeded energy metabolism of the seedling.

5.3.5 Positive Effects of Digestates on Soil

5.3.5.1 Effects on Nutrient Properties

The main positive effects of applying digestates to soil are enhancement of the soil's N and P contentS and improvement of the soil structure and water-holding capacity. The application of digestates increases the levels of mineral N, available P, and available K. This has been confirmed by evaluation of the nutrient properties after digestate application in short- and long-term experiments in both the laboratory and the field [43, 44].

Moreover, the OM content in arable soil plays a critical role in sustaining the fertilizer response. The recalcitrant molecules present in manure cannot be digested properly by AD, hence the application of digestate will substantially influence the OM content in soil and further contribute to soil fertility [45]. OM in soil is essential for good crop yields, because it not only provides nutrients for crops but also improves soil aggregation and facilitates the maintenance of structure, drainage, and aeration [46]. Consequently, soil application

Table 5.3.3 Examples of national regulations on nutrient loading on farmland [49, 50].

Country	Maximum nutrient load[a]	Required storage capacity	Compulsory season for spreading
Austria	100	6 mo	Feb 28–Oct 25
Denmark	170 (cattle), 140 (pig)	9 mo	Feb 1–harvest
Italy	170–500	90–180 d	Feb 1–Dec 1
Sweden	170	6–10 mo	Feb 1–Dec 1
Ireland	170	4 mo	Feb 1–Oct 14
UK	250–500	6–10 mo	Feb 1–Dec 1
Germany	170	6 mo	Feb 1–Oct 31 (arable land)
China	—	>30 d	—

[a]Maximum nutrient load unit: kg N $(ha\ y)^{-1}$.

of digestates makes soil less susceptible to erosion, easier to plow, and better at retaining nutrients, which indirectly leads to improved crop yield [46].

The results from a 112-day laboratory incubation study, conducted by applying raw manure, digested manure, and inorganic fertilizer to soil, indicate that the AD process does not substantially affect the content of bioavailable N and P of digestates compared to the nutrient supply from raw swine manure [15]. The levels of mineral N in burned soil increased with the application digestate from pig slurry, which suggests that the application of digestate may help to recover soil functions after a severe fire event [47]. Hence, the application of digestates has a positive agronomic effect on plant growth even if the nutrient properties in the soil tend to decrease with time [48].

Regulations set by some countries for the N loading rate, storage capacity, and application of digestates are summarized in Table 5.3.3. These regulations serve as a reminder of the fact that the amount of digestates applied to soil should be limited to prevent excess nutrient utilization, which might cause nutrient pollution in soil and water. The information included in the table indicates that European countries have gathered detailed information regarding digestate application, while more detailed regulation is needed in China.

5.3.5.2 Effects on Microbial Activity

Though digestates provide nutrients and OM to the soil, their application may potentially affect soil microbial activity. The microbial biomass in soil responds to changes in the surrounding environment. Hence, the dynamics of soil microbial parameters may reflect perturbations in soil quality induced by digestates [51]. Microorganisms present in the soil are stimulated after application of digestates. Furthermore, a broad range of soil functions are benefitted by the application of digestates. This can be confirmed by the increase in microbial biomass and soil enzymatic activities, including those of urease, alkaline phosphatase, and β-glucosidase. The increase in soil biological activity results from higher concentrations of mineral N and easily degradable carbon in digestates. The application of digestates may alter the bacterial community structure [52] and further affect the growth of the crop.

In contrast, the diversity of microbial community induced by the digestates seems to be smaller than that of cattle slurry, which would not influence the microbial functioning in soil [52]. However, the microbial activity declined in a 120-day incubation laboratory experiment [52]; this can be explained by the possibility that the acidic soil could increase the solubility of heavy metals such as Cd, Cu, Pb, and Zn from digestates. A study wherein four different biogas digestates and pig slurry were applied to soil resulted in both stimulatory and inhibitory short-term effects on microbial functions [53]. These four biogas digestates applied to soil would inhibit ammonia oxidation and denitrification, which could be an early warning of potentially hazardous substances in the digestates. This effect can also be regarded as positive, since it may reduce N losses.

5.3.5.3 Potential Negative Effects

Though digestates can be used as a fertilizer, they are not completely harmless to the soil or plant. They may contain heavy metals, organic pollutants, and pathogens, which can be incorporated into the soil ecosystem upon their application. This section will introduce some potential negative effects of digestate application, such as heavy metals [54].

The heavy metals present in digestates are likely to have adverse effects on the soil environment, but the various beneficial effects resulting from the use of digestates as fertilizer and soil amendments outweigh the potential risks of heavy metals. The accumulation of heavy metals in soil because of long-term land application of digestates and the potential risks in agriculture need to be taken seriously. The concentration of heavy metals in digestates is summarized in Table 5.3.4. The regulations set by different countries regarding the limits on heavy metals in soil are given in Table 5.3.5. Data in these two tables indicate that the levels of heavy metals in digestates are lower than the regulatory limits. Consequently, the environmental risk posed by heavy metals resulting from land application of digestates is mainly due to their potential accumulation in the long term, especially of Cu and Zn [67].

The origin of heavy metals in digestates can be traced to the metals (macroelements such as Cu, Zn, and Mn and microelements such as Ni, Cr, Pb, Cd, and As) present in the livestock feed and AD process [62]. As the bioavailability of these metals is not reduced during the AD process, the heavy metal contents are found to be higher when the soil is fertilized with digestates in comparison to equivalent nutrient inputs of manure or mineral fertilizer. Cu, Zn, and Mn in manure are usually accumulated more obviously in digestates after AD. After a micronucleus test, genotoxicity in *Vicia faba* was found to be strongly associated with the presence of Cu and Zn in digestate from pig slurry [68]. Exchangeable Zn and exchangeable Cu present in soil can be easily absorbed by plants and will accumulate in low pH conditions, which indicates the digestates are more adaptable to alkaline soils [69]. The accumulation of Mn in soil is toxic to most energy crops and vegetables, even at very low concentrations of 1 ppm [70]. However, the long-term fate of heavy metals resulting from land application of digestates, their release mechanism, species transformation, and the influence of heavy metals (especially Cu and Zn) on different types of soil are still not fully understood [54]. Hence, heavy metals need to be monitored because of their high mobility and potential risk to the soil environment.

Table 5.3.4 *Heavy metal contents in digestates.*

Feedstock	Cd	Pb	Hg	Ni	Cr	As	Zn	Cu	Reference
Cattle manure + sunflower	0.06–1.66	0.02–0.24	0.16–1.34	0.51–2.01	0.06–1.66		0.81–43	0.12–0.55	[55]
Maize silage + grass	3.1 ± 2.3	101.2 ± 2.5	ND	22.7 ± 3.0	26.5 ± 1.3		109.6 ± 1.2	109.6 ± 1.2	[56]
Pig manure	10.48	21.51	ND	72.72	60.93	23.58	2743.49	1158.28	[57]
Triticale + manure	ND	ND		0.7	0.4	1.0	31.1	10.9	[58]
Cattle manure + maize cereal	ND	ND		1.2	0.5	1.0	34.9	10.7	[58]
Maize + manure	0.3 – 0.7	2.6–5.1	ND	6.2–8.0	6.0–11.4		164–425	39–113	[59]
Sewage + strawberry	2 ± 1	ND		86 ± 1	85 ± 1		222 ± 1	101 ± 1	[60]
Cattle manure + OFMSW	0.6 ± 0.7	52.6 ± 23.6	0.5 ± 0.5	107.8 ± 61.4	96.9 ± 58.1		655.0 ± 234.5	159.9 ± 102.3	[61]
Biowaste	0–0.46	5–282		5–41	7.4–54		60–340	21–161	[62]
Pig manure	<0.001	0.02			0.17	0.032		2.71	Unpublished data
Pig manure	<0.001	0.01			0.14	0.027		2.49	Unpublished data
Pig manure	<0.001	<0.01			0.09	0.0096		0.62	Unpublished data

OFMSW, organic fraction of municipal solid wastes; ND, not detected. Heavy metals unit: mg kg^{-1} DM.

Table 5.3.5 *Regulatory guidelines for the limits on heavy metal in soil.*

	Cd	Pb	Hg	Ni	Cr	As	Zn	Cu	Resource
EU recommendations	20	750	16	300	1000		2500	1000	IEA Bioenergy [63]
EU recommendations starting 2015	5	500	5	200	600		2000	800	IEA Bioenergy [63]
EU recommendations starting 2025	2	300	2	100	600		1500	600	IEA Bioenergy [63]
UK	1.5	200	1	50	100		400	200	IEA Bioenergy [63]
Finland	1.5	100	1	100	300	25	1500	300	IEA Bioenergy [63]
USA	85	840	57	420		75	7500	4300	EPA [64]
Italy, fertilizer	1.5	140	1.5	100			500	230	[61]
Turkey	10	1200	25	400	1200		4000	1750	[65]
China, in acid soil (pH <6.5)	5	300	5	100	600	75	500	250	[57]
China, in alkali soil (pH ≥6.5)	20	1000	15	200	1000	75	1000	500	[57]
Sweden	2	100	2.5	50	100		800	600	[66]

Heavy metal unit: $mg\,kg^{-1}$ DM.

5.3.6 Conclusion

AD has been attracting considerable interest worldwide because of its potential for energy utilization of organic waste. However, large amounts of digestates are intensively generated as by-products in large-scale AD processes. Land application of digestates obtained from organic agricultural waste digestion has been shown to be the most suitable way of utilizing these resources, and the beneficial effects of digestates as fertilizers and soil amendments in agriculture have been validated. However, concerns have been raised regarding the risk of contamination, pathogenicity, and odor impact. In order to promote appropriate application and management of digestates, a comprehensive analysis of the benefits and risks they pose should be undertaken. The characteristics of digestates and the effects on soil of their land application were analyzed. This chapter could help in determining the safe and sustainable utilization of anaerobic digestates in agriculture and forward the sustainable development of scaled biogas plants.

References

1. National Development and Reform Commission. (2014). Annual Report of Comprehensive Utilization of Resources 2014. Available from: http://www.chinawaterrisk.org/research-reports/annual-report-on-comprehensive-utilization-of-resources-2014/ (accessed December 23, 2019).
2. Alburquerque, J.A., de la Fuente, C., Ferrer-Costa, A. et al. (2012). Assessment of the fertiliser potential of digestates from farm and agroindustrial residues. *Biomass and Bioenergy* **40**: 181–189.
3. Whitehead, T.R., Price, N.P., Drake, H.L., and Cotta, M.A. (2008). Catabolic pathway for the production of skatole and indoleacetic acid by the acetogen *Clostridium drakei, Clostridium scatologenes,* and swine manure. *Applied and Environmental Microbiology* **74** (6): 1950–1953.
4. Feng, H., Qu, G.F., Ning, P. et al. (2011). The resource utilization of anaerobic fermentation residue. In: *2nd International Conference on Challenges in Environmental Science and Computer Engineering* (ed. Q. Zhou), 1092–1099. New York: Procedia Environmental Sciences.
5. European Union. (1991). Council Directive 91/676/EEC of 12 December 1991 Concerning the Protection of Waters Against Pollution Caused By Nitrates from Agricultural Sources (Nitrate Directive). Official Journal of the European Communities 34(L 375 31/12/1991).
6. Yue, Z.B., Teater, C., Liu, Y. et al. (2010). Sustainable pathway of cellulosic ethanol production integrating anaerobic digestion with biorefining. *Biotechnology and Bioengineering* **105** (6): 1031–1039.
7. Singh, M., Reynolds, D.L., and Das, K.C. (2011). Microalgal system for treatment of effluent from poultry litter anaerobic digestion. *Bioresource Technology* **102** (23): 10841–10848.
8. Wang, M.Z., Wu, Y., Li, B.M. et al. (2015). Pretreatment of poultry manure anaerobic-digested effluents by electrolysis, centrifugation and autoclaving process for Chlorella vulgaris growth and pollutants removal. *Environmental Technology* **36** (7): 837–843.
9. Li, Y., Liu, Y., and Sun, C. (2013). Organic hydroponics using anaerobic fermentation liquid residue of animal manures from biogas generation. *Southwest China Journal of Agricultural Sciences* **26** (6): 2422–2429.
10. Jin, H., Chang, Z., Ye, X. et al. (2011). Physical and chemical characteristics of anaerobically digested slurry from large-scale biogas project in Jiangsu Province. *Transactions of the Chinese Society of Agricultural Engineering* **27** (1): 291–296.
11. Kobayashi, N., Noel, E.A., Barnes, A. et al. (2013). Characterization of three Chlorella sorokiniana strains in anaerobic digested effluent from cattle manure. *Bioresource Technology* **150**: 377–386.
12. Clemens, J., Trimborn, M., Weiland, P., and Amon, B. (2006). Mitigation of greenhouse gas emissions by anaerobic digestion of cattle slurry. *Agriculture, Ecosystems and Environment* **112** (2–3): 171–177.

13. Jia, Q., Xiang, W., Yang, F. et al. (2016). Low-cost cultivation of *Scenedesmus* sp. with filtered anaerobically digested piggery wastewater: biofuel production and pollutant remediation. *Journal of Applied Phycology* **28** (2): 727–736.
14. Garfí, M., Gelman, P., Comas, J. et al. (2011). Agricultural reuse of the digestate from low-cost tubular digesters in rural Andean communities. *Waste Management* **31** (12): 2584–2589.
15. Loria, E.R. and Sawyer, J.E. (2005). Extractable soil phosphorus and inorganic nitrogen following application of raw and anaerobically digested swine manure. *Agronomy Journal* **97** (3): 879–885.
16. Yetilmezsoy, K. and Sapci-Zengin, Z. (2009). Recovery of ammonium nitrogen from the effluent of UASB treating poultry manure wastewater by MAP precipitation as a slow release fertilizer. *Journal of Hazardous Materials* **166** (1): 260–269.
17. Lei, X., Sugiura, N., Feng, C., and Maekawa, T. (2007). Pretreatment of anaerobic digestion effluent with ammonia stripping and biogas purification. *Journal of Hazardous Materials* **145** (3): 391–397.
18. Alfa, M., Adie, D., Igboro, S. et al. (2014). Assessment of biofertilizer quality and health implications of anaerobic digestion effluent of cow dung and chicken droppings. *Renewable Energy* **63**: 681–686.
19. Schievano, A., D'Imporzano, G., Salati, S., and Adani, F. (2011). On-field study of anaerobic digestion full-scale plants (part I): an on-field methodology to determine mass, carbon and nutrients balance. *Bioresource Technology* **102** (17): 7737–7744.
20. Moller, H.B., Jensen, H.S., Tobiasen, L., and Hansen, M.N. (2007). Heavy metal and phosphorus content of fractions from manure treatment and incineration. *Environmental Technology* **28** (12): 1403–1418.
21. Rico, C., Rico, J.L., Tejero, I. et al. (2011). Anaerobic digestion of the liquid fraction of dairy manure in pilot plant for biogas production: residual methane yield of digestate. *Waste Management* **31** (9–10): 2167–2173.
22. Liu, W.K., Yang, Q.-C., and Du, L. (2009). Soilless cultivation for high-quality vegetables with biogas manure in China: feasibility and benefit analysis. *Renewable Agriculture and Food Systems* **24** (4): 300–307.
23. Li, X., Guo, J., Pang, C., and Dong, R. (2016). Anaerobic digestion and storage influence availability of plant hormones in livestock slurry. *ACS Sustainable Chemistry & Engineering* **4** (3): 719–727.
24. Gong, H., Yan, Z., Liang, K. et al. (2013). Concentrating process of liquid digestate by disk tube-reverse osmosis system. *Desalination* **326**: 30–36.
25. Scaglia, B., Pognani, M., and Adani, F. (2015). Evaluation of hormone-like activity of the dissolved organic matter fraction (DOM) of compost and digestate. *Science of the Total Environment* **514**: 314–321.
26. Cavalli, D., Cabassi, G., Borrelli, L. et al. (2016). Nitrogen fertilizer replacement value of undigested liquid cattle manure and digestates. *European Journal of Agronomy* **73**: 34–41.
27. Abubaker, J., Risberg, K., and Pell, M. (2012). Biogas residues as fertilisers – effects on wheat growth and soil microbial activities. *Applied Energy* **99**: 126–134.
28. Walsh, J.J., Jones, D.L., Edwards-Jones, G., and Williams, A.P. (2012). Replacing inorganic fertilizer with anaerobic digestate may maintain agricultural productivity at less environmental cost. *Journal of Plant Nutrition and Soil Science* **175** (6): 840–845.
29. Garg, R.N., Pathak, H., Das, D., and Tomar, R. (2005). Use of flyash and biogas slurry for improving wheat yield and physical properties of soil. *Environmental Monitoring and Assessment* **107** (1–3): 1–9.
30. Lošák, T., Zatloukalová, A., Szostková, M. et al. (2014). Comparison of the effectiveness of digestate and mineral fertilisers on yields and quality of kohlrabi (*Brassica oleracea*, L.). *Acta Universitatis Agriculturae et Silviculturae Mendelianae Brunensis* **59** (3): 117–122.
31. Schloter, M., Dilly, O., and Munch, J. (2003). Indicators for evaluating soil quality. *Agriculture, Ecosystems and Environment* **98** (1–3): 255–262.
32. Nkoa, R. (2014). Agricultural benefits and environmental risks of soil fertilization with anaerobic digestates: a review. *Agronomy for Sustainable Development* **34** (2): 473–492.
33. Tambone, F., Scaglia, B., D'Imporzano, G. et al. (2010). Assessing amendment and fertilizing properties of digestates from anaerobic digestion through a comparative study with digested sludge and compost. *Chemosphere* **81** (5): 577–583.

34. Kvasauskas, M. and Baltrenas, P. (2009). Research on anaerobically treated organic waste suitability for soil fertilisation. *Journal of Environmental Engineering and Landscape Management* **17** (4): 205–211.
35. Gutser, R., Ebertseder, T., Weber, A. et al. (2005). Short-term and residual availability of nitrogen after long-term application of organic fertilizers on arable land. *Journal of Plant Nutrition and Soil Science* **168** (4): 439–446.
36. Orzi, V., Scaglia, B., Lonati, S. et al. (2015). The role of biological processes in reducing both odor impact and pathogen content during mesophilic anaerobic digestion. *Science of the Total Environment* **526**: 116–126.
37. Keller, M. and Kollmann, J. (1999). Effects of seed provenance on germination of herbs for agricultural compensation sites. *Agriculture, Ecosystems and Environment* **72** (1): 87–99.
38. Möller, K. and Müller, T. (2012). Effects of anaerobic digestion on digestate nutrient availability and crop growth: a review. *Engineering in Life Sciences* **12** (3): 242–257.
39. Mehta, C.M. and Batstone, D.J. (2013). Nutrient solubilization and its availability following anaerobic digestion. *Water Science and Technology* **67** (4): 756–763.
40. Negre, M., Monterumici, C.M., Vindrola, D. et al. (2012). Horticultural and floricultural applications of urban wastes originated fertilizers. *Compost Science and Utilization* **20** (3): 150–155.
41. Gulyás, M., Tomocsik, A., Orosz, V. et al. (2012). Risk of agricultural use of sewage sludge compost and anaerobic digestate. *Acta Phytopathologica et Entomologica Hungarica* **47** (2): 213–221.
42. Abdullahi, Y., Akunna, J.C., White, N.A. et al. (2008). Investigating the effects of anaerobic and aerobic post-treatment on quality and stability of organic fraction of municipal solid waste as soil amendment. *Bioresource Technology* **99** (18): 8631–8636.
43. Riva, C., Orzi, V., Carozzi, M. et al. (2016). Short-term experiments in using digestate products as substitutes for mineral (N) fertilizer: agronomic performance, odours, and ammonia emission impacts. *Science of The Total Environment* **547**: 206–214.
44. Zerzghi, H., Gerba, C.P., Brooks, J.P., and Pepper, I.L. (2010). Long-term effects of land application of class B biosolids on the soil microbial populations, pathogens, and activity. *Journal of Environmental Quality* **39** (1): 402–408.
45. Sánchez, M., Gomez, X., Barriocanal, G. et al. (2008). Assessment of the stability of livestock farm wastes treated by anaerobic digestion. *International Biodeterioration and Biodegradation* **62** (4): 421–426.
46. Arthurson, V. (2009). Closing the global energy and nutrient cycles through application of biogas residue to agricultural land–potential benefits and drawback. *Energies* **2** (2): 226–242.
47. Cordovil, C., De Varennes, A., Pinto, R., and Fernandes, R. (2011). Changes in mineral nitrogen, soil organic matter fractions and microbial community level physiological profiles after application of digested pig slurry and compost from municipal organic wastes to burned soils. *Soil Biology and Biochemistry* **43** (4): 845–852.
48. Chiyoka, W.L., Zvomuya, F., and Hao, X. (2014). A bioassay of nitrogen availability in soils amended with solid digestate from anaerobically digested beef cattle feedlot manure. *Soil Science Society of America Journal* **78** (4): 1291–1300.
49. Holm-Nielsen, J.B., Al Seadi, T., and Oleskowicz-Popiel, P. (2009). The future of anaerobic digestion and biogas utilization. *Bioresource Technology* **100** (22): 5478–5484.
50. Birkmose, T.S. (ed.) (2009). *Nitrogen Recovery from Organic Manures: Improved Slurry Application Techniques and Treatment – The Danish Scenario*. York: International Fertiliser Society.
51. Sapp, M., Harrison, M., Hany, U. et al. (2015). Comparing the effect of digestate and chemical fertiliser on soil bacteria. *Applied Soil Ecology* **86**: 1–9.
52. Abubaker, J., Cederlund, H., Arthurson, V., and Pell, M. (2013). Bacterial community structure and microbial activity in different soils amended with biogas residues and cattle slurry. *Applied Soil Ecology* **72**: 171–180.
53. Abubaker, J., Risberg, K., Jönsson, E. et al. (2015). Short-term effects of biogas digestates and pig slurry application on soil microbial activity. *Applied and Environmental Soil Science* **2015**: 658542.

54. Govasmark, E., Stäb, J., Holen, B. et al. (2011). Chemical and microbiological hazards associated with recycling of anaerobic digested residue intended for agricultural use. *Waste Management* **31** (12): 2577–2583.
55. Demirel, B., Göl, N.P., and Onay, T.T. (2013). Evaluation of heavy metal content in digestate from batch anaerobic co-digestion of sunflower hulls and poultry manure. *Journal of Material Cycles and Waste Management* **15** (2): 242–246.
56. Quina, M., Lopes, D., Cruz, L. et al. (2015). Studies on the chemical stabilisation of digestate from mechanically recovered organic fraction of municipal solid waste. *Waste and Biomass Valorization* **6** (5): 711–721.
57. Zhu, N.-M. and Guo, X.-J. (2014). Sequential extraction of anaerobic digestate sludge for the determination of partitioning of heavy metals. *Ecotoxicology and Environmental Safety* **102**: 18–24.
58. Monlau, F., Francavilla, M., Sambusiti, C. et al. (2016). Toward a functional integration of anaerobic digestion and pyrolysis for a sustainable resource management. Comparison between solid-digestate and its derived pyrochar as soil amendment. *Applied Energy* **169**: 652–662.
59. Pabón-Pereira, C., De Vries, J., Slingerland, M. et al. (2014). Impact of crop–manure ratios on energy production and fertilizing characteristics of liquid and solid digestate during codigestion. *Environmental Technology* **35** (19): 2427–2434.
60. Serrano, A., Siles, J.A., Chica, A.F., and Martín, M.Á. (2014). Anaerobic co-digestion of sewage sludge and strawberry extrudate under mesophilic conditions. *Environmental Technology* **35** (23): 2920–2927.
61. Bonetta, S., Bonetta, S., Ferretti, E. et al. (2014). Agricultural reuse of the digestate from anaerobic co-digestion of organic waste: microbiological contamination, metal hazards and fertilizing performance. *Water, Air, and Soil Pollution* **225** (8): 2046.
62. Kupper, T., Bürge, D., Bachmann, H.J. et al. (2014). Heavy metals in source-separated compost and digestates. *Waste Management* **34** (5): 867–874.
63. Al Seadi, T., Lukehurst, C. (2012). Quality Management of Digestate from Biogas Plants Used as Fertiliser. Available from: https://www.iea-biogas.net/files/daten-redaktion/download/publi-task37/digestate_quality_web_new.pdf (accessed December 23, 2019).
64. Costa, A., Ely, C., Pennington, M., et al., eds. (2015). Anaerobic Digestion and its Applications. Available from: https://www.epa.gov/sites/production/files/2016-07/documents/ad_and_applications-final_0.pdf (accessed December 23, 2019).
65. Saruhan, V., Gul, I., Kusvuran, A., and Aydin, F. (2012). Effects of sewage sludge used as fertilizer on heavy metal contents of bird's-foot trefoil (*Lotus corniculatus* L.) and soil. *Asian Journal of Chemistry* **24** (2): 866–870.
66. Lehtomäki, A. and Björnsson, L. (2006). Two-stage anaerobic digestion of energy crops: methane production, nitrogen mineralisation and heavy metal mobilisation. *Environmental Technology* **27** (2): 209–218.
67. Win, A.T., Toyota, K., Ito, D. et al. (2016). Effect of two whole-crop rice (*Oryza sativa* L.) cultivars on methane emission and Cu and Zn uptake in a paddy field fertilized with biogas slurry. *Soil Science and Plant Nutrition* **62** (1): 99–105.
68. Marcato-Romain, C., Pinelli, E., Pourrut, B. et al. (2009). Assessment of the genotoxicity of Cu and Zn in raw and anaerobically digested slurry with the Vicia faba micronucleus test. *Mutation Research, Genetic Toxicology and Environmental Mutagenesis* **672** (2): 113–118.
69. Achiba, W.B., Gabteni, N., Lakhdar, A. et al. (2009). Effects of 5-year application of municipal solid waste compost on the distribution and mobility of heavy metals in a Tunisian calcareous soil. *Agriculture, Ecosystems & Environment* **130** (3–4): 156–163.
70. Lee, J. and Seong, D. (2015). Replacing conventional nutrient inputs for basal application with anaerobically digested pig slurry for bulb onion production. *Journal of Plant Nutrition* **38** (8): 1241–1253.

5.4

Enhancing Nutrient Use and Recovery from Sewage Sludge to Meet Crop Requirements

Ruben Sakrabani
Cranfield Soil and Agrifood Institute, Cranfield University, Cranfield, UK

5.4.1 Trends in Sewage Sludge Management in Agriculture

Over 9.4 million tons of sewage sludge are produced annually in the European Union (EU) [1, 2], and with a growing population, this amount will only increase in the future. Sewage sludge that has been de-watered and treated to a standard that is acceptable for agricultural use is often termed biosolids. More traditional routes of discharge to sea were banned by the EU Urban Waste Water Treatment Directive 91/271/EC (CEC 1991) or greatly restricted by the EU Landfill Directive 99/31/EC (CEC 1999), which required a 75% reduction in biodegradable waste going to landfill between 1995 and 2010. Alternative disposal routes such as incineration are believed to be unsustainable and have met strong public opposition at the planning stage [3]. The disposal of the ash from incinerated sewage may also contain high concentrations of heavy metals and other toxins that require careful disposal [4]. However, in a different situation, incineration is promoted as a solution in countries where sewage sludge application is banned, such as Switzerland, and incinerator ash is used to extract phosphorus [5]. The quantity of sewage sludge recycled to agriculture varies considerably between EU member states, with the United Kingdom recycling approximately 80% [1].

An alternative to disposal is to recycle sewage sludge as agricultural fertilizer, but this has a number of disadvantages. Sewage sludge (or biosolids) has a highly variable fertilizer

Biorefinery of Inorganics: Recovering Mineral Nutrients from Biomass and Organic Waste, First Edition.
Edited by Erik Meers, Gerard Velthof, Evi Michels and René Rietra.
© 2020 John Wiley & Sons Ltd. Published 2020 by John Wiley & Sons Ltd.

potential [6]; not only does the nutrient value vary, but the rate of nutrient release is difficult to predict, making it hard to manage from both an agronomic and an environmental perspective. The relatively low N:P ratio of biosolids can lead to a build-up of soil phosphorus when applied at a rate sufficient to meet crop nitrogen requirements [7]. Sewage sludge also contains substantial amounts of other important plant nutrients, including nitrogen, sulfur, magnesium, and micronutrients [8]. In addition, organic fertilizers have advantages over mineral fertilizers in that their use improves soil structure, drainage, available water, and organic matter (OM) content [9], which helps promote plant growth. A substantial proportion of P fertilizer is either sorbed on to colloids and mineral surfaces or precipitated. Subsequent solubilization is governed by soil physicochemical properties, pH, climate, the availability of other major nutrients, and the form in which the P was applied [10, 11]. Codling [12] reports that in a pot trial to establish wheat using soils that had received biosolids for 16–24 years, the soil still retained elevated levels of phytoavailable P, which indicates a big residual effect that could be tapped as a nutrient resource.

There are also concerns regarding heavy metal concentrations (such as Cu, Ni, and Zn) in sewage sludge and potential accumulation of these heavy metals in the soil with the potential for subsequent transfer to crops [13]. However, more recent work shows that heavy metals in soils derived from sewage sludge are not significant, though increased efforts are being made to investigate the effects of organic pollutants (such as pharmaceuticals and personal care products).

The use of amendments such as sewage sludge in agriculture supports the concept of sustainable intensification of agriculture. The definition of "sustainable agriculture," in its modern approach, can be traced back to the United States in the early 1980s, indicating a way of farming that should mimic natural ecosystems [14]. In the last few decades, in order to face the challenge of feeding 9 billion people by 2050, a concept called "sustainable intensification" has been discussed, which entails producing more food from the same area of land while reducing the environmental impacts [15–18]. In intensively managed agricultural land, sewage sludge can be a good source of OM, which is needed to preserve overall soil health in order to support sustainable intensification.

As phosphorus is an essential element for crop productivity, there are concerns that global scarcity of P has implications for food security [19]. Future agricultural P requirements may rely on renewable sources such as sewage sludge, livestock manure, digestates, food, and other organic wastes to alleviate these issues [20]. In addition, recycling organic waste will help offset the demise of finite P resources, which are under pressure from an expanding world population that needs feeding. There are various estimates of the global phosphate resource, but realistically the true figure is largely unknown. This is partly because of the sensitive nature of the information and partly because there may be phosphate rock yet to be discovered [19, 21]. Modeling contradicts the P scarcity issue by suggesting that phosphate rock reserves are sufficient to meet demand into the twenty-second century and can be extended well into the twenty-third with assessed use reduction and recycling measures [22].

The annual increment of phosphorus contained in the human population is estimated to be on the order of 1 million tons per year, which is a small proportion of the quantity mined [23]. Approximately 40 million tons of phosphate rock (P_2O_5 equivalent) are mined annually, of which 80–90% is used in fertilizers [24]. This suggests a clear requirement to ensure that phosphorus is recycled to a large extent, so that the rate of exhaustion of the reserves is significantly reduced. Dawson and Hilton [23] suggest that P will have to be managed in a new end-to-end way as functioning within a closed rather than an open system.

The aim of this chapter is to demonstrate through two case studies the efficacy of biosolids-derived fertilizers in supporting major crop yield and their impact on soil fertility.

5.4.2 Organomineral Fertilizer Use in Case Studies

This section provides an overview of the two types of organomineral fertilizers (OMFs) (labeled as OMF_{10} and OMF_{15}) and biosolids used in the case studies detailed later in the chapter. OMF_{10} and OMF_{15} indicate NPK formulations of 10 : 5 : 5 and 15 : 5 : 5, respectively. P was not added, so the P level of the OMF is equivalent to that of the biosolids. The agricultural benefits of biosolids are widely recognized, as previously discussed; however, in a study by Antille [25], biosolids were found to contain a total N insufficient to meet crop demands, as it was organically bound. Therefore, urea was added by the research team after the drying process, then pelletized. The OMF was a nutrient-balanced sludge-based product produced by drying digested sewage sludge cake at 80 °C in a tumbling evaporator, which produced sludge granules of between 3 and 6 mm diameter (Figure 5.4.1).

Supplementary mineral nutrients (such as urea and muriate of potash, as sources of nitrogen and potassium, respectively) were added to provide a higher proportion of available nutrients to the product. The proportion of these additional nutrients could be varied to take account of the variable nutrient content of the raw biosolids. The composition of biosolids used in this study is shown in Table 5.4.1. The supplementary mineral nutrients were added as a coating to the sludge granules by spray application of steam-melted urea granules in a rotary drum mixer. Further drying produced a c. 90% dry solid content. The product also contained 1.61% sulfur and 0.50% magnesium [26]. This prototype product did not yet incorporate potassium as a coating directly, so potassium was applied separately using muriate of potash (60% K_2O). The novelty of this product lay in the approach of blending

Figure 5.4.1 Organomineral fertilizer (OMF) shown as pellets.

Table 5.4.1 Initial characteristics of biosolids (mean ± standard error, n = 3).

	Concentration ± std error/% w/w
Biosolids parameters	
Total C%	29.3 ± 0.03
Total N%	5.47 ± 0.02
Total P%	5.65 ± 0.03
Water-soluble P (as P_2O_5)	0.15 ± 0.01
OMF_{10} parameters	
Total C%	28.4 ± 0.49
Total N%	9.45 ± 0.16
Total P%	5.39 ± 0.05
Water-soluble P (as P_2O_5)	0.10 ± 0.01
OMF_{15} parameters	
Total C%	27.4 ± 0.65
Total N%	16.05 ± 0.13
Total P%	5.45 ± 0.07
Water-soluble P (as P_2O_5)	0.07 ± 0.01

two sources of fertilizers (i.e. sewage sludge and an inorganic fertilizer such as urea or muriate if potash).

Biosolids fertilizers provide a slow release of phosphate supply to crops [27], while OMFs provide additional readily available nutrients from mineral sources. Urea was chosen as the preferred source of additional N, being less expensive and of lower risk compared to ammonium nitrate, which would be potentially explosive when mixed with sludge and thermally dried [25]. When added to soil, urea hydrolyzes to release ammonium and hydroxide ion, which then nitrifies to produce nitrate and protons [28]. The change in pH brought about by the mineralization of urea is likely to increase the mineralization of soil OM.

Application of OMF can lead to a build-up of P within the soil [25]. It was part of the purpose of the case studies to demonstrate whether any build-up would P will occur, but this could not be done (see later). Application of P at rates greater than required by the crop represents a wasted resource and potential environmental hazard. More information is required regarding the long-term fate and release of P in sludge-treated agricultural soils, in order to assess the efficiency of crop P utilization [29]. Sustainable farming practice aims to balance nutrient inputs with outputs. An ideal fertilizer would release nutrients at a suitable time to meet crop requirements [30] with minimal environmental impacts associated with nutrient losses from over-application or poorly timed applications. Understanding the ability to adjust the nutrient values of biosolids with supplementary mineral fertilizers is critical to reducing wastage, minimizing environmental impacts, and improving crop yields [29].

5.4.3 Case Study 1: Field Trial Using OMF (Broxton)

An experimental field site was established at Broxton, north-west England, in 2008. The site had a generally flat topography with a slight (<1°) incline in a westerly direction and was located in an area with a typical annual rainfall of 1250–1500 mm and an average annual maximum temperature of 13–14 °C. The soil within the field belonged to the Clifton Association [31] and consisted of slowly permeable seasonally waterlogged reddish fine and coarse loamy soils, defined as a Stagnic Luvisol soil under the World Reference Base (WRB) classification system. Soil pH was 7.1, with adequate phosphorus levels (24 mg l^{-1}; P index 2) but low levels of potassium (115 mg l^{-1}; K index 1), sulfur (5 mg l^{-1}), and manganese (39 mg l^{-1}). However, calcium (2204 mg l^{-1}) levels were above the recommended level of 1600 mg l^{-1}. The cation exchange capacity (CEC) of the soil at the start of the trial was 14.1 meq 100 g^{-1}, which indicates a slightly low nutrient holding capacity. This soil type is typically used for cereals and grassland. 1.6 ha of a 5-ha field was marked out into 48 plot areas, each of 288 m^2. The rotation sequences and crop varieties chosen were typical of the geoclimatic region in which the experiment was undertaken. This choice enabled the experimental plots to be used in demonstration events hosted for local farmers, as the trial had direct relevance to their farming systems.

The crop rotations included winter wheats Einstein and Solstice (*Triticum aestivum*); spring wheat Tybolt (*T. aestivum*); all oilseed rape Ability (*Brassica napus*); winter barley Sequal (*Hordeum vulgaris*); spring beans Fuego (*Phaseolus vulgaris*); and forage maize ES Ballade (*Zea mays*). The remaining field area was sown with perennial rye grass mix in 2008. The arable plots were laid out between tramlines (24 m wide), which were used to traffic the field during routine crop treatment such as spraying of herbicides and fungicides. Two fertilizer treatments – with OMF and conventional fertilizers – were applied across the experimental plots. This plot design was used consistently throughout the project. Each strip of crop (crop rows A–D) was subdivided into twelve 12 m-wide plots between a series of secondary tramlines at right angles to the main lines. Each strip had six replicates of each fertilizer treatment, which were randomly allocated down the strip but not between strips to reduce the amount of trafficking over the field area. Both fertilizer treatments were applied using a 12 m pneumatic boom spreader. Further details of this experiment can be obtained from Deeks et al. [32].

Combinations of single nutrient mineral fertilizers were used to provide the recommended nutrient requirements as a comparator to OMF. Nitrogen and potassium were applied as ammonium nitrate (Nitram 34% N) and muriate of potash (60% K_2O), respectively. No additional phosphorus fertilizer was added as the soil contained adequate levels (24 mg l^{-1} Olsen-P; P index 2). One-way analysis of variance (ANOVA) was used to test the significant difference between treatments for crop yields and soil properties using a 95% confidence level.

According to Deeks et al. [32], with the exception of one crop, winter wheat (year 2008/2009), treatment differences in yield were observed between biosolids and OMF but were not significant (at a 95% confidence level). Figure 5.4.2 shows the absence of effect on yield of the OMF and biosolids treatments. The mean yields were generally compatible with typical yields for the United Kingdom [33], though oilseed rape crops were slightly lower than UK average yields and the spring oilseed rape crop in 2009/2010 failed due to adverse weather conditions. Spring bean and winter barley yields were also slightly lower than UK averages in 2010/2011. For winter wheat (year 2008/2009), where significant differences were observed between treatments, the mean yield from crops receiving OMF treatment was approximately 20% lower than the crop yield from the conventional fertilizer (ammonium nitrate) plots. However, comparison of the weight of 1000 grains of wheat (44.1 g conventional treatment and 42.0 g OMF per 1000 grains)

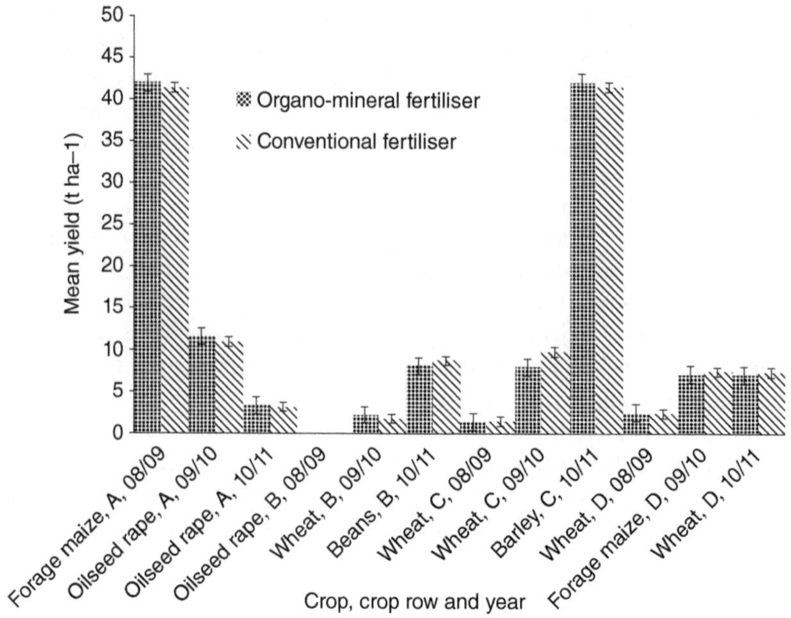

Figure 5.4.2 *Comparison of mean crop yields between the two fertilizer treatments of conventional fertilizer and organomineral fertilizer (OMF) over the 3-year trial (2008–2011).*

showed no significant difference ($p = 0.379$). The timing of nitrogen application and its availability in soil to meet crop demand is known to affect yield [34]. As similar amounts of nitrogen were applied using OMF and conventional fertilizer treatments, the rate and timing of release of the nutrient may have been the influencing factor. Since OMF has biosolids as a component, it provides OM, which can help in retaining the nitrate needed for crops and also facilitates water retention, which explains the lack of yield difference between treatments.

5.4.4 Case Study 2: Field Trial Using OMF (Silsoe)

A 3-year field-scale experiment was established to compare two formulations of OMF (with different N : P : K ratios) to biosolids and urea application. The experiment was situated at Cranfield University's farm (UK), grid reference TL 07423 35708. The soil was from the soil series Bearsted (Typical Brown Earth), which is a sandy loam soil (70.6% sand, 18.5% silt, 10.9% clay), OM% $= 4.2 \pm 0.06$ and = pH 7.4 ± 0.03. Historically, the field was under arable rotation for commercial crop production. The rotation consisted of set-aside (fallow after harvest) (2004–2005), oilseed rape (2005–2006), and winter wheat (2006–2007). The experimental plots (60 plots, dimensions $2\,m \times 5\,m$) used in the study were established with ryegrass (*Lolium perenne* L.) in February 2007 [25]. A randomized experimental design was incorporated for the fertilizer type (biosolids, OMF_{10}, OMF_{15}, and urea) and application rate (0, 50, 100, 150, 200, and 250 kg N ha^{-1}). The biosolids were in granulated (6 mm) form. They were mixed with urea (46%) as an N source and with muriate of potash (60% K_2O) as a K source, resulting in an OMF consisting of biosolids with additional N and K. A detailed description of the production and characteristics of the biosolids and OMF are presented by Antille [25]. The chemical characteristics of the biosolids are shown in Table 5.4.1. Further details of this experiment can be obtained from Pawlett et al. [35]. One-way ANOVA was used to test the significant difference between treatments for crop yields and soil properties using a 95% confidence level.

The fertilizer, application rates, and replication combinations resulted in a 60-plot experiment in a big rectangular field site with individual plots with tramlines for easy access to a combine harvester.

According to Pawlett et al. [35], ryegrass yield increased for all fertilizers up to an application rate of 100 kg N ha^{-1}, with a maximum yield of 1390 ± 57 kg dry matter (DM) ha^{-1} in 2011 (Figure 5.4.3). Thereafter, increasing the fertilizer application rate did not significantly increase yield, as observed in 2012 and 2013. The crop yield for all fertilizers was similar up to 200 kg N ha^{-1} in 2013, but in 2012 and 2011 biosolids had a lower yield than OMF. This demonstrates that crop yields were not compromised by the OMF formulations up to 200 kg N ha^{-1} (2013), but in addition there was no yield benefit from incorporating higher amounts of urea into the biosolids. This study demonstrates reduced yields where urea alone was applied at 250 kg N ha^{-1} (2011 yield). This was not observed for the other fertilizers. The reduced yield associated with urea may be a phytotoxic response. Leaf-tip necrosis ("tip-scorching") of ryegrass was reported by Watson and Miller [36] where urea was applied at 100 kg N ha^{-1}, with increasing tip-scorch with application rate. Delayed urea hydrolysis (to NH_4^+ and CO_2) enabled urea uptake into plants, with subsequent leaf necrosis [36, 37]. Urea fertilizers have also been associated with poor seed germination, seedling growth, and early plant growth due to the ammonia produced through rapid urea hydrolysis [38].

It is important to note that the yield overall was reduced compared to previous years [25]. The peak of growth throughout the year for ryegrass is between April and July [39]. During this period in 2013, no rainfall was recorded, and the peak temperature was 18.4 °C, with a maximum ryegrass yield of 1390 kg DM ha^{-1}. This represents a reduction of yield compared to the two previous years. In 2012, the average rainfall for the same period was 70.6 mm, with a mean temperature of 19.3 °C and a yield equivalent of 7800 kg DM ha^{-1} (two ryegrass cuts), whereas in 2011 the average rainfall was 152.8 mm, with a mean temperature of 19.5 °C and a yield equivalent of 1400 kg dry weight (DW) ha^{-1} (three ryegrass cuts) (data not shown). It is likely that yields were influenced by a lack of rainfall and modest temperatures during the key stages of the ryegrass growth cycle. Soil-water availability is often considered the main factor that affects ryegrass yields [40].

One of the main challenges of this project was the energy cost of drying the OMF pellets to 90% dry solids. Energy is expensive in the United Kingdom. Though some attempts were made to recycle the waste heat in the granulation plant, this was not enough. Hence, the larger-scale production that was envisaged at the start of the project never took off. The work to date has proved the efficacy of OMF as a fertilizer and its impact on crop yield and soil fertility. The manufacturing and engineering aspects of OMF production still remain to be fully resolved.

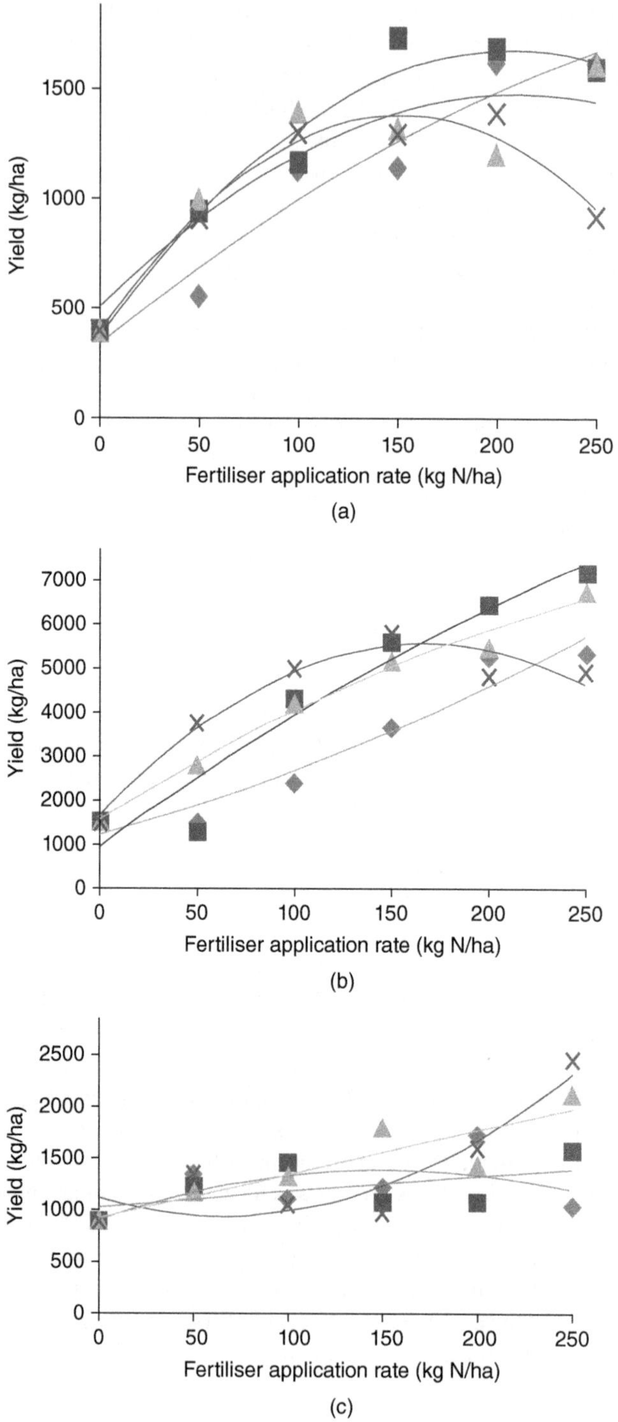

Figure 5.4.3 *Ryegrass yield (mean ± SE) for (a) 2011, (b) 2012, and (c) 2013. Diamond and blue, biosolids; square and red, OMF_{10}; triangle and green; OMF_{15}; cross and black, urea.*

5.4.5 Conclusion

While sewage sludge contains variable amounts of nutrients, making its use as a fertilizer unpredictable, transforming it through combination with mineral fertilizers to an OMF product has been shown to be a promising step forward. The novelty of producing this wholesome fertilizer from a renewable resource without compromising crop yield is a key feature of our work, which also aims to support sustainable intensification of agriculture. This chapter has demonstrated that OMF derived from biosolids and urea is able to produce a comparable yield to urea at $150\,kg\,N\,ha^{-1}$.

The chapter also set the scene for future experimental work looking at the conversion of waste into resource. Challenges include ensuring the heterogeneity of the waste material, which will influence the reproducibility of the final product (an alternative fertilizer in this case), as well as the energy costs related to drying of pellets. A small amount of waste heat was used to supplement the use of fossil fuel in order to dry the pellets. Future work is needed to increase the use of waste heat and minimize that of fossil fuels in order to truly claim that OMF pellets are a greener alternative to chemical fertilizers. In the current project, since the wastewater treatment plants (WWTPs) were stringent in ensuring the influent came from certain sources, some control was exerted on the treatment process and finally on the sludge that was produced. Regulations and policies also help provide a greater control on the quality of sludge that is being produced. Collectively, these factors influence the quality and consistency of the OMF derived from sludge material.

Acknowledgments

This research was funded by the European Union Seventh Framework Programme (FP7-ENV.2010.3.1.1-2 ENV) under grant agreement n° 265269. http://www.end-o-sludg.eu/. Additional thanks are due to Dr. Mark Pawlett, Dr. Lynda Deeks, Mr. Bob Walker, Mr. Rob Read, and Ms. Abby Wallwork for field work and Dr. Diogenes Antille and Prof. Dick Godwin for contributing to the background work on the trials at Silsoe.

References

1. European Environment Agency (2005) Sewage sludge – a future waste problem? Indicator Fact Sheet Signals 2001. Available from: http://www.eea.europa.eu/data-and-maps/indicators/generation-and-treatment-of-sewage-sludge (accessed December 23, 2019).
2. Laturnus, F., von Arnold, K., and Gron, C. (2007). Organic contaminants from sewage sludge applied to agricultural soils. *Environmental Science and Pollution Research* **14** (1): 53–60.
3. Petts, J. (1994). Incineration as a waste management option. In: *Waste Incineration and the Environment* (eds. R.E. Hester and R.M. Harrison). Cambridge: Royal Society of Chemistry.
4. Wei, Y., Van Houten, R.T., Borger, A.R. et al. (2003). Minimisation of excess sludge production for biological wastewater treatment. *Water Research* **37**: 4453–4467.
5. Herzel, H., Krüger, O., Hermann, L., and Adam, C. (2015). Sewage sludge ash – a promising secondary phosphorus source for fertilizer production. *Science of the Total Environment* **542**: 1136–1143.
6. Sommers, L.E. (1977). Chemical composition of sewage sludges and analysis of their potential use as a fertilizers. *Journal of Environmental Quality* **6**: 225–232.
7. Edge, D. (1999). Perspectives for nutrient removal from sewage sludge and implications for sludge strategy. *Environmental Technology* **20**: 759–763.

8. MAFF (2000) Fertilizer Recommendations for Agricultural and Horticultural Crops (RB209). Norwich: Ministry of Agriculture, Fisheries and Food.
9. Environment Agency (2004) The State of Soils in England and Wales. Bristol: Environment Agency.
10. Blake, L., Mercik, S., Koerschens, M. et al. (2000). Phosphorus content in soils, uptake by plants and balance in three European long-term field experiments. *Nutrient Cycling in Agroecosystems* **56**: 263–275.
11. Kirk, G. (2002). Use of modelling to understand nutrient acquisition by plants. *Plant and Soil* **247**: 123–130.
12. Codling, E.E. (2014). Long-term effects of biosolid-amended soils on phosphorus, copper, manganese and zinc uptake by wheat. *Soil Science* **179** (1): 21–27.
13. Johnston, A.E. (2008). *Resource or Waste: The Reality of Nutrient Recycling to Land*. York: International Fertilizer Society.
14. Gomiero, T., Pimentel, D., and Paoletti, M.G. (2011). Is there a need for a more sustainable agriculture? *Critical Reviews in Plant Sciences* **30** (1–2): 6–23.
15. Pretty, J. (2002). *Agri-Culture: Reconnecting People, Land and Nature*. London: Earthscan.
16. Pretty, J. (2008). Agricultural sustainability: concepts, principles and evidence. *Philosophical Transactions of the Royal Society B* **363**: 447–465.
17. Royal Society of London (2009) Reaping the benefits: science and the sustainable intensification of global agriculture. RS Policy Document 11/09. Issued: October 2009 RS1608. Available from: https://royalsociety.org/topics-policy/publications/2009/reaping-benefits/ (accessed December 23, 2019).
18. Godfray, H.C.J., Beddington, J.R., Crute, I.R. et al. (2010). Food security: the challenge of feeding 9 billion people. *Science* **327**: 812–818.
19. Cordell, D., Drangert, J.O., and White, S. (2009). The story of phosphorus: global food security and food for thought. *Global Environmental Change* **19**: 292–305.
20. Ashley, K., Cordell, D., and Mavinic, D. (2011). A brief history of phosphorus: from the philosopher's stone to nutrient recovery and reuse. *Chemosphere* **84**: 737–746.
21. Hilton, J.K., Johnston, A.E., and Dawson, C.J. (2010). *The Phosphate Life-Cycle: Rethinking the Options for a Finite Resource*. York: International Fertilizer Society.
22. Koppelaar, R.H.E.M. and Weikard, H.P. (2013). Assessing phosphate rock depletion and phosphorus recycling options. *Global Environmental Change* **23**: 1454–1466.
23. Dawson, C.J. and Hilton, J. (2011). Fertilizer availability in a resource-limited world: production and recycling of nitrogen and phosphorus. *Food Policy* **36**: S14–S22.
24. Defra (2009). Review of the feasibility of recycling phosphates at sewage treatment plants in the UK. A Report by the Advisory Committee on Health and Safety.
25. Antille, D.L. (2011). Formulation, utilization and evaluation of organo-mineral fertilizers. EngD Dissertation: Cranfield University.
26. Gedara, S., Le, M.S., Murray C, et al. (2009) Farm scale crop trials of SMART-PTM Organo mineral fertilizers. Proceedings of the 14th European Biosolids and Organic Resources Conference and Exhibition.
27. Gendebien, A., Davis, B., Hobson, J., et al. (2010). Environmental, Economic and Social Impacts of the Use of Sewage Sludge on Land. Final Report. Project III. Project Interim Reports. EC DG ENV.G.4/ETU/2008/0076r.
28. Hofman, G. and Van Cleemput, O. (2004). *Soil and Plant Nitrogen*. Paris: International Fertilizer Industry Association.
29. Smith, S.R. (2008) The Implications for Human Health and Environment of Recycling Biosolids onto Agricultural Land. Centre for Environmental Control and Waste Management, 76-7.
30. Trinchera, A., Allegra, M., Rea, E. et al. (2011). Organo-mineral fertilizers from glass-matrix and organic biomasses: a new way to release nutrients. A novel approach to fertilisation based on plant demand. *Journal of the Science of Food and Agriculture* **91**: 2386–2393.

31. Ragg, J.M., Beard, G.R., George, H., et al. (1984) Soils and Their Use in Midland and Western England. Soil Survey of England and Wales, Bulleting No. 12, Harpenden.
32. Deeks, L.K., Chaney, K., Murray, C. et al. (2013). A new sludge-derived organo-mineral fertilizer gives similar crop yields as conventional fertilizers. *Agronomy for Sustainable Development* **33**: 539–549.
33. Nix, J. (2011). *Farm Management Pocketbook 41*. London: Imperial College, The Anderson Centre.
34. Alley, M.M., Scharf, P., Brann, D.E., et al. (2009) Nitrogen Management for Winter Wheat: Principles and Recommendations. Virginia Cooperative Extension, Publication 424–026. Available from: https://www.pubs.ext.vt.edu/content/dam/pubs_ext_vt_edu/424/424-026/424-026.pdf (accessed December 23, 2019).
35. Pawlett, M., Deeks, L.K., and Sakrabani, R. (2015). Nutrient potential of biosolids and urea derived organo-mineral fertilizers in a field scale experiment using ryegrass (*Lolium perenne* L.). *Field Crops Research* **175**: 56–63.
36. Watson, C.J. and Miller, H. (1996). Short-term effects of urea amended with urease inhibitor N-(n-butyl) thiophosphoric triamide on perenial ryegrass. *Plant and Soil* **184**: 33–45.
37. Saggar, S., Singh, J., Giltrap, D.L. et al. (2013). Quantification of reductions in ammonia emissions from fertilizer urea and animal urine in grazed pastures with urease inhibitors for agriculture inventory: New Zealand as a case study. *Science of the Total Environment* **465**: 136–146.
38. Bremner, J.M. and Krogmeier, M.J. (1988). Elimination of the adverse effects of urea fertilizer on seed germination, seedling growth, and early plant growth in soil [urease inhibitor/phenylphosphorodiamidate/N-(n-butyl)thiophosphoric triamide]. *Proceedings of the Natural Academy of Sciences of the United States of America* **85** (13): 4601–4604.
39. Anslow, R.C. and Green, J.O. (1967). The seasonal growth of pasture grasses. *Journal of Agricultural Science (Cambridge)* **68**: 109–122.
40. Doyle, C.J., Ridout, M.S., Morrison, J., and Edwards, C. (1986). Predicting the response of perennial ryegrass to fertilizer nitrogen in relation to cutting interval, climate and soil. *Grass and Forage Science* **41**: 303–310.

5.5

Application of Mineral Concentrates from Processed Manure

Gerard Velthof[1], Phillip Ehlert[1], Jaap Schröder[2], Jantine van Middelkoop[3], Wim van Geel[2], and Gerard Holshof[3]

[1] *WENR, Wageningen University & Research, Wageningen, The Netherlands*
[2] *WPR, Wageningen University & Research, Wageningen, The Netherlands*
[3] *WLR, Wageningen University & Research, Wageningen, The Netherlands*

5.5.1 Introduction

Manure is a valuable source of nutrients for crops. However, in regions with intensive livestock farming systems and limited agricultural land, the amount of nutrients available exceeds the nutrient demand of crops. In these regions, the surplus of nitrogen (N) and phosphorus (P) results in an increase in soil nutrient status and high emissions of N and P to groundwater, surface water, and the atmosphere [1]. A series of policies and measures have been implemented in the European Union (EU) to decrease emissions of N and P from agriculture to the environment [2]. Processing of manure is considered as an option to increase the nutrient use efficiency of manure [3]. One treatment method is separation of livestock slurry into a solid fraction (SF) and a liquid fraction (LF) followed by reverse osmosis (RO) of the LF [4, 5]. The RO decreases the volume of the LF (Chapter 4.3), resulting in a concentrated N–potassium (K) solution ("mineral concentrate"), in which most of the N is present as ammonium (NH_4^+). The SF is rich in organic matter (OM) and P, and can be used as a soil amendment. The water removed by RO has low concentrations of nutrients

Biorefinery of Inorganics: Recovering Mineral Nutrients from Biomass and Organic Waste, First Edition.
Edited by Erik Meers, Gerard Velthof, Evi Michels and René Rietra.
© 2020 John Wiley & Sons Ltd. Published 2020 by John Wiley & Sons Ltd.

and can be discharged to sewer or surface water [6]. The reduction of the volume by RO increases the ability to transport mineral concentrates from areas with high livestock density to arable farming areas. This chapter presents an overview of the nitrogen fertilizer replacement value (NFRV) of mineral concentrates from processed manure, based on a series of studies carried out in the Netherlands. In Section 5.5.2, an assessment is made of the fertilizer value of manure concentrates on the basis of the composition. Section 5.5.3 presents N fertilizer values obtained in pot and field experiments. Section 5.5.4 deals with N losses from mineral concentrates as ammonia (NH_3), nitrous oxide (N_2O), and nitrate (NO_3^-).

5.5.2 Product Characterization

Table 4.3.1 shows the average composition of mineral concentrates obtained from slurry treatment based on separation and RO. The average total N content of the concentrates is 8.15 g N kg^{-1} product. The N in mineral concentrates is mainly found in the NH_4^+ form (on average, 90% of total N in the concentrate). The remaining N is organically bound. The pH of mineral concentrates is high (about pH 8), thus it is likely that NH_4^+ partly occurs in the form of NH_3 in mineral concentrates.

The efficiency of N in mineral concentrates as a fertilizer depends on the presence and degradability of organic N and the gaseous N losses (as NH_3 and via denitrification) during and after application [7]. The NFRV of an organic fertilizer is the percentage of the applied N that has the same effect on crop N yield as mineral N fertilizer. In the Netherlands, NFRV is generally determined by comparison with broadcast mineral fertilizer calcium ammonium nitrate (CAN), which is the most commonly used mineral N fertilizer in the country.

Part of the N in mineral concentrates becomes available for the crop via N mineralization. According to fertilizer recommendations in the Netherlands (www.bemestingsadvies.nl; www.kennisakker.nl), it is assumed that the NFRV of organic N in manure amounts to 20–60% during the first 12 months after application. The NH_3 emission from surface-applied slurry amounts to 69–74% of applied NH_4^+-N and that from slurry injected in the soil (including injection) is 2–26% [8]. Assuming that these figures also hold for mineral concentrates, it is estimated that the NFRV of surface-applied mineral concentrates is 25–30% and that of injected slurry 70–90% compared to CAN. This theoretical approximation of NFRV has been tested in experiments, the results of which are presented in Section 5.5.3.

The P content in mineral concentrates is generally low (<0.2 g P kg^{-1}; Chapter 4.3), and therefore mineral concentrates have no agronomic value as P fertilizers.

The K content in mineral concentrates is about 8 g K kg^{-1} (Chapter 4.3). The exact chemical form in which it occurs in mineral concentrates is not known, but based on chemical analysis it is assumed that it is found bound to bicarbonate, chloride, and sulfate and in fatty acids [9]. Therefore, it is likely that the K in mineral concentrates is fully available to the crop. The supply of K in mineral concentrate reduces the need for other mineral K fertilizers. This is particularly advantageous for crops with a high K demand such as potato and maize. The K demand of grassland is also high, but is partly met when cattle manure is produced on a farm. An excess supply of K to cattle can cause health problems (grass tetany). The amounts of K in feed, fertilizer, and manure should hence be taken into consideration when importing mineral concentrates to a dairy farm.

Mineral concentrates also contain other nutrients, including calcium (Ca), magnesium (Mg), sulfur (S), sodium (Na), and trace elements. If a concentrate is applied at common N and K application rates, the supply of most other nutrients is not of agronomic importance. However, the levels of Na in mineral concentrates are approximately 20–25% (w/w) that of K [9]. When using a mineral concentrate as an N or K fertilizer, a significant amount of Na is applied (20–40 kg Na ha^{-1}). Na has value in animal feeding, and some arable crops (e.g. sugar beet) respond positively to its application. S is also a valuable component of mineral concentrate, but the average total S application rate is low (about 4 kg per 100 kg N as mineral concentrate, of which about 3 kg is in the form of sulfate-S). The availability of S for the crop in mineral concentrates is unknown. The average CL concentration is 3 g kg^{-1} mineral concentrate [9]. Harmful effects to crops of excess Cl are not an issue when using mineral concentrates, as long as the supply of Cl with other fertilizers is taken into account.

The contents of the heavy metals Cd, Cr, Ni, Pb, and As and of organic contaminants such as dioxins, non-ortho PCBs, mono-ortho PCBs, indicator PCBs, organochlorine pesticides residues, polyaromatic hydrocarbons (PAHs), and mineral oil in mineral concentrates are low, and often below the detection limit. These contents meet the standards in the Fertilizer Act of the Netherlands [10]. Consequently, it is unlikely that the use of a mineral concentrate as fertilizer will lead to an unacceptable loading of soil with heavy metals and organic contaminants.

5.5.3 Agronomic Response

5.5.3.1 Pot Experiments

Pot experiments to test the NFRV of mineral concentrates, using grass as a test crop and CAN as a reference fertilizer, have been carried out by Ehlert et al. [11], Klop et al. [12], and Rietra and Velthof [13]. CAN consists of 27% N, of which half is nitrate and half is ammonium. In these pot experiments, the NFRV of injected mineral concentrate compared to broadcast CAN was on average 91%, and higher than that of injected pig slurry (75%; Table 5.5.1). These findings are in agreement with the theoretical NFRV estimated from the chemical composition (Section 5.5.2).

In Klop et al.'s experiment [12], grass yields of surface-applied mineral concentrate were low, partly due to scorching of the grass after surface application of the mineral concentrate.

Table 5.5.1 *Average nitrogen fertilizer replacement value (NFRV) of injected mineral concentrate and injected pig slurry compared to calcium ammonium nitrate (CAN) (in %) in pot experiments.*

References	Crop	NFRV, % of CAN			
		Mineral concentrate		Pig slurry	
		Injected	Surface-applied	Injected	Surface-applied
[12]	Grass	96	50	79	41
[11]	Grass	86	—	74	—
[11]	Swiss chard	87	—	71	—
[13]	Grass	93	72	76	—
Average		91	—	75	—

The same held with surface-applied pig slurry. Scorching did not occur after injection of mineral concentrate or pig slurry or after surface-application of CAN. Deposition of urine during grazing has also been shown to induce scorching of grass [14]. Probably, salt, NH_3, and volatile fatty acids concentrations near the grass roots were too high after surface application of mineral concentrate and pig slurry but not after injection. Part of the difference between surface application and injection is due to differences in NH_3 emissions. The NFRV of pig slurry was only 41% after surface application, and increased to 79% when injected [12]. Measurements showed that NH_3 emission was much lower from injected concentrate than from surface-applied concentrate. Emission of N_2O from mineral concentrate was higher than from CAN, but lower than from pig slurry.

In Ehlert et al.'s experiment [11], the NFRV of mineral concentrate was tested with perennial rye grass and Swiss chard and with different types of mineral N fertilizer. The NFRV of mineral concentrate compared to CAN was on average 87%. The NFRV of liquid ammonium nitrate (AN), ammonium sulfate (AS), and ammonium chloride was on average 100% compared to CAN. This indicates that the efficiency of solid ammonium fertilizers was higher than that of mineral concentrate in this experiment. The NFRV of urea was somewhat lower (except when applied to grass on a sandy soil), probably due to NH_3 emission [15].

In Rietra and Velthof's experiment [13], the effects of soil moisture content and acidification of mineral concentrate on NFRV were tested. Acidification is a measure to decrease NH_3 emission [16]. The NFRV of injected concentrates (84–93%, with the highest NFRV at the highest moisture content) was significantly higher than that of surface-applied concentrate (64–79%). The NFRV of acidified concentrate was similar to that of CAN. Measurements showed that acidification minimized NH_3 emission.

5.5.3.2 Field Experiments

Field experiments in the Netherlands show NFRVs of injected (to a depth of 5 cm) mineral concentrates ranging from 54 to 84% compared to broadcast CAN (Table 5.5.2). The lowest NFRV, 54%, was observed on grassland in 2009. An explanation for this relative outlier is as yet lacking. Averaged over all experiments, the NFRV of mineral concentrate compared to CAN was 79% on arable land and 71% on grassland. These values are lower than those obtained in the pot experiments (Table 5.5.2) and at the lower end of the theoretically estimated NFRV values of 70–90% (Section 5.5.2). Van Geel et al. [21] also determined the NFRV in field experiments, with a less detailed set-up than the experiments described in Table 5.5.2. Their results showed a wide range in NFRV (0–130%). In 20 experiments, the NFRV of mineral concentrate was similar to CAN, in 10 it was lower than CAN, and in 1 it was higher than CAN [21].

The NFRV of mineral concentrates was higher (79–117%, average 93%) when compared to liquid ammonium nitrate injected with the same equipment as mineral concentrate. Clearly, the application method and form of fertilizer affect N use efficiency; that is, the N use efficiency of an injected liquid N fertilizer (liquid ammonium nitrate and mineral concentrate) was lower than that of broadcast CAN prills. The distribution of N in the soil differed between broadcast-applied CAN and injected liquid fertilizers, and this could be a factor in the differences in N use efficiency between CAN and the liquid fertilizers.

Table 5.5.2 *Average nitrogen fertilizer replacement value (NFRV) of injected mineral concentrate compared to calcium ammonium nitrate (CAN) or liquid ammonium nitrate (AN) (in %) in field experiments.*

Crop	Year	Soil type	NFRV, %		References
			Compared to CAN	Compared to liquid AN	
Potato	2009	Clay	76		[17]
Potato	2009	Sand	84		[17]
Potato	2010	Clay	75	117	[17, 18]
Potato	2010	Sand	81		[17]
Maize	2010	Sand	72		[19]
Maize	2011	Sand	84		[19]
Grassland	2009	Sand/clay	54[a]	86	[20]
Grassland	2010	Sand/clay	71[a]	102	[20]
Grassland	2011	Sand	80[a]	79	[20]
Grassland	2012	Sand	81[a]	83	[20]

[a]For each year, the average NFRV of two to four experiments is included.

The results of the experiments indicate that there is scope to increase NFRV in the field by optimizing the use of mineral concentrate via low NH_3 emission application techniques and by decreasing the organic N content of mineral concentrate.

5.5.4 Risk of Nitrogen Losses

5.5.4.1 Ammonia Emission

Mineral concentrate is an NH_4^+-containing fertilizer with a high pH (about 8), and therefore it carries a risk of NH_3 emission. Injection into the soil is a well-known NH_3 emission abatement technique [22]. A review by Hou et al. [23] showed that emissions of NH_3 from slurries following band spreading, incorporation, and injection were 55% (range: 37–67%), 70% (50–82%), and 80% (72–86%) lower than those from surface applied manures, respectively.

In a series of incubation studies, the NH_3 emissions from untreated pig slurry, mineral concentrate, mineral fertilizers, and the SF of separated slurry were quantified [24]. The products were both surface-applied and injected in the soil at 5 cm depth. Surface application of mineral concentrate, pig slurry, and urea resulted in high NH_3 emissions (Figure 5.5.1). The NH_3 emission from injected mineral concentrate was low, similar to that of surface-applied CAN (Figure 5.5.1). Averaged over three incubation tests, the NH_3 emission from injected mineral concentrate was significantly lower than that of injected pig slurry [24].

In a review, Hou et al. [23] found significantly lower NH_3 emissions (reduction with 18% based on 44 observations) for separated LF relative to untreated slurry. The NH_3 emissions from mineral concentrates and LF were not determined in the same experiment, so it is not clear if NH_3 emission from mineral concentrates differs from that from LF under the same conditions. Differences in NH_3 emission may be expected because of the higher NH_4^+ concentration of mineral concentrates compared to the LF or the lower water content of

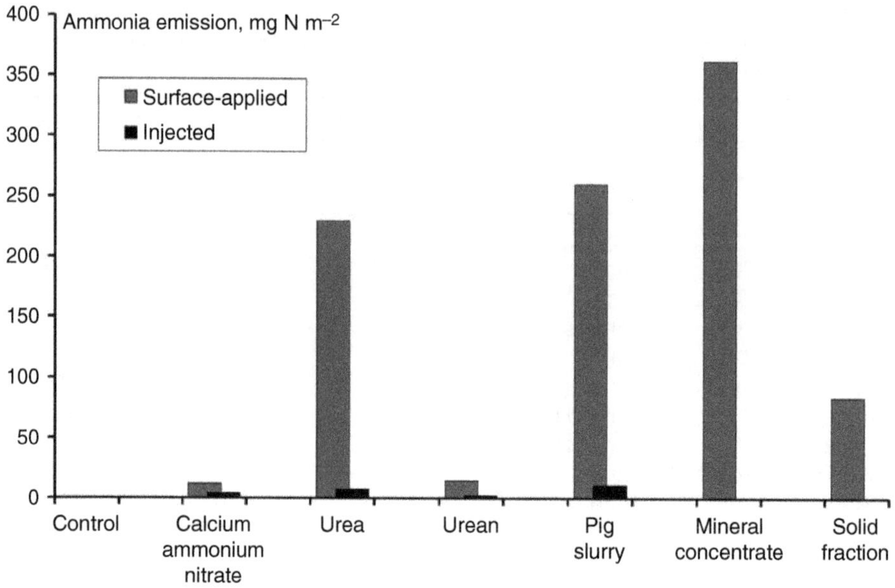

Figure 5.5.1 *NH_3 emissions from calcium ammonium nitrate (CAN), urea, urean (liquid mixture of urea and ammonium nitrate), pig slurry, mineral concentrate, and the SF of separated slurry applied at equal amounts of total N. All fertilizers and manures were both surface-applied and injected (indicated as incorporated). Source: Results of an incubation experiment with arable sandy soil by Velthof and Hummelink [24].*

mineral concentrates, as a result of which the infiltration rate into the soil may be reduced. Because of these differences, the NH_4^+ concentration gradient at the liquid–air interface will be greater after application of mineral concentrate than after application of LF.

Field experiments in 2010 showed that the NH_3 emission after sod injection in cereals was 3% of the applied NH_4^+-N in the mineral concentrate, or 12% when applied via trailing hoses [25]. The NH_3 emission from mineral concentrate applied to grassland via sod injection averaged 8% of the applied NH_4^+-N. These measurements were carried out in just one year, and thus the emission factors cannot be generally applied, as weather conditions have a major effect on NH_3 emission [26].

The risk of NH_3 emission from applied mineral concentrate is probably higher when applied to soils containing lime than to neutral or acidic soils, as is the case for any other NH_4^+-based mineral fertilizer [15]. Additional NH_3 abatement techniques may be applied to decrease NH_3 emission and increase N efficiency. Rietra and Velthof [13] showed in a pot experiment that acidification of mineral concentrates minimized NH_3 emission. The NFRV of acidified mineral concentrate was equal to that of CAN.

5.5.4.2 Nitrous Oxide Emission

In a series of incubation studies, Velthof and Hummelink [24] quantified N_2O emissions from untreated pig slurry, mineral concentrate, and mineral fertilizers (see Figure 5.5.2 for the results of the experiment with grassland soils). The average N_2O emission of injected mineral concentrate was higher than the N_2O emission from a similar N rate of

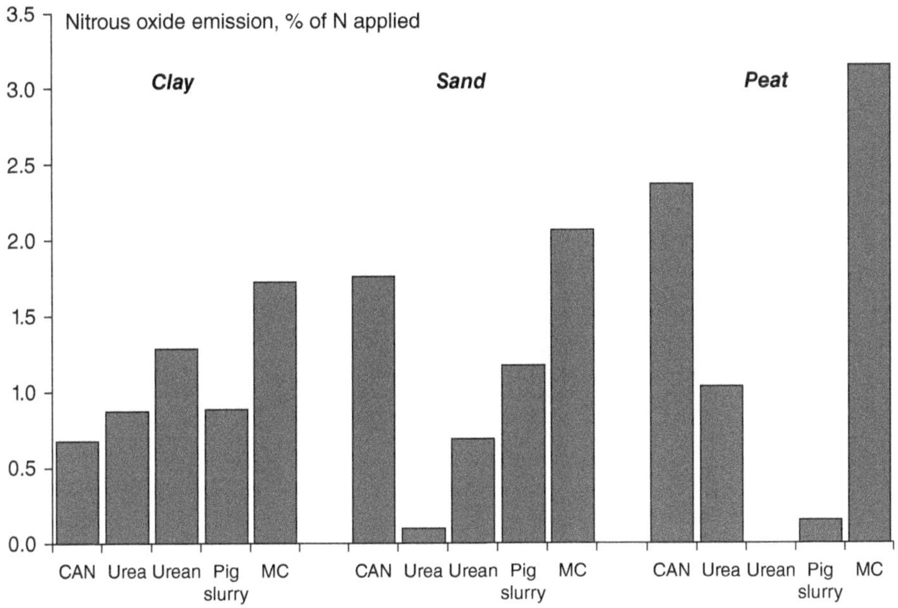

Figure 5.5.2 *N_2O emission factors as per cent of N applied for calcium ammonium nitrate (CAN), urea, urean, pig slurry, and mineral concentrate (MC) applied to intact grassland columns (PVC columns with a diameter of 10 cm and height of 10 cm) from clay, sand, and peat soils. The application rate was equivalent to 170 kg N ha^{-1}. CAN, urea, and urean were surface-applied and the pig slurry and mineral concentrates were injected in a row at 5 cm depth with a knife. The soil moisture content was kept at field capacity and incubation was carried out at 20 °C. Source: Results of a 14-day experiment under controlled conditions by Velthof and Hummelink [24].*

surface-applied CAN [24]. The N_2O emission from mineral concentrate was approximately 1.5-fold higher than that from untreated pig slurry, averaged over all studies and application techniques [24]. The injection of mineral concentrate and pig slurry resulted in higher N_2O emissions than surface application.

Both nitrification and denitrification may be sources of N_2O after application of mineral concentrates to a soil. Application of mineral concentrates may strongly increase concentrations of NH_3 in the soil. Ammonia is toxic for nitrifying organisms. A high NH_3 concentration in soil may thus inhibit nitrification, leading to the production of nitrite and N_2O [27, 28]. These effects are likely to be similar to those found in urine patches [29] and after application of anhydrous ammonia as fertilizer [30]. As for denitrification-related N_2O production, it must be noted that mineral concentrates contain organic carbon, including volatile fatty acids [11]. When degradable OM is applied to a nitrate containing soil under wet conditions, denitrifying bacteria may use the carbon as an energy source, and nitrate can be transformed into gaseous N_2O and N_2. Paul and Beauchamp [31] showed that volatile fatty acids are effective energy sources for denitrifiers. Accordingly, Ehlert et al. [11] found that application of mineral concentrates to soil increased potential denitrification. The higher N_2O emission from injected mineral concentrate compared with surface-applied mineral concentrate is probably related to the lower oxygen concentrations in the soil, and the higher N concentrations after injection [32].

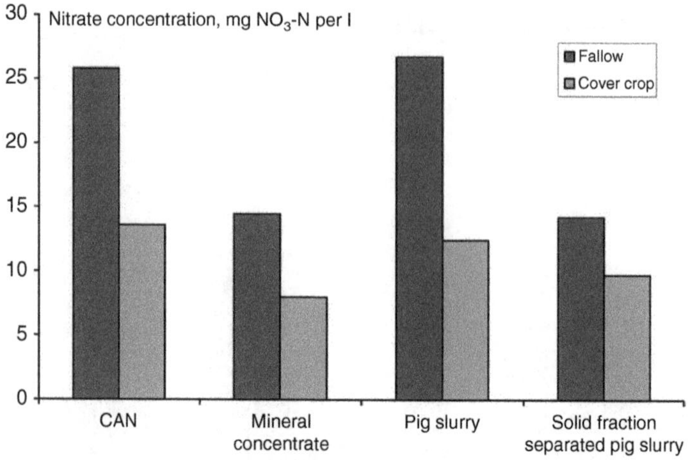

Figure 5.5.3 *Average NO_3^--N concentration (mg NO_3^--N per l^{-1}) in the upper groundwater in a field experiment with maize on sandy soil carried out in 2010 and 2011. Calcium ammonium nitrate (CAN), mineral concentrate, pig slurry, and the solid fraction (SF) of separated pig slurry were applied at 150 kg N ha^{-1}. The experiment was carried out with and without a rye winter crop [19].*

The amount of N lost via N_2O emission is low (usually less than 2% of the N applied as fertilizer or manure; [33]). Similar amounts will be lost in the form of NO_x [34]. Emissions of N_2 can be much higher than those of N_2O, especially under wet conditions [35]. The total gaseous N losses as N_2O, NO_X, and N_2 by nitrification and denitrification may significantly affect the N efficiency of mineral concentrates.

5.5.4.3 Nitrate Leaching

Mineral N in the soil in the fall is an indicator for the risk of NO_3^- leaching in winter in the Netherlands [36]. Measurements of mineral N contents in the soil after harvest in fall showed overall no differences between CAN and mineral concentrates in experiments with potato, maize, and grassland [17–20]. These results suggest that the use of mineral concentrate does not increase the risk of NO_3^- leaching compared to CAN.

Measurements in 2010 and 2011 of the NO_3^- concentration in the groundwater of the field experiments with maize showed that it was on average lower in the plots to which mineral concentrate was applied than in those that received CAN and pig slurry (Figure 5.5.3). A winter cover crop significantly reduced leaching of NO_3^-. In these maize experiments, the NFRV of mineral concentrate was 72–84% that of CAN. The lower NFRV did not increase N leaching losses from concentrates compared to CAN, however. Immobilization of mineral N probably didn't occur either [11]. A reduced availability of N from mineral concentrate due to an incomplete mineralization is also not likely, given the low content (5–10%) of organic N in mineral concentrates. These results indicate that the lower NFRV of mineral concentrates compared to CAN in these experiments was most likely related to gaseous N losses by denitrification and NH_3 emission.

Measurements of NO_3^- concentrations in the upper groundwater in a grassland experiment in 2012 showed no clear differences in nitrate concentration between mineral concentrate and CAN [20]. In an experiment by Schils et al. [37], NO_3^- concentration in upper groundwater was measured in 10 maize and 20 grassland farm fields. In one part of each field, mineral fertilizer and cattle slurry were applied, while in another part, mineral fertilizer was replaced by mineral concentrate at a comparable total N rate. The variation in NO_3^- concentration was large, and concentrations were higher in maize land than in grassland. There was no statistically significant difference in NO_3^- concentration between the two nutrient-management treatments.

These studies show that the risk of NO_3^- leaching from applied mineral concentrates is similar to or lower than that with CAN, for both grassland and arable land. The N leaching losses from slurries were higher than those from mineral concentrate, probably due to release of mineral N by mineralization outside the growing period of the crop. The organic N contents of mineral concentrates were much lower than those of untreated slurries.

5.5.5 Conclusion

Mineral concentrates are N-K fertilizers produced by RO of the LF of separated livestock slurry. About 90% of the N in mineral concentrate is present as ammonium, the other 10% as organic N. Pot experiments showed that the NFRV of injected mineral concentrate was on average 91% that of CAN, and higher than that of injected pig slurry (75%). In field experiments in the Netherlands, NFRVs of injected mineral concentrates ranged from 54 to 84% compared to broadcast CAN. Injection of mineral concentrate into the soil strongly decreased NH_3 emission in incubation experiments, but increased N_2O emission. Field measurements show that the risk of nitrate leaching from applied mineral concentrates is similar to or lower than that from CAN and untreated manure. Obviously, NH_3 emission and denitrification are the dominant N-loss pathways after application of mineral concentrate. Losses of N after such applications can be decreased and NFRV can be increased by using low-NH_3-emission application techniques, acidification, and a reduction of the organic N content of the concentrate.

References

1. Velthof, G.L., Lesschen, J.P., Webb, J. et al. (2014). The impact of the nitrates directive on nitrogen emissions from agriculture in the EU-27 during 2000–2008. *Science of the Total Environment* **15**: 468–469.
2. Oenema, O., Bleeker, A., Braathen, N.A. et al. (2011). Nitrogen in current European policies. In: *The European Nitrogen Assessment* (eds. M.A. Sutton, C.M. Howard, J.W. Erisman, et al.), 62–81. Cambridge: Cambridge University Press.
3. Burton, C.H. (2007). The potential contribution of separation technologies to the management of livestock manure. *Livestock Science* **112**: 208–216.
4. Thörneby, L., Persson, K., and Trägårdh, G. (1999). Treatment of liquid effluents from dairy cattle and pigs using reverse osmosis. *Journal of Agricultural Engineering Research* **73**: 159–170.
5. Ledda, C., Schievano, A., Salati, S., and Adani, F. (2013). Nitrogen and water recovery from animal slurries by a new integrated ultrafiltration, reverse osmosis and cold stripping process: a case study. *Water Research* **47**: 6157–6166.

6. Hoeksma, P., de Buisonjé, F.E. (2012) Mineral Concentrates from Animal Slurry. Monitoring of the Pilot Plants in 2011 [Dutch]. Report 626, Livestock Research, Wageningen.
7. Birkmose, T.S. (2009). *Nitrogen Recovery from Organic Manures: Improved Slurry Application Techniques and Treatment; the Danish Scenario*. York: International Fertilizer Society.
8. Huijsmans, J.F.M. and Schils, R.L.M. (2009). *Ammonia and Nitrous Oxide Emissions Following Field-Application of Manure: State of Art Measurements in the Netherlands*. York: International Fertilizer Society.
9. Ehlert, P.A.I., Hoeksma, P. (2011) Agronomic and Environmental Perspectives of Mineral Concentrates. Study in the Framework of the Pilot Mineral Concentrates [Dutch]. Report 2185, Alterra, Wageningen.
10. Ehlert, P.A.I., Hoeksma, P., Velthof, G.L. (2009) Inorganic and Organic Pollutants in Mineral Concentrates. Results of First Screening [Dutch]. Report 256, Animal Sciences Group, Wageningen.
11. Ehlert, P.A.I., Nelemans, J.A., Velthof G.L. (2012) Nitrogen Efficiency and Losses by Denitrification and Nitrogen Immobilisation Obtained under Controlled Conditions [Dutch]. Report 2314, Alterra, Wageningen.
12. Klop, G., Velthof, G.L., and van Groenigen, J.W. (2012). Application technique affects the potential of mineral concentrates from livestock manure to replace inorganic nitrogen fertilizer. *Soil Use and Management* **28**: 468–477.
13. Rietra, R.P.J.J., Velthof, G.L. (2014) Nitrogen Efficiency of Mineral Concentrate under Controlled Conditions. Effects of Acidification, Moisture, and Application Technique [Dutch]. Report 2518, Alterra, Wageningen.
14. Lantinga, E.A., Keuning, J.A., Groenwold, J., and Deenen, P.J.A.G. (1987). Distribution of excreted nitrogen by grazing cattle and its effect on sward quality herbage production and utilization. In: *Animal Manure on Grassland and Fodder Crops: Fertilizer or Waste?* (eds. H.G. Van der Meer, R.J. Unwin, T.A. Van Dijk and G.C. Ennik), 103–117. Dordrecht: Martinus Nijhoff.
15. Sommer, S.G., Schjoerring, J.K., and Denmead, O.T. (2004). Ammonia emission from mineral fertilizers and fertilized crops. *Advances in Agronomy* **82**: 557–622.
16. Kai, P., Pedersen, P., Jensen, J.E. et al. (2008). A whole-farm assessment of the efficacy of slurry acidification in reducing ammonia emissions. *European Journal of Agronomy* **28**: 148–154.
17. Schröder, J.J., de Visser, W., Assinck, F.B.T. et al. (2014). Nitrogen fertilizer replacement value of the liquid fraction of separated livestock slurries applied to potatoes and silage maize. *Communications in Soil Science and Plant Analysis* **45**: 73–85.
18. Van Geel W., van den Berg, W., van Dijk, W. (2012) Nitrogen Efficiency of Mineral Concentrates Applied to Potatoes. Report of Field Research in 2009 and 2010 [Dutch]. Report 475, PPO, Wageningen.
19. Schröder, J.J., de Visser, W., Assinck, F.B.T., and Velthof, G.L. (2013). Effects of short-term nitrogen supply from livestock manures and cover cropping on silage maize production and nitrate leaching. *Soil Use and Management* **29**: 151–160.
20. Holshof G., van Middelkoop, J.C. (2014) Nitrogen Efficiency of Mineral Concentrates Applied to Grassland. Field Experiments 2012 and Overall Analysis [Dutch]. Report 769, Livestock Research 769, Wageningen.
21. Van Geel, W, van den Berg, W., van Dijk, W., Wustman, R. (2012) Additional Research Mineral Concentrates 2009–2011 on Arable Land and Grassland. Summary of the Results of the Field Experiments and Determination of the Nitrogen Use Efficiency [Dutch]. Report 476, PPO, Wageningen.
22. Bittman, S., Dedina, M., Howard, C.M. et al. (eds.) (2014). *Options for Ammonia Mitigation: Guidance from the UNECE Task Force on Reactive Nitrogen*. Edinburgh: Centre for Ecology and Hydrology.
23. Hou, Y., Velthof, G.L., and Oenema, O. (2015). Mitigation of ammonia, nitrous oxide and methane emissions from manure management chains: a meta-analysis and integrated assessment. *Global Change Biology* **21**: 1293–1312.

24. Velthof, G.L., Hummelink, E. (2011) Ammonia and Nitrous Oxide Emission after Application of Mineral Concentrates. Results of Laboratory Studies as Part of the Pilot Mineralen Concentrates [Dutch]. Report 2180, Alterra, Wageningen.
25. Huijsmans, J.F.M., Hol, J.M.G. (2011) Ammonia Emission after Application of Mineral Concentrate to Arable Land and Grassland [Dutch]. Report 398, Plant Research International, Wageningen.
26. Søgaard, H.T., Sommer, S.G., Hutchings, N.J. et al. (2002). Ammonia volatilization from field-applied animal manure – the ALFAM model. *Atmospheric Environment* **36**: 3309–3319.
27. Chalk, T.J. and Smith, C.J. (1983). Chemodenitrification. In: *Gaseous Loss of Nitrogen from Plant-Soil Systems*, Developments in Plant and Soil Sciences, vol. **9** (eds. J.R. Freney and J.R. Simposn), 65–89. Dordrecht: Springer.
28. Venterea, R.T. (2007). Nitrite-driven nitrous oxide production under aerobic soil conditions: kinetics and biochemical controls. *Global Change Biology* **13**: 1798–1809.
29. Oenema, O., Velthof, G.L., Yamulki, S., and Jarvis, S.C. (1997). Nitrous oxide emissions from grazed grassland. *Soil Use and Management* **13**: 288–295.
30. Bremner, J., Breitenbeck, G., and Blackmer, A. (1981). Effect of anhydrous ammonia fertilization on emissions of nitrous oxide from soils. *Journal of Environmental Quality* **1**: 77–80.
31. Paul, J.W. and Beauchamp, E.G. (1989). Effect of carbon constituents in manure on denitrification in soil. *Canadian Journal of Soil Science* **69**: 49–61.
32. Velthof, G.L. and Mosquera, J. (2011). The impact of slurry application technique on nitrous oxide emission from agricultural soils. *Agriculture, Ecosystems & Environment* **140**: 298–308.
33. Velthof, G.L., Mosquera, J. (2011). Calculation of Nitrous Oxide Emission from Agriculture in the Netherlands. Update of Emission Factors and Leaching Fraction. Report 2151, Alterra, Wageningen.
34. Stehfest, E. and Bouwman, L. (2006). N2O and NO emission from agricultural fields and soils under natural vegetation: summarizing available measurement data and modelling of global annual emissions. *Nutrient Cycling in Agroecosystems* **74**: 207–228.
35. Weier, K.L., Doran, J.W., Power, J.F., and Walters, D.T. (1993). Denitrification and the dinitrogen/nitrous oxide ratio as affected by soil water, available carbon, and nitrate. *Soil Science Society of America Journal* **57**: 66–72.
36. Ten Berge H. (2002) A Review of Potential Indicators for Nitrate Loss from Cropping and Farming Systems in the Netherlands. Report 31, Plant Research International, Wageningen.
37. Schils, R., Geerts, R., Oenema, J., et al. (2014) Effects of Application of Mineral Concentrate on the Nitrate Concentration of Groundwater. Study in Frame of the Pilot Mineral Concentrates [Dutch]. Report 2570, Alterra, Wageningen.

5.6

Liquid Fraction of Digestate and Air Scrubber Water as Sources for Mineral N

Ivona Sigurnjak, Evi Michels, and Erik Meers

Laboratory of Analytical Chemistry and Applied Ecochemistry, Faculty of Bioscience Engineering, Ghent University, Ghent, Belgium

5.6.1 Introduction

The need to meet the rapidly increasing demand for mineral nitrogen (N) while reducing dependence on fossil fuels (2% of world energy use is dedicated to the Haber–Bosch process; [1]) has been driving widespread attention toward the recuperation and reuse of nutrients present in digestate and manure, through the generation of various end- and side-products of potential use as sustainable (biobased) alternatives.

As a first step in digestate processing, mechanical separation is often applied, resulting in a phosphorus (P)-rich solid fraction (SF) and a P-poor, nitrogen–potassium (N-K)-rich liquid fraction (LF). The latter, with a relatively high mineral N content, where NH_4-N makes up more than 75% of the total N [2], might be of particular interest for P-saturated soils [3, 4]. A low C/N ratio and an alkaline pH of LF digestate may cause ammonia (NH_3) volatilization during inappropriate storage [5]. This N loss from digestate facilities and animal housing can be recovered by using an acidic air scrubber, where NH_3 in the air is captured in sulfuric acid. As such, air scrubber water (ASW) is generated with high potential to be used as a formulated nitrogen–sulfur (N-S) fertilizer [2, 5, 6]. However, insufficient knowledge about the properties and the impact of these products on soil quality and crop yield, combined with current legal constrains, limits their use. Therefore, this chapter

Biorefinery of Inorganics: Recovering Mineral Nutrients from Biomass and Organic Waste, First Edition.
Edited by Erik Meers, Gerard Velthof, Evi Michels and René Rietra.
© 2020 John Wiley & Sons Ltd. Published 2020 by John Wiley & Sons Ltd.

investigates the performance of LF digestate and ASW as N fertilizers, by examining their impact on soil and yield production in silage maize and lettuce cultivation.

In order to evaluate the environmental impact of using these products in agriculture, physicochemical soil properties, including soil total and available N (NO_3-N) pool, pH, and soil electrical conductivity (EC), were evaluated. Moreover, yield production with crop N uptake was determined in order to calculate the N use efficiency (NUE) of the proposed strategies. It is hypothesized that the use of ASW and LF digestate will not cause significant differences in crop yield and soil properties compared to the conventional fertilization practice of using mineral N in horticulture and using animal manure (AM) enriched with mineral N in arable crops. For ASW, the hypothesis was based on the fact that the product contains 100% mineral N. In the case of LF digestate, our recent research (Sigurnjak, 2017) has shown that LF digestate with a high NH_4-N/N_{total} ratio (≈0.80) might provide mineral N in a similar fashion as mineral N fertilizer, despite the small presence of organic N.

5.6.2 Materials and Methods

5.6.2.1 Experimental Design

Experimental research was carried out at two sites in Flanders (Belgium), a temperate marine region with an average annual precipitation of 800 mm and an average annual air temperature of 9.0 °C [7].

Site 1 was an experimental field run on a sandy soil for three consecutive years (2011, 2012, and 2013) in Wingene (51° 3′ 0″N, 3° 16′ 0″E). The main characteristics of the soil in April 2011 were: pH-KCl 7.0, organic carbon 1.9%, available P (0–23 depth) 816 mg kg^{-1} dry weight (DW), and NO_3-N (0–90 cm depth) 40 kg ha^{-1}. Silage maize (*Zea mays*, cv. Atletico KWS [FAO Ripeness Index: 280] in 2011, Fernandez [FAO Ripeness Index: 260] in 2012, and Millesim [FAO Ripeness Index: 240] in 2013) was grown as a test crop. The preceding crop was fodder maize. During the experimental period, Italian rye grass was sown as an intercrop. Based on the soil characteristics and crop demand, the fertilizer dosage was formulated at 150 kg effective N ha^{-1}, 180 kg K_2O ha^{-1}, and 30 kg MgO ha^{-1} in 2011, and 135 kg effective N ha^{-1}, 250 kg K_2O ha^{-1}, and 60 kg MgO ha^{-1} in 2012 and 2013. The effective N is the amount of N from the applied product that is expected to be available for crop uptake in the season of application [8]. In accordance with Flemish legislation, the effective N for organic fertilizers was set at 60% of the total N-content, whereas for ASW it was set at 100%. The maximum allowable rate of 80 kg P_2O_5 ha^{-1} for the cultivation of silage maize was respected [9]. Experimental treatments were tested in three replicated subplots ($n = 3$) of 9 × 7.5 m, randomized in the field in order to minimize the potential influence of variable soil conditions on the results. Product N doses for all fertilization years can be found in Table 5.6.1. Treatment CAN + AM presents the reference treatment, where only AM and mineral fertilizers (N as calcium ammonium nitrate [CAN], K_2O as potassium sulphate [PAT]) were used. In treatments CAN + AM + ASW and CAN + AM + LF, only a partial replacement of CAN took place, due to the fact that 25–33 kg was still applied as a mineral start N fertilizer. Finally, in treatments AM + ASW and AM + ASW + LF (tested only in 2012 and 2013), complete replacement of mineral N was achieved by fulfilling crop N demand completely with ASW or LF digestate in combination with AM. It should be remarked that the actual doses were sometimes different than the intended doses and

Table 5.6.1 *Product dosage of total nitrogen (kg ha^{-1}) applied for proposed fertilization treatments as compared to a conventional fertilization regime in silage maize (CAN + AM) and lettuce (CAN) cultivation; additional application of synthetic K_2O (kg ha^{-1}) in order to satisfy crop nutrient requirements; P_2O_5 contribution from applied products.*

Test plant	Treatments	Synthetic N (CAN)	Animal manure N (AM)	Air scrubber water N (ASW)	LF digestate N (LF)	Synthetic K_2O	P_2O_5 contribution	Total N applied
Maize	CAN + AM	54–60	97–163			78–161	45–108	157–217
	CAN + AM + ASW	25–30	97–163	29–46		78–161	45–108	157–217
	CAN + AM + LF	25–33	87–143		60–110	33–121	45–108	185–282
	AM + ASW		97–163	54–92		78–161	49–117	157–217
	AM + ASW + LF		87–143	33–46	65–110	66–121	49–117	152–299
Lettuce	CAN	210				240	0	210
	ASW			210		240	0	210
	LF				210	173	34	210

In the lettuce trial, additional synthetic P fertilizer (triple superphosphate, TSP) was added up to 125 kg P_2O_5 ha^{-1}. For the maize trial, applied dosage is presented within a range of all three experimental years and synthetic P fertilizer was not applied.

sometimes higher than the maximum allowable level due to differences in manure, ASW, and LF digestate composition over time. This variability especially affected the addition of P_2O_5, which in 2011 and 2013 exceeded the maximum allowable rate of 80 kg P_2O_5 ha^{-1} (Table 5.6.1).

Site 2 was an experimental greenhouse trial run on a sandy soil for 34 days (June 13 – July 17, 2013) at the Vegetable Research Centre (PCG) in Kruishoutem (50° 54′ 0″N, 3° 31′ 0″E). Prior to the experiment, soil flushing was performed as a common practice among farmers, especially between crop cycles during the summer months or after soil disinfection. In this study, soil flushing concerned application of water where 6 hours of irrigation were spread over a period of 8 days: 1 minute of irrigation equals 180 ml m^{-2}. As such, a homogenous nutrient-poor soil was obtained in June 2013: pH 6.3, EC 0.4 dS m^{-1}, total P 1.7 mg kg^{-1}, NO_3-N 1.7 mg kg^{-1}. This allowed higher application rates of ASW and LF digestate, leading to more accurate assessment of the their impact on soil quality and crop production as compared to CAN. Lettuce (*Lactuca sativa*, cv. Cosmopolia) was grown as a test crop with nutrient requirements of 210 N, 125 P_2O_5, and 240 K_2O kg ha^{-1}. The experiment was set up according to a fully randomized block design with four replicate plots of 10 m^2 (4 m × 2.5 m) per treatment. Three different fertilization treatments were established (Table 5.6.1). Treatment CAN was used as a reference where mineral N was applied along with triple superphosphate (TSP) and PAT. In the following two treatments, a single replacement of CAN took place, where ASW or LF digestate were used as an N fertilizer in combination with synthetic P_2O_5 and K_2O fertilizer.

5.6.2.2 Fertilizer Sampling

The LF digestate was sampled from an anaerobic codigestion plant at the site of Sap Eneco Energy (Merkem, Belgium). The feed of the installation consisted of 30% AM, 30% silage maize, and 40% organic biological waste produced by the food industry. The LF digestate underwent an obligatory hygienization step (1 hour at 70 °C) and separation step (sieve band press). AM was collected at the pig farm of Huisman (Aalter, Belgium). ASW for the silage maize field trial was collected at the piggery of Ladevo BVBA, Ruiselede (2011) and Sap Eneco Energy, Aalter, Belgium (2012 and 2013). In the greenhouse trial, ASW from a stable air washing installation at a pig farm belonging to Casier (Merkem, Belgium) was used. All products were collected in polyethylene bottles (2 l), stored (<4 °C), and characterized to determine the required application rate for the different cultivation treatments based on the total N crop demand (Table 5.6.2). Because the pH of ASW was low (pH 2–3), it was neutralized before application to the field. In 2011, the pH adjustment was carried out by adding NaOH (1 l NaOH per 200 l acidic waste water), whereas in 2012 and 2013 the acidic ASW was mixed with alkaline ASW from the same site [2, 10]. A pH adjustment is advised to prevent the reduction of soil pH, corrosion of application instruments, and leaf burning [2]. However, it is challenging to obtain a neutral pH of 7. Therefore, the pH of adjusted ASW applied in the 3-year maize trial varied between 6.9 and 9.0. The pH adjustment did not occur in the lettuce trial, since the application dosage of the product was lower (only 2 l per plot), it was applied manually, and it was diluted with tap water prior to fertilization. Methods of product physicochemical characterization and application to the

Table 5.6.2 *Physicochemical characteristics of biobased products applied as fertilizers in silage maize and lettuce cultivation.*

	Maize field trial (2011–2013)			Lettuce greenhouse trial (2013)	
Parameter	AM	LF digestate	ASW (pH-adjusted)	LF digestate	ASW (no pH-adjustment)
DW (%)	4.3–11	2.5–4.3	ND	3.3	33
EC (dS m^{-1})	31–35	33–34	135–208	41	262
pH	7.7–7.8	7.4–7.8	6.9–9.0	8.6	2.4
Total N (g kg^{-1} FW)	5.3–8.3	3.6–7.2	27–64	5.3	86
NH_4-N (g kg^{-1} FW)	3.2–5.6	2.8–6.2	27–64	4.6	86
P_2O_5 (g kg^{-1} FW)	2.4–5.4	0.57–1.7	<0.06	0.86	0.11
K_2O (g kg^{-1} FW)	2.9–6.7	3.0–4.5	0.2	4.4	0.2
NH_4-N/N_{total}	0.60–0.67	0.78–0.86	100	0.87	100

Data presented on fresh weight (FW) basis within a range of all three experimental years for long-term field trial.
AM, animal manure; LF, liquid fraction; ASW, air scrubber water; DW, dry weight; EC, electrical conductivity; FW, fresh weight; ND, not determined.
[a]50 vol. % digestate and 50 vol. % LF digestate in 2011, 40 vol. % digestate and 60 vol. % LF digestate in 2012 and 2013.

field are reported in Vaneeckhaute et al. [2, 10] and Sigurnjak et al. [6] for the maize and the lettuce experiment, respectively.

5.6.2.3 Plant and Soil Sampling

During the maize trial, samples of soils and plants were taken in October (at harvest) and November 2011 and 2013 and in November (at harvest) 2012. At each sampling moment, homogeneous soil samples were taken per subplot at three depths (0–30 cm, 30–60 cm, 60–90 cm) using a soil core sampler. The maize was harvested with a maize chopper. The crop fresh weight (FW) yield (i.e. aboveground plant) was determined at the field by hand harvesting 6.5 m^2, 10 m^2, and 7.5 m^2 per plot in 2011, 2012, and 2013, respectively. Methods of soil sampling and physicochemical analysis of soil and plant samples are reported in Vaneeckhaute et al. [2, 10].

In the lettuce trial, harvest was performed on July 17, 2013, after the commercial weight of lettuce (500 g fresh yield) was reached. Representative plant material samples were obtained by harvesting 12 random plants per plot, and soil samples were obtained as a mixture of five random sample points per plot, taken from the 0–30 cm soil layer. Methods of soil and plant analysis are reported in Sigurnjak et al. [6].

5.6.2.4 Statistical Analysis

Statistical analyses were performed with SPSS 22.0 for Windows. A one-way analysis of variance (ANOVA) was used to determine the effect of the applied fertilizers on soil quality and fertility, along with the effect on crop yield and nitrogen uptake, based on the obtained physicochemical data. Additional post hoc assessment was performed using Tukey's Test ($p < 0.05$, $n = 3$), when significant differences between means were observed. The condition of normality was checked using the Shapiro–Wilk test. Significant parameter correlations were determined using the Pearson correlation coefficient (r).

5.6.2.5 Nitrogen Use Efficiency

Based on the physicochemical data, the nitrogen use efficiency (NUE) of each treatment was determined using the following equation:

$$NUE\ (\%) = \frac{nitrogen\ crop\ uptake\ (kg\ ha^{-1})}{nitrogen\ applied\ through\ fertilization\ (kg\ ha^{-1})} \tag{5.6.1}$$

Furthermore, adjusted nitrogen use efficiency (ANUE) was obtained as the ratio of the NUE of treatments with biobased products (ASW and LF digestate) and the conventional fertilization regime (reference):

$$ANUE\ (\%) = \frac{NUE\ biobased\ treatment}{NUE\ reference\ treatment} \tag{5.6.2}$$

by assuming that conventional fertilization practice of using synthetic fertilizer in horticulture and AM additionally supplied with synthetic fertilizer in arable crops is 100% efficient. ANUE above 100% implies a relative deficit and ANUE below 100% a relative surplus of applied N.

5.6.3 Impact of Fertilization Strategies on Crop Production

In the first experimental year of the maize trial, Vaneeckhaute et al. [2] reported a statistically significant effect ($p < 0.05$) of fertilizer type on FW yield at harvest time when treatment CAN + AM + LF, the partial substitution of mineral N by LF digestate, resulted in significantly ($p < 0.05$) higher yield than other tested treatments, including conventional fertilization (CAN + AM). This effect, however, was not validated in the following 2 years (Figure 5.6.1), since in that period no significant differences ($p > 0.05$) were detected between proposed sustainable fertilization strategies and conventional fertilization regime (CAN + AM). The lowest FW and DW maize yields were observed in 2012 as compared to the other experimental years as a result of wet weather conditions [11] and delayed planting, which can lead to delayed leaf area index (LAI) development. Interestingly, the decrease in FW yield did not influence the DW content (%), which at harvest time was 28 ± 1% in 2011, 29 ± 0% in 2012, and 28 ± 1% in 2013. Hence, though each year a different cultivar was grown (Atletico KWS in 2011, Fernandez in 2012, and Millesim in 2013), the differences in the maize FW yield during the 3-year trial can mainly be attributed to weather conditions. Finally, a significant correlation was found between the FW maize yield and crop N uptake ($r = 0.911$; $p = 0.00$). It is assumed that heavy rainfall at the beginning of the maize cropping period in 2012 influenced the crop uptake by making N less available, which led to lower FW yields and reduced maize N uptake for all treatments, including the reference.

Similar trends regarding the crop yield were observed in the lettuce trial, where no significant statistical differences ($p > 0.05$) were detected among the three different fertilization treatments with respect to lettuce FW yield (Figure 5.6.2). However, it was observed that in

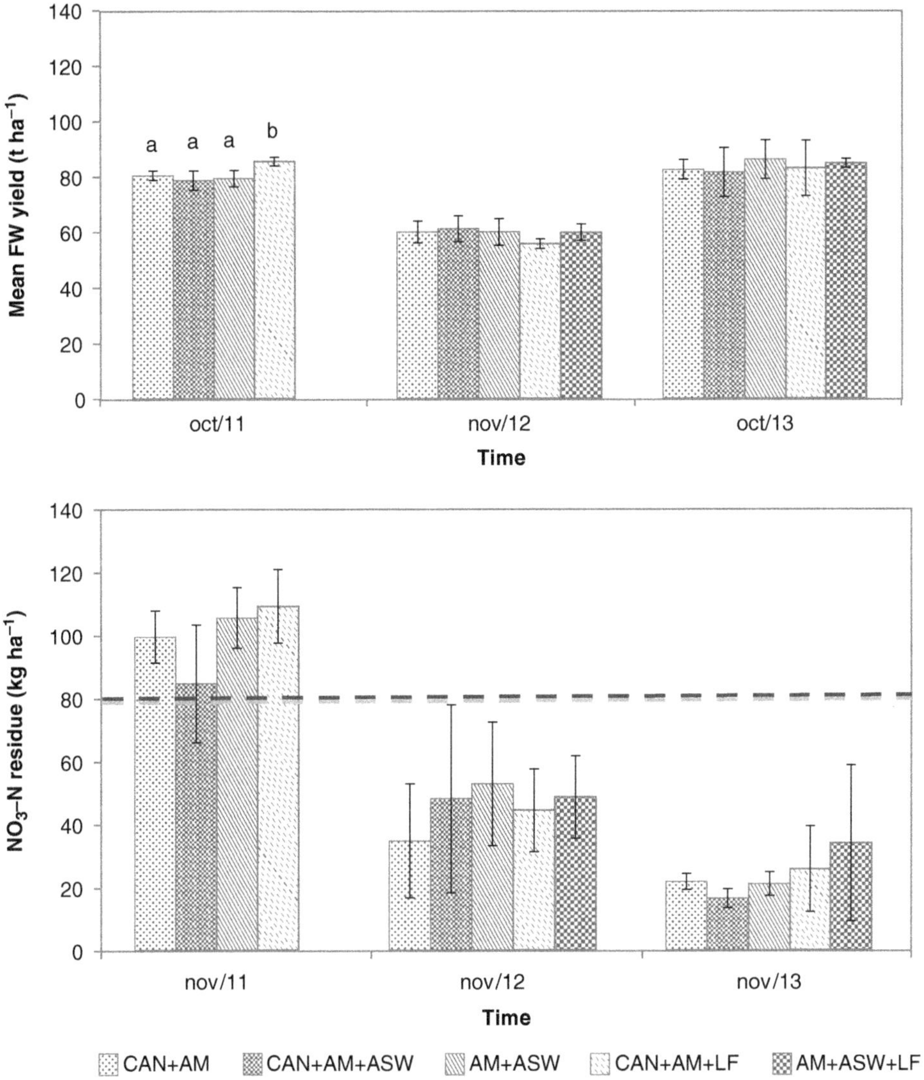

Figure 5.6.1 *Fresh weight (FW) maize yield (t ha^{-1}) at harvest time and NO_3-N residue (kg ha^{-1}) in the soil layer 0–90 cm for the five different cultivation treatments in November of each experimental year (2011, 2012, 2013). The line indicates the maximum allowable level of nitrate residue in soil (80 kg NO_3-N ha^{-1} for sandy soil in 2012 and 2013) between October 1 and November 15 according to Flemish environmental standards. In 2011, the maximum allowable level of nitrate residue in soil was 90 kg NO_3-N ha^{-1}. Error bars indicate standard deviations (n = 3) and lower-case letters (a and b) indicate significant different means between fertilizer treatments within a single year.*

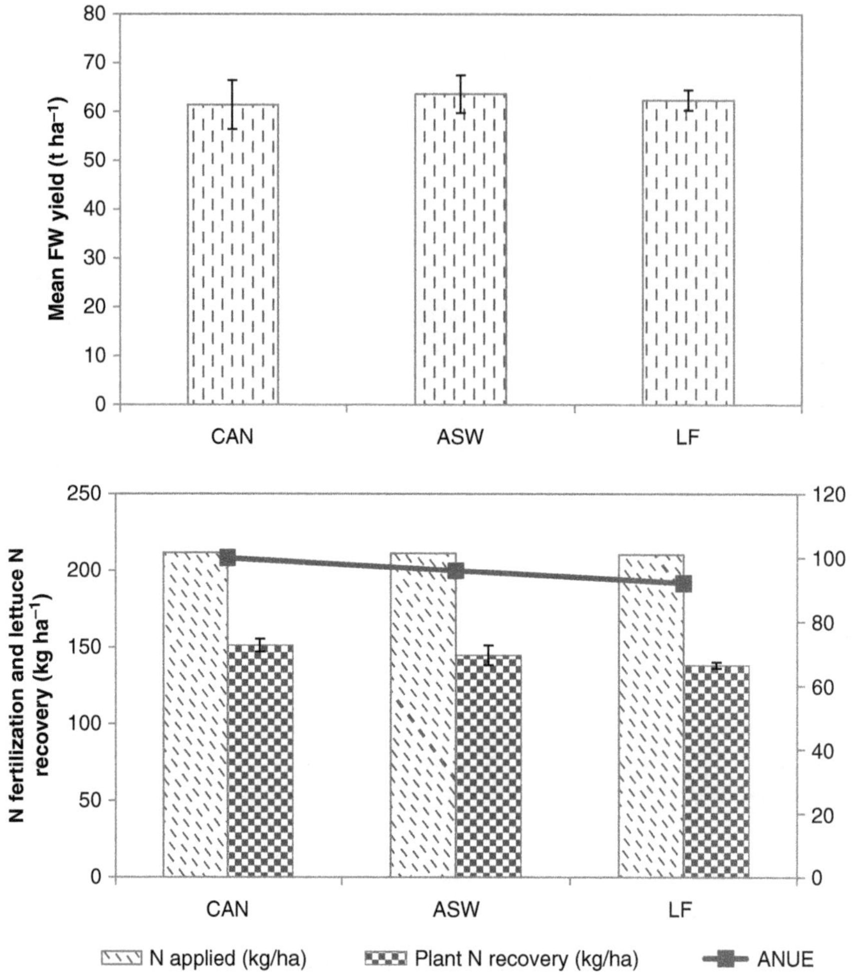

Figure 5.6.2 *Fresh weight (FW) lettuce yield (t ha^{-1}) at harvest time and N fertilization supply and lettuce N recovery in relation to adjusted nitrogen use efficiency (ANUE). Error bars indicate standard deviations (n = 4).*

LF treatment lettuce had a difficult start, which was visible in crop volume and uniformity (data not shown; [6]). This is usually the result of two crucial variables in LF application, namely EC and presence of ammonia-N in the product. Nonetheless, with soil EC values of 0.6 ± 0.2, 0.8 ± 0.2, and $0.8 \pm 0.2\,\mathrm{dS\,m^{-1}}$ for CAN, ASW, and LF treatments, respectively, there is no clear evidence that the lettuce uniformity was affected by high salt content. Up to now, contradictory results concerning digestate phytotoxicity have been reported. Some authors implicate NH_4-N and organic acid concentrations in digestate as limiting factors in plant growth [12], and others report that plant growth might be inhibited not only by ammonia volatilization but also to a lesser extent by ethylene oxide in AM [13]. These findings may give indications about the difficult start of the lettuce that occurred in treatments with LF digestate. It is, however, important to remark that in this study, no significant differences as compared to a reference were observed at harvest time with respect to crop yield and volume. The negative influence was only visible in lettuce uniformity, which was less homogeneous in comparison to the reference.

5.6.4 Impact of Fertilization Strategies on Soil Properties

Soil quality and fertility may be threatened by high nutrient inputs, which could negatively affect the environment through leaching. In order to assure efficient N use, N losses should not exceed those of synthetic fertilization if the aim is to (partially) substitute them in the future. In Flanders, the level of NO_3-N residue in the soil profile (0–90 cm) during the period October 1–November 15 is used as an indicator for unwanted leaching losses to surface and groundwater. In the first experimental year of the maize study, Vaneeckhaute et al. [2] reported an exceedance of the maximum allowable NO_3-N level of 90 kg ha^{-1} for all treatments, including the reference (CAN + AM), due to unfavorable weather conditions and late sowing of the intercrop, Italian rye grass (Figure 5.6.1). In the following 2 years, this limit was reduced even further to 80 kg NO_3-N ha^{-1}, as described by the Flemish Manure Regulation [14], and another decrease is expected in the near future. Despite the stringent environmental legislation, the NO_3-N residue for all treatments in 2012 and 2013 was below the legally stipulated limit and non-significant between treatments.

Though nitrate leaching is environmentally more relevant in open-air cropping, nutrient losses in greenhouse settings are also undesirable. At harvest time, NO_3-N residue (0–30 cm) amounted to 31 ± 11, 30 ± 10, and 36 ± 28 mg kg^{-1} for CAN, ASW, and LF treatment, respectively. Hence, no significant differences in the use of ASW or LF as a replacement for CAN were detected from an environmental point of view.

5.6.5 Adjusted Nitrogen Use Efficiency

Figure 5.6.3 gives more insight into the nutrient use efficiency of each treatment as compared (by ANUE) to the conventional fertilization regime in arable crops (being 100%). In 2011, CAN + AM + LF treatment gave the highest ANUE of 105%, due to having the highest mean value of crop N uptake. This effect, however, was not validated in the following 2 years, in which period CAN + AM + LF had the lowest ANUE and AM + ASW (ASW was used for the complete replacement of synthetic N fertilizer) had a significantly higher ANUE as compared to other treatments, including the reference (CAN + AM). It should be remarked that ANUE largely accounts for differences in N application rate [15], which in this study were caused by variability in nutrient composition between the intended and actually applied N dosage. Hence, regarding the fact that crop N uptake was mostly equal for all treatments, the lower ANUE of LF in treatments (CAN + AM + LF and AM + ASW + LF) can be attributed to its higher variability in nutrient composition, which may lead to overfertilization and a subsequent decrease in ANUE. Nevertheless, taking into consideration that recovery of N from fertilizers is about 30–70% of applied N in the year following application [16], fertilization strategies where LF digestate is used as an N fertilizer have resulted in similar agronomic (yield) and environmental (no accumulation of NO_3-N residue) values to conventional fertilization using synthetic N and AM.

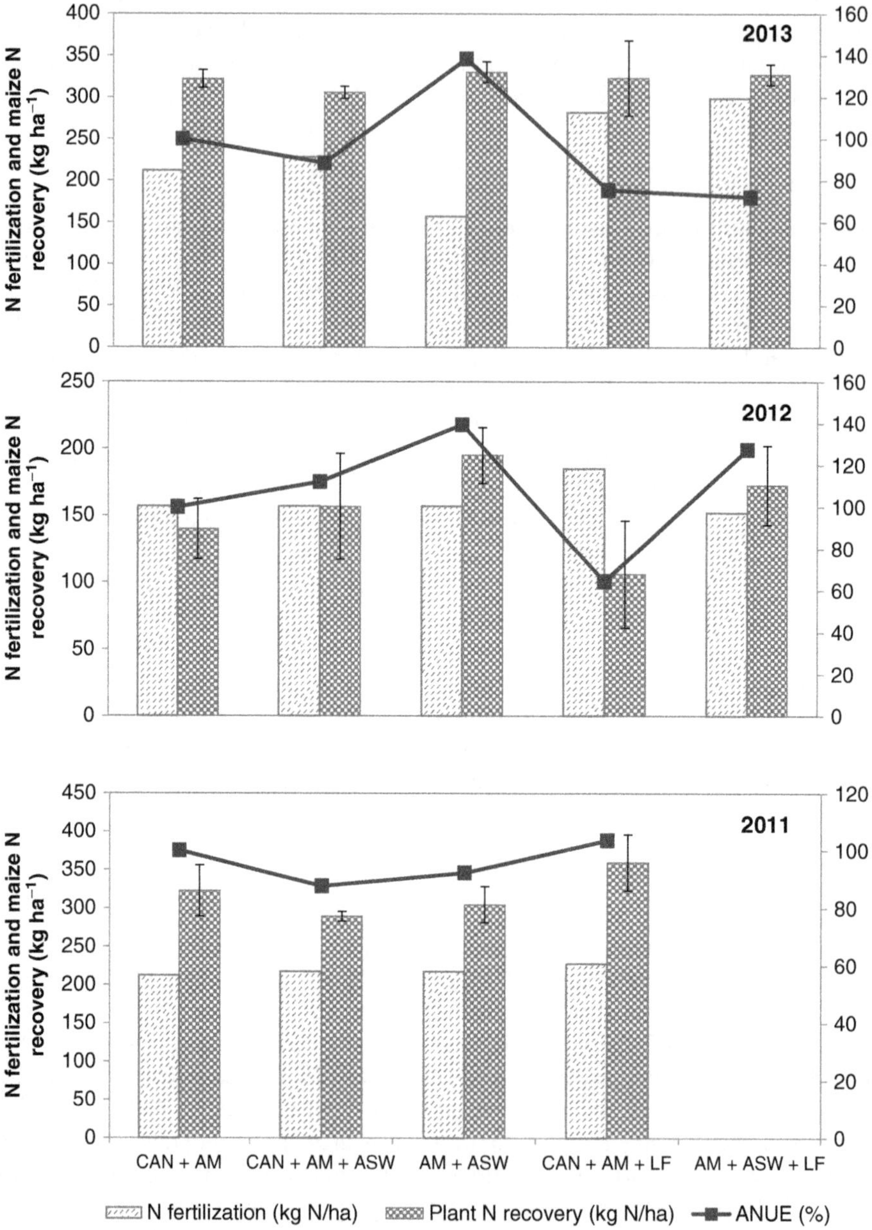

Figure 5.6.3 *N fertilization supply (kg ha^{-1}) and maize N recovery (kg ha^{-1}) in relation to adjusted nitrogen use efficiency (ANUE). Error bars indicate standard deviations (n = 3).*

Finally, similar results regarding ANUE were obtained in the greenhouse trial, where ASW and LF were added in combination with synthetic P_2O5 and K_2O fertilizers (Figure 5.6.2). For both sustainable alternatives, ANUE was above 90% as compared to the conventional fertilization regime.

5.6.6 Conclusion

A long-term (3-year) field trial and a short-term (<1-year) greenhouse experiment were conducted to assess the N fertilizer value of ASW and LF of digestate in silage maize (*Z. mays*) and lettuce (*L. sativa*) cultivation. Their N fertilizer potentials were compared to the conventional fertilization practice of using mineral N fertilizer in horticulture and AM supplemented with mineral N fertilizer in arable crops. At harvest time, no significant differences in crop fresh yield were observed in either tested crop. Also, no significant differences were detected with respect to the amount of nitrate left in the soil after harvest. NUE compared to conventional fertilization was higher when ASW was used as N fertilizer and lower when LF of digestate was applied. The lower efficiency of LF digestate was mainly due to its higher variability in nutrient composition.

The experiments described in this chapter demonstrate that mineral N fertilizers could be safely substituted by ASW and LF without adversely affecting productivity. Also, there is no clear evidence that the use of ASW or LF is associated with significantly greater nitrate losses compared to the use of a conventional mineral N fertilizer.

References

1. M.A. Sutton, A. Bleeker, C.M. Howard, et al. Our Nutrient World: The Challenge to Produce Moore Food and Energy with Less Pollution. Global Overview of Nutrient Management, Report. Edinburgh: Centre for Ecology and Hydrology (on behalf of the Global Partnership on Nutrient Management and the International Nitrogen Initiative); 2013.
2. Vaneeckhaute, C., Meers, E., Michels, E. et al. (2013). Closing the nutrient cycle by using bio-digestion waste derivatives as synthetic fertilizer substitutes: a field experiment. *Biomass and Bioenergy* **55**: 175–189.
3. Bomans, E., Fransen, K., Gobin, A. et al. (2005). *Addressing Phosphorus Related Problems in Farm Practice*. Leuven-Heverlee: Soil Service of Belgium.
4. Nkoa, R. (2014). Agricultural benefits and environmental risks of soil fertilization with anaerobic digestates: a review. *Agronomy for Sustainable Development* **34** (2): 473–492.
5. V. Lebuf, F. Accoe, C. Vaneeckhaute, et al. Nutrient Recovery from Digestate: Techniques and End-Products. Fourth International Symposium on Energy from Biomass and Waste, Venice. 2012.
6. Sigurnjak, I., Michels, E., Crappé, S. et al. (2016). Utilization of derivatives from nutrient recovery processes as alternatives for fossil-based mineral fertilizers in commercial greenhouse production of *Lactuca sativa* L. *Scientia Horticulturae* **198**: 267–276.
7. RMI, Climatology Database. Belgium: Royal Meteorological Institute of Belgium (RMI). Available from: www.meteo.be (accessed December 23, 2019).
8. Webb, J., Sørensen, P., Velthof, G. et al. (2010). *Study on Variation of Manure N Efficiency Throughout Europe*. Harwell: AEA Technology.
9. FMD, Flemish Manure Decree, Decree Concerning the Protection of Water against Nitrate Pollution from Agricultural Sources. May 13, 2011. Decree No.: BS13.05.2011-MAP4.
10. Vaneeckhaute, C., Ghekiere, G., Michels, E. et al. (2014). Assessing nutrient use efficiency and environmental pressure of macronutrients in biobased mineral fertilizers: a review of recent advances and

best practices at field scale. In: *Advances in Agronomy* (ed. L.S. Donald), 137–180. Cambridge, MA: Academic Press.
11. Boerenbond, Vlaams Landbouwinkomen 2012 blijft ondermaats. Leuven: Flemish Farmers Organization; 2012. pp. 1–13.
12. Möller, K. and Müller, T. (2012). Effects of anaerobic digestion on digestate nutrient availability and crop growth: a review. *Engineering in Life Sciences* **12** (3): 242–257.
13. Wong, M.H., Cheung, H., and Cheung, Y.H. (1983). The effects of ammonia and ehtylene oxide in animal manure and sewage sludge on the seed germination and root elongation. *Environmental Pollution Series A* **30**: 109–123.
14. VLM, Support Measures 2013 at a High Nitrate Residue: Sampling Campaign 2012/Packages of Measures in 2013. Brussels: Flemish Land Agency (VLM); 2013.
15. Klop, G., Velthof, G.L., and van Groenigen, J.W. (2012). Application technique affects the potential of mineral concentrates from livestock manure to replace inorganic nitrogen fertilizer. *Soil Use and Management* **28** (4): 468–477.
16. Bock, B.R. (1984). Efficient use of nitrogen in cropping systems. In: *Nitrogen in Crop Production* (ed. R.D. Hauck), 273–294. Madison, WI: ASA-CSSA-SSSA.

5.7

Effects of Biochar Produced from Waste on Soil Quality

Kor Zwart

Environmental Research, Wageningen University & Research, Wageningen, The Netherlands

5.7.1 Introduction

Biochar has become tremendously popular since the discovery of Dark Amazon Earth or "Terra Preta do Indio" soils by the Dutch Scientist Wim Sombroek in the Amazon river basin, and even more so since his discovery was picked up by the German scientists Johannes Lehman and Bruno Glaser during the late 1990s [1]. The latter two suggested that the presence of large amounts of char in these Terra Preta soils, next to organic and mineral household waste, explained why they were far more fertile than the surrounding soils, which lacked such human-made amendments. It was discovered that this char had already been present for periods over thousands of years [2], and it was concluded that it must be a very stable carbon source that led to the positive effect on soil fertility seen in the soils.

The presence of char in the fertile Terra Preta soils has led to the idea that the application of charred biomass to low-productive soils anywhere in the world could convert them into high-productive ones (identified as Terra Preta Nova). This would then contribute to successful regeneration of degraded land and further prevent soil desertification and loss of fertile soils [3]. For this to occur, biochar would have to restore critical soil functions and functionalities such as higher organic matter (OM) content and associated improvement of water holding capacity (WHC) and water storage or buffering and release of nutrients to enhance soil fertility and crop productivity via enhanced cation exchange capacity (CEC) of the soils. Further improvements to soil quality would be based on biochar providing places

Biorefinery of Inorganics: Recovering Mineral Nutrients from Biomass and Organic Waste, First Edition.
Edited by Erik Meers, Gerard Velthof, Evi Michels and René Rietra.
© 2020 John Wiley & Sons Ltd. Published 2020 by John Wiley & Sons Ltd.

to live and survive for soil organisms [4], and thereby enhancing the microbial-based benefits to plants (e.g. mineralization of OM and residues completing the cycling of nutrients and interventions in any soil-based pathogen populations and soil repressiveness) [5].

Char(coal) is produced from biomass in a process of pyrolysis. Pyrolysis heats up biomass (range of temperatures 250–900 °C) in a closed and anaerobic environment (see also Chapter 4.3). It produces char and energy-rich gasses and liquids. Thus, pyrolysis results not only in production of bioenergy but also in a highly stabile organic carbon fraction. The latter is believed to improve soil quality and productivity, and also acts as a carbon sink to sequester CO_2 and store it as organic carbon in the soil semi-permanently. The term "biochar" was introduced to identify the materials that would provide this range of functions. Biochar is a specific form of char that is produced especially to be applied to soils [6]. The prefix "bio" is very much a marketing term, though some may claim that it excludes soil application of char from materials of non-biological origin such as plastics or rubber tires, as such materials may also be pyrolyzed with char as a residue.

The idea to apply biochar to soils in order to simultaneously enhance soil fertility and mitigate climate change has generated a vast number of scientific and non-scientific papers over the last 20 years, and the number of contributions is still increasing. An exploration on a scientific search engine such as Web of Science of peer-reviewed papers on biochar may lead to over 4300 hits (June 2017); if non-scientific papers are included, a search on Google produces several hundred thousand. Many of them support the idea of a quadruple-win situation if biochar is produced from waste biomass: waste reduction, energy production, carbon sequestration, and improved soil fertility. Recently, next to the pyrolysis of wood and other relatively pure materials (see also Chapter 5.8), pyrolysis of organic waste has received a lot of attention. An analysis on Web of Science shows that 64 out of 94 papers with pyrolysis of waste in general in their title and 54 out of 100 on animal manure (see also Chapter 4.3) have been published since 2014.

However, not every application of biochar to every type of soil has resulted in increased productivity or improved soil quality. In fact, negative non-anticipated impacts and effects have been reported frequently. Recent reviews have shown that application of biochar only has a positive effect on crop yields in poor tropical soils and not in relatively fertile moderate climate soils [7, 8].

In this chapter, we identify and explain our current understanding of biochar effects: where, when, and how positive effects of biochar may be expected in soils and where and when they may not, on the basis of current scientific knowledge of both biochar and soils. This is done first by describing the properties of biochar with a focus on biochar from waste, then by discussing soil fertility, especially in relation to soil organic matter (SOM). Finally, the chapter concludes by examining where the two may or may not match.

5.7.2 Biochar Production and Properties

5.7.2.1 Pyrolysis

Biochar is produced by heating biomass under anaerobic conditions (i.e. in the absence of oxygen) in a so-called kiln during pyrolysis. Pyrolysis temperatures may range from 250 to over 900 °C. In the lower temperature range up to 300 °C, the process is called torrefaction, while above 600 °C in the presence of certain oxygen levels, it is referred to as gasification. Fast pyrolysis involves a rapid heating up time and a short biomass residence time in the kiln. Slow pyrolysis, in contrast, uses a long residence times and a slow heating up process. Hydrothermal conversion (HTC) is pyrolysis under wet conditions, for example using sewage sludge as a feedstock. Gasification is the heating of biomass in the presence of certain amounts of oxygen; thus, more biomass is converted into energy-rich gasses and CO_2 and less biochar is produced [9]. So, there are several conditions under which pyrolysis takes place, and since the processing time, pressure, and other elements of the process may vary under each of these conditions, it is not surprising that pyrolysis of even a single feedstock may result in the production of biochar with a wide range of different properties.

5.7.2.2 Biochar Feedstock

Several sources of biomass can be used to produce biochar. Biomass encompasses materials that originate from plants and animals. It includes processed biomass like compost and sewage sludge, animal bones, and digestate from anaerobic fermentation processes. This short list of processed biomass types already gives an indication that the composition of biochar feedstocks may vary widely, which becomes even more clear from Tables 5.7.1 and 5.7.2.

The composition of the biomass feedstock from which biochar is produced may vary widely in, for example, dry matter (DM) content, ash content, elementary composition (C, N, P, K, etc.), and cellular structure.

Even if biochar is produced from a single feedstock, its composition and thus its properties may vary widely, depending on the pyrolysis conditions during biochar production (Sections 5.7.2.1 and 5.7.2.2). Clearly, this variation even increases if a mixture of feedstocks is included.

This implies that though the term "biochar" is used widely, there is in fact not one biochar. Instead, there are many different types of biochar with a broad range of individual properties. Biochar properties depend on the type of biomass feedstock used and the circumstances under which it has been carbonized. Thus, it is very difficult to provide general statements

Table 5.7.1 *Biomass composition (%) of various feedstocks used for biochar production.*

Biomass	Water	Ash	C	H	O	N	P	References
Grass		6.9	48.6	7.25	44.1	0.64		[69]
Pine wood		1.2	50.6	6.68	42.7	0.05		
Pig manure	68.3		36.6			3.24		[74]
Saw dust	8.12		46.5			0.22		
MSW compost	39.9	50	21.3	—	—	1.2	0.32	[75]
Sewage sludge (AD)		34.62	32.04	51		5.466		[76]
Peanut hull		3.3	50.7	6.1	38.1	1.7	—	[77]
Pecan shell		1.6	51.6	5.7	41	0.3	—	
Poultry litter		24.4	36.2	4.8	24.4	4.1	—	
Switch grass		2.3	48.3	6.2	42.7	0.51	—	
Animal bone		15	58.0	8.3	19.7	11.55	1.6	[78]

MSW, municipal solid waste; AD, anaerobic digestion.

Table 5.7.2 Composition (%) of various types of biochar.

Biomass	Temperature, °C	Ash	C	H	O	N	P	References
Wheat straw	500–600	17	69			9.6		[79]
Wheat straw	400	10	65.79	3.43	20.4	0.21		[80]
Grass	500	15.4	82.2	3.32	13.4	1.04	—	[69]
Pine wood	500	2.1	81.9	3.45	14.5	0.06	—	
Poultry manure	400	25	53.54	2.8	15	3.71		[80]
Sewage sludge (AD)	500	74.21	17.4	0.7	10.5	1.54		[81]
Peanut hull	500	9.3	81.8	2.9	3.3	2.7	0.26	[77]
Pecan shell	700	5.2	91.2	1.5	1.5	0.51	0.05	
Poultry litter	700	52.4	44	0.3	0.3	2.8	4.28	
Switch grass	500	7.8	84.4	2.4	4.3	1.07	0.24	
Animal bones	700–800	> 60	13.1			1.87	15.2	[82]

AD, anaerobic digestion.

regarding the effects of biochar in soils, especially because biochar may behave differently in different soil types. The large variation in biochar types, together with their different effects in different soils, may explain the wide variation in, for example, crop yields that has been found after biochar application [8].

5.7.2.3 Biochar Composition

Biochar contains in general three different major fractions: a stable organic fraction, a labile organic fraction, and an ash fraction. In this section, we will describe the properties of each in more detail.

During pyrolysis, the composition of the biomass changes with time and temperature [10]. Fresh biomass is predominantly built up by the elements carbon (C), hydrogen (H), and oxygen (O), and during pyrolysis not only does the total biomass change with increasing temperature, but so too do the O : C (Figure 5.7.1) and H : C (Figure 5.7.1) ratios.

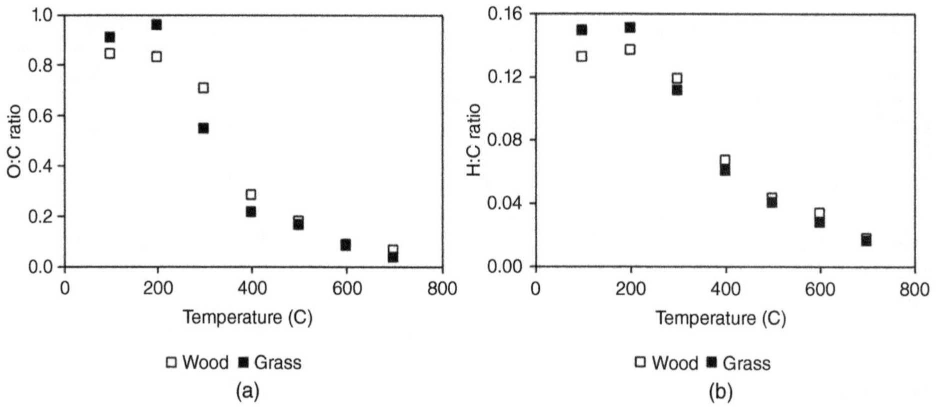

Figure 5.7.1 (a) O : C and (b) H : C ratios in wood and grass biomass with increasing pyrolysis temperature. Source: After Keiluweit et al. [69].

Figure 5.7.1 demonstrates that these changes are very similar in grass and wood, despite the different structures of these feedstocks.

The C, H, and O released during pyrolysis are mainly in the form of the energy-rich gasses H2, CO, and CH_4, organic liquids, plus water and CO_2. Other important biomass elements are nitrogen (N), phosphorous (P), potassium (K), calcium (Ca), and sulfur (S). While ammonia is removed by scrubbers [11], a large fraction of N also disappears into the atmosphere during pyrolysis [12]; the other elements remain mostly in the ashy fraction of biochar. The same holds for metals. While lead, cadmium, and zinc are considered relatively volatile, chromium, copper, and nickel are the least volatile heavy metals [13]. Thus, with increasing temperature, biochar becomes more enriched in C, P, K, Ca, and a number of metals, including toxic heavy metals. The ash content of biochar not only depends on the ash content of the feedstock, but also increases with increasing pyrolysis temperatures. Fertilizer production (ammonium sulfate) due to removal of ammonia from the flue gas after biomass gasification and use of acid scrubbers is potentially possible but requires a rather high NH_3 selling price [14]. Biochar from waste materials can also contain high contents of heavy metals and organic contaminants, which can hamper its use as fertilizer. The content of contaminants and various aspects thereof have recently been reviewed [15–17].

The molecular structure of the remaining biochar is characterized by highly condensed aromatic carbon rings containing other heteroatoms like O, N, and S. The H : C and O : C ratios decrease with increasing temperatures (Figure 5.7.1). The chemical and biological stability of biochar is largely determined by the level of condensation and aromatization [18] and the O : C [19] and H : C ratios [20]. According to Spokas [19], biochar with a molar O : C ratio below 0.6 should be considered as stable, and that above 0.6 as labile. Rittl [21] has found that biochar with a large aryl fraction decomposed rapidly after soil application of biochar from a traditional kiln to Brazilian savannah soils. Both biochar properties and soil properties play a role in biochar stability. Wang et al. [22] have performed a meta-analysis on biochar decomposition and found that it decreases with the clay content of soils.

There are some international initiatives to tackle the problems regarding biochar definition, but no consensus has yet been reached, with maybe one exception: charcoal should only be named "biochar" if it is used for any purpose that does not involve its rapid mineralization to CO_2 and may eventually become a soil amendment [23].

5.7.2.4 Biochar Structure

Biochar is a porous material with pore sizes ranging from the nanometer range to over 20 mm in diameter. The original cell and fibrous structure of plants can often still be seen in plant-derived biochar (Figure 5.7.2). Its porous nature enables biochar to take up water one to two times its own weight. However, in contrast with a sponge or with peat and clay, the swelling capacity of biochar is very limited, and addition of biochar to clay soils may even decrease their swelling capacity as compared to controls without addition [24]. Biochar is a rigid material, and the comparison with a sponge is in fact incorrect. Peat, which does have swelling capacities, may take up water 10–100 times its own weight [25] and is as such far more effective than biochar [26].

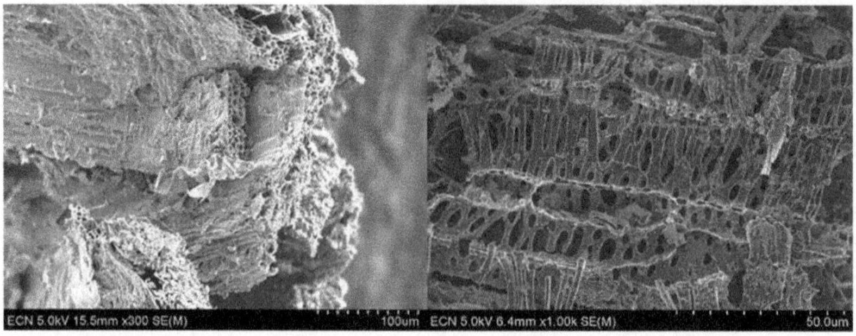

Figure 5.7.2 Scanning electron microscope (SEM) photographs of biochar from wood (left) and sweet pepper (right), clearly showing its porous structure. Source: Courtesy of Dr. H.J.M. Visser, ECN, Petten, the Netherlands.

Due to its porous structure, biochar has a large specific surface area (SSA) that may reach several hundreds of square meters per gram. SSA, which increases with increasing pyrolysis temperature [27], explains at least partly the sorption capacity of biochar for certain nutrients and organics.

5.7.2.5 Functional Groups

Several functional groups are linked to the biochar carbon matrix, including carboxyl, hydroxyl, and alcohol groups [28]. Many of these are negatively charged, but some (e.g. quaternary ammonium groups) may have a positive charge. The total number of functional groups and their charges determine some important biochar properties. Biochar's capacity to adsorb cations like calcium, potassium, and ammonium ions, also referred to as its CEC, is determined by the fraction of negatively charged functional groups. Its anion exchange capacity (AEC), on the other hand, is determined by the fraction of positively charged groups. Biochar properties regarding water also depend on these functional groups: highly charged biochar is hydrophilic, while biochar with few functional groups is hydrophobic. The functional groups often contain oxygen and hydrogen molecules, and the O : C and H : C ratios decrease with increasing pyrolysis temperature (Figure 5.7.1), with low-temperature biochar being more hydrophilic and having a higher CEC than that produced at high temperatures.

5.7.3 Effect of Biochar on Soil Fertility

This section looks at the possible effects of biochar on soil fertility, beginning with a brief discussion of the factors that determine soil fertility in general.

5.7.3.1 Factors Determining Soil Fertility

Soil fertility is primarily determined by water and nutrient availability to plants. Soil texture, soil structure, and SOM content and quality all affect these direct soil fertility factors, and hence can be considered as secondary factors regarding soil fertility. Soil texture determines

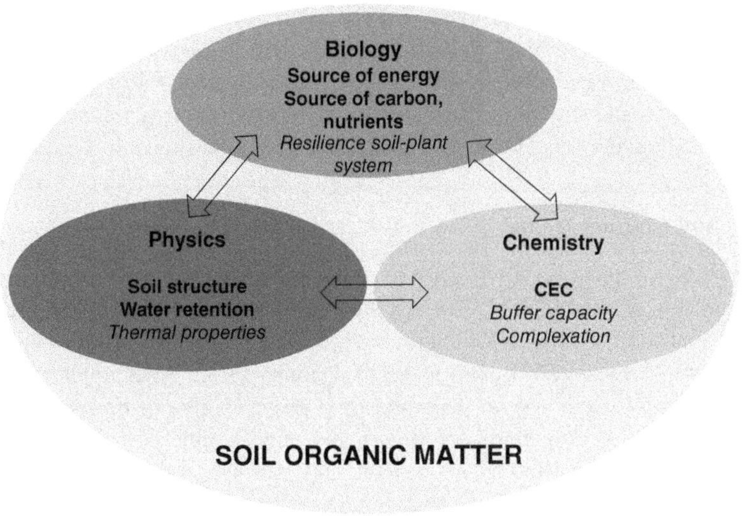

Figure 5.7.3 *Simplified schematic representation of factors determining soil fertility in relation to soil organic matter (SOM). Source: Based upon relations in e.g. Baldock and Skjemstad [70].*

among other things the soil WHC, clay soils having a higher WHC than sandy soils [29]. Also, water availability is higher in clay soils. Soil texture has a great effect on CEC (see Section 5.7.3.5), with clay soils again having a much higher CEC than sandy soils. Soil structure strongly determines the root depth plants can reach, the degree of aeration, and the WHC. Mineralization of SOM determines nutrient availability, and SOM also has effects on soil structure and water retention. Figure 5.7.3 shows the effects of SOM on CEC and soil chemistry. SOM directly and indirectly affects water and nutrient availability and indirectly affects water availability via its effect on soil structure. Thus, SOM plays an important role in soil fertility.

Soil fertility is also determined by biological factors. Decomposition of SOM by soil organisms is associated with the release of mineral nutrients. Many soil organisms (e.g. earthworms and fungi) play a role in soil structure formation. Moreover, it is believed that soil organisms determine soil health both positively and negatively. Pathogenic soil organisms may cause crop diseases, while antibiotics produced by soil bacteria and fungi may suppress them.

5.7.3.2 Effects of Biochar on Soil Fertility Factors

5.7.3.2.1 Soil Texture and Structure

The soil texture (i.e. the fraction of clay, silt, and sand of soil) is not affected by any biochar, perhaps unless extreme amounts of high-ash-containing biochar are applied over several decades or centuries, which seems very unlikely to occur. Biochar is lighter in specific density than the minerals in soils. Thus, application of biochar theoretically will result in a decrease in soil bulk density, as confirmed by a number of publications [30]. However, biochar does not always affect bulk density [31]. In some cases, this may be due to the fact that it was applied to the soil as very small particles or was rapidly transformed into very small particles in the soil. Soils with decreased bulk density have an increased pore space, resulting in a higher WHC and a lower energy requirement for soil tillage. Biochar may indirectly lead to an increase in the formation of water-stable soil aggregates [32] and thus to a decrease in bulk density. However, this effect is again not shown in all studies [32].

5.7.3.2.2 Soil Organic Matter

SOM is the general name for all OM (dead and alive) present in soils. It originates from the decomposition of microbial, plant, and animal residues and comprises a mixture of very different organic components, including proteins, carbohydrates (cellulose, hemicellulose), lipids, and lignin, each in a different stage of decomposition. After application of fresh organic materials to a soil, several decomposition stages can be distinguished. The most rapidly decomposable compounds are sugars and other simple carbohydrates, proteins, and lipids. These are converted by microorganisms into new microbial cells. Complexes of lignin and cellulose (ligno-cellulose) are more resistant to decomposition. Soil formation is partly associated with the decomposition of organic compounds. Soil particles are glued together by sticky components from bacteria and fungi, and thus large soil aggregates are formed.

In almost every publication regarding biochar, it is mentioned that it can be used to increase soil organic carbon content. This claim is undoubtedly correct in view of the high and stable carbon content of biochar. Implicitly, however, by suggesting that organic carbon means SOM, it is then also claimed that biochar therefore improves soil fertility. However, biochar carbon is not in every respect similar to SOM-carbon, and biochar differs from SOM in many ways.

Biochar contains a large stable carbon fraction; it is resistant to decomposition. There are exceptions where the carbon fraction decomposes rapidly [21], but such cases are likely due to incomplete carbonization as a result of too low pyrolysis temperatures or too short residence times in the kiln.

The labile organic fraction of biochar is decomposed within a period of a few weeks to decades after application of biochar to soils, while the stabile fraction shows an extremely low decomposition rate and remains present for several centuries, like in the Terra Preta soils. It will be clear that any labile component of the char supplied to Terra Preta soils centuries ago will by now be completely decomposed. This implies that only the stabile fraction of char will currently affect the soil fertility of such soils. For that reason, one might expect that regarding its effect on soil fertility, a great deal of attention would be paid to the role of the stabile organic fraction. In practice, that does not occur. Biochar is considered as a whole, and the effect of the distinctive fractions is seldom studied.

Data on the mean residence time (MRT) for biochar (556 years) [22, 33, 34], fresh OM (0.4 years) [35, 36], and coarse wood debris (decay rates: 0.0025–1.09 per year) [37–39] show that biochars are more refractory than their precursors. The MRT of SOM varies strongly (1–1000 years) [40–42] according to soil type, vegetation, and soil depth [22, 40]. The higher decomposition rate of surface SOM as compared to biochar is in line with its higher O : C and H : C ratios and in many cases higher C : N ratio. The C : N ratio of SOM in fertile soils is generally between 10 and 30 [43], whereas that of biochar is usually higher than 100 (Table 5.7.2). Humic substances (fulvic acids and humines) are the end-products of SOM decomposition and play an important role in CEC, for example (see Section 5.7.3.5).

It is clear from this section that biochar organic components differ in many respects from the organic components present in SOM. Thus, application of biochar organics will probably not have similar effects on soil fertility to application of organics from compost or of the organic compounds in SOM itself.

5.7.3.2.3 *Water Availability*

The effect of biochar on water availability to crops is limited and very much depends on the type of biochar in combination with the soil type [26]. The pore size distribution of biochar is largely dependent on the pyrolysis temperature. Biochar produced at relatively low temperatures largely maintains its original biological structure, in which, for instance, cells can still be recognized [44]. These relatively large pores (up to 20 μm in diameter) can store water, which can easily be extracted by plants. Biochar produced at high temperatures is characterized by the presence of mostly pores with a diameter in the nanometer range [45]. Since plants can only extract water from pores larger than 200 nm, it is clear that though the WHC of such biochar may be increased, water from these nanometer pores is not available to plants. Thus, it is not surprising that in the relatively few water-retention curves that have been published for soils in the presence of biochar, the largest effects are seen in the wettest regions [46]. The results would have been more interesting if biochar had increased the WHC between soil field capacity and the wilting point of crops; that is, the total amount of water available to crops.

Biochar may have a positive effect on water retention in soils with a very low water retention capacity of their own (i.e. coarse sandy soils with a very low clay and silt content and a low SOM content). In soils containing clay, silt, and SOM, the effects of texture and SOM are far greater than the effect of biochar. The effect of biochar on the total amount of water available to crops is mostly limited. There may be a positive effect below the field capacity (i.e. when the soil already contains a large amount of water). At high soil moisture contents, soils with biochar often contain more water than controls without. However, such water can freely drain out of the topsoil to deeper layers out of the reach of plant roots. At a high matric potential close to the wilting point of plants, the biochar effect is mostly very limited to completely absent, meaning that biochar has little effect under dry conditions. In smaller pores, capillarity becomes too large for plants to extract water.

The positive effect of SOM on WHC has been widely published [19, 36]. There is a positive non-linear correlation between SOM content and plant water availability. The effect decreases with higher SOM contents and is rather limited [29]. An increase in water availability from −10 to 30% has been observed after long-term amendment of organic residues [47].

5.7.3.2.4 *Nutrient Availability*

Next to a small fraction of water and a large organic fraction, biochar is composed of ash (i.e. mineral fraction). The ratio between the organic and the ash fraction varies with the type of biochar. Plant-derived biochars generally have a low ash content, while waste-derived biochar generally has a much higher one (Table 5.7.2). The ash content also depends on the pyrolysis conditions (i.e. higher temperatures result in higher ash contents). The ash fraction contains mostly calcium, magnesium, potassium, and phosphate, alongside small amounts of other mineral elements. The ash fraction may directly contribute to soil fertility by supplying nutrients and via its liming effect.

These effects are limited in biochar with a low ash content. In contrast, a high ash content is associated with a high pH and a high liming effect [48], mostly due to the presence of the carbonates Ca, Mg, and K. However, a high ash content (i.e. a high nutrient content) does not automatically mean a high nutrient availability. Nitrogen may not be immediately available since it may be bound to the organic fraction and decomposition of this fraction is needed to release it. Moreover, as already described, the labile biochar fraction is readily decomposable. That, in combination with a high C : N ratio of this fraction, may result in nutrient immobilization rather than nutrient mineralization, and, indeed, several cases of nitrogen immobilization after biochar application have been reported in the literature [49]. Biochar from oak wood may adsorb ammonium from digestate, the residue of biogas production, albeit to a much lesser degree than for the clay mineral clinoptilolite [50, 51]. Co-composting of biochar with fresh compost may result in an organic coating on the char, leading to a higher (but still very low) nitrate content ($2\,g\ NO_3\text{-}N\,kg^{-1}$) as compared to pristine biochar [52]. The phosphate content in digestate solution was decreased in the presence of biochar, probably due to a precipitation reaction [51]. Potassium, on the other hand, was increased in the solution, indicating its ready release from the biochar.

Results regarding phosphorous availability may vary to a large degree. P in the feedstock was fully recovered in biochars from a dairy manure–wood mixture and from a biosolid–wood mixture, and these biochars were effective as P-fertilizers, though P uptake was less than from fertilizer phosphate [22]. Warren et al. [53] showed that 0–73% of phosphate from pyrolyzed animal bones (bone char) in soil could be extracted with NaOH; they concluded that the P-fertilizer capacity of animal bone char was between that of rock phosphate and that of phosphate fertilizer [53]. Bone char is therefore considered an effective P-fertilizer [54]. However, one may question whether bone char should be considered as a biochar. According to the European Biochar Certificate, biochar should contain at least 50% carbon [23], whereas bone char, which was studied extensively during the EU Project REFERTIL, mostly consists of CaO (40%) and P_2O_5 (30%) [55]. Phosphorous availability varied between biochar produced from corn stover, Ponderosa pine, and Switch grass, being highest in Ponderosa pine wood residual and lowest in Switch grass biochar, which had the highest P-sorption capacity [56]. When the latter was applied to calcareous soils, however, P-availability decreased and P-sorption capacity increased. Biochar from chicken manure applied to an acid soil at low rates ($5\,g\,kg^{-1}$, approximately $20\,tons\,ha^{-1}$) increased its P-availability. At high application rates ($40\,g\,kg^{-1}$), P levels became at risk for leaching [57].

5.7.3.2.5 Cation Exchange Capacity

Apart from the direct nutrient availability from the ash fraction and mineralization from biochar, CEC may play a role in more indirect nutrient availability. Several papers indicate that application of biochar to soils may increase their overall CEC [56, 58, 59]. However, a comparison of the CECs reported for biochar with those of soils and soil fractions shows that the impact of biochar application to soils on overall CEC is rather small. Figure 5.7.4

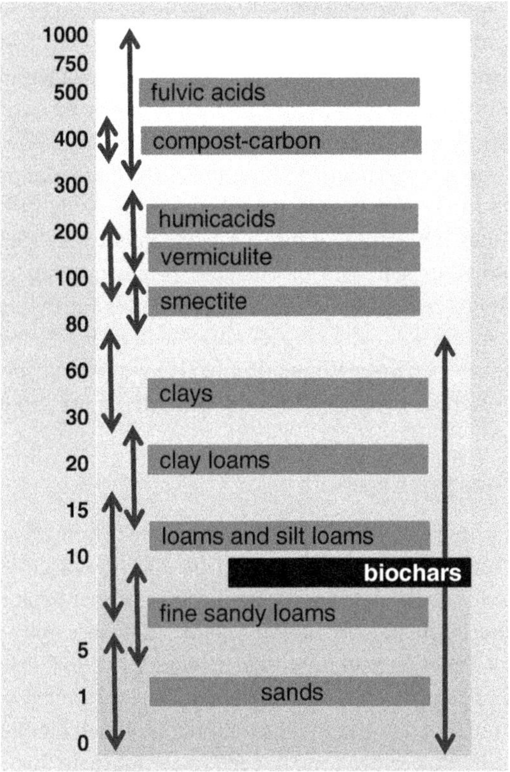

Figure 5.7.4 *CECs (meq 100 g^{-1}) of soils, soil fractions [71], and biochar [72, 73].*

compares the CECs of several soil fractions, including clay and organic fractions, with the range of CECs of biochar reported. It is evident that many soil fractions have a higher CEC than biochar, which means that only application to certain sandy soils low in clay and SOM content will lead to an increase in CEC. It is reported that "aging" of biochar may lead to an increase in CEC [60]. However the mechanism behind this is not completely clear and it cannot be excluded that the increase is resulting from adhering soil particles or humic substances.

5.7.3.3 Biochar as a Fertilizer or Soil Conditioner

The high initial expectations regarding the application of pyrolyzed biomass to soils in order to improve soil properties and increase plant production to levels analogous to Amazonian Terra Preta soils need to be reconsidered after a few decades of fundamental and practical biochar research. As clearly shown, positive effects on plant production are restricted to highly weathered soils with a low pH and low nutrient content [8]. Even there, the positive effects are more due to the liming effect of biochar (i.e. its ash content) than to an effect on SOM of the carbon fraction. In view of current knowledge regarding the origin of Amazonian Terra Preta soils, this lack of a positive effect may not be very surprising.

Terra Preta soils certainly do have an artificial origin, but charred biomass was not the only material applied to them. Inorganic wastes like ash, (fish) bones, and kitchen cutlery and organic wastes like manure, excrement, and urine have also been identified [1]. Thus, it can be concluded that the high fertility of Terra Preta soils cannot be attributed to the effect of applied charred biomass alone. In that respect, Terra Preta soils show a great similarity to the plaggen soils of Northern Germany, Belgium, and the Netherlands [61].

For that reason, and in view of the properties of biochar as described in this chapter, it is also not surprising that application of biochar alone is not very successful as a fertilizer or soil conditioner. Whether activated biochar can produce better results in soils in the temperate regions remains to be seen. Currently, draft rules for the use of waste materials such as biochar as fertilizers are under consideration with a view to their possible inclusion into the revised European Fertiliser Regulation [62]. Switzerland (not an EU member) is the only nation within Europe that has allowed biochar to be used in agriculture [23].

5.7.4 Trends in Biochar Research

Broadly, the following trend can be found in biochar research regarding its potential to improve soil quality. Initially, during the first period after the discovery of the Terra Preta soils, scientific attention was focused on the application of raw fresh biochar.

The presence of other remnants, including organic remnants, may explain why biochar application alone is not effective in many cases. Professor Bruno Glaser (pers. comm.) considers the application of raw biochar as completely useless in most situations. Stimulated by Glaser, a new trend in biochar research has started, in which biochar is "activated" prior to its application to soils. Raw biochar is "loaded" with nutrients by mixing it with animal manure or by co-composting it with plant and animal residues [63, 64]. An improvement in crop production through the application of such co-composted biochar has been shown, where application of the raw product resulted in a production decrease as compared to a control without biochar [64]. The effect of co-composted biochar might be explained by an increased capture and subsequent release of nitrogen [52]. Also, combined applications of biochar and chemical fertilizers may have a stimulating effect on crop production [65].

Next to activation with manure, fertilizer, or during composting, another type of chemical modification of biochar may help improve its effects. Fang et al. [66] have successfully modified biochar by adding magnesium in order to adsorb phosphate from pig wastewater.

The next trend may come from the observation that the biochar content of many Terra Preta soils is in fact not very high. The increase in soil carbon as compared to non-Terra Preta soils is rather the effect of an increase in SOM, rather than of charcoal itself. This has led to the theory that SOM decomposition is retarded in the presence of biochar, for instance by sorption of humic substances, which subsequently are protected from further decomposition. [67, 68]. However, since such a process may need hundreds of years to become effective, one might wonder if farmers will be convinced by arguments for its application.

Many experiments in which biochar is tested in soils have a simple set-up and lack a good control. Biochar is produced from a certain feedstock, applied to a soil (often in a range of quantities), and has its effects on soils or crops measured and described. The results obtained vary widely, because different feedstocks pyrolyzed under different conditions produce biochars with different properties. Experiments testing the different fractions in biochar, namely the labile fraction, the stabile carbon fraction, and the ash fraction, have seldom been carried out. The former two may be difficult to obtain, but the ash fraction can be extracted from biochar. Experiments in which the effects of the ash fraction on soil properties are compared to those with a complete biochar would provide indications whether the carbon fraction of biochar has any effects at all. For example, Jeffery et al. [8] observed that in many of the cases where biochar had a positive effect on crop growth, this could be explained by an increase in pH. This is to a large extent due to the liming effect of the ash in biochar.

Clearly, there is still little understanding of the specific effects of the stable and labile organic fractions and the ash fraction in biochar on the physical, chemical, and biological properties of soils of different textures and compositions, and subsequently on crop growth. Future research should focus on these mechanisms and on the specific effects of the different biochar fractions in order to determine whether or not biochar can be used as a soil amendment to improve soil quality and crop production.

References

1. Glaser, B. and Birk, J.J. (2012). State of the scientific knowledge on properties and genesis of Anthropogenic Dark Earths in Central Amazonia (*terra preta de Índio*). *Geochimica et Cosmochimica Acta* **82**: 39–51.
2. Pessenda, L., Gouveia, S., and Aravena, R. (2001). Radiocarbon dating of total soil organic matter and humin fraction and its comparison with 14 C ages of fossil charcoal. *Radiocarbon* **43** (2B): 595–601.
3. Glaser, B., Lehmann, J., and Zech, W. (2002). Ameliorating physical and chemical properties of highly weathered soils in the tropics with charcoal – a review. *Biology and Fertility of Soils* **35** (4): 219–230.
4. Lehmann, J., Rillig, M.C., Thies, J. et al. (2011). Biochar effects on soil biota – a review. *Soil Biology and Biochemistry* **43** (9): 1812–1836.
5. Bonanomi, G., Ippolito, F., and Scala, F.A. (2015). "Black" future for plant pathology? Biochar as a new soil amendment for controlling plant diseases. *Journal of Plant Pathology* **97** (2): 223–234.
6. Verheijen, F., Jeffery, S., Bastos, A. et al. (2009). *Biochar Application to Soils, a Critical Scientific Review of Effects on Soil Properties, Processes and Functions*. Luxembourg: Office for the Official Publ. of the European Communities.
7. Jeffery, S., Verheijen, F.G., van der Velde, M., and Bastos, A.C. (2011). A quantitative review of the effects of biochar application to soils on crop productivity using meta-analysis. *Agriculture, Ecosystems & Environment* **144** (1): 175–187.
8. Jeffery, S., Abalos, D., Prodana, M. et al. (2017). Biochar boosts tropical but not temperate crop yields. *Environmental Research Letters* **12** (5): 053001.
9. Nachenius, R., Ronsse, F., Venderbosch, R., and Prins, W. (2013). Biomass pyrolysis. In: *Advances in Chemical Engineering*, vol. **42** (ed. D.Y. Murzin), 75–139. Amsterdam: Elsevier.
10. Cha, J.S., Park, S.H., Jung, S.-C. et al. (2016). Production and utilization of biochar: a review. *Journal of Industrial and Engineering Chemistry* **40**: 1–15.
11. Sarıoğlan, A., Durak-Çetin, Y., Okutan, H., and Akgün, F. (2017). Decomposition of ammonia: the effect of syngas components on the activity of zeolite Hβ supported iron catalyst. *Chemical Engineering Science* **171**: 440–450.
12. Van Der Drift, A., Van Doorn, J., and Vermeulen, J.W. (2001). Ten residual biomass fuels for circulating fluidized-bed gasification. *Biomass and Bioenergy* **20** (1): 45–56.
13. Vervaeke, P., Tack, F.M.G., Navez, F. et al. (2006). Fate of heavy metals during fixed bed downdraft gasification of willow wood harvested from contaminated sites. *Biomass and Bioenergy* **30** (1): 58–65.

14. Andersson, J. and Lundgren, J. (2014). Techno-economic analysis of ammonia production via integrated biomass gasification. *Applied Energy* **130**: 484–490.
15. Bucheli, T.D., Hilber, I., and Schmidt, H.-P. (2015). Polycyclic aromatic hydrocarbons and polychlorinated aromatic compounds in biochar. In: *Biochar for Environmental Management*, Science, Technology and Implementation (eds. J. Lehmann and S. Joseph), 595–624. London: Routledge.
16. Hilber, I., Bastos, A.C., Loureiro, S. et al. (2017). The different faces of biochar: contamination risk versus remediation tool. *Journal of Environmental Engineering and Landscape Management* **25** (2): 86–104.
17. Beesley, L., Moreno-Jimenez, E., Fellet, G. et al. (2015). Biochar and heavy metals. In: *Biochar for Environmental Management*, Science, Technology and Implementation (eds. J. Lehmann and S. Joseph), 563–594. London: Routledge.
18. Crombie, K., Mašek, O., Sohi, S.P. et al. (2013). The effect of pyrolysis conditions on biochar stability as determined by three methods. *GCB Bioenergy* **5** (2): 122–131.
19. Spokas, K.A. (2010). Review of the stability of biochar in soils: predictability of O: C molar ratios. *Carbon Management* **1** (2): 289–303.
20. Kuhlbusch, T. and Crutzen, P. (1995). Toward a global estimate of black carbon in residues of vegetation fires representing a sink of atmospheric CO_2 and a source of O_2. *Global Biogeochemical Cycles* **9** (4): 491–501.
21. Rittl, T.F. (2015). *Challenging the Claims on the Potential of Biochar to Mitigate Climate Change*. Wageningen: Wageningen University.
22. Wang, J., Xiong, Z., and Kuzyakov, Y. (2016). Biochar stability in soil: meta-analysis of decomposition and priming effects. *GCB Bioenergy* **8** (3): 512–523.
23. European Biochar Foundation. European Biochar Certificate – Guidelines for a Sustainable Production of Biochar. Available from: http://www.european-biochar.org/en/download (accessed December 23, 2019).
24. Zhao, D., Huang, S., and Huang, J. (2015). Effects of biochar on hydraulic parameters and shrinkage-swelling rate of silty clay. *Transactions of the Chinese Society of Agricultural Engineering* **31** (17): 136–143.
25. Rezanezhad, F., Price, J.S., Quinton, W.L. et al. (2016). Structure of peat soils and implications for water storage, flow and solute transport: a review update for geochemists. *Chemical Geology* **429**: 75–84.
26. Atkinson, C. (2018). How good is the evidence that soil-applied biochar improves water-holding capacity? *Soil Use and Management* **34** (2): 177–186.
27. Trigo, C., Cox, L., and Spokas, K. (2016). Influence of pyrolysis temperature and hardwood species on resulting biochar properties and their effect on azimsulfuron sorption as compared to other sorbents. *Science of the Total Environment* **566**: 1454–1464.
28. Li, X., Shen, Q., Zhang, D. et al. (2013). Functional groups determine biochar properties (pH and EC) as studied by two-dimensional 13C NMR correlation spectroscopy. *PLoS One* **8** (6): e65949.
29. Minasny, B. and McBratney, A. (2018). Limited effect of organic matter on soil available water capacity. *European Journal of Soil Science* **69** (1): 39–47.
30. Sun, Z., Moldrup, P., Elsgaard, L. et al. (2013). Direct and indirect short-term effects of biochar on physical characteristics of an arable sandy loam. *Soil Science* **178** (9): 465–473.
31. Sun, Z. (2015). *Biochar Effects on Soil Physical and Biological Characteristics*. Aarhus: Aarhus University.
32. Soinne, H., Hovi, J., Tammeorg, P., and Turtola, E. (2014). Effect of biochar on phosphorus sorption and clay soil aggregate stability. *Geoderma* **219**: 162–167.
33. Lehmann, J., Abiven, S., Kleber, M. et al. (2015). Persistence of biochar in soil. In: *Biochar for Environmental Management*, Science, Technology and Implementation (eds. J. Lehmann and S. Joseph), 233–280. London: Routledge.
34. Paustian, K., Lehmann, J., Ogle, S. et al. (2016). Climate-smart soils. *Nature* **532** (7597): 49–57.

35. Dungait, J.A.J., Hopkins, D.W., Gregory, A.S., and Whitmore, A.P. (2012). Soil organic matter turnover is governed by accessibility not recalcitrance. *Global Change Biology* **18** (6): 1781–1796.
36. Mondini, C., Cayuela, M.L., Sinicco, T. et al. (2017). Modification of the Roth C model to simulate soil C mineralization of exogenous organic matter. *Biogeosciences* **14** (13): 3253–3274.
37. Freschet, G.T., Weedon, J.T., Aerts, R. et al. (2012). Interspecific differences in wood decay rates: insights from a new short-term method to study long-term wood decomposition. *Journal of Ecology* **100** (1): 161–170.
38. Zell, J., Kändler, G., and Hanewinkel, M. (2009). Predicting constant decay rates of coarse woody debris – a meta-analysis approach with a mixed model. *Ecological Modelling* **220** (7): 904–912.
39. Russell, M.B., Fraver, S., Aakala, T. et al. (2015). Quantifying carbon stores and decomposition in dead wood: a review. *Forest Ecology and Management* **350**: 107–128.
40. Ohno, T., Heckman, K.A., Plante, A.F. et al. (2017). 14C mean residence time and its relationship with thermal stability and molecular composition of soil organic matter: a case study of deciduous and coniferous forest types. *Geoderma* **308**: 1–8.
41. Trumbore, S.E. (1993). Comparison of carbon dynamics in tropical and temperate soils using radiocarbon measurements. *Global Biogeochemical Cycles* **7** (2): 275–290.
42. Tiessen, H., Cuevas, E., and Chacon, P. (1994). The role of soil organic matter in sustaining soil fertility. *Nature* **371** (6500): 783–785.
43. Springob, G. and Kirchmann, H. (2003). Bulk soil C to N ratio as a simple measure of net N mineralization from stabilized soil organic matter in sandy arable soils. *Soil Biology and Biochemistry* **35** (4): 629–632.
44. Hyväluoma, J., Kulju, S., Hannula, M. et al. (2017). Quantitative characterization of pore structure of several biochars with 3D imaging. *Environmental Science and Pollution Research* **25** (26): 1–11.
45. Klasson, K.T., Uchimiya, M., and Lima, I.M. (2015). Characterization of narrow micropores in almond shell biochars by nitrogen, carbon dioxide, and hydrogen adsorption. *Industrial Crops and Products* **67**: 33–40.
46. Ruysschaert, G., Nelissen, V., Postma, R. et al. (2016). Field applications of pure biochar in the North Sea region and across Europe. In: *Biochar in European Soils and Agriculture: Science and Practice* (eds. S. Shackley, G. Ruysschaert, K. Zwart and B. Glaser), 121–157. London: Routledge.
47. Eden, M., Gerke, H.H., and Houot, S. (2017). Organic waste recycling in agriculture and related effects on soil water retention and plant available water: a review. *Agronomy for Sustainable Development* **37** (2): 11.
48. Smider, B. and Singh, B. (2014). Agronomic performance of a high ash biochar in two contrasting soils. *Agriculture, Ecosystems & Environment* **191**: 99–107.
49. Nelissen, V., Rütting, T., Huygens, D. et al. (2015). Temporal evolution of biochar's impact on soil nitrogen processes – a 15N tracing study. *GCB Bioenergy* **7** (4): 635–645.
50. Kocatürk-Schumacher, N.P., Zwart, K., Bruun, S. et al. (2017). Does the combination of biochar and clinoptilolite enhance nutrient recovery from the liquid fraction of biogas digestate? *Environmental Technology* **38** (10): 1313–1323.
51. Kocatürk, N.P. (2016). *Recovery of Nutrients from Biogas Digestate with Biochar and Clinoptilolite*. Wageningen: Wageningen University.
52. Hagemann, N., Joseph, S., Schmidt, H.-P. et al. (2017). Organic coating on biochar explains its nutrient retention and stimulation of soil fertility. *Nature Communications* **8** (1): 1089.
53. Warren, G.P., Robinson, J.S., and Someus, E. (2009). Dissolution of phosphorus from animal bone char in 12 soils. *Nutrient Cycling in Agroecosystems* **84** (2): 167–178.
54. Zwetsloot, M.J., Lehmann, J., Bauerle, T. et al. (2016). Phosphorus availability from bone char in a P-fixing soil influenced by root–mycorrhizae–biochar interactions. *Plant and Soil* **408** (1–2): 95–105.
55. Someus E. Reducing Mineral Fertilisers and Chemicals Use in Agriculture by Recycling Treated Organic Waste as Compost and Bio-char Products (REFERTIL). Project ID 289785 FP7-KBBE. Terra Humana Ltd, editor. Hungary: Biofarm Manor; 2016.

56. Chintala, R., Schumacher, T.E., McDonald, L.M. et al. (2014). Phosphorus sorption and availability from biochars and soil/biochar mixtures. *CLEAN – Soil, Air, Water* **42** (5): 626–634.
57. Hass, A., Gonzalez, J.M., Lima, I.M. et al. (2012). Chicken manure biochar as liming and nutrient source for acid Appalachian soil. *Journal of Environmental Quality* **41** (4): 1096–1106.
58. Hansen, V., Müller-Stöver, D., Munkholm, L.J. et al. (2016). The effect of straw and wood gasification biochar on carbon sequestration, selected soil fertility indicators and functional groups in soil: an incubation study. *Geoderma* **269**: 99–107.
59. Zhang, Q.-Z., Wang, X.-H., Du, Z.-L. et al. (2013). Impact of biochar on nitrate accumulation in an alkaline soil. *Soil Research* **51** (6): 521–528.
60. Mukherjee, A., Zimmerman, A., Hamdan, R., and Cooper, W. (2014). Physicochemical changes in pyrogenic organic matter (biochar) after 15 months of field aging. *Solid Earth* **5** (2): 693.
61. Wiedner, K. and Glaser, B. (2015). *Traditional Use of Biochar*. London/New York: Taylor & Francis Group.
62. Huygens D, Saveyn H, Eder P, Delgado Sancho L. Draft Nutrient Recovery Rules for Recovered Phosphate Salts, Ash-Based Materials and Pyrolysis Materials in View of Their Possible Inclusion as Component Material Categories in the Revised Fertiliser Regulation. Interim Report. Circular Economy and Industrial Leadership Unit Directorate B – Growth and Innovation. JRCEC, editor; 2017.
63. Busch, D. and Glaser, B. (2015). Stability of co-composted hydrochar and biochar under field conditions in a temperate soil. *Soil Use and Management* **31** (2): 251–258.
64. Kammann, C.I., Schmidt, H.-P., Messerschmidt, N. et al. (2015). Plant growth improvement mediated by nitrate capture in co-composted biochar. *Scientific Reports* **5**: 11080.
65. Glaser, B., Wiedner, K., Seelig, S. et al. (2015). Biochar organic fertilizers from natural resources as substitute for mineral fertilizers. *Agronomy for Sustainable Development* **35** (2): 667–678.
66. Fang, C., Zhang, T., Li, P. et al. (2014). Application of magnesium modified corn biochar for phosphorus removal and recovery from swine wastewater. *International Journal of Environmental Research and Public Health* **11** (9): 9217–9237.
67. Lorenz, K. and Lal, R. (2014). Biochar application to soil for climate change mitigation by soil organic carbon sequestration. *Journal of Plant Nutrition and Soil Science* **177** (5): 651–670.
68. Courtier-Murias, D., Simpson, A.J., Marzadori, C. et al. (2013). Unraveling the long-term stabilization mechanisms of organic materials in soils by physical fractionation and NMR spectroscopy. *Agriculture, Ecosystems & Environment* **171**: 9–18.
69. Keiluweit, M., Nico, P.S., Johnson, M.G., and Kleber, M. (2010). Dynamic molecular structure of plant biomass-derived black carbon (biochar). *Environmental Science & Technology* **44** (4): 1247–1253.
70. Baldock, J.A. and Skjemstad, J.O. (2000). Role of the soil matrix and minerals in protecting natural organic materials against biological attack. *Organic Geochemistry* **31** (7): 697–710.
71. Rowell, D.L. (2014). *Soil Science: Methods & Applications*. London: Longman Scientific and Technical.
72. Bakshi, S., Aller, D.M., Laird, D.A., and Chintala, R. (2016). Comparison of the physical and chemical properties of laboratory and field-aged biochars. *Journal of Environmental Quality* **45** (5): 1627–1634.
73. Shenbagavalli, S. and Mahimairaja, S. (2012). Production and characterization of biochar from different biological wastes. *International Journal of Plant, Animal and Environmental Sciences* **2** (1): 197–201.
74. Huang, G., Wu, Q., Wong, J., and Nagar, B. (2006). Transformation of organic matter during co-composting of pig manure with sawdust. *Bioresource Technology* **97** (15): 1834–1842.
75. He, X.-T., Logan, T.J., and Traina, S.J. (1995). Physical and chemical characteristics of selected US municipal solid waste composts. *Journal of Environmental Quality* **24** (3): 543–552.
76. Tay, J., Chen, X., Jeyaseelan, S., and Graham, N. (2001). A comparative study of anaerobically digested and undigested sewage sludges in preparation of activated carbons. *Chemosphere* **44** (1): 53–57.
77. Novak, J.M., Busscher, W.J., Laird, D.L. et al. (2009). Impact of biochar amendment on fertility of a southeastern coastal plain soil. *Soil Science* **174** (2): 105–112.

78. ECN. Phyllis2 Database for Biomass and Waste. Available from: https://phyllis.nl/ (accessed December 23, 2019).
79. O'Toole, A., Knoth de Zarruk, K., Steffens, M., and Rasse, D. (2013). Characterization, stability, and plant effects of kiln-produced wheat straw biochar. *Journal of Environmental Quality* **42** (2): 429–436.
80. Sun, K., Ro, K., Guo, M. et al. (2011). Sorption of bisphenol A, 17α-ethinyl estradiol and phenanthrene on thermally and hydrothermally produced biochars. *Bioresource Technology* **102** (10): 5757–5763.
81. Chen, T., Zhang, Y., Wang, H. et al. (2014). Influence of pyrolysis temperature on characteristics and heavy metal adsorptive performance of biochar derived from municipal sewage sludge. *Bioresource Technology* **164**: 47–54.
82. Siebers, N. and Leinweber, P. (2013). Bone char: a clean and renewable phosphorus fertilizer with cadmium immobilization capability. *Journal of Environmental Quality* **42** (2): 405–411.

5.8

Agronomic Effect of Combined Application of Biochar and Nitrogen Fertilizer: A Field Trial

Wei Zheng and Brajendra K. Sharma
Illinois Sustainable Technology Center, University of Illinois at Urbana-Champaign, Champaign, IL, USA

5.8.1 Introduction

Biochar is a carbon-rich material generated when biomass is heated in a closed container, under either an oxygen-starved or an oxygen-free environment. Biochar production and its soil application are being considered as a potential win-win strategy for simultaneously sequestering carbon, immobilizing contaminants, and improving soil quality and fertility [1, 2]. Thermochemical conversion of biomass to biochar results in carbons from the atmosphere–biosphere pool being transferred to a slower cycling form (i.e. biochar), which may sequester these carbons into soils for millennia [3]. Moreover, biochar has the potential to mitigate emissions of greenhouse gases from soils, including carbon dioxide (CO_2), methane (CH_4), and nitrous oxide (N_2O) [4–6]. Also, the addition of biochar to soil has been shown to help retain chemical contaminants, and so prevent their loss by runoff and leaching to surrounding watershed and groundwater [7, 8]. For example, it has been reported that dairy-manure biochar can simultaneously immobilize lead and atrazine in contaminated soils [9], effectively retarding their transport through the soils.

The use of biochar as a soil amendment originated over 2000 years ago in the Amazon Basin. At that time, biochar known as charcoal was applied to soil to improve soil quality and fertility. The history of the use of biochar in North American agriculture dates back to

Biorefinery of Inorganics: Recovering Mineral Nutrients from Biomass and Organic Waste, First Edition.
Edited by Erik Meers, Gerard Velthof, Evi Michels and René Rietra.
© 2020 John Wiley & Sons Ltd. Published 2020 by John Wiley & Sons Ltd.

the beginning of modern science [10, 11]. According to the earliest publications, charcoal could not only be utilized as a pesticide to prevent some diseases in early US agriculture [10, 11], but also was used as a soil amendment to replenish nutrients in fields through combination with ammonia [10]. Recent studies have hypothesized that biochar used as a soil amendment may offer a series of benefits capable of improving soil quality, including raising soil pH, increasing moisture holding capacity, attracting more beneficial fungi and microbes, improving cation exchange capacity (CEC), and retaining nutrients in the soil [4, 12].

Given that biochar has the potential to improve soil quality and fertility, it is expected that biochar as a soil amendment would increase agronomic yields. One report showed that some crop yields could be increased by up to 200~300% with biochar additions [13], which may be attributed to higher nutrient uptake by plants in biochar amended soils [14]. Moreover, some studies have further shown that the increase of productivity derived from biochar application in agricultural fields does not impact product quality [15, 16]. Several reviews have summarized and evaluated the effects of biochar soil application on agronomic yield responses [17–19]. One quantitative review estimated through a meta-analysis that biochar as a soil amendment would increase crop productivity by approximately 10% [17]. However, biochar application in agricultural fields has not always resulted in positive effects on agronomic production. Spokas et al. [18] reviewed previously published articles and found that approximately 50% reported yield increases after biochar soil application, 30% observed no significant differences, and 20% showed negative effects on crop yield or plant growth. They further noted that negligible and negative agronomic yields might be greater than these suggested numbers, since the publication of positive results is typically favored [18]. In general, the positive effects of biochar on agronomic yields have been reported in nutrient-poor and acidic soils [19, 20]. For highly fertile soils, the benefits of biochar on crop productivity are often very limited [19, 21, 22]. Therefore, the responses of biochar soil application among agronomic yields are variable, depending on soil properties, climate conditions, crop plants, and biochar type [17, 18]. A recent study further indicated that the effect of biochar as a soil amendment is more dependent on-site characteristics (soil and climate) than on crop or biochar type [22].

From the standpoint of increasing agronomic yields, it is crucial to add proper biochars to fitting soils for appropriate plant growth under suitable environmental conditions. In the United States, biochar is commonly applied to highly weathered or nutrient-poor soils [23, 24]. For fertile soils, combined applications of biochar and chemical fertilizers or compost can be effective and feasible in improving plant growth and increasing crop yields [25–27]. Our previous greenhouse experiments showed that utilizing combined applications of biochar and nitrogen fertilizer in fertile soils had a synergetic effect in increasing agronomic yields [28]. Moreover, this combined application reduced the amount of chemical fertilizer needed [28]. This could help to partially offset the cost of biochar, which is currently considered one of the main barriers to its widespread use in agricultural fields. The objective of the present study was to conduct a field trial in fertile Illinois agricultural soils and assess the potential effects of the combined application of biochar and nitrogen fertilizer on agronomical production.

5.8.2 Materials and Methods

5.8.2.1 Biochars

Two biochars obtained from two different feedstocks were used in the field trial. Biochar A was produced from corn cobs by a lab-scale pyrolysis unit developed at the Illinois Sustainable Technology Center. The pyrolysis system consists of a batch reactor equipped with a programmable temperature controller (up to >900 °C) and a cooling system (to collect gas and bio-oil) [29]. The corn cobs were collected from a local farm, air dried, and then hammer milled to approximately 5 mm particle size. The ground feedstock was weighed into a batch reactor and pyrolyzed at 450 °C under oxygen-limited conditions. After pyrolysis, biochar in the reactor was allowed to cool overnight to room temperature, when the sample mass was recorded. Biochar B was obtained from BEST Energies, Inc. of Madison, Wisconsin. It was produced from hardwood chips using their slow pyrolysis pilot plant with a pyrolysis temperature range around 450~550 °C.

Both biochars were characterized for physical and chemical properties. Elemental analyses were conducted by the Microanalysis Laboratory at the University of Illinois using a CE 440 CHN analyzer (Exeter Analytical). The N_2-BET surface areas of the biochar samples were measured in a static volumetric apparatus (Monosorb, Quantachrome Corporation). Ash content was measured by heating the samples at 800 °C for 4 hours. The physicochemical properties of the biochars are shown in Table 5.8.1.

5.8.2.2 Soil and Site Description

The field trial site was located on a farm with more than 400 acres of agricultural fields near Monticello, Illinois (40°11′ N, 88°17′ W). The fields were typical no-till fields. Corn and soybean had been cultivated there for many years. It had never previously had any biochar additions. The soil was a silt loam comprising 34% sand, 54% silt, and 12% clay. Its pH was ~5.5 and its organic matter (OM) content ~4% in the upper 10 cm, decreasing with depth.

5.8.2.3 Field Experimental Design

The experiments were conducted during the corn-growing season in Illinois (from April to October 2010). The field experiment was designed to test whether the use of biochar as a soil amendment could reduce the application rates of nitrogen fertilizer while at the same time maintaining the same or higher corn yield. Three nitrogen fertilizer application rates were investigated. The field experimental design is shown in Figure 5.8.1. Six rows (each

Table 5.8.1 *Physicochemical properties of biochars.*

Biochar	SSA ($m^2\ g^{-1}$)	% C	% H	% N	% O[a]	% Moisture	% Ash
Biochar A	7.8	77.8	2.9	0.9	11.5	1.3	5.6
Biochar B	66.3	71.0	2.7	0.1	19.7	1.7	4.8

SSA, specific surface area.
[a]By difference.

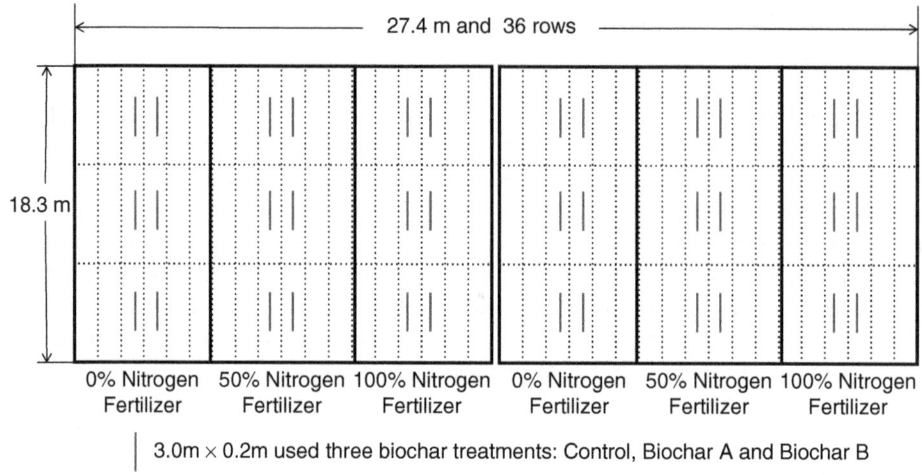

Figure 5.8.1 *Schematic diagrams showing biochar field experimental design.*

0.64 m in width and 18.3 m in length) were used for each of the three nitrogen fertilizer application rates. They included three biochar applications: control (no biochar), biochar A, and biochar B. Therefore, there were a total of nine treatments. The treatments were run in duplicate so that there was a total of 36 rows and 18 plots in the field trial. As shown in Figure 5.8.1, the biochars were randomly applied in 3 m segments in rows 3 and 4 for each fertilizer treatment in the plots.

The fields were prepared and managed by the farmer overseeing them according to standard corn-growing practices, including cultivation, fertilization, and weed and insect control. In the middle of April 2010, both biochars were completely mixed with the topsoil at a depth of approximately 10 cm according to the previously described experimental design. The biochar application rate was 40 t ha^{-1}, which corresponds to the levels used in previous research [30]. After 2 weeks, the corn seeds were planted in the designed fields using a regular corn sower with six row positions. Meanwhile, a pesticide (Capture Liquid Insecticide) was applied at 0.46 kg ha^{-1} in the furrow of each row. A month after seeding, a commercial Solution Nitrogen (28%) was side-dressed on the field plots according to three designed nitrogen application rates: no nitrogen fertilizer (0 kg N ha^{-1}), 50% nitrogen fertilizer application rate (107.6 kg N ha^{-1}), and 100% nitrogen fertilizer application rate (215.2 kg N ha^{-1}). In the middle of October 2010, the corn in the experimental fields was harvested.

5.8.2.4 Measurements and Analyses

Topsoil samples, 10 cm in depth, were collected from each plot before the experiment and at harvest. For each of the collected soil samples, properties including soil pH, soil OM, available phosphorus (P_1 weak Bray and P_2 strong Bray), nitrogen as nitrate, exchangeable potassium, magnesium, and calcium, and CEC were determined by Midwest Laboratories in Nebraska.

For each treatment, corn ears were harvested from the field plot, transferred to the greenhouse, and peeled to obtain the dried grains. The crop yields were calculated from the weight of grains for each treatment. The results were subjected to statistical analyses using a multivariate ANOVA test [31].

5.8.3 Results and Discussion

5.8.3.1 Effect of Biochar Application on Agronomic Yields

The effects of biochar A and biochar B used as soil amendments on corn yields are shown in Figure 5.8.2 and Table 5.8.2. On the control fields in the absence of biochar, the use of 50% nitrogen fertilizer significantly enhanced corn yields (Figure 5.8.2). However, further increase of the fertilizer application rate to 100% did not increase crop yields as expected, suggesting that the application rate of nitrogen fertilizer could be reduced by half in these highly fertile agricultural soils.

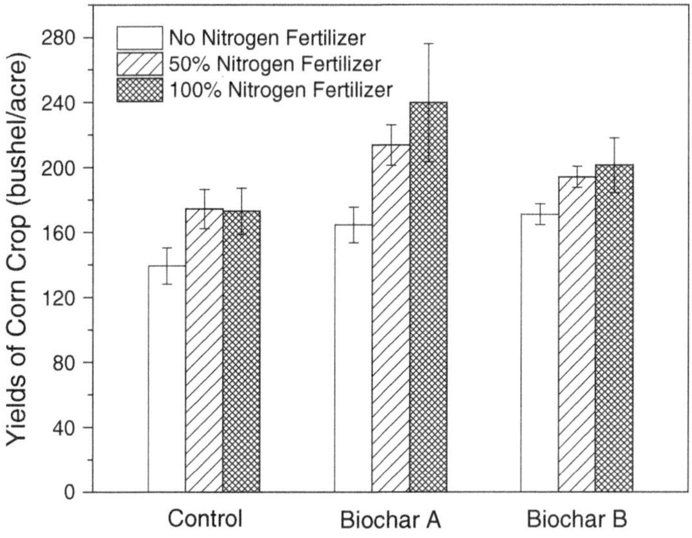

Figure 5.8.2 *Effect of biochar on corn yields under different nitrogen fertilizer application rates.*

Table 5.8.2 *Effect of nitrogen fertilizer application rates on corn yields (bushel/acre) with and without biochar additions.*

Nitrogen fertilizer	0	50%	100%
No biochar	139.3[a]	174.3[a]	173.0[a]
Biochar A	164.6[b]	213.7[b]	239.8[b]
Biochar B	170.9[b]	194.2[b]	201.3[a]

Crop yields between no biochar applications and biochar applications followed by different letters are significantly different ($\alpha = 0.5$) by Tukey's mean separation test.

In the absence of nitrogen fertilizer, the use of biochar as a soil amendment significantly increased crop yields (Table 5.8.2). For example, corn yields increased approximately 18 and 23% for biochar A and biochar B, respectively, compared to the fields without any treatments. This implies that the addition of biochar may improve soil quality and thereby help plant growth. The actual mechanisms responsible for the crop yield increase brought about by biochar application are very complicated and not yet completely understood. Previous studies indicated that some biochars could not only supply nitrogen and improve nutrient uptake, but also increase the mineralization of native soil nitrogen (as a result of the priming effect) [30, 32]. In addition, several researchers have shown that biochar application increases soil water content and enhances growth of microorganisms [16, 18], which would help to drive a substantial increase in agronomic productivity. Further research is needed to more precisely and sufficiently elucidate the mechanisms of these effects.

For the combined applications of biochar and nitrogen fertilizer, more significant increases in corn yields were observed compared to fields with only fertilizer applications (Figure 5.8.2). In the absence of biochar, for example, an approximate 25% increase in yield was obtained when 50 or 100% nitrogen fertilizers were used. When the use of biochar was integrated with fertilizer, the crop yield increased by approximately 54 and 39% for biochar A and biochar B, respectively, in the 50% fertilizer treatment and by approximately 72 and 44% in the 100% fertilizer treatment. Similar synergetic effects have also been reported in previous field [33] and greenhouse [30, 32] experiments. Such effects may be attributable to enhanced availability of nutrients for plant uptake, since biochar can efficiently hold ammonium ions in order to produce a slow-released fertilizer in soils, which can reduce the runoff of nitrogen fertilizer and inhibit its nitrification [28, 34]. In addition, the crop yield resulting from the combined application of biochar A and 100% fertilizers was significantly higher than that from the combined use of biochar B and 100% fertilizers (Table 5.8.2). The difference may be attributed to different biomass feedstocks and biochar production conditions.

As shown in Figure 5.8.2 and Table 5.8.2, the crop yields in the fields treated with biochar and 50% fertilizer were much higher than those in the fields with 100% fertilizer application only. It may be inferred that the use of biochar as a soil amendment is thus able to reduce fertilizer use while at the same time maintaining high crop yield, even though an increase in crop yield did not occur with increasing fertilizer application rates in the absence of biochar in this study. The reduction in the amount of chemical fertilizer needed may decrease the fertilizer cost, which would partly offset the cost of biochar use and thus facilitate the wider application of biochar in fertile soils.

5.8.3.2 Effect of Biochar as a Soil Amendment on Soil Quality

The addition of biochar into soils may cause a significant alteration in soil physical and chemical characteristics [16]. The results of soil analysis before the field experiments and after harvest are shown in Table 5.8.3. Before the experiments, the ranges of soil properties in the experimental plots were 3.2~4.7% for soil OM, 5.2~6.0 for soil pH, 12~27 mg kg^{-1} for available phosphorus P_1 and 17~43 mg kg^{-1} for P_2, 19~75 mg kg^{-1} for nitrogen as nitrate, and 15.4~20.8 meq/100 g for CEC. After harvest, the soil OM, soil pH, available phosphorus P_1 and P_2, and CEC generally increased in the field plots treated with biochar. The increase in soil OM and CEC showed that fairly large amounts of carbon and

Table 5.8.3 *Selected soil properties before and after experiments.*

Treatments	Soil organic matter (%)	Phosphorus		Neutral ammonium acetate (exchangeable)			pH	Cation exchange capacity (CEC) (meq 100 g^{-1})	Nitrate-N (mg kg^{-1})
		P_1 (mg kg^{-1})	P_2 (mg kg^{-1})	K (mg kg^{-1})	Mg (mg kg^{-1})	Ca (mg kg^{-1})			
Before experiment									
No fertilizer and no biochar	3.2	15	22	206	345	1953	5.8	16.3	19
No fertilizer with biochar-A	3.6	19	23	326	274	1808	5.7	15.4	39
No fertilizer with biochar-B	4.1	20	28	218	349	2046	5.7	17.3	35
50% fertilizer and no biochar	4.2	19	32	180	438	2340	5.6	20.7	32
50% fertilizer with biochar-A	4.7	27	42	358	406	2474	5.6	19.0	33
50% fertilizer with biochar-B	4.0	21	31	251	362	1986	5.2	20.8	67
100% fertilizer and no biochar	4.6	12	17	172	527	2617	6.0	21.1	34
100% fertilizer with biochar-A	4.6	22	43	195	449	2422	5.9	19.7	75
100% fertilizer with biochar-B	3.3	15	24	149	345	1878	5.4	17.6	74
After experiment (at harvest)									
No fertilizer and no biochar	3.9	20	30	198	310	1906	5.4	15.2	21
No fertilizer with biochar-A	5.0	33	46	259	318	2105	6.1	16.1	12
No fertilizer with biochar-B	4.5	19	31	166	329	1915	5.8	20.1	21
50% fertilizer and no biochar	4.5	22	38	176	366	2234	5.5	18.3	32
50% fertilizer with biochar-A	5.5	40	74	342	313	2298	6.2	19.0	29
50% fertilizer with biochar-B	4.8	29	47	179	324	2077	6.1	21.3	25
100% fertilizer and no biochar	4.2	14	21	175	456	2398	5.8	17.7	58
100% fertilizer with biochar-A	5.1	35	60	239	403	2308	6.5	22.6	13
100% fertilizer with biochar-B	4.9	27	39	221	376	2247	5.9	20.9	56

exchangeable cations were introduced by biochar application. The high level of available phosphorus P_1 and P_2 after biochar application indicated that the use of biochar as a soil amendment led to a high retention of nutrients in the soil. These results are consistent with many previous studies, in which biochar as a soil amendment can improve soil quality by increasing soil OM, pH, and CEC, and holding nutrients in soils.

By contrast, the contents of nitrate-N in these biochar-amended plots were significantly reduced even when they underwent nitrogen fertilizer application as compared to non-biochar-amended soils (Table 5.8.3). This confirms that biochar can sorb nitrogen fertilizers and inhibit their nitrification, leading to a decrease in the nitrate concentrations in fields amended with biochar. From the perspective of pollution prevention, the enhancement of nitrogen utilization could minimize excess nutrient loading into nearby water bodies. In addition, soil analysis after harvest revealed that soil OM, soil pH, available phosphorus P_1 and P_2, and CEC were generally higher after the application of biochar and fertilizer than after the application of fertilizer only (Table 5.8.3). These results support the idea that the combined application of biochar and chemical fertilizers can increase nutrient retention in fertile soils and enhance their plant uptake, resulting in synergetic effects on agronomic yields.

Acknowledgments

This study was supported by an Illinois Department of Agriculture Sustainable Agriculture Grant and the Russell and Helen Dilworth Memorial Fund. The authors would like to acknowledge BEST Energies, Inc. for providing biochar for the field trial. The authors also thank Dr. Edward Zaborski for his assistance in conducting the field work.

References

1. Laird, D.A. (2008). The charcoal vision: a win–win–win scenario for simultaneously producing bioenergy, permanently sequestering carbon, while improving soil and water quality. *Agronomy Journal* **100**: 178–181.
2. Lehmann, J. and Joseph, S. (2008). *Biochar for Environmental Management Science and Technology*. Sterling, VA: Earthscan.
3. Spokas, K.A. (2010). Review of the stability of biochar in soils: predictability of O:C molar ratios. *Carbon Management* **1**: 289–303.
4. Lehmann, J., Gaunt, J., and Rondon, M. (2006). Bio-char sequestration in terrestrial ecosystems – a review. *Mitigation and Adaptation Strategies for Global Change* **11**: 403–427.
5. Spokas, K.A. and Reicosky, D.C. (2009). Impacts of sixteen different biochars on soil greenhouse gas production. *Annals of Environmental Science* **3**: 179–193.
6. Cayuela, M.L., van Zwieten, L., Singh, B.P. et al. (2014). Biochar's role in mitigating soil nitrous oxide emissions: a review and meta-analysis. *Agriculture Ecosystems & Environment* **191**: 5–16.
7. Uchimiya, M., Cantrell, K.B., Hunt, P.G. et al. (2012). Retention of heavy metals in a Typic Kandiudult amended with different manure-based biochars. *Journal of Environmental Quality* **41**: 1138–1149.
8. Cabrera, A., Cox, L., Spokas, K. et al. (2014). Influence of biochar amendments on the sorption-desorption of aminocyclopyrachlor, bentazone and pyraclostrobin pesticides to an agricultural soil. *Science of the Total Environment* **470**: 438–443.
9. Cao, X.D., Ma, L.N., Gao, B., and Harris, W. (2009). Dairy-manure derived biochar effectively sorbs lead and atrazine. *Environmental Science & Technology* **43**: 3285–3291.

10. Allen, R.L. (1847). *A Brief Compend of American Agriculture*. New York: C. M. Saxton.
11. Haldeman, S.S. (1852). Devoted to agriculture, horticulture, and rural economy. *Pennsylvania Farm Journal* **1**: 57–58.
12. Lehmann, J. (2007). Bio-energy in the black. *Frontiers in Ecology and the Environment* **5**: 381–387.
13. Glaser, B., Lehmann, J., and Zech, W. (2002). Ameliorating physical and chemical properties of highly weathered soils in the tropics with charcoal – a review. *Biology and Fertility of Soils* **35**: 219–230.
14. Husk, B., Major, J. (2011). Biochar commercial agriculture field trial in Québec, Canada – Year three: Effects of biochar on forage plant biomass quantity, quality and milk production. Available from https://www.researchgate.net/profile/Barry_Husk/publication/273467636_Commercial_Scale_Agricultural_Biochar_Field_Trial_in_Quebec_Canada_Over_two_Years_Effects_of_Biochar_on_Soil_Fertility_Biology_and_Crop_Productivity_and_Quality/links/550455ab0cf2d60c0e664255/Commercial-Scale-Agricultural-Biochar-Field-Trial-in-Quebec-Canada-Over-two-Years-Effects-of-Biochar-on-Soil-Fertility-Biology-and-Crop-Productivity-and-Quality.pdf (accessed December 23, 2019).
15. Vaccari, F.P., Baronti, S., Lugato, E. et al. (2011). Biochar as a strategy to sequester carbon and increase yield in durum wheat. *European Journal of Agronomy* **34**: 231–238.
16. Genesio, L., Miglietta, F., Baronti, S., and Vaccari, F.P. (2015). Biochar increases vineyard productivity without affecting grape quality: results from a four years field experiment in Tuscany. *Agriculture Ecosystems & Environment* **201**: 20–25.
17. Jeffery, S., Verheijen, F.G.A., van der Velde, M., and Bastos, A.C. (2011). A quantitative review of the effects of biochar application to soils on crop productivity using meta-analysis. *Agriculture Ecosystems & Environment* **144**: 175–187.
18. Spokas, K.A., Cantrell, K.B., Novak, J.M. et al. (2012). Biochar: a synthesis of its agronomic impact beyond carbon sequestration. *Journal of Environmental Quality* **41**: 973–989.
19. Crane-Droesch, A., Abiven, S., Jeffery, S., and Torn, M.S. (2013). Heterogeneous global crop yield response to biochar: a meta-regression analysis. *Environmental Research Letters* **8**: 044049.
20. Hass, A., Gonzalez, J.M., Lima, I.M. et al. (2012). Chicken manure biochar as liming and nutrient source for acid Appalachian soil. *Journal of Environmental Quality* **41**: 1096–1106.
21. Liu, X.Y., Zhang, A.F., Ji, C.Y. et al. (2013). Biochar's effect on crop productivity and the dependence on experimental conditions – a meta-analysis of literature data. *Plant and Soil* **373**: 583–594.
22. Vaccari, F.P., Maienza, A., Miglietta, F. et al. (2015). Biochar stimulates plant growth but not fruit yield of processing tomato in a fertile soil. *Agriculture Ecosystems & Environment* **207**: 163–170.
23. Ippolito, J.A., Laird, D.A., and Busscher, W.J. (2012). Environmental benefits of biochar. *Journal of Environmental Quality* **41**: 967–972.
24. Lentz, R.D. and Ippolito, J.A. (2012). Biochar and manure affect calcareous soil and corn silage nutrient concentrations and uptake. *Journal of Environmental Quality* **41**: 1033–1043.
25. Otterpohl, R. (2012). Boosting compost with biochar and bacteria. *Nature* **486**: 187–188.
26. Alburquerque, J.A., Salazar, P., Barron, V. et al. (2013). Enhanced wheat yield by biochar addition under different mineral fertilization levels. *Agronomy for Sustainable Development* **33**: 475–484.
27. Schmidt, H.P., Kammann, C., Niggli, C. et al. (2014). Biochar and biochar-compost as soil amendments to a vineyard soil: influences on plant growth, nutrient uptake, plant health and grape quality. *Agriculture Ecosystems & Environment* **191**: 117–123.
28. Zheng, W., Holm, N., and Spokas, K.A. (2016). Research and application of biochar in North America. In: *Agricultural and Environmental Applications of Biochar: Advances and Barriers* (eds. M. Guo, Z. He and M. Uchimiya), 475–494. Madison, WI: SSSA Special Publication.
29. Zheng, W., Guo, M.X., Chow, T. et al. (2010). Sorption properties of greenwaste biochar for two triazine pesticides. *Journal of Hazardous Materials* **181**: 121–126.
30. Chan, K.Y., Van Zwieten, L., Meszaros, I. et al. (2007). Agronomic values of greenwaste biochar as a soil amendment. *Australian Journal of Soil Research* **45**: 629–634.

31. Papiernik, S.K., Dungan, R.S., Zheng, W. et al. (2004). Effect of application variables on emissions and distribution of fumigants applied via subsurface drip irrigation. *Environmental Science & Technology* **38**: 5489–5496.
32. Chan, K.Y., Van Zwieten, L., Meszaros, I. et al. (2008). Using poultry litter biochars as soil amendments. *Australian Journal of Soil Research* **46**: 437–444.
33. Yamato, M., Okimori, Y., Wibowo, I.F. et al. (2006). Effects of the application of charred bark of Acacia mangium on the yield of maize, cowpea and peanut, and soil chemical properties in South Sumatra, Indonesia. *Soil Science and Plant Nutrition* **52**: 489–495.
34. Spokas, K.A., Koskinen, W.C., Baker, J.M., and Reicosky, D.C. (2009). Impacts of woodchip biochar additions on greenhouse gas production and sorption/degradation of two herbicides in a Minnesota soil. *Chemosphere* **77**: 574–581.

Section VI

Economics of Biobased Products and Their Mineral Counterparts

6.1

Economics of Biobased Products and Their Mineral Counterparts

Jeroen Buysse and Juan Tur Cardona
Department of Agricultural Economics, Faculty of Bioscience Engineering, Ghent University, Ghent, Belgium

6.1.1 Introduction

Fertilizers are a key resource for modern agriculture. The production and application of fertilizers has changed drastically over the last hundred years. Until the twentieth century, soils, crop rotation with nitrogen (N)-fixating crops, and animal and human waste were the main sources of nutrients for crop production. The development of the Haber–Bosch process to produce reactive nitrogen from N_2 changed the agricultural production system in many regions of the world. The abundant availability of reactive N for fertilization has not only contributed to increased yields and production, it has also led to a change in production systems where the practice of recycling waste and the use of N-fixating crops has declined. Meanwhile, the availability of other nutrients such as in phosphorous (P) and potassium (K) has increased due to the developments P and K mining.

The objective of this chapter is to explain why and how fossil-based fertilizers have replaced the traditional organic fertilizers and how this process could be reverted using new legislation and technology. A key element to understanding this process is the knowledge that not only has use by farmers changed, but so too have the market and interactions between different market participants.

The first part of the chapter focuses on the demand side by comparing and explaining the different fertilizers used in different parts of the world. Simple economic theories and agronomic growth functions can to a large extent explain why the use of fossil fertilizers

Biorefinery of Inorganics: Recovering Mineral Nutrients from Biomass and Organic Waste, First Edition.
Edited by Erik Meers, Gerard Velthof, Evi Michels and René Rietra.
© 2020 John Wiley & Sons Ltd. Published 2020 by John Wiley & Sons Ltd.

is high in some regions compared to others. Reasons for the changed use of mineral and biobased fertilizers are also discussed.

The chapter continues by explaining supply-side evolutions in the fertilizer sector in relation to both fossil-based and organic sources. A key difference between the two supply chains is the scale of the firms and the logistic processes involved. Fossil-based fertilizers are traded by multinational companies internationally, while organic fertilizers are traded more locally by smaller companies. This section discusses the market power of the international fertilizer industry and the possible impact this has on organic fertilizers.

The final section concludes by describing the policy and mental shifts required to steer a transition toward the use of more organic-based sources.

6.1.2 Fertilizer Demand

Fertilizers are a main input for agriculture. Nowadays, fertilization is mainly based on mineral fertilizers, but the dominance of chemical fertilizers over traditional organic nutrients has only taken hold in the last 50 years (see Figure 6.1.1). In 1960, the proportion of nutrients applied by chemical fertilizers was still less than half of the quantity applied from organic fertilizers [1]. Since then, the world consumption of chemical fertilizers has increased by 15% every year.

Indeed, since the green revolution in the 1950, the consumption of mineral fertilizers has grown rapidly. The big expansion in fertilizer use took place mainly in developing economies, as shown in Figure 6.1.2. Except for a temporal decrease in fertilizer use due to the collapse of the Soviet Union around 1990, total fertilizer consumption is still increasing. In 2014, the total worldwide chemical fertilizer consumption in the world reached 186 million metric tons (Mt) of nutrients, comprising 113 million Mt N, 42 million Mt P_2O_5, and 31 million Mt K_2O [1]. Demand has evolved differently in different regions. In developed regions such as North America and Western Europe, fertilizer consumption has dropped compared to the level of the 1980s, due to a more efficient use of nutrients

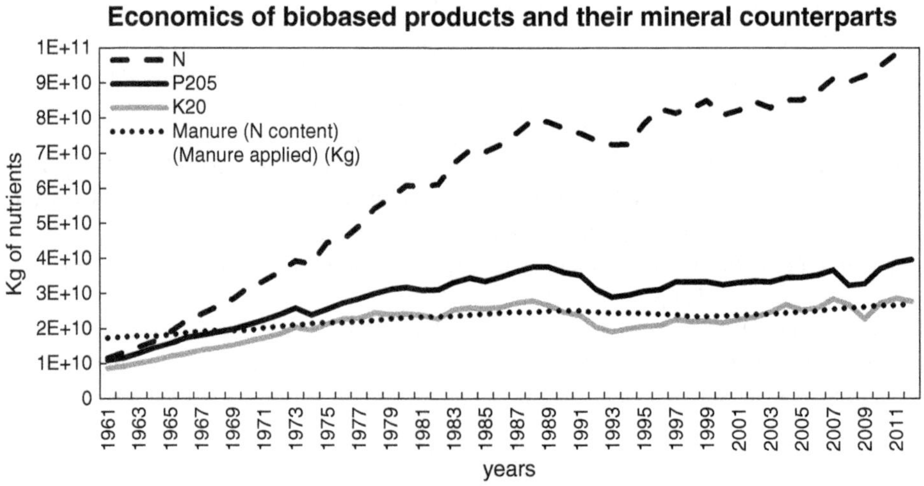

Figure 6.1.1 World nutrient application in agricultural crops [1].

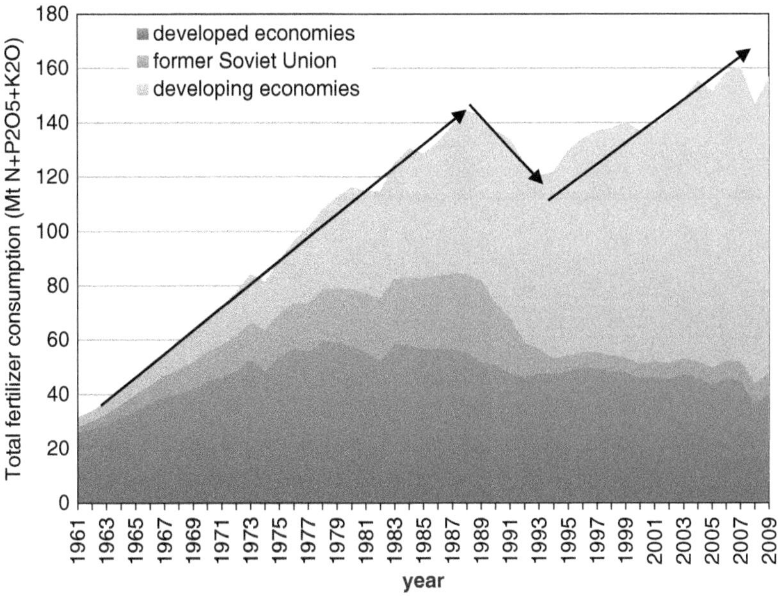

Figure 6.1.2 *Evolution of total fertilizer consumption worldwide from 1961 to 2009. Source: Based on IFA Agriculture.*

(in both crop and livestock production) as a result of increased technical efficiency and specific agricultural policies. Developing countries, on the other hand, have a booming and still-growing demand.

Within these developing economies, Asia is the continent with the highest demand for N mineral fertilizers. Globally, it is followed by America and Europe (see Figure 6.1.3). Africa and Oceania account for the lowest regional consumption of the world's total.

The higher increase in Asia is a result of a population increase and the support instruments put in place there, including investment in fertilizer production, improved marketing

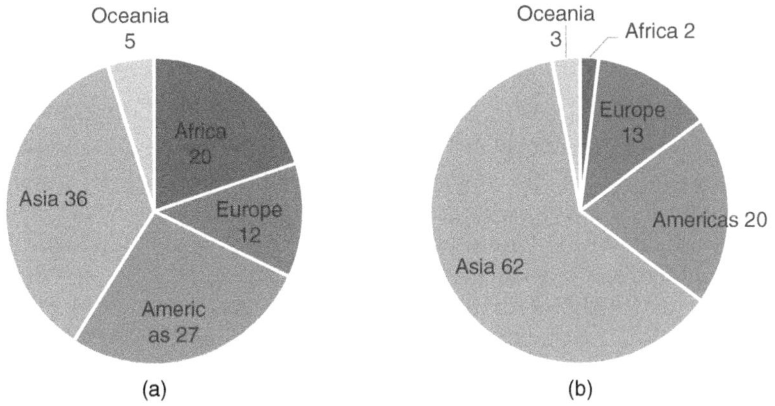

Figure 6.1.3 *Share of consumption of (a) organic nitrogen fertilizers and (b) mineral fertilizer. Total for organic fertilizer =* 97×10^9 *kg N, for inorganic nitrogen =* 111×10^9 *kg [2].*

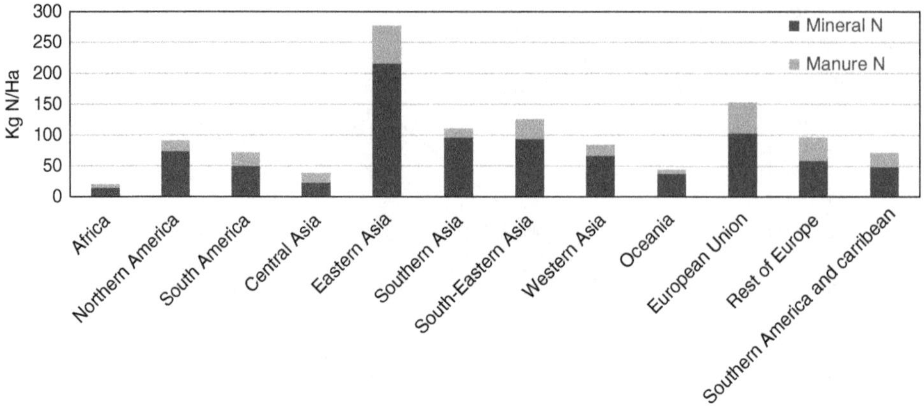

Figure 6.1.4 *Application rates per hectare per region.*

and distribution infrastructure, fertilizer and credit subsidies, and price-support programs. A slightly lower increase in fertilizer consumption is expected for Latin America, since production gains from improvements in the efficiency of fertilizer application are still high compared to the acquisition of new land. On the other hand, the current and forecasted demand for fertilizers in sub-Saharan Africa remains low. Sub-Saharan Africa still faces structural problems including poor physical infrastructure, lack of foreign exchange, lack of credit possibilities for farmers, and low crop prices, which together with the removal of fertilizer subsides explain the low chemical fertilizer use there [3]. It also has limited access to manure and other nutrient sources [4].

The uneven geographical distribution of mineral fertilizer and manure application can also be observed in the application rates per hectare. Figure 6.1.4 shows the amounts of N applied by chemical fertilizers and manure proportional to the total agricultural land of the regions. Eastern Asia has the highest application rates from both mineral and manure supplies. While Europe has a limited share in the world demand (9%), the application per hectare remains quite high. The low total use of nitrogen fertilizers in Africa and Oceania corresponds to low application of fertilizers in these regions.

6.1.2.1 Crop Demand

Unlike with the total demand for global consumption of fertilizers, detailed data on fertilizer consumption by crop is not often available. Heffer [5] discusses fertilizer consumption in 2010–2011, covering 94% of the world fertilizer demand divided among 13 crop groups (see Figure 6.1.5). Over half of this consumption (50.8%) is applied to cereals, followed by maize (the main fertilizer consumer crop), then wheat, then rice.

However, Heffer [6] also observes that, at a national level, the diversity of application rates of total fertilizer consumption per crop depends highly on the crop mix and the level of intensification. In some countries, for instance, almost all fertilizer use is allocated to a single crop (e.g. rice in Bangladesh, oil palm in Malaysia). This indicates that farmers' decision to apply mineral fertilizer to different crops is guided by the relative price of fertilizer compared to the price of the crop and not by the absolute price of mineral fertilizer [7].

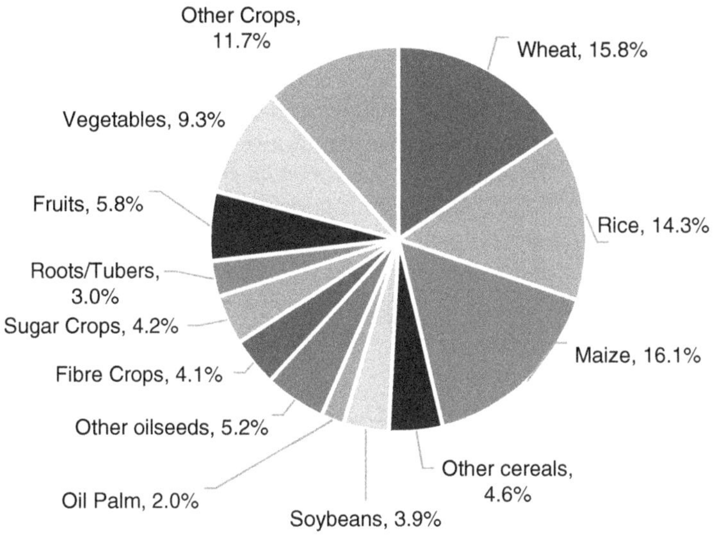

Figure 6.1.5 *Total global fertilizer use by crop. Source: Heffer [5].*

6.1.2.2 Drivers of the Increased Use of Mineral Fertilizers

The global demand for fertilizer is driven by the need to produce ever more food and energy. Population increases, changing diets, and the new demand for alternative sources of energy all contribute to a continuous need for production growth [8]. An increase in production requires a global increase either in the proportion of agricultural area or in the production per unit land. According to the US Food and Agriculture Organization (FAO) [9], an agricultural land increase is still possible, mainly in Africa and Latin America, but the small production prospective there due to low yields on the land in question makes it unlikely to compensate for the high investment, environmental, and social cost of bringing it into production. Therefore, the challenge of growth in the future depends on agricultural intensification [10]: obtaining higher yields on the same agricultural area. Fertilizer application has been one of the key aspects in this regard. Organic manure and guano enabled an increased production per hectare in East Asia in the last century and thus supported a larger population there [11]. But the always increasing demand for agricultural crops contributed to a drive for efficient and readily available chemical fertilizers. The growth in production since has been significant, to the point that it has been estimated that 48% of the population is fed from nitrogen manufactured in the Haber–Bosh process [12].

The early utilization of chemical fertilizers aimed merely to cover the gap created by the limited availability of nutrients from recycled organic waste, but they were quickly taken up by the expanding agricultural production system and used to replace organic fertilizers [13]. Chemical fertilizers also allowed for a more efficient nutrient application, since the nutrients were not organically bound, resulting in very robust economic return on use.

At the same time, the agricultural sector experienced a restructuring in its production system. A similar trend was observed in the concentration of population in cities and agricultural areas. Specialization in the farming sector resulted in spatial concentration in regions with mainly arable farmland or with a focus on livestock production. This disconnection, together with the availability of chemical fertilizers, changed farmers' views and reliance on manure and other leftovers from farming activities [14], altering the nutrient cycle. Also, the cost of labor increased, which decreased the attractiveness of labor-intensive manure management on and between farms. Some farmers considered manure a waste product, despite its nutrient content [15], to the extent that the positive effects of organic waste streams were often ignored [16]. The transport of manure could also contribute to the spreading of pathogens and weeds [17], and the use of manure entailed management challenges with regard to environmental problems such as greenhouse gas emissions [18, 19] and leaching of nutrients to the surface and groundwater [20]. As a consequence of these environmental issues and the evolution of farm-sector production requirements, farmers today often have a preference for chemical fertilizers as a source of nutrients [21].

6.1.2.3 Drivers of Biobased Fertilizer Demand

Despite the preference for chemical fertilizers, the use of nutrients from other sources is still important in agriculture, and comes with certain benefits. Manure has an important role in replenishing organic carbon in soils [22, 23], especially in intense agricultural systems [24], and helps improve their physical structure [14].

Lately, there has been increasing interest in nutrient recycling. The main driver for this is increased mineral fertilizer prices, caused either by increased fuel prices, depletion of nutrient stocks [25–27], or the market power of the fertilizer industry (see later). These increased prices create welfare problems for farmers given the continued squeezing margins in the agricultural sector.

Another driver is the spatial concentration of nutrient production and related environmental problems with regard to the disposal of nutrients. Due to the accumulation of nutrients in the agricultural system because of the import of external nutrient sources (chemical fertilizers, imported animal feed, etc.) and the disappearance of mixed farming systems, there is a need for more processing of excess nutrients, leading to new biobased fertilizers. In north-western European areas with overproduction of nutrients, farmers face stringent rules and limits on N and P application. Here, a maximum application rate of total N provides incentives for the optimization of nutrient use efficiency from all sources. In these areas, the processing and reuse of nutrients from manure are supposed to both decrease operating costs for the growing crops and simplify the transport of these nutrients over long distances. This transition in nutrient management will protect the environment and, at the same time, reintroduce recycling for manure management [28]. Progress in processing activities and technological change have allowed for a more efficient nutrient application, thereby achieving a necessary integration of chemical and biobased fertilizer applications [24]. Especially in cases where the total N application per hectare is limited, there will be interest in trying to utilize nutrients from biobased sources, especially when they are available at affordable prices. More and more attention toward nutrient recycling is included in novel strategies for sustaining food security and the environment [29].

The demand for biobased products is expected to further increase in the future, driven by the need for affordable nutrients and to increase the sustainability of the nutrient flow through the agricultural system. The use of biobased fertilizers thus seems set to play an important role in the future. On one hand, the import of chemical fertilizers has contributed and keeps contributing to the (excess) availability of nutrients in agricultural production. On the other, the increase in fertilizer prices and nutrient stock depletion will likely contribute to the reduction of chemical fertilizer use, thereby shifting the demand from chemical fertilizers to other sources of nutrients, where available.

As with the application of chemical fertilizers, application of biobased fertilizers to specific crops may differ by country. In the United States, corn production accounts for over half of all land spread with manure, making corn the largest single crop in the region, followed by the likes of hay, grasses, and soybeans [30]. Velthof et al. [31] describe different manure applications from housing and storage in Europe, mainly applied to fodder or roughage crops (grass, maize) or to non-fodder crops at different rates, distinguishing between those with high manure application rates (potato, sugar beet, barley), those with moderate rates (wheat, rye, oat, grain maize), and those where in general no manure is applied (fruits, citrus, oil crops). In this case, the application of biobased fertilizers will strongly depend on the specific nutrient crop requirements and the availability of the regional biobased or animal manure products. The low nutrient concentration of biobased products often limits the economic possibility of transport over long distances. Thus, availability often depends on livestock systems, manure type, crop rotation, and distribution. At the same time, the specific requirements and the uncertainty and variability in NPK content can greatly influence the yields of crops.

6.1.2.4 Importance of Fertilizer Use in the Cost of Production

The European Union and United States have systematic farm accounts that make it possible to determine the importance of the cost of fertilizers to the total gross value of a crop. Figure 6.1.6 shows the evolution of the different operational costs of cereal production in the European Union. In this type of agriculture, the share of fertilizers is the highest of the total operational costs; in 2013, the cost of fertilizers represented more than 25% of all such costs. This is lower than United States Department of Agriculture (USDA) estimates for corn operational costs, where the cost of fertilizers was estimated at 37% in 2016 and at 35% in 2017. Both EU and US statistics show that, for some crops, the cost of fertilizers is the most important of all operational costs in terms of total production. In addition, Figure 6.1.6

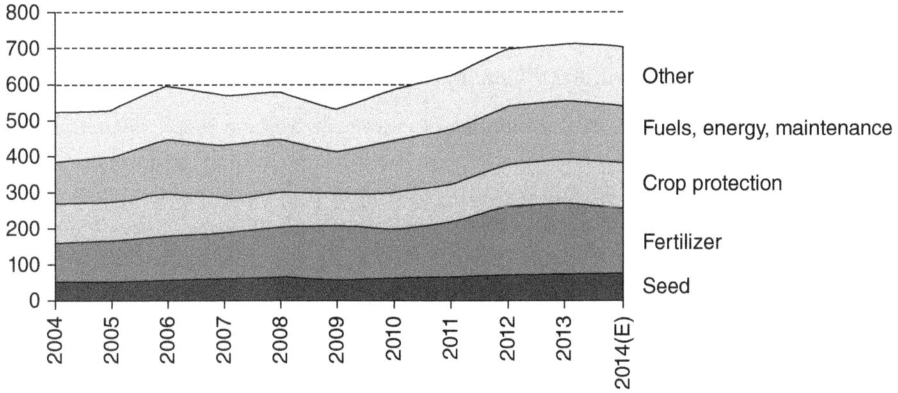

Figure 6.1.6 *Cost of production of cereals in the European Union (euro ha^{-1}). Source: Based on Farm Accountancy Data Network European Commission [32].*

shows that fertilizer costs are increasing more than other operational costs, and further that they are increasing more than the value of crop outputs. Some explanations for these trends are provided in the next sections.

6.1.3 Fertilizer Supply

This section first illustrates the current state of production of fertilizers, next looks at the link between food, fertilizer, and fuel prices, and finally examines the concentration of market power in the fertilizer industry. Insight into these characteristics is important in comprehending the future outlook of the biobased fertilizer industry and how the supply will meet the growing demand for fertilizers.

6.1.3.1 Global Production: Statistics and Regional Distribution

The first fertilizer factories were established in the United Kingdom in the early nineteenth century. Soon, other firms entered the market. Business began to boom after World War 2, partly because a lot of factories using nitrate to make explosives converted to the fertilizer industry.

The global total nutrient capacity ($N + P_2O_5 + K_2O$) was 284 million tons in 2014, with a total supply of 240 million tons. In that same year, the total capacity was expected to increase by 2.9%, and supply to grow by 1.6% [1]. The evolution based on FAO statistics suggests somewhat different results, yet the trend is clear: the production of all NPK fertilizers increased. Compared to 2002, the production of phosphate fertilizers in 2014 increased by 45%, potash fertilizer production by 40%, and nitrogen fertilizer production by 33% (Table 6.1.1).

As with consumption, production is not equal worldwide. Figure 6.1.7 shows the geographical production variation for ammonia and phosphoric acid. The biggest producer (37%) of Ammonia is East Asia. Eastern Europe, Central Asia, North America, and South Asia each produce between 12 and 15% . Between 2000 and 2012, the geographical distribution of ammonia production remained similar. That is not the case for phosphoric acid, where East Asia tripled its share to 42% in 2012, while North America's halved to 20%.

6.1.3.2 Link Between Food, Fertilizer, and Fuel Prices

The food price crisis in 2007 had a number of causes. It was the result of meteorological conditions coinciding with rising oil prices, the increase in the world population, the changing diet of the expanding middle-class population in Asia, and the competition for agricultural land due to the food versus fuel tension. Some authors question the link between fuel and food [33, 34]. A link between fertilizer and fuel seems obvious because of the very energy-intensive Haber-Bosch process required for ammonia fertilizer production and the energy needed for the mining of potassium and phosphate fertilizers. Yet, energy prices do not show the same spikes in 1973 and 2008 as do fertilizer prices. Gnutzmann and Śpiewanowski [35] believe this lack of co-integration between energy and food suggests that earlier studies may have identified an indirect impact of energy on food, channeled through increased fertilizer prices. This can be seen by the similar spikes occurring in 1973

Table 6.1.1 *Global production of fertilizer by nutrient, 2002–2014 [2].*

1000 tons nutrients	2002	2003	2004	2005	2006	2007	2008	2009	2010	2011	2012	2013	2014
Nitrogen fertilizers (N total nutrients)	85.348	87.459	94.823	96.202	95.476	100.189	99.189	97.196	104.102	105.983	107.965	109.341	113.310
Phosphate fertilizers (P205 total nutrients)	36.847	39.078	41.198	42.141	41.474	43.988	46.153	43.500	49.822	53.237	52.553	52.321	53.300
Potash fertilizers (K20 total nutrients)	29.610	32.954	34.851	33.673	35.658	36.915	35.064	31.027	39.301	40.147	38.949	38.979	41.372

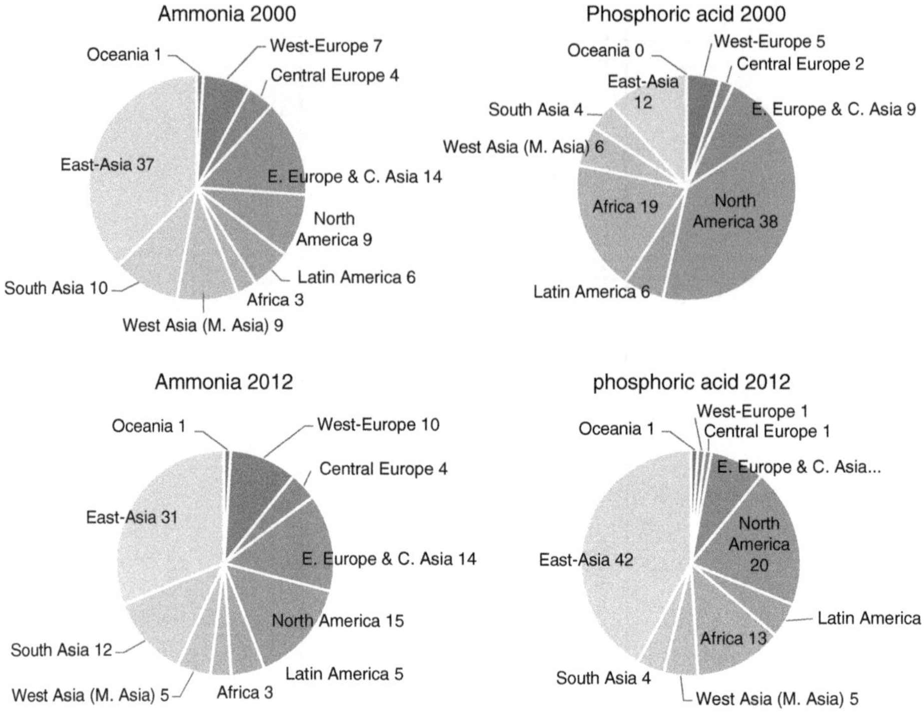

Figure 6.1.7 *Production share of ammonia and phosphoric acid worldwide. Source: KNOEMA statistics.*

and 2008 for both fertilizers and food. Indeed, Gnutzmann and Śpiewanowski [35] do find robust evidence for the co-integration of food, fuel, and fertilizer prices when they include fertilizer in their analysis. They emphasize the importance of incorporating fertilizer into equilibrium models of agriculture commodity prices.

Though there is an obvious link between energy and fertilizer prices, with crisis levels in both 1973/1974 and 2007/2008, fertilizer prices were much higher in those years than one would expect from energy prices at the time. Thus, these fertilizer crisis levels cannot be sufficiently explained by increasing energy prices. Recently, some authors have investigated to what extent international export cartels or market concentration might be the explaining factor [36–38].

6.1.3.3 Concentration and Market Power

The fertilizer industry is getting more consolidated and market-oriented. In the past, it was affected by state funds driving investments from a food security point of view rather than from a business one, and by weak fertilizer companies that existed as part of government-owned enterprises or conglomerates. As state involvement has declined and conglomerates have cleaned up their portfolios, there has been a trend toward consolidation and more financial discipline across the whole industry.

The high and increasing levels of market concentration and trade have two causes. First, industry consolidation is related to the concentration of availability of raw materials. In the case of nitrogen, a supply of energy is needed to fuel the Haber–Bosch process, in order to convert inert air nitrogen into reactive nitrogen forms such as ammonia and urea. Concentration of the industry is driven by access to natural gas, whether it be in gas-producing regions, in harbors with liquefied natural gas terminals, or in locations with pipeline connections. In the case of phosphorous and potash, access to raw material is mainly related to the limited number of countries where mines of phosphate rock or potash deposits exist.

A second reason for the concentration of production is economies of scale. Increased scale benefits the logistics, size, and efficiency of processing installations. Humber [36] provides an overview of the developments of the industry in the United States. In 1976, there were over 50 ammonia producers, in 2000, there were 27, and as of 2010, there were just 13. The top four nitrogen fertilizer producers by capacity, Koch, PCS Nitrogen, Agrium, and CF Industries, made up 43% of the total US capacity in 2000, but as of 2010 they represent nearly 74%.

Taylor and Moss [37] state that the fertilizer industry is fulfilling the conditions that make an industry prone to cartel formation and tacit agreements: (i) easy communication between firms; (ii) conditions of mutual monitoring; (iii) a small number of sellers; (iv) similar relative sizes of sellers; (v) difficult conditions of entry; (vi) ease of expansion through buying smaller firms; (vii) a similar cost structure among sellers; (viii) structure of the buyer side of the market; (ix) industry conditions; (x) the nature of the product; and (xi) industry history and sociology.

But this consolidation has also been noticed worldwide. Hernandez and Torero [38] provide an overview of the most important fertilizer-producing countries and discuss the share of market power of the five biggest firms (see Table 6.1.2). For ammonia, urea, and NPK fertilizers, the share of the five biggest companies is 50% or more. For phosphorous

Table 6.1.2 *Concentration of world fertilizer production capacity, 2008–2009 [38].*

Fertilizer	Top five countries (% of world in parenthesis)	Top five capacity (000 MT)	Top five share (% of world)
Ammonia	China (22.8), India (8.9), Russia (8.5), United States (6.5), and Indonesia (3.9)	84 183	50.6
Urea	China (33.1), India (13.1), Indonesia (5.4), Russia (4.2), and United States (4.1)	95 802	59.9
DAP/MAP	China (23.3), United States (21.2), India (11.4), Russia (6), and Morocco (4)	22 896	65.9
Phosphoric acid	United States, (20.9), China (19.3), Morocco (9.6), Russia (6.2), and India (5.3)	28 274	61.3
Potash	Canada (37.6), Russia (13.2), Belarus (9.9), Germany (8.2), and China (7.7)	39 687	76.7
NPK	China (29.3), India (8.2), Russia (6), France (4), and Turkey (3)	47 186	50.4

fertilizer, this is up to 65%, and for potash, it is 77%. The top five countries in terms of P and K fertilizer production are highly correlated with the availability of raw materials.

This concentration of the industry over time and its potential impact on market power are increasingly attracting the attention of the scientific literature. Hernandez and Torero [38] calculate different market concentration indexes for the case of the urea industry and use panel data to link this with fertilizer prices. They use the CR4 (i.e. the total market share of the top four producers) and the Herfindahl–Hirschman index (HHI) to measure the size of firms in relation to the industry. Their estimation suggests that the higher the level of concentration in this market, the higher the prices. It seems that the expected cost-efficiency due to economies of scale in production (and procurement) does not result in lower prices. This indicates that there is indeed some market-power exertion and tacit collusion among firms.

Taylor and Moss [37] explain the underestimation of the HHI index as follows: an industry with a small number of companies with equal market shares will have a low HHI index, indicating low market power, based on the estimation of Lerner indices for three big fertilizers companies. The Lerner index calculates the market power of each firm based on the difference between market price and marginal production costs divided by market price. The overall average of the dynamic Lerner index for all three companies and all three nutrients is 0.29 over the 2008–2012 period, meaning that profits for those companies are 29% above the cost of goods sold. Based on these calculations, the authors describe different periods of market power in the fertilizer industry. Between 1998 and 2007, there was a wave of consolidation. In 2008, prices rose spectacularly and the market power and profits were enormous, as suggested by dynamic Lerner index analysis. This had consequences for the demand of phosphorus and potash worldwide. Manufacturers' inventories grew to the point where they temporarily shut down production. Yet, prices did not decrease substantially afterwards, which could be an indication for cartel formation and tacit anticompetitive behavior. A third period from 2009 to 2012 is characterized by a drop in prices, explained by a "combination of factors: a weakening of a possible 'super' cartel, incumbent firms lowering prices to dissuade potential entrants, or to quiet the public outcry for an antitrust investigation" [37].

Humber [36] has investigated the price spike of nitrogenous fertilizer in early 2010, at the moment of a natural gas price decline. He assumed that a merger between CF and Terra industries in 2010 contributed to this inconsistent market behavior. A counterfactual fertilizer price series analysis concluded that the merger raised prices by roughly 75%.

The influence of an export cartel on the food spike prices of 2007/2008 has also been suggested [39]. Taylor and Moss [37] identified strong cartel market power in the period 2008–2012. A dummy cartel variable for these years was added to a regression analysis, where the author estimated the price for potash fertilizer based on energy prices. According to their estimates, the cartel directly led to a 60% overcharge in the fertilizer market, which through cost pass-through to agriculture translated into a 25% increase in food prices.

6.1.3.4 Impact of a Strong Fertilizer Industry on the Production of Biobased Fertilizers

This high market power of the fertilizing industry did not go unnoticed by the public. Different complaints were raised in different countries. From a political point of view, the focus has been on national antitrust laws that safeguard the competition within one country's borders. Some countries have explicit exemptions on these laws for export-related activities when this does not restrict competition on the domestic market. The best known is the Webb–Pomerance Act (1918) in the United States, which enables US firms to develop sufficient market power to countervail that of foreign firms on the world market. The situation changed when many firms became international companies [8]. Other countries have implicit exemptions where antitrust laws only reach to the domestic market. Taylor and Moss [37] and Jenny [8], among others, emphasize the need for an international elimination of the explicit exemption on competition law.

Little is known about the impact of the market power of the mineral fertilizer industry on the small but growing biobased fertilizer sector. One would expect producers of biobased fertilizer to profit from the current high prices, installed by some of the export cartels. But when more public resistance and more antitrust complaints are raised, more international sanctions and laws will be installed. This could induce the end of these cartels, leading to more competition worldwide and thus lower fertilizer prices. This would make it even harder for biobased fertilizer producers to compete with mineral fertilizers.

However, the fertilizer industry can also use its market power to apply a pricing strategy to stop new entrants. Temporarily lowering prices could diminish the profitability of new business models in the biobased fertilizer industry and hamper its development.

Our hypothesis is that a more likely development is the extension of the portfolio of the current fertilizer industries. Through the extension of the business toward biobased fertilizers, the fertilizer industry can guarantee the sourcing of nutrients, improve its sustainability image, and develop pricing strategies where the prices of biobased and other fertilizers are adjusted to maximize profitability.

Price is not the only factor influenced by the strong market power of the fertilizer industry. It is also a stakeholder in the discussions on standards and regulations around other fertilizers, such as the biobased ones. These have different characteristics, but they are recycled nutrient products with a clear ecological benefit compared to the energy- and resource-intensive production of mineral fertilizers. That is why Taylor and Moss [37] see more gain in the development of policies stimulating recycling and believe that this will indirectly affect cartel pricing.

6.1.4 Conclusion

Since the late 1950s, the use of mineral fertilizers has increased rapidly. Over the last several decades, the increase in mineral fertilizer use has boomed, especially in developing countries. Though the prices have been rising drastically since 2008, the trend continues, because of a growing world population and changing diet patterns. Several governments of developing countries have introduced supportive measures for mineral fertilizer use. In several areas with very intensive livestock farming, however, a nutrient excess is looming, leading to environmental problems and increasing the need for nutrient recovery.

In 2008, prices for chemical fertilizer spiked, and though this is linked to the energy crisis, several authors found some discrepancy between the spike prices of energy and those of fertilizers. In that regard, it became clear that there is a lot of consolidation in the mineral fertilizer industry, and indications exist of export cartel formation with tacit price-setting.

On the other hand, could this also create an opportunity for the development of the biobased fertilizer industry, which will be better able to compete since it currently has higher production costs compared with mineral fertilizers.

From an environmental point of view, in order to slow the depletion of mineral resources and the accumulation of nutrients in agricultural areas, governments should in parallel rethink the rules on export cartels and price fixing and impose measures to support nutrient recycling.

More detailed research is needed to assess the impact of the market power of the fertilizer industry on biobased fertilizer production. Simulation studies of future scenarios could give insight into the development paths of the biobased fertilizer industry. Scenarios with continued strong cartel formation versus weakening or decomposed cartels and policies advantaging nutrient recycling could be developed.

References

1. FAOSTAT (2016) FAOSTAT statistical database. Available from: http://www.fao.org/faostat/en/#home (accessed December 23, 2019).
2. FAOSTAT (2017) FAOSTAT statistical database. Available from: http://www.fao.org/faostat/en/#home (accessed December 23, 2019).
3. Bumb, B.L., Baanante, C.A. (1996). Role of fertilizers in food security and protecting the environment to 2020. IFPRI Discussion Paper.
4. Garrity, D.P., Akinnifesi, F.K., Ajayi, O.C. et al. (2010). Evergreen agriculture: a robust approach to sustainable food security in Africa. *Food Security* **2** (3): 197–214.
5. Heffer, P. (2013). Assessment of Fertiliser Use by Crop at the Global Level 2010–2010/11. International Fertiliser Industry Association (IFA).
6. Heffer, P. (2013). Global status and challenges of fertilizer use world fertilizer demand. Evolution of world fertilizer demand. IFA Global Soil Partnership Technical Workshop, pp. 1–14.
7. Jensen, L.S., Schjoerring, J.K., Van der Hock, K.W. et al. (2011). Benefits of nitrogen for food, fiber and industrial production. In: *The European Nitrogen Assessment* (eds. M.A. Sutton, C.M. Howard, J.W. Erisman, et al.), 32–61. Cambridge: Cambridge University Press.
8. Jenny, F. (2012). Export cartels in primary products: the potash case in perspective. In: *Trade, Competition and the Pricing of Commodities* (eds. S.J. Evenett and F. Jenny), 99–132. London: Centre for Economic Policy Research.
9. Food and Agriculture Organization (2009). *The State of Food and Agriculture 2009: Livestock in the Balance*. Rome: FAO.
10. Tilman, D., Cassman, K.G., Matson, P.A. et al. (2002). Agricultural sustainability and intensive production practices. *Nature* **418** (6898): 671–677.
11. King, F.H. (1911). *Farmers of Forty Centuries, or Permanent Agriculture in China, Korea and Japan*. Madison, WI: Mrs. F.H. King.
12. Erisman, J.W., Sutton, M.A., Galloway, J. et al. (2008). How a century of ammonia synthesis changed the world. *Nature Geoscience* **1**: 636–639.
13. Smil, V. (2001). *Feeding the World: A Challenge for the Twenty-First Century*. Cambridge, MA: MIT Press.
14. Wadman, W.P., Sluijsmans, C.M.J., and De La Lande Cremer, L.C.N. (1987). Value of animal manures: changes in perception. In: *Animal Manure on Grassland and Fodder Crops. Fertilizer or Waste?* SE-1, vol. **30** (eds. H.G. Van Der Meer, R.J. Unwin, T.A. Van Dijk and G.C. Ennik), 1–16. Amsterdam: Springer.

15. Schröder, J. (2005). Revisiting the agronomic benefits of manure: a correct assessment and exploitation of its fertilizer value spares the environment. *Bioresource Technology* **96** (2): 253–261.
16. Lory, J.A., Russelle, M.P., and Peterson, T.A. (1995). A comparison of two nitrogen credit methods: traditional vs. difference. *Agronomy Journal* **87** (4): 648–651.
17. Gessel, P.D., Hansen, N.C., Goyal, S.M. et al. (2004). Persistence of zoonotic pathogens in surface soil treated with different rates of liquid pig manure. *Applied Soil Ecology* **25** (3): 237–243.
18. Anderson, N., Strader, R., and Davidson, C. (2003). Airborne reduced nitrogen: ammonia emissions from agriculture and other sources. *Environment International* **29** (2–3): 277–286.
19. Copeland, C. (2010). Air quality issues and animal agriculture: a primer. In: *Animal Agriculture Research Progress* (eds. K.B. Tolenhoff, G.S. Becker, C. Copeland and C.H. Vincent). New York: Nova Science.
20. Hooda, P.S., Edwards, A.C., Anderson, H.A., and Miller, A. (2000). A review of water quality concerns in livestock farming areas. *Science of the Total Environment* **250** (1–3): 143–167.
21. Tur Cardona, J., Buysse, J. (2015). Farmers' reasons to accept bio-based fertilizers: a choice experiment in 8 different European countries. 13th International Conference on Renewable Resources and Biorefineries.
22. Gollany, H.T., Rickman, R.W., Liang, Y. et al. (2011). Predicting agricultural management influence on long-term soil organic carbon dynamics: implications for biofuel production. *Agronomy Journal* **103** (1): 234–246.
23. Stockmann, U., Adams, M.A., Crawford, J.W. et al. (2013). The knowns, known unknowns and unknowns of sequestration of soil organic carbon. *Agriculture, Ecosystems and Environment* **164**: 80–99.
24. Matson, P.A., Parton, W.J., Power, A.G., and Swift, M.J. (1997). Agricultural intensification and ecosystem properties. *Science* **277** (5325): 504–509.
25. Öborn, I., Andrist-Rangel, Y., Askegaard, M. et al. (2005). Critical aspects of potassium management in agricultural systems. *Soil Use and Management* **21** (1): 102–112.
26. Dawson, C.J. and Hilton, J. (2011). Fertiliser availability in a resource-limited world: production and recycling of nitrogen and phosphorus. *Food Policy* **36** (Suppl. 1): S14–S22.
27. Cooper, J., Lombardi, R., Boardman, D., and Carliell-Marquet, C. (2011). The future distribution and production of global phosphate rock reserves. *Resources, Conservation and Recycling* **57** (January): 78–86.
28. Martinez, J., Dabert, P., Barrington, S., and Burton, C. (2009). Livestock waste treatment systems for environmental quality, food safety, and sustainability. *Bioresource Technology* **100** (22): 5527–5536.
29. Hossain, M. and Singh, V.P. (2000). Fertilizer use in Asian agriculture: implications for sustaining food security and the environment. *Nutrient Cycling in Agroecosystems* **57** (2): 155–169.
30. USDA – Economic Research Service (2006). Use of manure as a crop fertilizer. Manure Use for Fertilizer and for Energy. Report to Congress by the United States Department for Agriculture. pp. 7–17.
31. Velthof, G.L., Oudendag, D., Witzke, H.P. et al. (2009). Integrated assessment of nitrogen losses from agriculture in EU-27 using MITERRA-EUROPE. *Journal of Environmental Quality* **38** (2): 402–417.
32. European Union. (2016). EU Cereal Farms. Report based on 2013 FADN data.
33. Nazlioglu, S. and Soytas, U. (2011). World oil prices and agricultural commodity prices: evidence from an emerging market. *Energy Economics* **33** (3): 488–496.
34. Zhang, Z., Lohr, L., Escalante, C., and Wetzstein, M. (2010). Food versus fuel: what do prices tell us? *Energy Policy* **38** (1): 445–451.
35. Gnutzmann, H. and Śpiewanowski, P. (2016). Fertilizer fuels food prices: identification through the oil-gas spread. *Social Science Research Network* https://doi.org/10.2139/ssrn.2808381.

36. Humber, J. (2014). A Time Series Approach to Retrospective Merger Analysis: Evidence in the US Nitrogen Fertilizer Industry Abstract. Available from: http://ageconsearch.umn.edu/bitstream/170667/2/AAEA%20Paper.pdf (accessed December 23, 2019).
37. Taylor, C.R., Moss, D.L. (2013). The Fertilizer Oligopoly: The Case for Global Antitrust Enforcement. Available from: https://www.antitrustinstitute.org/work-product/the-fertilizer-oligopoly-the-case-for-global-antitrust/ (accessed December 23, 2019).
38. Hernandez, M.A. and Torero, M. (2013). Market concentration and pricing behavior in the fertilizer industry: a global approach. *Agricultural Economics* **44**: 723–734.
39. Gnutzmann, H., Śpiewanowski, P. (2014). Did the Fertilizer Cartel Cause the Food Crisis? Available from: https://papers.ssrn.com/sol3/papers.cfm?abstract_id=2534753 (accessed December 23, 2019).

Section VII

Environmental Impact Assessment on the Production and Use of Biobased Fertilizers

7.1

Environmental Impact Assessment on the Production and Use of Biobased Fertilizers

Lars Stoumann Jensen, Myles Oelofse, Marieke ten Hoeve, and Sander Bruun

Department of Plant and Environmental Sciences, University of Copenhagen, Copenhagen, Denmark

7.1.1 Introduction

There are a wide range of environmental impacts associated with the production and use of synthetic (mineral) fertilizers and biobased fertilizers. Some of the major ones are consumption of valuable natural resources, eutrophication, acidification, and global warming. Use of biobased fertilizers holds the advantage that nutrients from waste products are recycled and their production is associated with lower-energy and non-renewable resource consumption compared to synthetic fertilizers. However, there are also some disadvantages. Gaseous emissions to the environment occur throughout the production and use chain of biobased fertilizers, and a variable fraction of the nitrogen (N) is usually in organic form, which may lead to a lower crop utilization efficiency and nitrate leaching when N mineralization occurs outside the growing season.

The environmental impacts of biobased fertilizers are of a different nature than those of mineral fertilizers, and they occur at different life-cycle stages. For this reason, it is pertinent to ask whether biobased fertilizers are better for the environment than their mineral counterparts based on synthetically fixed atmospheric N and mining of earth minerals

Biorefinery of Inorganics: Recovering Mineral Nutrients from Biomass and Organic Waste, First Edition.
Edited by Erik Meers, Gerard Velthof, Evi Michels and René Rietra.
© 2020 John Wiley & Sons Ltd. Published 2020 by John Wiley & Sons Ltd.

for phosphorous (P), potassium (K), and most micronutrients. It may also be relevant to compare different ways of producing biobased fertilizers in order to choose the most environmentally benign one.

Previous sections of this book have outlined the scope and potential of emerging technologies available for mineral recovery from different waste streams (Sections III and IV) and the quality and agricultural performance of the mineral products developed from wastes (Sections V and VI). The aim of this chapter is to elucidate the environmental effects, both negative and positive, that occur at the different stages of the biobased fertilizer production process, as well as following land application, when benchmarked against fossil-based synthetic fertilizers. Untreated wastes can generally be divided into two broad categories, namely animal manures and urban and industrial solid wastes or wastewaters, both of which can serve as a source of minerals and energy. The environmental impact of a biobased fertilizer will therefore depend on the type of waste in question, coupled with the treatment technology and resultant use of the outputs.

The acquisition of a full understanding of the environmental impacts of treatment technologies and of land application requires an approach that can capture the full system effects, focusing on a range of environmental indicators. A methodology that can answer this kind of environmental question, and which is increasingly being applied to do so, is life-cycle assessment (LCA). LCA has been used to analyze the impacts of processes producing biobased fertilizers, including anaerobic digestion (AD) [1, 2] and composting [3].

We therefore begin this chapter with a summary of the LCA methodology, presenting an introduction to the method that is specifically relevant for the environmental analysis of biobased fertilizer production and use. Following this, we further review a range of integrative environmental assessments and comparisons of biobased fertilizers with conventional synthetic fertilizers. In the following Chapter 7.2, we explore potential environmental impacts as well as potential benefits that occur during the treatment technology step and during and after field application. Finally, we provide a case study example of an LCA applied to a manure treatment. In addition, Chapter 7.3 gives a more extensive example of an LCA of selected processes of nutrient recovery from digestate.

7.1.2 Life Cycle Assessment of Biobased Fertilizer Production and Use

In this section, we present a general introduction to LCA and a simplified case study to clarify the methodology. The objective is to give readers the necessary understanding of the LCA methodology to understand what it can be used for, and how it can analyze systems for the production of biobased fertilizers. This will allow them to understand and evaluate LCA papers and reports on the subject.

7.1.2.1 Life Cycle Assessment

LCA is a methodology that can be used to compare the environmental impacts of products providing the same service. The LCA framework and methodology are standardized in ISO standards ISO 14040 [4] and ISO 14044 [5], which provide guidelines for how to conduct LCAs. In addition, there are numerous detailed guidelines, such as the ILCD handbook [6] and standard textbooks [7].

LCA has two main features that distinguish it from other types of environmental analysis. First, it investigates the complete life cycle of a product or service, as the name implies. This entails that resource use and emissions are analyzed from raw material extraction through manufacturing, transportation, and use to final disposal. Furthermore, secondary processes related to the life cycle of the product or service are also included. Examples are diesel production for transportation and electricity production for machinery. Considering the complete life cycle is necessary to avoid burden shifting from one stage of the life cycle to another. Second, LCA attempts to include all natural resources needed by and all emissions to the environment associated with a product or service. This is necessary to avoid reaching false conclusions. An example: Product A has a low impact on global warming and a very high impact on acidification and eutrophication, while the opposite is the case for product B. If only global warming (the carbon footprint) is included in the analysis, the false conclusion that A is better than B will be reached. Often, however, it is difficult to include all impacts, as it requires measurement of all relevant emissions and well-developed impact assessment methods. This can be difficult to obtain for some impact categories (e.g. in relation to biobased fertilizers, impacts like odor, noise, and soil quality).

7.1.2.2 The Four Phases of LCA

According to the ISO standards, there are four phases you have to go through when conducting an LCA: goal and scope, inventory analysis, impact assessment, and interpretation. In reality, conducting an LCA often includes iteration; for example, when the practitioner in an impact assessment finds out that a certain process has a major influence on the final results, he or she will go back to the inventory phase and collect more accurate data on that process.

7.1.2.2.1 Goal and Scope

In the goal and scope phase, the methodological framework of the study is described, boundaries are set, and the results are interpreted in relation to the goal and scope of the study. The goal of an LCA should be defined as clearly as possible, including the intended application and the context in which the study is performed. The intended application defines which products or services to compare. As an example, a study could compare different biobased fertilizers produced from the same waste material or different waste materials used to produce a biobased fertilizer. Stakeholders who might have an interest in the results could include farmers, producers, or politicians. This type of information is important in order to avoid misinterpretation of the results.

The scope of a study includes the definition of multiple aspects: the functional unit, the system boundaries, co-product allocation, and the temporal, geographical, and technological scope.

7.1.2.2.1.1 Functional Unit. The functional unit represents the function that the product or service provides. For biobased fertilizers, a functional unit could be "treatment of 1 ton of a certain kind of waste" or "production of 1 ton of biobased fertilizer." The functional unit is used to ensure that the services provided by the analyzed systems are identical.

7.1.2.2.1.2 System Boundaries. There are two kinds of system boundaries that need to be defined: boundaries between the technical system and the natural system and boundaries between processes that are included in and excluded from the assessment.

Environmental impacts occur only when substances cross the system boundary; that is, when a substance leaves the technical system and enters the natural system. An LCA is not concerned about what happens in the technical system, but as soon as substances enter the natural system, the environment might be affected. Similarly, when a substance leaves the natural system and enters the technical system, it constitutes a consumption of resources in the natural system. If the natural reserve of these resources is limited and non-renewable, use of the resources affects the environment. Obviously, it is important to define the system boundary precisely, otherwise it is impossible to quantify environmental emissions and non-renewable resource consumptions.

The second kind of boundary is the technical delimitation: which processes to include in and exclude from the assessment, a decision that must be taken by the LCA practitioner. The general rule is that all processes that make a significant contribution from an environmental perspective need to be included. It is possible to define some rigorous cut-off criteria, but in practice the LCA practitioner will just exclude processes that he or she deems unimportant. An LCA study needs to clearly state which processes are included in the product system model and which are left out. This is often done with a flowchart (see Section 7.1.5).

7.1.2.2.1.3 Allocation Problems. In the scope definition, decisions about how to handle co-products need to be made. Co-products result from processes with multiple outputs. Many systems designed to produce biobased fertilizers also have other by-products. An example could be AD, which in addition to the digestate also produces biogas and thus energy. If the digestate is to be compared with another biobased fertilizer, such as a compost, a functional unit could be “1 ton of N in biobased fertilizer”. The complication then arises that the two systems do not provide the same service, as the AD scenario produces biogas, while the other biobased fertilizer scenario does not. The way in which the LCA practitioner will deal with this extra product depends on the approach taken. One method is to allocate environmental impacts to the different products. This could be done based on the mass, economical value, or energy content of the different co-products, for example. However, the preferred way of solving this problem is to use system expansion. This is done by identifying which product the co-product would replace and then modeling the consequences of avoiding the production of the replaced product. In the AD example, this would mean that the same amount of electricity needs to be produced conventionally (e.g. by coal in a combined heat and power [CHP] plant) in the other scenario(s) as is produced when the biogas from the AD scenario is used for electricity production.

7.1.2.2.1.4 Consequential Versus Attributional Modeling. Two different approaches to LCA modeling are usually distinguished: the attributional approach should and the consequential approach. In the attributional approach, co-products are dealt with either by allocation or by system expansion. In the attributional approach, system expansion is done by using weighted averages of current sources. In the consequential approach, allocation is not used and system expansion always is. The important difference lies in the way that system expansion is carried out. In the consequential approach, the expansion is done by modeling the actual consequences of putting the by-product on the market. Somehow, the product that is not produced because of the introduction of this by-product has to be identified. Logic tells us that it will be the least competitive product that will be replaced. In our AD example, the biogas can be used to generate electricity, which is sold by the AD plant. In the attributional approach, it would be assumed that a mix of electricity would be saved, for example from coal-fired plants, nuclear power, natural gas, hydropower, and windmills, according to the current production in the region. In the consequential approach, on the other hand, the consequences of the introduction of the additional electricity are modeled. The electricity produced is not likely to replace that produced from wind because the production from windmills will not depend on the increased supply. It is more likely that it will replace electricity produced from coal or natural gas, as electricity production from CHP plants running on these fuels can be adjusted quickly. In the longer term, of course, an increased supply of electricity may affect the need and willingness to invest in windmills, so identification of the consequentially replaced product is not easy and depends on scale and time frame.

7.1.2.2.2 Inventory Analysis

In inventory analysis, all of the emissions to the environment and all of the non-renewable resource consumption associated with the production of one functional unit are calculated. This is done by collecting input (non-renewable resources and products from other processes) and output (emissions to the environment and products to other processes) data for all the processes in the system. This is illustrated for a single process in Figure 7.1.1. Here, the input of products from other processes could be the feedstock, which is fed to a composting facility each day, the output of products could be the compost, the emissions to air could be methane and other gases emitted, and emissions to soil could be nutrients leached from the compost heap.

For every process, relative amounts of input and output flows and all emissions need to be known. For example, they could all be given in units per year or per ton of product entering.

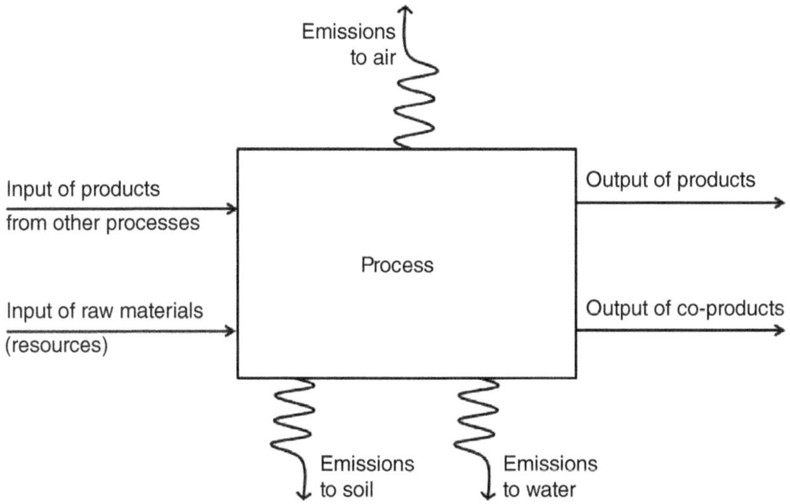

Figure 7.1.1 *General representation of a process receiving inputs of products and resources and delivering other products, resulting in emissions to air, soil, and water.*

Data for the processes have to be compiled from available sources such as measurements, scientific papers, reports, and databases. Some emissions may also be calculated based on the principles of conservation of energy, mass, and elements. Finally, models can be used to estimate emission factors. Examples are numerous in the literature [8, 9].

When these data are collected for all processes, they have to be connected to produce a flow model and related to the functional unit. This means that all processes have to be scaled so that the inputs, outputs, and emissions correspond to the functional unit. Subsequently, the model can be used to make a total inventory of environmental exchanges per functional unit.

The inventory analysis is usually carried out using LCA software, which makes the calculations easier and provides a database with inventory data for some of the most common processes. Among the most widely used LCA softwares are GaBi (www.gabi-software.com/LCA) and SimaPro (www.pre-sustainability.com), which are commercially available, and openLCA (www.openlca.org), which is open-access.

7.1.2.2.3 Impact Assessment

The third phase of LCA is the impact assessment, in which the potential impacts of the emissions and resource consumptions quantified in the inventory analysis are assessed.

7.1.2.2.3.1 The Nature of Environmental Impact Potentials. Before describing how to conduct or understand the impact assessment, we will give an explanation of what environmental impact potentials are. The reason that LCA works with impact *potentials* instead of *real* impacts is that it is uncertain what the real environmental effect of a specific emission is, as this depends on different factors. The first factor is location. If a substance is emitted close to vulnerable ecosystems or in highly populated areas, its impact will probably be higher than if it is emitted far from vulnerable ecosystems or in deserted areas. The second factor is the time when the emission is taking place. All substances have a specific half-life, and when the amount of a substance in an environmental compartment (soil, water, or air) gets too high, a tipping point may be reached, with a detrimental impact. The third factor is related to the second: it is the total mass of the substance emitted at once. Again, for this factor a tipping point can be reached. Finally, emissions can show synergies with other substances (emitted at the same time or already present in the environmental compartment), enhancing the impact of the emissions.

An impact happens whenever there is something we treasure that is somehow threatened or destroyed. There are three general areas of protection: human health, natural environment, and natural resources. There is general agreement that these areas are worth protecting and are threatened by human actions through various environmental mechanisms. An example of how human actions can threaten areas of protection is leaching of mercury to the aquatic environment. The mercury accumulates in the upper levels of the aquatic food chain and may change the ecosystem, affecting the area of natural environment. If fish from this ecosystem are caught for human consumption, it may also have an impact on human health. Often, the cause–effect chains leading to impacts in the areas of protection are very complicated. Consider, for example, an emission of carbon dioxide (CO_2). When CO_2 is emitted to the atmosphere, the radiative forcing is increased slightly, making it more difficult for long-wave radiation to escape from the earth. We can determine fairly well the degree to which this occurs via experiments coupled with knowledge about the ability of the CO_2 molecule to absorb radiation of different wavelengths and its residence time in the atmosphere. However, this knowledge does not seem to be directly relevant for the impacts that occur on the three areas of protection. We know that increased radiative forcing is causing the temperature of the earth to increase, but this is still not relevant for the quantification of the impacts on the areas of protection. Increases in global temperature lead to rising sea levels and changed weather patterns, and subsequently to destroyed houses, failed crop yields, starvation, wars, and so on. It is extremely difficult to predict these things, though they are beginning to be very relevant for the areas of protection.

Environmental impacts can be assessed at different points along the cause–effect chain. In the example with an emission of CO_2, we can assess how much radiative forcing will rise as a consequence of this emission. This will be considered a mid-point impact because it is assessed in the middle of the cause–effect chain. A common way of doing this is to use CO_2 equivalents, so that the effect of emissions of other substances also contributing to global warming is benchmarked against CO_2. In contrast, an end-point method would assess the impact on human health, ecosystem diversity, and resource availability directly. This could be measured in disability-adjusted loss of life years, loss of species during a year, and increased cost of resource extraction. Obviously, it is much easier to assess impacts at the mid-point, so estimates are more precise, but it is typically also less relevant for the area we are interested in protecting.

7.1.2.2.3.2 Calculation of Impacts. There are a range of different impact assessment methods that define impact categories such as global warming, eutrophication, acidification, and human toxicity potential and resource consumption. Some methods are based on calculating impacts at the mid-point, some at the end-point, and some use a combination of both. The most frequently applied impact assessment methods include EDIP 2003 [10], IMPACT 2002+ [11], Stepwise [12] ReCiPe [13], and the ILCD method [6]. These are often included in LCA software or can be imported into it so that the calculations are performed automatically.

After choosing one or more impact assessment methods, the next step is characterization. This means that all emissions contributing to the same impact category are added together. As different substances will contribute with ranging severity to an impact category, we need to use characterization factors that represent this. Impact assessment methods provide such characterization (or equivalency) factors, which are a measure of how much an emission of a given substance contributes to the impact potential in each impact category:

$$IP(j) = \sum_{i=1}^{n} Q_i CF(j)_i \qquad (7.1.1)$$

where $IP(j)$ is the impact potential for impact category j, Q_i is the emission of substance i to a certain compartment (e.g. air, soil, water), and $CF(j)_i$ is the characterization factor for emission of substance i to category j.

A well-known example of a characterization factor is the global warming factors defined by the Intergovernmental Panel on Climate Change (IPCC) [14]. In a 100-year perspective, emissions of 1 kg methane will cause as much global warming as 25 kg CO_2, while 1 kg nitrous oxide will cause as much as 298 kg CO_2. Thus, if the production of a biobased fertilizer corresponding to one functional unit results in an emission of 32 kg CO_2,

0.26 kg N_2O, and 12 kg CH_4 then the total global warming impact potential is: *IP*(global warming) = 32 kg × 1 kg CO_2 eq. kg^{-1} + 0.26 kg × 298 kg CO_2 eq. kg^{-1} + 12 kg × 25 kg CO_2 eq. kg^{-1} = 409 kg CO_2 eq.

7.1.2.2.3.3 Normalization and Weighting. To make the results easier to interpret, there are some optional steps that can be taken following characterization. These are normalization and weighting. In normalization, the impacts are normalized with respect to a reference point. This could be the impact of an average person in a chosen target group or area, for example. By using normalization, we can evaluate the magnitude of the impact potentials related to one functional unit compared with other contributions to the impact category. The normalization procedure can be described by the equation:

$$NIP(j) = \frac{IP(j)}{NR(j)} \tag{7.1.2}$$

where *NIP(j)* is the normalized impact potential in impact category *j* and *NR(j)* is the normalization reference.

The normalized results do not say anything about the importance of an impact contribution in terms of the seriousness of the potential environmental impacts. This means that it can still be difficult to compare two products, because one may have high impacts in one category and the other in another. To solve this problem, weighting can be used. If *W(j)* is the weighting factor for category *j*, then the weighted impact potential is calculated as:

$$WIP(j) = NIP(j)W(j) \tag{7.1.3}$$

and then the weighted impact potentials in the different impact categories can be summed to give a single value for the product, which can be compared with an alternative:

$$WIP = \sum_{j=1}^{m} WIP(j) \tag{7.1.4}$$

The use of weighting is often criticized because determination of weighting factors is associated with much uncertainty and because their estimation often seems very subjective. Some of the problems can be avoided by using end-point methods, which calculate the impacts related to the three areas of protection. However, it should be noted that this is also associated with uncertainty and subjectivity. In addition, one may still end up in a situation where one has to choose between a product that has a high impact in one category and one that has a high impact in another. Then, one must still rely on weighting or some other subjective approach. By using a specific weighting method, an arbitrary but mutually agreed approach is achieved.

Most studies are published without standardized weighting but with an individual interpretation, which enables the results to be more context-specific. To give an example, imagine two products. Product A has a high global warming potential (GWP) and low acidification potential, while product B has the opposite. Assume that acidification is not seen as a major environmental threat worldwide, but it is in the region on which this LCA is focused. Product A might have the highest weighted impact potential when using weighting factors from an impact assessment method, but potential damage in the region of focus will probably be higher when product B is produced. Whether to choose one's own weighting method or one that is a part of an impact assessment method depends on personal preferences. In the LCA community these days, there is a general trend toward own interpretation.

7.1.2.2.4 Interpretation

In the final phase of the LCA, the results are interpreted. Here, the findings of the inventory analysis and impact assessment are combined in order to draw conclusions and recommendations in accordance with the goal and scope. The results can be interpreted at very different levels, all the way from a detailed investigation of where in the life cycle emissions of a particular compound are emitted to a comparison of a single weighted impact potential for two products.

An important part of the interpretation is to test the robustness of the results and conclusion for as many parameters as possible through sensitivity analysis. For example, if the emission factor for nitrate after application of a biobased fertilizer is estimated to be 20% of N reaching the field, but the value might be 15 or 30%, the effects of decreasing the value to 15% and increasing it to 30% can be tested in a sensitivity analysis. If the lower and higher estimates do not change the ranking of the biobased fertilizers, our conclusions can be considered robust.

7.1.3 Environmental Impacts from the Production and Use of Biobased Fertilizers

In this section, the various direct and indirect environmental impacts of fertilizer production, distribution, and use are described. Though the focus is on biobased fertilizers, we use synthetic fertilizers as a benchmark to evaluate their environmental impacts, and therefore the differences and similarities in impacts between the two are described.

7.1.3.1 Climate Change and Global Warming Potential

Global warming or climate change covers the climatic changes following from an accumulation of greenhouse gases (long-lived gases that absorb infrared radiation from the earth) in the atmosphere. The resulting heating of the atmosphere propagates to the continents and oceans, leading to a warming of the earth, to rises in sea level, and to changes in the regional and global climate. The main greenhouse gas is CO_2, when it is formed by combustion of fossil fuels (e.g. for transport, electricity production, or ammonia synthesis for use in mineral fertilizer); CO_2 produced from the biological decomposition of organic waste under aerobic (e.g. composting) and anaerobic (e.g. AD) treatments does not provide net contributions to the atmospheric content of CO_2, since such waste degrades naturally and releases its carbon content as CO_2 anyway. However, methane (CH_4) emitted during the production or use of biobased fertilizers (e.g. through leakage from AD) is a much stronger greenhouse gas than CO_2 (25 times on a weight basis according to the IPCC [14]). Nitrous oxide (N_2O), the strongest among the naturally occurring greenhouse gases (about 298 times stronger than CO_2) may be released during the production, and especially after the field application, of synthetic as well as biobased fertilizers.

In general, the GWP from synthetic fertilizer production derives from increased greenhouse gas emissions occurring mainly from: (i) the fossil fuel use in (a) ammonia production by industrial N_2 fixation (the Haber–Bosch process), (b) fossil energy use for the extraction of P and K from minerals, and (c) transportation of raw materials (e.g. P and K minerals) and fertilizer products; and (ii) the nitrous oxide emissions in nitric acid production [15].

For production of biobased fertilizers, GWP stems mainly from the energy used to drive the nutrient recovery processes (separation, filtration, drying, etc.), but also from direct emissions of greenhouse gases from the treatment processes (e.g. leaks of CH_4 from biogas reactors or digestate storage, CH_4 and N_2O emissions from composting, etc.). The primary impact of synthetic as well as biobased fertilizers on GWP during land application is N_2O production from soil. Nkoa [16] reviewed studies of N_2O from anaerobic digestates versus raw manures, and found significant reductions: 17–71% lower N_2O emissions from digested versus raw manures applied to sandy and loamy soils. However, fertilizer use can also increase carbon storage in the soil due to increased biomass production [17] and greater return of crop residues such as roots, stubble, and straw, in principle lowering atmospheric CO_2 concentrations. This may also lead to a substantial net saving of C if the increase in straw or other crop residues biomass is used for biofuel, which can lead to an overall substitution of fossil fuel consumption.

7.1.3.2 Eutrophication

N and P are macronutrients for higher plants and algae, and the release of compounds of these two elements therefore has the ability to fertilize and perturb the ecological balance of natural aquatic and terrestrial ecosystems, increasing their primary production, but also changing their species composition and diversity. In lakes and coastal waters, this causes algal blooms followed by oxygen depletion in the bottom-near strata.

For synthetic fertilizer production, the main impact on eutrophication is caused by dispersion during phosphate production and accumulation of phosphogypsum [15]. For biobased fertilizer production, direct impacts will be mainly on terrestrial eutrophication through gaseous emissions of NH_3 from the treatment or nutrient recovery plant, which can be deposited on nearby land and vegetation, and subsequently may also contribute to acidification (see later) as well as indirectly to aquatic eutrophication, if the deposited N is leached.

For both synthetic and biobased fertilizer use, an increasing fertilizer application rate has an impact on terrestrial eutrophication through volatilization of NH_3, while aquatic eutrophication derived from nitrate leaching mainly occurs at N fertilizer rates above the economic optimum for the specific crop (see Section 7.1.5). However, for the biobased fertilizers with a high proportion of organically bound N, the relative contribution to aquatic eutrophication will typically be higher, due to mineralization of organically bound N over longer time spans and during periods when there is no crop with a sufficient N uptake to prevent leaching [18]. Möller [19] reviewed studies of the effects of AD on soil nitrate leaching and concluded that AD in itself does not appear to have a significant direct influence on leaching, but as on-site biogas plants tend to influence cropping system compositions and changes in the import of dedicated energy crops, municipal biowastes, biowastes from the food industry, and so on for the biogas plant, this has a stronger effect on leaching risk compared to the chemical changes induced in the biowaste during AD; that is, the systems effects are much stronger.

7.1.3.3 Acidification

Emissions of nitrogen oxides (NO_x, the sum of NO, nitric oxide, and nitrogen dioxide [NO_2]), ammonia (NH_3), and sulfur oxides (SO_x) lead to the release of hydrogen ions when the gases are deposited in terrestrial or aquatic environments. The protons contribute to acidification of soils and lakes when they are released in areas where the buffering capacity is low, resulting in forest decline and dead lakes.

The main source of SO_x and NO_x is combustion of fossil fuels during fertilizer production and transport processes; emissions will therefore typically be higher for synthetic than for biobased fertilizers. In contrast, NH_3 emissions occur mainly after field application, and synthetic fertilizers typically show low losses at such times, apart from of urea, where NH_3 volatilization of up to 30% of applied N may occur under adverse conditions [20]. However, during production of biobased fertilizers, potentially large NH_3 emissions may occur during production or storage, especially when feedstocks are high in ammonium or the treatment process produces fractions rich in ammonium (e.g. mineral concentrates produced by ultra-filtration [UF] or reverse osmosis [RO] separation of animal manures) and appropriate measures to reduce NH_3 losses are not taken. Composting and, especially, drying of manure and biowaste solids are particularly prone to NH_3 volatilization, as the combination of elevated ventilation and high temperature promotes NH_3 formation and transfer to the gaseous phase. When manures and biowaste are anaerobically digested for biogas, NH_3 loss from stored digestate slurry can be high – often higher than that of undigested manure, due to an increased pH, though this depends on temperature and storage conditions. When applied to fields, the combined effect of increased pH and increased ammonium concentration in digestates potentially increases NH_3 volatilization risk [16]. However, the literature reports contradictory results, some papers showing reduced NH_3 loss (perhaps due to the generally lower viscosity of digestates, promoting rapid soil infiltration), others no effect, and some increases [19].

7.1.3.4 Eco- and Human Toxicity

Ecotoxicity covers impacts on an ecosystem, damaging individual species and changing the structure or function of the ecosystem range from death over reproductive damages to behavioral changes. Ecotoxicity is a composite category that includes all substances with a direct effect on the health of the ecosystem, and involves many different mechanisms of toxicity; in that sense, it in principle partly overlaps with eutrophication impacts, but nutrient and organic matter (OM) enrichment and consequent perturbation of the ecological balance of the natural ecosystem is normally considered under eutrophication and not part of the ecotoxicity impact. The most important contributions to ecotoxicity from fertilization activities therefore come from toxic metals and persistent organic pollutants.

Toxic impacts on humans occur through inhalation with air, ingestion with food and water, and penetration of the skin after contact with polluted surfaces. There are thousands of substances that have the potential to exert human toxicity, but the most important exposure from fertilization activities comes from (i) exposure to particles emitted from transportation (fossil energy combustion) and ammonia emissions and (ii) exposure to toxic metals, persistent organic pollutants, and pathogenic organisms.

For LCA of eco- and human toxicity, there is a general scientific consensus that the USEtox model endorsed by the UNEP/SETAC Life Cycle Initiative for the characterization of the human and ecotoxicological impacts of chemicals is currently the best approach [21]. USEtox includes a database of characterization factors, comprising the fate, exposure, and effect parameters for a very broad range (over 3000) of potentially toxic compounds. However, there is still controversy over the fate modeling – and thus estimated characterization factors – for heavy metals in especially the soil compartment, as well as a need for improved spatial differentiation (e.g. local soil characteristics and environmental parameters) [22]. Smolders et al. [23] showed the toxicity of a range of heavy metals (incl. Cd, Cu, Co, Ni, Pb, and Zn) decreased up to 100-fold with aging of field-contaminated soils, as compared to freshly amended soils. Furthermore, it has been shown that metals in soils amended with sewage sludge are typically less available compared with those in soils spiked with soluble metal salts, which has been attributed to both lower availability in the original sludge matrix (complexation) and aging reactions in the soil [24].

For synthetic fertilizers, the main concern is over the contribution to heavy metal accumulation, primarily for ecotoxicity, but potentially also for human toxicity through food-chain transfers. In particular, P-containing fertilizers may be prone to cadmium (Cd) contamination due to the inherent presence of Cd, particularly in secondary rock phosphates. However, fertilizer regulations with low thresholds for trace metal content at national and transnational (e.g. European Union) levels are supposed to effectively prevent this; recent estimates of the average EU Cd mass balance show this to have become negative, as compared with earlier positive balances estimated in 2002 [25].

For biobased fertilizers, similar concerns over heavy metals concentrate on Cd, Zn, Cu, Ni, Pb, Cr, and Hg. Smith [26] reviewed international data on the total concentrations of heavy metals in municipal solid waste (MSW) and green-waste composts and found that all types of compost contain larger concentrations of heavy metals than the background values present in the soil, irrespective of the source. Waste compost and sewage sludge additions to agricultural and other soils will therefore raise the soil content and the availability of heavy metals for transfer into crop plants. Source segregation and processing will lower this to a minimum, and concentrations have been declining in most EU sewage sludge samples in the past decades. Soil availability depends on the soil concentration, the soil pH, the nature of the chemical association between the metal, the organic residual, and the soil matrix, and the plant's ability to regulate uptake of the particular element. Aerobic composting processes increase the complexation of heavy metals, and metals are strongly bound to the compost OM matrix (Pb most strongly, Ni least so, with Zn, Cu, and Cd showing intermediate sorption characteristics), limiting their solubility and potential bioavailability in soil. There is generally no tangible evidence of negative impacts of heavy metals applied to soil in compost on soil microbial processes or biomass, and experimental evidence has demonstrated reduced bioavailability and crop uptake of metals from composted biosolids compared to other types of unprocessed sewage sludge [26]. Composting is therefore overall likely to contribute to lowering the availability of metals in amended soil compared to other waste biostabilization techniques. However, risks to the environment, human health, and soil fertility from heavy metals in source-segregated MSW, green-waste, or sewage sludge composts need to be monitored closely with strict quality criteria and best management practices in order to avoid metal content becoming a barrier to the end-use of compost products from these sources. For manure-based fertilizers, the situation is different, as these are often high in Cu and Zn, used as animal feed additives. Long-term accumulation of Zn and Cu from manure-based digestates derived from biogas plants therefore tends to be greater than that from the use of mineral fertilizers [16]; here again, quality criteria and best management practices need to be set.

Organic chemicals discharged in urban wastewater and solid waste from industrial and domestic sources are predominantly lipophilic in nature and therefore become concentrated in sewage and processing sludge, with potential implications for the agricultural use of sludge as a soil improver. Biodegradation occurs to varying degrees during treatment processes, and is generally higher for aerobic treatments like composting than for anaerobic ones like those found in a biogas plant. However, residues will usually still be present to some extent, or metabolites with similar or even higher toxicity may be formed. Nevertheless, a review by Smith [27] indicates that the potential environmental and health impacts of organic contaminants in sludge are generally negligible or very small when the material is recycled to farmland, though certain groups of compounds or organisms still require further investigation (i.e. chlorinated parafins, phthalates, triclosan, certain pharmaceuticals and endocrine-disrupting compounds, and antibiotic-resistant microorganisms).

Attempts have been made to include the pathogen risk to human health in LCA for wastewater and sludge management systems, though this is commonly omitted due to methodological limitations. Heimersson et al. [28] found that for such waste management systems, pathogen risk could constitute an important part (up to 20%) of the total life-cycle impact on human health. The risk increases when animal manures are included as feedstock, since several outbreaks of gastroenteritis (Salmonella, Yersinia, Campylobacter, and the protozoa Giardia and Cryptosporidium) have been linked to livestock operations [16]. Well-managed composting is effective in controlling most of these, while AD is often reported to be less effective, but pathogen risk depends on feedstocks, the configuration features of the digester (i.e. pre- and post-treatments such as pasteurization), digestion temperature (mesophilic or thermophilic), pH, ammonia concentration, hydraulic retention time, and storage conditions, among other factors [16].

7.1.3.5 Resource Use

Resources can be evaluated in relation to their depletion (consumption related to geological/natural reserve), scarcity (economic availability), and criticality (where a resource is scarce and also crucial for society). A natural resource is either extracted permanently from the natural environment (in the case of abiotic resources like minerals and fossil fuels, regeneration of which is negligible on a human time scale) or subject to varying but limited regeneration rates (in the case of renewable, biotic resources) [29]. A considerable range of methodologies for assessing resource depletion in LCA have been proposed, but there is not yet a consensus, though most are based on the notion that extraction of a resource from the natural environment leads to a decrease in its future availability for human use [29].

In relation to fertilizers, the major issues relate to the use of fossil energy and rock minerals, notably P, which is an essential and non-substitutable nutrient for plants and animals. Rock phosphate is the main material used for the production of P fertilizers. However, its ore reserves are finite and its availability is estimated to decline in the next 100–400 years [30, 31], so increased efficiency of use and enhanced recycling of P-containing waste streams are increasingly encouraged [32]. When comparing biobased and synthetic fertilizers, the replacement of the abiotic resource rock P with recycling of organic fertilizer P (e.g. digestate from a biogas plant as a fertilizer) can reduce resource consumption, but it may also have other impacts than rock P extraction, for example in terms of eutrophication and acidification.

7.1.3.6 Land Use: Direct and Indirect Land Use Change

The physical impacts imposed on land often lead to more severe ecosystem damage than the emission-related impacts just described. In LCA, physical disruption of habitats is typically dealt with under the impact category "land use," which is quantified as the product of the size of the affected area, the duration of the impact, and the quality change caused by the impact.

In the context of fertilizer use, the main effect on land use change (LUC) is directly in terms of intensification (higher yield per area) with increasing fertilizer inputs up to the point of no further marginal increase, and hence a lesser need for land area for a certain production of crop yield (see Section 7.1.5). However, indirect land use change (iLUC) may also occur in connection with biobased fertilizer production, especially for those involving plant-based biomass (e.g. combined biofuel production). The production of plant biomass for this purpose may cause increased land use somewhere, for the production of replacement crop products; this is termed iLUC. However, iLUC can be very difficult to quantify, as one has to decide where in the world the displaced crop will be produced (this will affect marginal biome impacts), and furthermore the supply–demand response on the world crop market will affect and be affected by crop prices [33]. Agricultural land expansion for crop cultivation has been shown to typically affect forest biomes and potential grassland/steppe [34], and hence has significant impacts in terms of soil C loss from the transformation of these C-rich ecosystems.

Tonini et al. [35] conducted an LCA of four biomass-to-energy conversion technologies (anaerobic co-digestion of manure and plant biomass, thermal gasification, combustion in small-scale CHP plant and co-firing in large-scale coal-fired CHP plants) using different perennial crops and including iLUC and soil C changes in the impact assessment. They showed that global warming was the determinant impact, with only co-firing in a large-scale CHP yielding improvement over the fossil fuel reference, and iLUC impact represented as much as 41% of the induced greenhouse gas emissions from soil C loss.

7.1.3.7 Other Impacts, Including Odor

Some of the classical LCA impact categories like stratospheric ozone depletion and photochemical oxidant formation (smog) are typically not of any major importance in LCAs of fertilizer products, whether synthetic or biobased, and therefore are often omitted. However, other impacts of great perceived importance in relation to waste management, like work environment (safety, noise, etc.) and odor, are also not commonly included in LCAs. Emissions of certain gases from manure represent a special impact, in terms of both odor nuisance and work environment. Both ammonia (NH_3) and hydrogen sulfide (H_2S) are abundant gases in livestock environments and can be a health risk if inhaled. Both gases have a low odor detection limit but can paralyze the olfactory gland and represent a substantial health risk at higher concentrations.

However, odor is a highly subjective sensory perception and its quantification is by no means a straightforward task, because the sense of smell differs between people and because many components contribute to odor – for manures, more than 100 different odorous compounds have been identified. Odor is mainly considered to be a nuisance rather than a health hazard, since epidemiological evidence of adverse health effects of odor per se is relatively weak [36]. Still, odor emissions from industry, livestock enterprises, and waste treatment systems are a common concern for citizens residing nearby, and incidences of odor annoyance are increasing. With production of biobased fertilizers deriving from organic waste streams like animal manures, organic household waste, sewage, and industrial sludges, gaseous odor emissions from their collection, processing, transport, and field application are more or less inevitable, though appropriate measures can be taken to minimize them and their impacts on human nuisance. On the other hand, these waste streams are not "produced" as a raw material for biobased fertilizer production, but would need to be disposed in any case (or fed into another valorization chain), and thus odor emissions must be compared between business-as-usual and the production of biobased fertilizers.

Therefore, it is increasingly important to be able to compare technologies in terms of odorous emissions and impacts on humans. However, odor is not yet a standardized impact category in LCA. Peters et al. [37] found that though many authors have made use of odor threshold concepts in odor management studies, very few have tried to implement them in an LCA approach. Their literature review identified only a few case studies in which odor from organic waste handling or processing had actually been assessed in an LCA framework. They therefore proposed an overall framework for incorporating odor into LCA, and developed a mid-point indicator that goes beyond current approaches, including new characterization factors to operationalize the method [37]; however, studies for manure or waste-based processes are yet to be published applying this approach.

7.1.4 Benefits and Value of Biobased Fertilizers in Agricultural and Non-Agricultural Sectors

In contrast to the preceding section on environmental impacts, we here describe the benefits and potential positive effects of biobased fertilizer production and use. Again, synthetic fertilizers are commonly used as a reference or benchmark.

7.1.4.1 Crop Yield, Nutrient Use Efficiency, and Substitution of Mineral Fertilizers

Organic waste-based fertilizers typically contain several plant macronutrients, primarily N and P, some K, and varying amounts of micronutrients such as B, Cu, Fe, Mn, Mo, and Zn. The amounts and ratios of nutrients in the original waste types often do not match plant demands in the same way as synthetic fertilizers are designed to do, but this can to some extent be adjusted during biobased fertilizer production, either by combining different waste types and process derivatives or by supplementing missing nutrients in pure chemical form to provide the proper ratios needed for crop production [38].

The fertilizer value, or mineral fertilizer equivalent (MFE, defined as yield or nutrient uptake per unit of nutrient applied, relative between the biobased and typically a commercial mineral fertilizer reference) value, of biobased fertilizers ranges from negligible or even negative to 100%, particularly if the residual release of nutrients in subsequent years is included [39]. The potential MFE value for N in most animal manures is closely related to the proportion of ammonium-N in total N; for that reason, composts should be considered slow-release fertilizers and commonly have rather low potential MFE values in the year of application, but as the organically bound nutrients are mineralized, the accumulated MFE value over more than 10 years may reach 40–70% [40]. While most manure-based fertilizers have a relatively high MFE value for P, fertilizers based on Fe-precipitated sludges or incineration ash, where the P is very strongly bound, will be much less available. Digestates from anaerobic co-digestion of animal manures and agro-industrial residues generally have a high potential MFE value in the range of 60–80%, mainly associated with their high content of plant-available ammonium-N, and the AD process typically enhances the MFE value by 10–15% compared to the feedstocks. However, recycling of digestates as crop fertilizers may be restricted by their Cu and Zn contents, high salinity, low biological stability, or phytotoxicity and hygiene characteristics, which must be addressed to make them suitable biobased fertilizers [41]. Similarly, mineral concentrates produced by UF, evaporation, or RO from liquid manures or waste streams typically have a potential MFE value around 100% within the year of application, but may also have a high loss potential (e.g. due to a high pH, inducing high ammonia loss risk) [40]. Even though such treated and concentrated products may cause fewer environmental emissions and are cheaper and easier to transport, facilitating the natural nutrient cycle by substituting mineral fertilizers, low demand (and consequently low market prices compared to their real MFE value) and regulatory barriers often prohibit the realization of their true fertilizing and humus value [42].

7.1.4.2 Substitution of Peat-Based Products

Organic waste-based fertilizers may substitute not only nutrients in synthetic fertilizers, but with certain applications also other resources, such as topsoil when used for landscaping [43] and peat-derived growth media when used for potted plants, greenhouse production, or even private gardening [44, 45]. Use of compost for landscaping has several benefits, apart from replacing the often expensive, energy-intensive, and environmentally detrimental procurement of suitable topsoils, including greater resilience and water holding capacity, improved biological activity, and reduced labor requirements during construction and maintenance [43]. Compost replacement of peat in the production of growth media avoids the emissions that occur during peat extraction and subsequent C mineralization under aerobic conditions, releasing peat CO_2, which would otherwise have remained sequestered in the peat bog from where it was excavated [44]. In an LCA of compost substitution of peat-based growth media, Boldrin et al. [45] showed that the savings in global warming impact from compost substitution of peat were much greater than the savings caused by C sequestration through application to soil only and in the same order of magnitude as the savings in eutrophication potential from the substitution of mineral fertilizer by compost nutrients in the growth media. However, when growth media and composts are used in private gardens, the substitution rate becomes important. Andersen et al. [46] showed through a survey study of private garden owners that 1 ton of compost substituted only 0.19–0.22 tons of growth media.

7.1.4.3 Soil Quality Enhancement

Biobased fertilizers with a content of stable organic C will contribute to soil humus build-up and C sequestration. A study by Petersen et al. [47] showed that inclusion of soil C changes in an LCA can constitute a major contribution to the total greenhouse gas emissions per crop unit for plant products, and the results will greatly depend on the time perspective chosen. However, such biobased fertilizers will also enhance the quality, fertility, and crop productivity of soils [38, 43]. A review by Martinez-Blanco et al. [48] identified eight significant and quantifiable environmental benefits apart from nutrient supply and fertilizer substitution, namely carbon sequestration, pest and disease suppression, soil workability, biodiversity, crop nutritional quality, crop yield potential, soil erosion prevention, and soil moisture retention. So far, however, LCA impact assessment methods are available only for fertilizer nutrient substitution and soil C sequestration (reducing global warming impact), and these are increasingly included in LCA studies. When further impact assessment methods that include, for example, improved soil workability, crop yield potential improvements, or soil erosion prevention are developed, these could provide improved assessment of the multiple benefits of biobased fertilizers.

7.1.5 Integrative Comparisons of Synthetic and Biobased Fertilizers

This section reviews a number of recent LCA studies of various biobased fertilizer production technologies and use scenarios. Astrup et al. [49] recently reviewed a wide range of LCAs of thermal waste-to-energy technologies (250 studies in total) and concluded that very few of the published studies provided full and transparent descriptions of goal and scope definitions, technology parameters, or modeling principles. In many cases, this prevented an evaluation of the validity of the results, limiting the applicability of the data and results in other contexts, as well as prohibiting comparisons across studies. A similar conclusion can be reached for the so far more limited number of LCA studies of biobased fertilizer production and use, or what we could term waste-to-fertilizer technologies and applications. In the following section, we will therefore not attempt to make cross-studies comparisons, but shall only review and discuss study findings individually. Astrup et al. [49] provide a detailed overview of assumptions and modeling choices in the existing literature and recommendations for best practices in state-of-the-art LCA; hopefully, future LCA studies of biobased fertilizers will gradually adhere more appropriately to such best practices.

7.1.5.1 Synthetic Fertilizers

Since most studies use synthetic fertilizers as a benchmark to evaluate the environmental impacts of the biobased fertilizers in focus, it is important to first understand the general environmental effects relating to mineral fertilizer production and use. The production of fertilizers causes high values in the impact categories climate change, resource depletion, and acidification, whereas resource depletion is dominant for transportation and delivery.

For the application of fertilizer, eutrophication is the most important impact category [50]. Table 7.1.1 provides an overview of the main environmental impacts from synthetic fertilizer production and delivery. The largest differences between synthetic fertilizer types are clearly with respect to climate change impacts, where GWP is highest for N fertilizers due to the high energy consumption for industrial N fixation. However, recent improvements in best available technologies in the industry can cut this impact to half the value. The traditional LCA assessment methodology for GWP has recently been challenged by Laratte et al. [53], who indicate a much higher overall GWP from fertilizer production.

Production is the main contributor to climate change impact, with less contribution after application to the crop [50]. However, nitrous oxide emissions are often related exponentially to N fertilizer application rate; once the crop N needs have been met, the N_2O emission rates appear to increase non-linearly [54]. Nevertheless, IPCC methodologies [14] assume a linear response, typically amounting to 1.0–1.2% of the N applied, but real loss rates are highly variable and can vary by at least an order of magnitude. For eutrophication impacts, losses from field application contribute much more to the life cycle impacts than production and transport. As N application rates strongly affect the eutrophication potential, and to some extent the GWP and consequently LCA results, it is essential that the right amounts of N are used and that for N fertilizer production the best available technique is installed. Crop production with insufficient N input can, however, also lead to increases in overall life cycle impacts due to increased land use for crop yield loss compensation [51]. A good environmental performance in cereal production can be achieved by (i) maintaining optimum yields, in order to use land most efficiently, (ii) applying N according to crop demand in order to minimize nitrate-leaching, (iii) using N fertilizers with low ammonia volatilization rates in order to keep acidification and terrestrial eutrophication potentials low, and (iv) reducing N_2O emissions during nitrate fertilizer production (scrubbing techniques) in order to reduce the GWP [55]. Therefore, mineral fertilizer application rates that are either too high or too low increase the environmental impacts of crop production, mainly due to eutrophication in the former case and to inefficient land use in the latter, as illustrated in Figure 7.1.2. The lowest aggregated impact occurs at N fertilization levels around, or somewhat below, the economically optimal N rate.

Ahlgren et al. [56] investigated the land use, environmental impact, and fossil energy use when biogas (produced by AD of ley grass and maize) was used as an alternative to natural gas in the production of N fertilizers. Their analysis showed that 1 ha of agricultural land (south-west Sweden) can produce 1.7 metric ton N yr^{-1} (ammonium nitrate) from ley grass, or 3.6 ton N yr^{-1} from maize. GWP was lower when producing N fertilizer from biomass compared with natural gas, but eutrophication and acidification potential were higher. The greatest advantage of the biomass systems, however, lies in their potential to reduce agriculture's dependency on fossil fuels; only 2–4 MJ kg^{-1} N of primary fossil energy was required in the biomass scenarios, compared to 35 MJ kg^{-1} N when utilizing natural gas (similar to Table 7.1.1).

7.1.5.2 Unprocessed Animal Manures

The use of animal manures, more or less unprocessed, as fertilizers has been assessed in a few LCA studies. Nemecek et al. [57] conducted an LCA comparison of intensive and

Table 7.1.1 *Main environmental impacts from the production, packaging, and delivery of the main synthetic fertilizer types, per nutrient unit [15, 51, 52].*

Fertilizer product[a]	Unit (kg)	Primary energy consumption (MJ)	Global warming potential (GWP) (kg CO_2 eq)			Eutrophication potential (g PO_4^{3-} eq.)	Acidification potential (g SO_2 eq.)	Abiotic resource use (g Sb eq.)
			CO_2	N_2O	Total			
AN	N	40.0[a]/29.8[b]	2.34[a]/1.77[b]	3.69[a]/0.83[b]	6.2[a]/2.74[b]	0.5	4.7	23
Urea	N	51.6[a]/44.1[b]	1.39[a]/0.98[b]	0	1.59[a]/1.13[b]	0.54	5.3	23
CAN	N	42.6[a]/31.4[b]	2.49[a]/1.89[b]	3.66[a]/0.83[b]	6.3[a]/2.83[b]	0.55	5.3	21
AS	N	42	l.d.	l.d.	3	0.52	5.3	20
TSP	P	30.25	1.6[a]	0	1.66[a]	0.74	8.1	15
SSP	P	13	l.d.	l.d.	0.6	0.57	6.6	16
MOP	K	10.06	0.58[a]	0	0.60[a]	0.30	7.2	3.9
L	Ca	2.3	l.d.	l.d.	0.15	0.26	1.6	2.4

AN, ammonium nitrate; CAN, calcium ammonium nitrate; AS, ammonium sulfate; TSP, triple super phosphate; SSP, single super phosphate; MOP, muriate of potash; L, limestone; l.d., lack of data.

[a] Production (European average) at plant gate.

[b] Production (BAT) at plant gate.

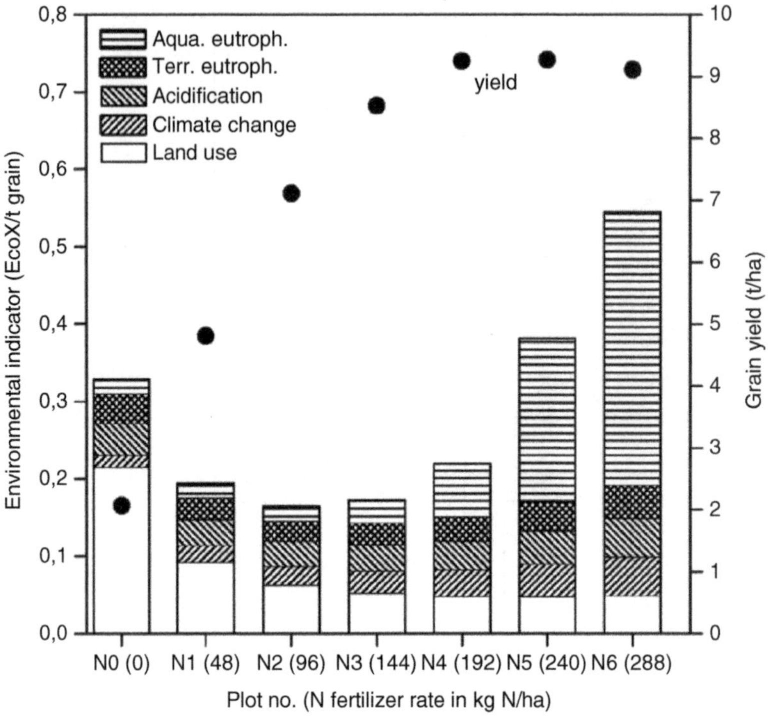

Figure 7.1.2 *Aggregated environmental indicator values (EcoX) per ton of grain (stacked bars) and yields (t/ha, dots) at increasing N fertilizer rates [55].*

extensive agricultural systems, and as part of this analyzed the effect of replacing synthetic fertilizers with farmyard manure as a special form of extensification (lowering of energy and resource inputs). With crop yield as the functional unit, their results indicated significant reductions in resource depletion and energy demand, improved soil quality, and slightly increased eutrophication potential but significantly increased acidification potential.

In a UK study by Sandars et al. [58], the functional unit was a certain amount of animal product (pork meat). The system boundaries only included manure storage and the field application system (and not the animal housing or upstream feed supply), and impact weighing based on the UK environmental emissions inventory was applied. The authors concluded that though low-emission slurry application techniques (band spreaders and injection) reduced impacts significantly (64–71%) compared to the standard application method (splash plate), the aggregated (weighed) environmental impact reduction was only around a maximum of 10% compared to mineral fertilizer and was not significant when taking uncertainty into account.

Griffing et al. [59] compared two animal manure treatment systems (deep pit storage under the animal house and storage in an anaerobic lagoon), with urea ammonium nitrate (UAN) synthetic fertilizer as the reference. Here, the functional unit was the fertilization needed for a certain crop yield. The manure fertilization showed lower resource depletion and a similar eutrophication potential to the synthetic fertilizer, but a higher GWP and acidification potential for manure than for synthetic fertilizer. However, CH_4 emission factors from housing and storage were critical to this conclusion. Furthermore, the authors pointed out that conclusions were very sensitive to the allocation method and the boundary between animal production (pork) and crop fertilization, especially when land was limiting and there was a resultant surplus of manure P and N.

7.1.5.3 Mechanically Separated and Processed Animal Manures

When intensive livestock production results in a surplus of nutrient that cannot be adequately utilized on adjacent land, mechanical separation and processing may facilitate nutrient redistribution. AN LCA study by ten Hoeve et al. [60] used a functional unit of 1 ton of animal slurry and compared four different separation and processing technologies (screw press, screw press + composting of solids, decanter centrifuge, and decanter centrifuge + NH_3 stripping liquids, all with local application of liquids and 100 km transport of solids before land application) with conventional local application of raw slurry as the reference. Separation, especially by decanter centrifuge, was found to result in lower environmental impact potentials than application of untreated slurry to adjacent land. Composting and ammonia stripping either slightly increased or slightly decreased the environmental impact potentials, depending on the impact category considered. Differences were relatively small and no single treatment option for geographically redistributing slurry nutrients could therefore be identified as clearly superior with respect to potential environmental impacts. Thus, the choice of appropriate technology should depend on other considerations, such as local environmental policy, cost, odor, or practicality. Sensitivity analyses furthermore showed that the emission factors used for NH_3, N_2O, and P emissions from the field influenced the conclusions greatly, and thus emission factors for the specific context should be considered.

Surplus nutrients from high-intensity livestock production were also in focus in a study by Lopez-Ridaura et al. [61], who conducted an LCA comparing manure treatment (centrifuge separation, biological N removal by nitrification–de-nitrification of the liquid fraction, composting of solids, transport, and soil application) with transfer of excess unprocessed slurry (avg. 40 km) and injection in cropland areas with a nutrient demand and no surplus of manure. The functional unit was the amount of treated or managed manure. The LCA clearly indicated lower environmental impacts (all categories) from the transfer and crop injection than from the biological treatment, as the former produced fewer gaseous emissions and consumed less non-renewable energy than the latter, even when including energy consumption for transport of the slurry before application to cropland. Brockmann et al. [62] conducted a more in-depth analysis of these two management options, which reached similar conclusions, and revealed that direct field emissions from organic fertilizer application and the amount of avoided mineral fertilizers (which was much lower for the biological treatment, due to loss of available N) dominated the environmental impacts.

Ten Hoeve et al. [63, 64] examined the reduction in environmental impacts from slurry acidification, applied either in the animal house or during field application and in combination with mechanical separation, and found that the most effective solution depended on the environmental regulations (fertilizer and manure N and P application limits) under which it operated; strict N regulations favored combinations of in-house acidification and separation, whereas strict P regulations favored separation alone.

7.1.5.4 Manure-Based Digestates and Post-Processing Products

AD of animal manures for biogas production is rapidly becoming the method of choice for processing of manure [16, 19], due to the multiple benefits achieved at the farm system and societal levels: renewable energy production, generally reduced greenhouse gas emissions (if fugitive emissions from reactor and storage are avoided), and higher crop fertilizing value of the digestate. Therefore, several recent LCA studies have focused on assessing manure-based AD scenarios. Hamelin et al. [1] compared three scenarios involving different slurry separation technologies (decanter centrifuge, screw press, and screw press + drying and pelletizing of solids) and the use of these solid fractions as a co-substrate for AD. The results showed that the environmental benefits of manure-based AD are highly dependent upon the efficiency of the separation technology used to concentrate volatile solids in the solid fraction used as AD feedstock. The AD scenario involving the most efficient separation technology resulted in a dry matter (DM) separation efficiency of 87% into the solids and allowed a net reduction of the GWP of 40% compared to the reference slurry management. This figure covered the whole slurry life cycle, including the flows bypassing the AD plant, and included soil carbon balances and the changes in yield resulting from increased N availability and thus mineral fertilizer displacement.

Hamelin et al. [65] analyzed a range of different co-substrate alternatives for manure-based AD, including maize energy crop, residual cereal straw, household biowaste, commercial biowaste, garden waste, and separated manure solids, with a reference scenario of manure AD without co-substrate. Due to the large overall benefits of AD of manure, as mentioned earlier, the scenario with separated manure solids showed the greatest environmental benefits in the LCA analysis, as this allowed a higher proportion of manure per functional unit (1 ton of animal manure excreted) to be put through the AD plant. LUC emissions from energy crops and soil carbon changes were included, and this made the maize energy crop the co-substrate with the worst environmental performance. Next to manure solids, straw and all biowaste categories should therefore be prioritized as co-substrates for manure AD. More complex treatment scenarios, including some with AD, were analyzed using LCA by Prapasponga et al. [66]. The most effective environmental impact reduction was found for treatment technology systems aiming at combined energy recovery with high nutrient recovery and control of greenhouse gas, ammonia, and nitrate emissions at every handling stage, and AD achieved the highest impact reduction because of its high efficiencies in energy and nutrient recovery with restricted GWP and eutrophication potential.

Post-processing of digestate was assessed in an LCA study by Rehl and Müller [67], including separated solids processing by composting, belt dryer, drum dryer, solar drying, or thermal concentration and liquid processing by physical-chemical treatment (micro-filtration, RO, and ion exchange) and compared against a conventional digestate management (storage and land application) reference. Overall, solar drying, composting, and physical-chemical treatment were identified as the most suitable options for educing the use of resources and environmental impacts compared to conventional digestate management. Belt drying turned out to be the handling process with the highest energy demand, GWP, and acidification potential among the compared options. Nitrogen-related emissions from digestate treatment, storage, and field application have a major influence on environmental impacts. Other important aspects are the amount and kind of fuel used for heat supply (biogas, natural gas) and the procedure chosen for allocation among heat and power.

7.1.5.5 Municipal Solid Waste and Wastewater Biosolids Processed by AD or Composting

Management and treatment of urban waste streams has been the subject of many more LCA studies than manure management and treatment, simply because technologies for the former have been available and applied for several decades, whereas those for the latter have only appeared more recently. Laurent et al. [68] reviewed 222 published LCA studies of solid waste management systems and, like, Astrup et al. [49] found little consistency in methodological approaches and data management, and thus not surprisingly little agreement in the conclusions of the studies. Only a limited number were focused on management of organic waste components (29 studies in total), but overall most of these pointed toward positive environmental impacts from recycling through AD or composting (12 studies), whereas thermal treatment (e.g. incineration with recovery of energy) was only found superior in five.

In their LCA comparison of multiple biogas production and utilization pathways, including both agricultural residues (manure, straw), energy crops, organic fraction of MSW (OFMSW), and food industry organic wastes, Poeschl et al. [69] found that co-digestion of OFMSW with agricultural and food industry residues resulted in the lowest overall environmental impacts for both small- and large-scale biogas plants. Righi et al. [70] conducted a more specific LCA of anaerobic co-digestion of de-watered sewage sludge and OFMSW and found that small AD plants combined with composting post-treatment may offer an environmentally sustainable waste management option for small communities. This was achieved mainly by a strong reduction in the distances and volumes transported by road, the low energy requirement for the process itself, energy savings from the CHP production and the energy and nutrient resources saved from the compost fertilizer produced by the digested matter. However, Lyng et al. [71] conducted an LCA comparison of OFMSW or manure processing by AD, including different applications of biogas and digestate and different assumptions regarding which products should be substituted, and found the results for GWP impact were less sensitive for the different choices than all the other environmental indicators, which were highly sensitive for them. Tonini et al. [72] conducted an LCA comparison of alternatives for resource recovery from OFMSW treatment (waste refinery: AD + solids composting vs. incineration, mechanical-biological treatment (MBT) with or without energy recovery, landfilling with or without energy recovery). Overall, the waste refinery provided global warming savings comparable to efficient incineration and MBT with energy recovery, but the main environmental benefits from waste refining were from changes in resource use (i.e. improved P recovery [about an 85% increase] and increased renewable electricity production [by 15–40% compared with incineration]), albeit at the potential expense of additional toxic emissions to the soil. Further, Tonini et al. [73] estimated that this waste refinery concept had the potential for recovery of N, P, K, and biogenic C equaling 81–89% of the input, but found that the quality of the digestate or compost may be critical with respect to use on land, and hence source input quality should be monitored closely.

Many LCAs of wastewater treatment plant (WWTP) systems have focused on the removal of nutrients and contaminants from the water stream and the protection of the aquatic environment, and only recently have studies including quantification of the environmental impacts and benefits of nutrient recovery and utilization as biobased fertilizers begun to appear. In a comparative LCA of different types of WWTPs (combinations of pure chemical vs. biological + chemical precipitation of P; AD vs. aerobic mineralization of the sludge; land application vs. incineration of residual sludge), representative of mainstream treatment options in Denmark, Niero et al. [74] found that for the GWP and fossil resource depletion impact categories, WWTPs with AD and recycling P to agricultural soils appear to be more sustainable compared to the incineration of sludge. However, for eutrophication- and toxicity-related categories, WWTP with AD and sludge incineration yielded higher environmental impacts compared to WWTPs treating the sludge via aerobic stabilization prior to agricultural application; these conclusions were very sensitive to assumptions and parameter choices, however. In a later study of similar WWTP options, Yoshida et al. [75] found that, generally, the scenario with incineration of residual sludge performed better than or comparably to the different scenarios involving land application. However, their sensitivity analysis also showed that the results were sensitive to soil, crop, and climatic factors as well as local conditions of marine eutrophication. Overall, these studies highlight the importance of including all sludge treatment stages and of conducting a detailed, contextual N flow analysis, since the emission of reactive N into the environment is often the major driver for almost all non-toxic impact categories.

Linderholm et al. [76] also conducted an LCA study on WWTP options, but with a different scope and hence functional unit, namely the provision of a certain quantity of P to agricultural land (11 $kg\,P\,ha^{-1}$, the average crop requirement in Sweden). They compared recycled P sources (certified sewage sludge, struvite [$NH_4MgPO_4{\cdot}6H_2O$] precipitated from wastewater treatment, and P recovered from ash produced by incineration of sewage sludge) with synthetic fertilizer P (triple superphosphate, TSP) and found that sewage sludge had the lowest energy consumption, resource depletion, and GWP but increased ecotoxicity due to cadmium in sludge, whereas sludge incineration and P recovery had the highest energy use and GWP. Similar findings were made by Vaneeckhaute et al. [77], who evaluated the P use efficiency of recovered biobased fertilizers (struvite and Fe-precipitated sludge, both from wastewater treatment, as well as manure and anaerobic digestate based on manure + maize + foodwaste) and compared them to synthetic fertilizer (TSP). Struvite demonstrated potential as a slow-release mixed fertilizer, while the Fe-sludge showed very low P use efficiency in the short term; digestate was superior to manure and had the added benefit of lowering environmental impacts and providing renewable energy. Spångberg et al. [78] assessed the environmental impacts of recycling nutrients in human excreta (source separation of urine and excreta) to agriculture compared with traditional wastewater treatment (nutrient removal) and found the former to have lower energy consumption and GWP than the latter, but higher eutrophication and acidification potentials, due to large impacts by the ammonia emitted during storage and after spreading of the fertilizers.

While AD may often be the optimal solution for high moisture-containing or liquid organic wastes, composting will often be the method of choice for recycling of low-moisture, solid organic waste. Hansen et al. [79] conducted an LCA of the potential environmental impacts from agricultural land application of source-separated OFMSW processed by either composting or AD, using the waste-LCA model EASEWASTE.

The case study included compost and digestate fertilizer applied to two different soil types (loamy and sandy) under relatively humid, temperate climatic conditions (western Denmark). The local agricultural conditions such as soil type and climate, as well as the composition of the processed MSW, showed large influence on the normalized environmental impacts, with the main ones being global warming (0.4–0.7 PE), acidification (−0.06 [saving] to 1.6 PE), eutrophication (−1.0 [saving] to 3.1 PE), and toxicity potential. The environmental impacts were only markedly different between composting and AD for a few impact categories; for acidification and eutrophication potentials, AD had a much larger impact than compost. This was expanded on in a more recent study by Boldrin et al. [80], who also applied the EASEWASTE LCA model to quantify potential environmental effects from biological treatment of OFMSW, using composting, AD, and combinations thereof in relation to mass flows, energy consumption, gaseous emissions, biogas recovery, and compost/digestate utilization. In contrast to Hansen et al. [79], their results indicated significant benefits to GWP in scenarios including AD, though the application of digestate to crops may increase nutrient leaching and hence give a somewhat more negative impact on eutrophication potential than composting. More recently, Nielsen et al. [81] showed that the environmental emission factors and associated impacts from land application of different types of compost produced from green/garden park waste depend entirely on the local conditions, such as the soil type, precipitation regime, and general fertilization level, with the composition of the materials being of lesser importance. In contrast, the soil C sequestration factor was almost unaffected by the environmental conditions but depended to a large extent on the stability or degradability of the added material.

The effect of applying composted OFMSW for the production of tomatoes was compared against the use of synthetic fertilizer by Martinez-Blanco et al. [82]. Compost generally performed better than mineral fertilizer, particularly with respect to GWP and eutrophication potential, though it also had a higher impact in terms of acidification potential and photochemical oxidation. Industrial-scale composting versus home composting of OFMSW was compared in an LCA by Quiros et al. [83] looking at utilization of the compost for horticultural production of cauliflower crops, with synthetic fertilizer as the reference. The home composting had the better environmental performance, with the lowest impact in all categories assessed except resource depletion and eutrophication potential (which was lower for industrial compost). The industrial compost had the highest environmental impact in five of the seven categories assessed, while the mineral fertilizer reference had the highest eutrophication and GWP.

Several of the studies on compost, in particular those of Hansen et al. [79] and Boldrin et al. [80], point out that a range of benefits, mainly related to improved soil quality from long-term application of the processed OFMSW, are generally not quantified with the current commonly used LCA impact categories; therefore, these should be considered specifically in conjunction with the interpretation of LCA impact results. In line with this, Martinez-Blanco et al. [48] reviewed studies of compost use for agriculture with the idea of including more benefits in LCA; apart from substitution of synthetic fertilizer nutrients and carbon sequestration, pest and disease suppression, soil workability, biodiversity, crop nutritional quality, crop yield potential, soil erosion prevention, and soil moisture retention should also be included, and LCA impact assessment methods should be developed to fully account for these.

7.1.5.6 Mineral Concentrates, Extracts, Precipitates, Chars, and Ashes from Organic Wastes

A range of new and improved technologies for recovering nutrients and OM from biowaste and recycling them to the soil as fertilizer have appeared in recent years [40]. These include mineral concentrates (in particular N- and K-rich) and precipitates (in particular P-rich) from liquid waste streams and ashes and chars from thermal processing of solid organic wastes.

The use of UF and nano-filtration, RO, membrane distillation, and air stripping for the recovery of nutrients, especially N, from liquid waste (separated manures, digestates) is gradually being developed [84], with a number of full-scale commercial plants operational in the Netherlands and Belgium [85]. De Vries et al. [86] assessed the environmental consequences of manure separation and dewatering using RO and found the processing increased overall environmental impacts for GWP, acidification potential, particulate matter formation, and fossil fuel depletion when compared to current practice (direct application of manure), though inclusion of AD reduced GWP and fossil fuel depletion. Results were highly sensitive to methane emissions from manure storage, ammonia emissions from processing, and application and the replacement value of N fertilizer by the mineral concentrate, which may be highly variable depending on the application technique, as also shown by Klop et al. [87], due to the ammonia volatilization potential of the mineral concentrates.

Thermal treatment by pyrolysis or thermal gasification of nutrient-rich biowaste results in the production of a char product, which can be applied to soils as a fertilizer or amelioration product, and is then termed "biochar." This may have several co-benefits, including in biowaste management, renewable energy generation, long-term C sequestration from stable C in the biochar, and soil quality amelioration and fertility improvement. Roberts et al. [88] conducted an LCA of biochar systems based on feedstocks of crop residues (maize stovers), yard waste, and switchgrass energy crops, and assessed fossil fuel consumption and GWP (including the soil C sequestration effect). They found that GWP was negative (savings) from the crop residues and yard waste (mainly due to soil C sequestration) but increased from the energy crop due to iLUC; the results were also highly sensitive to the transport distances of the biomass.

7.1.6 Conclusion

LCA provides a comprehensive, systematic, and coherent framework for the assessment and comparison of the overall environmental impacts and benefits from products or processes delivering the same or similar services (in this context, synthetic and biobased fertilizers), and is therefore rapidly becoming the standard method of choice for such assessments. LCA could thus be an important tool for decision-making by important stakeholders, such as farmers, agricultural advisors, environmental technology developers, environmental authorities, and politicians.

However, application of the LCA approach to biobased fertilizer production and use is still relatively sparse. LCA also includes many different assumptions, method selections, data sources, and interpretation choices, all of which are highly influential for the outcome of the assessment. From our review of current LCA studies of waste-to-fertilizer technologies and applications, we can conclude that they are generally not systematic and coherent; goal and scope definitions are very different between studies, and most do not provide full and transparent descriptions of methodological choices, technology parameters, or modeling principles. This lack of consistency prohibits proper evaluation of the validity of results and limits the applicability of data and results in other contexts, as well as prohibiting comparisons across studies. For future LCA studies of biobased fertilizers, this situation needs to be improved; recommendations from the review of thermal waste-to-energy LCA studies by Astrup et al. [49] could be taken as a point of departure, but they will need to be adapted to biobased fertilizer production and use.

In spite of this, some overall tendencies can still be drawn from the review. Generally, biobased versus synthetic fertilizers result in reduced fossil resource depletion and energy consumption, while increasing eutrophication and acidification potential, due to increased N and P leaching and NH_3 losses, respectively. GWP varies, depending on one hand on the energy intensity of the processing and on the other hand on the possible co-production of energy from the process, as in AD, for example. In particular, scenarios including AD induce lower or even negative GWP impacts (savings) due to both fossil energy substitution and reduced greenhouse gas emissions from the entire waste-to-fertilizer chain. Generally, very processing-intensive technologies do not provide substantially lower impacts, unless the alternative solutions for, say, surplus manure includes very long transport distances or high emissions. Composting only scores positively in LCA when gaseous emissions are kept to a minimum and the use on land includes soil C sequestration; soil quality improvements and other benefits from composts are currently not covered by standard LCA impact categories, and hence these benefits are not well represented in LCA. For urban wastes, in particular, the contents of potential contaminants influencing the toxicity potentials of biobased fertilizer products play a more important role; however, there are indications that their importance is declining due to lower contaminant load in waste streams and that improvements in resource substitution will become more dominant.

Acknowledgments

This chapter was written with financial support from the People Programme (Marie Curie Actions) of the European Union's Seventh Framework Programme under the *ReUseWaste* Initial Training Network project (REA grant agreement no. 289887) as well as from the Danish Council for Strategic Research under the *CleanWaste* (no. 2104-09-0056) and *BioChain* projects (no. 12-132631).

References

1. Hamelin, L., Wesnæs, M., Wenzel, H., and Petersen, B.M. (2011). Environmental consequences of future biogas technologies based on separated slurry. *Environmental Science & Technology* **45**: 5869–5877.
2. Croxatto-Vega, G.C., ten Hoeve, M., Birkved, M. et al. (2014). Choosing co-substrates to supplement biogas production from animal slurry – a life cycle assessment of the environmental consequences. *Bioresource Technology* **171**: 410–420.
3. van Haaren, R., Themelis, N.J., and Barlaz, M. (2010). LCA comparison of windrow composting of yard wastes with use as alternative daily cover (ADC). *Waste Management* **30**: 2649–2656.
4. ISO 14040 (2006). Environmental Management – Life Cycle Assessment – Principles and Framework. European Standard.
5. ISO 14044 (2006). Environmental Management – Life Cycle Assessment – Requirements and guidelines. European Standard.

6. European Commission – Joint Research Centre (2010). International Reference Life Cycle Data System (ILCD) Handbook – General Guide for Life Cycle Assessment – Detailed Guidance. EUR 24708 EN. Luxembourg: Publications Office of the European Union.
7. Hauschild, M., Rosenbaum, R.K., and Olsen, S. (eds.) (2018). *Life Cycle Assessment – Theory and Practice*. New York: Springer.
8. Yoshida, H., Nielsen, M.P., Scheutz, C. et al. (2015). Long-term emission factors for land application of treated organic municipal waste. *Environmental Modeling and Assessment* **21**: 111–124.
9. ten Hoeve, M., Bruun, S., Naroznova, I. et al. (2018). Life cycle inventory modeling of phosphorus substitution, losses and crop uptake after land application of organic waste products. *International Journal of Life Cycle Assessment* http://dx.doi.org/10.1007/s11367-017-1421-9.
10. Hauschild, M.Z. and Potting, J. (2005). *Spatial Differentiation in Life Cycle Impact Assessment – the EDIP 2003 Methodology*. Copenhagen: Danish Ministry of the Environment, Environmental Protection Agency.
11. Jolliet, O., Margni, M., Charles, R.L. et al. (2003). IMPACT 2002+: a new life cycle impact assessment methodology. *International Journal of Life Cycle Assessment* **8**: 324–330.
12. Weidema, B.P., Hauschild, M.Z., and Jolliet, O. (2008). Preparing characterisation methods for endpoint impact assessment. Annex II. In: *Environmental Improvement Potentials of Meat and Dairy Products* (eds. B.P. Weidema, M. Wesnæs, J. Hermansen, et al.), 151–170. Seville: European Commission, Joint Research Centre, Institute for Prospective Technological Studies (EUR 23491 EN).
13. Goedkoop, M., Heijungs, R., Huijbregts, M. et al. (2009). *ReCiPe 2008: A Life Cycle Impact Assessment Method Which Comprises Harmonised Category Indicators at the Midpoint and the Endpoint Level*. Amsterdam: Ruimte en Milieu.
14. Forster, P., Ramaswamy, V., Artaxo, P. et al. (2007). Changes in atmospheric constituents and in radiative forcing. In: *Climate Change 2007: The Physical Science Basis. Contribution of Working Group I to the Fourth Assessment Report of the Intergovernmental Panel on Climate Change* (eds. S. Solomon, D. Qin, M. Manning, et al.). Cambridge: Cambridge University Press.
15. Skowrowska, M. and Filipek, T. (2014). Life cycle assessment of fertilizers: a review. *International Agrophysics* **28**: 101–110.
16. Nkoa, R. (2014). Agricultural benefits and environmental risks of soil fertilization with anaerobic digestates: a review. *Agronomy for Sustainable Development* **34**: 473–492.
17. Tian, G., Chiu, C.-Y., Franzluebbers, A.J. et al. (2015). Biosolids amendment dramatically increases sequestration of crop residue-carbon in agricultural soils in western Illinois. *Applied Soil Ecology* **85**: 86–93.
18. Sørensen, P. and Jensen, L.S. (2013). Nutrient leaching and runoff from land application of animal manure and measures for reduction. In: *Animal Manure Recycling – Treatment and Management* (eds. S.G. Sommer, M.L. Christensen, T. Schmidt and L.S. Jensen), 195–210. Chichester: Wiley.
19. Möller, K. (2015). Effects of anaerobic digestion on soil carbon and nitrogen turnover, N emissions, and soil biological activity. A review. *Agronomy for Sustainable Development* **35**: 1–21.
20. Sommer, S.G. and Hutchings, N.J. (2001). Ammonia emission from field applied manure and its reduction – invited paper. *European Journal of Agronomy* **15**: 1–15.
21. Rosenbaum, R.K., Bachmann, T.M., Gold, L.S. et al. (2008). USEtox – the UNEP-SETAC toxicity model: recommended characterisation factors for human toxicity and freshwater ecotoxicity in life cycle impact assessment. *International Journal of Life Cycle Assessment* **13**: 532–546.
22. Henderson, A.D., Hauschild, M.Z., van de Meent, D. et al. (2011). USEtox fate and ecotoxicity factors for comparative assessment of toxic emissions in life cycle analysis: sensitivity to key chemical properties. *International Journal of Life Cycle Assessment* **16**: 701–709.
23. Smolders, E., Oorts, K., van Sprang, P. et al. (2009). Toxicity of trace metals in soil as affected by soil type and aging after contamination: using calibrated bioavailability models to set ecological soil standards. *Environmental Toxicology and Chemistry* **28**: 1633–1642.

24. Smolders, E., Oorts, K., Lombi, E. et al. (2012). The availability of copper in soils historically amended with sewage sludge, manure, and compost. *Journal of Environmental Quality* **41**: 506–514.
25. Smolders, E. (2013). *Revisiting and Updating the Effect of Phosphorus Fertilisers on Cadmium Accumulation in European Agricultural Soils*. York: International Fertiliser Society.
26. Smith, S.R. (2009). A critical review of the bioavailability and impacts of heavy metals in municipal solid waste composts compared to sewage sludge. *Environment International* **35**: 142–156.
27. Smith, S.R. (2009). Organic contaminants in sewage sludge (biosolids) and their significance for agricultural recycling. *Philosophical Transaction of the Royal Society A* **367**: 4005–4041.
28. Heimersson, S., Harder, R., Peters, G.M., and Svanström, M. (2014). Including pathogen risk in life cycle assessment of wastewater management. 2. Quantitative comparison of pathogen risk to other impacts on human health. *Environmental Science and Technology* **48**: 9446–9453.
29. Klinglmair, M., Sala, S., and Brandão, M. (2014). Assessing resource depletion in LCA: a review of methods and methodological issues. *International Journal of Life Cycle Assessment* **19**: 580–592.
30. Withers, P.J.A., Elser, J.J., Hilton, J. et al. (2015). Greening the global phosphorus cycle: how green chemistry can help achieve planetary P sustainability. *Green Chemistry* **17**: 2087–2099.
31. Karunanithi, R., Szogi, A.A., Bolan, N. et al. (2015). Phosphorus recovery and reuse from waste streams. *Advances in Agronomy* **131**: 173–250.
32. Withers, P.J.A., van Dijk, K.C., Neset, T.-S.S. et al. (2015). Stewardship to tackle global phosphorus inefficiency: the case of Europe. *Ambio* **44** (Suppl. 2): 193–206.
33. Kløverpris, J., Baltzer, K., and Nielsen, P.H. (2010). Life cycle inventory modelling of land use induced by crop consumption. Part 2: Example of wheat consumption in Brazil, China, Denmark, and the USA. *International Journal of Life Cycle Assessment* **15**: 90–103.
34. Kløverpris, J. (2009). Identification of biomes affected by marginal expansion of agricultural land use induced by increased crop consumption. *Journal of Cleaner Production* **17**: 463–470.
35. Tonini, D., Hamelin, L., Wenzel, H., and Astrup, T. (2012). Bioenergy production from perennial energy crops: a consequential LCA of 12 bioenergy scenarios including land use changes. *Environmental Science and Technology* **46**: 13521–13530.
36. Sommer, S.G. and Feilberg, A. (2013). Gaseous emissions of ammonia and malodorous gases. In: *Animal Manure Recycling: Treatment and Management* (eds. S.G. Sommer, M.L. Christensen, T. Schmidt and L.S. Jensen), 131–151. Chichester: Wiley.
37. Peters, G.M., Murphy, K.R., Adamsen, A.P.S. et al. (2014). Improving odour assessment in LCA – the odour footprint. *International Journal of Life Cycle Assessment* **19**: 1891–1900.
38. Holm, P.E., McLaughlin, M.J., and Jensen, L.S. (2010). Utilization of biologically treated organic waste on land. In: *Solid Waste Technology & Management* (ed. T.H. Christensen), 665–682. Chichester: Wiley.
39. Jensen, L.S. (2013). Animal manure fertiliser value, crop utilisation and soil quality impacts. In: *Animal Manure Recycling – Treatment and Management* (eds. S.G. Sommer, M.L. Christensen, T. Schmidt and L.S. Jensen), 295–328. Chichester: Wiley.
40. Jensen, L.S. (2013). Animal manure residue upgrading and nutrient recovery in biofertilisers. In: *Animal Manure Recycling – Treatment and Management* (eds. S.G. Sommer, M.L. Christensen, T. Schmidt and L.S. Jensen), 271–294. Chichester: Wiley.
41. Alburquerque, J.A., de la Fuente, C., Ferrer-Costa, A. et al. (2012). Assessment of the fertiliser potential of digestates from farm and agroindustrial residues. *Biomass and Bioenergy* **40**: 181–189.
42. Golkowska, K., Vázquez-Rowe, I., Lebuf, V. et al. (2014). Assessing the costs and value of output products in digestate treatment systems. *Water Science and Technology* **69**: 656–662.
43. Carlsbæk, M. (2010). Use of compost in horticulture and landscaping. In: *Solid Waste Technology & Management* (ed. T.H. Christensen), 651–664. Chichester: Wiley.
44. Boldrin, A., Andersen, J.K., Møller, J. et al. (2009). Composting and compost utilisation: accounting of greenhouse gases and global warming contributions. *Waste Management Research* **27**: 800–812.

45. Boldrin, A., Hartling, K.R., Laugen, M., and Christensen, T.H. (2010). Environmental inventory modelling of the use of compost and peat in growth media preparation. *Resources, Conservation and Recycling* **54**: 1250–1260.
46. Andersen, J.K., Christensen, T.H., and Scheutz, C. (2010). Substitution of peat, fertiliser and manure by compost in hobby gardening: user surveys and case studies. *Waste Management* **30**: 2483–2489.
47. Petersen, B.M., Knudsen, M.T., Hermansen, J.E., and Halberg, N. (2013). An approach to include soil carbon changes in life cycle assessments. *Journal of Cleaner Production* **52**: 217–224.
48. Martinez-Blanco, J., Lazcano, C., Christensen, T.H. et al. (2013). Compost benefits for agriculture evaluated by life cycle assessment. A review. *Agronomy for Sustainable Development* **33**: 721–732.
49. Astrup, T.F., Tonini, D., Turconi, R., and Boldrin, A. (2015). Life cycle assessment of thermal waste-to-energy technologies: review and recommendations. *Waste Management* **37**: 104–115.
50. Hasler, K., Bröring, S., Omta, S.W.F., and Olfs, H.-W. (2015). Life cycle assessment (LCA) of different fertilizer product types. *European Journal of Agronomy* **69**: 41–51.
51. Brentrup, F. and Pallière, C. (2008). *Energy Efficiency and Greenhouse Gas Emissions in European Nitrogen Fertilizer Production and Use*. York: International Fertiliser Society.
52. Williams, A.G., Audsley, E., and Sandars, D.L. (2010). Environmental burdens of producing bread wheat, oilseed rape and potatoes in England and Wales using simulation and system modelling. *International Journal of Life Cycle Assessment* **15**: 855–868.
53. Laratte, B., Guillaume, B., and Birregah, J.K.B. (2014). Modeling cumulative effects in life cycle assessment: the case of fertilizer in wheat production contributing to the global warming potential. *Science of the Total Environment* **481**: 588–595.
54. Gregorich, E., Janzen, H.H., Helgason, B., and Ellert, B. (2015). Nitrogenous gas emissions from soils and greenhouse gas effects. *Advances in Agronomy* **132**: 39–74.
55. Brentrup, F., Küsters, J., Lammel, J. et al. (2004). Environmental impact assessment of agricultural production systems using the life cycle assessment (LCA) methodology II. The application to N fertilizer use in winter wheat production systems. *European Journal of Agronomy* **20**: 265–279.
56. Ahlgren, S., Bernesson, S., Nordberg, Å., and Hansson, P.-A. (2010). Nitrogen fertiliser production based on biogas – energy input, environmental impact and land use. *Bioresource Technology* **101**: 7181–7184.
57. Nemecek, T., Huguenin-Elie, O., Dubois, D. et al. (2011). Life cycle assessment of Swiss farming systems: II. Extensive and intensive production. *Agricultural Systems* **104**: 233–245.
58. Sandars, D.L., Audsley, E., Cañete, C. et al. (2003). Environmental benefits of livestock manure management practices and technology by life cycle assessment. *Biosystems Engineering* **84**: 267–281.
59. Griffing, E.M., Schauer, R.L., and Rice, C.W. (2014). Life cycle assessment of fertilization of corn and corn–soybean rotations with swine manure and synthetic fertilizer in Iowa. *Journal Environmental Quality* **43**: 709–722.
60. ten Hoeve, M., Hutchings, N.J., Peters, G. et al. (2014). Life cycle assessment of pig slurry treatment technologies for nutrient redistribution in Denmark. *Journal of Environmental Management* **132**: 60–70.
61. Lopez-Ridaura, S., van der Werf, H., Paillat, J.M., and Le Bris, B. (2009). Environmental evaluation of transfer and treatment of excess pig slurry by life cycle assessment. *Journal of Environmental Management* **90**: 1296–1304.
62. Brockmann, D., Hanhoun, M., Negri, O., and Helias, A. (2014). Environmental assessment of nutrient recycling from biological pig slurry treatment: impact of fertilizer substitution and field emissions. *Bioresource Technology* **163**: 270–279.
63. ten Hoeve, M., Nyord, T., Peters, G.M. et al. (2016). A life cycle perspective of slurry acidification strategies under different nitrogen regulations. *Journal of Cleaner Production* **127**: 591–599.
64. ten Hoeve, M., Gómez-Muñoz, B., Jensen, L.S., and Bruun, S. (2016). Environmental impacts of combining pig slurry acidification and separation under different regulatory regimes – a life cycle assessment. *Journal of Environmental Management* **181**: 710–720.

65. Hamelin, L., Naroznova, I., and Wenzel, H. (2014). Environmental consequences of different carbon alternatives for increased manure-based biogas. *Applied Energy* **114**: 774–782.
66. Prapaspongsa, T., Christensen, P., Schmidt, J.H., and Thrane, M. (2010). LCA of comprehensive pig manure management incorporating integrated technology systems. *Journal of Cleaner Production* **18**: 1413–1422.
67. Rehl, T. and Müller, J. (2011). Life cycle assessment of biogas processing treatment technologies. *Resources, Conservation and Recycling* **56**: 92–104.
68. Laurent, A., Bakas, I., Clavreul, J. et al. (2014). Review of LCA studies of solid waste management systems – Part I: Lessons learned and perspectives. *Waste Management* **34**: 573–588.
69. Poeschl, M., Ward, S., and Owende, P. (2012). Environmental impacts of biogas deployment – Part II: Life cycle assessment of multiple production and utilization pathways. *Journal of Cleaner Production* **24**: 184–201.
70. Righi, S., Oliviero, L., Pedrini, M. et al. (2013). Life cycle assessment of management systems for sewage sludge and food waste: centralized and decentralized approaches. *Journal of Cleaner Production* **44**: 8–17.
71. Lyng, K.-A., Modahl, I.S., Møller, H. et al. (2015). The BioValueChain model: a Norwegian model for calculating environmental impacts of biogas value chains. *International Journal of Life Cycle Assessment* **20**: 490–502.
72. Tonini, D., Martinez-Sanchez, V., and Astrup, T.F. (2013). Material resources, energy, and nutrient recovery from waste: are waste refineries the solution for the future? *Environmental Science and Technology* **47**: 8962–8969.
73. Tonini, D., Dorini, G., and Astrup, T.F. (2014). Bioenergy, material, and nutrients recovery from household waste: advanced material, substance, energy, and cost flow analysis of a waste refinery process. *Applied Energy* **121**: 64–78.
74. Niero, M., Pizzol, M., Bruun, H.G., and Thomsen, M. (2014). Comparative life cycle assessment of wastewater treatment in Denmark including sensitivity and uncertainty analysis. *Journal of Cleaner Production* **68**: 25–35.
75. Yoshida, H., ten Hoeve, M., Christensen, T.H. et al. (2018). Life cycle assessment of sewage sludge management options including long-term impacts after land application. *Journal of Cleaner Production* **174**: 538–547.
76. Linderholm, K., Tillman, A.M., and Mattsson, J.E. (2012). Life cycle assessment of phosphorus alternatives for Swedish agriculture. *Resources, Conservation and Recycling* **66**: 27–39.
77. Vaneeckhaute, C., Janda, J., Meers, E., and Tack, F.M.G. (2015). Efficiency of soil and fertilizer phosphorus use in time: a comparison between recovered struvite, $FePO_4$-sludge, digestate, animal manure, and synthetic fertilizer. In: *Nutrient Use Efficiency: From Basics to Advances* (eds. A. Rakshit, H.B. Singh and A. Sen), 73–85. New Delhi: Springer.
78. Spångberg, J., Tidåker, P., and Jönsson, H. (2014). Environmental impact of recycling nutrients in human excreta to agriculture compared with enhanced wastewater treatment. *Science of the Total Environment* **493**: 209–219.
79. Hansen, T.L., Bhander, G.S., Christensen, T.H. et al. (2006). Life cycle modelling of environmental impacts of application of processed organic municipal solid waste on agricultural land (EASEWASTE). *Waste Management Research* **24**: 153–166.
80. Boldrin, A., Neidel, T.L., Damgaard, A. et al. (2011). Modelling of environmental impacts from biological treatment of organic municipal waste in EASEWASTE. *Waste Management* **31**: 619–630.
81. Nielsen, M.P., Yoshida, H., Raji, S.G. et al. (2019). Deriving environmental life cycle inventory factors for land application of garden waste composts under European conditions. *Environment Modeling and Assessment* **24**: 21–35.
82. Martinez-Blanco, J., Munoz, P., Anton, A., and Rieradevall, J. (2009). Life cycle assessment of the use of compost from municipal organic waste for fertilization of tomato crops. *Resources, Conservation and Recycling* **53**: 340–351.

83. Quiros, R., Villalba, G., Muñoz, P. et al. (2014). Environmental and agronomical assessment of three fertilization treatments applied in horticultural open field crops. *Journal of Cleaner Production* **67**: 147–158.
84. Zarebska, A., Nieto, D.R., Christensen, K.V. et al. (2015). Ammonium fertilizers production from manure: a critical review. *Critical Reviews in Environmental Science and Technology* **45**: 1469–1521.
85. Hoeksma, P., de Buisonjé, F.E., Aarnink, A.A. (2012) Full-scale production of mineral concentrates from pig slurry using reverse osmosis. Proceeding of the International Conference of Agricultural Engineering, CIGR-AgEng2012, Valencia.
86. De Vries, J.W., Groenestein, C.M., and de Boer, I.J.M. (2012). Environmental consequences of processing manure to produce mineral fertilizer and bio-energy. *Journal of Environmental Management* **102**: 173–183.
87. Klop, G., Velthof, G.L., and van Groenigen, J.W. (2012). Application technique affects the potential of mineral concentrates from livestock manure to replace inorganic nitrogen fertilizer. *Soil Use and Management* **28**: 468–477.
88. Roberts, K.G., Gloy, B.A., Joseph, S. et al. (2010). Life cycle assessment of biochar systems: estimating the energetic, economic, and climate change potential. *Environmental Science and Technology* **44**: 827–833.

7.2

Case Study: Acidification of Pig Slurry

Lars Stoumann Jensen, Myles Oelofse, Marieke ten Hoeve, and Sander Bruun

Department of Plant and Environmental Sciences, University of Copenhagen, Copenhagen, Denmark

7.2.1 Introduction

In this chapter, we present a simple case study applying life cycle assessment (LCA) to the technology of pig slurry acidification for mitigation of ammonia emissions. The basic LCA concepts are described in Chapter 7.1. A more advanced case study focused on LCA of environmental impacts of different digestate treatment and nutrient recovery systems can be found in Chapter 7.3.

In this case study, a conventional scenario for pig slurry management is compared to a management scenario with acidification of the pig slurry, an environmental technology developed to limit the emission of ammonia to the atmosphere [1]. Slurry acidification is currently applied at commercial scale by farmers in Denmark and other parts of Scandinavia.

In the acidification scenario, a strong acid is added to the slurry, which is temporarily stored in the pits below the slated floor of the pig housing, while nothing is added to the non-acidified slurry. For the rest, management in the two scenarios is identical; both involve outdoor storage in a covered tank, and the same field application method is used. Figure 7.2.1 provides a graphical overview of the two scenarios. In the following sections, we have taken emission and substitution factors from more detailed and extensive LCA studies of slurry acidification technology [2, 3].

Biorefinery of Inorganics: Recovering Mineral Nutrients from Biomass and Organic Waste, First Edition.
Edited by Erik Meers, Gerard Velthof, Evi Michels and René Rietra.
© 2020 John Wiley & Sons Ltd. Published 2020 by John Wiley & Sons Ltd.

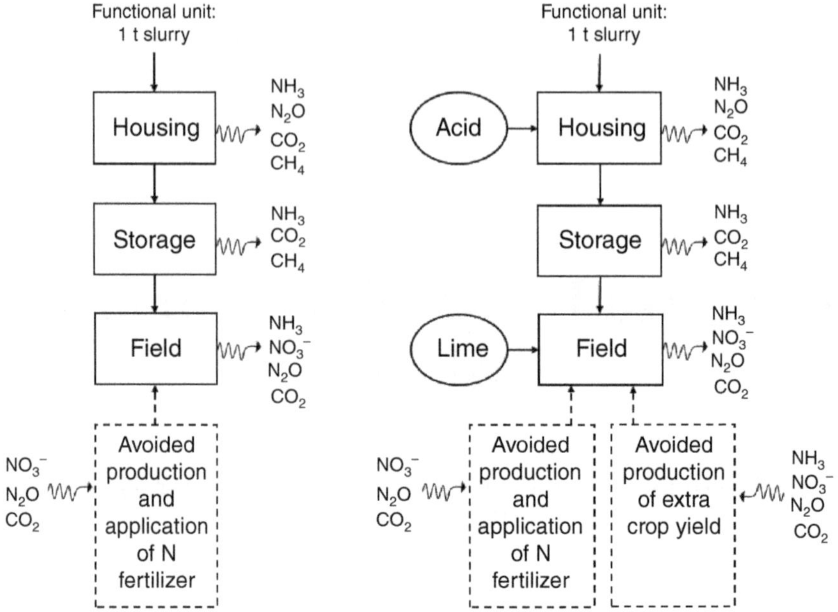

Figure 7.2.1 *Overview of the conventional scenario and the acidification scenario for the handling of 1 ton of slurry (functional unit). Avoided processes are indicated by dashed lines.*

In both scenarios, we define the functional unit as treatment of 1 ton of pig slurry. For the treatment of pig slurry, both with and without acidification, natural resources are needed and emissions to the environment take place. For example, ammonia (NH_3) and carbon dioxide (CO_2) are emitted from the housing during storage and after field application, while nitrate (NO_3) is emitted to the aquatic environment after field application. These emissions affect the environment in different ways: NH_3 contributes to terrestrial acidification, CO_2 to climate change, and NO_3^- to marine water eutrophication.

The use and treatment of slurry do not only have a negative effect on the environment, but also a positive one, since slurry is a valuable fertilizer, supplying nitrogen (N) to the crops. The N applied in slurry means that the farmer can reduce the amount of mineral N fertilizer used. We can model this by using system expansion. This is done by identifying the N fertilizer that will not be used and then running the production and application processes in reverse. When a process is run in reverse, all inputs become outputs and outputs become inputs. For the production of N fertilizer, fossil resources are needed and CO_2 is emitted. When we avoid the production of fertilizer, as we are using pig slurry, we save fossil resource use and CO_2 emissions. So, instead of having fossil resources as an input to fertilizer production and CO_2 emissions as an output, we model it as if CO_2 were taken from the atmosphere (input) and fossil resources were produced (output).

Slurry acidification is done to reduce NH_3 emissions to the environment [1]. When N in the slurry is not emitted as NH_3 during slurry storage in pig housing and in outdoor storage tanks, it remains in the slurry. This implies that a larger amount of N per functional unit can be used to fertilize agricultural fields, and therefore that crop yields are higher. This needs to be accounted for, which is done by assuming that the extra yield does not need to

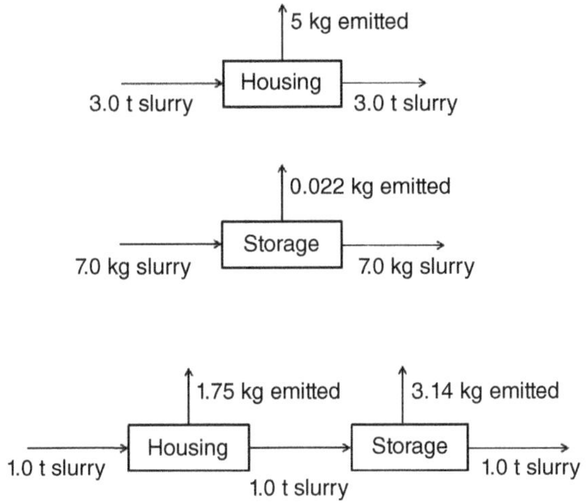

Figure 7.2.2 *Example illustrating how to adjust two processes to the functional unit of 1 ton of slurry as input.*

be produced elsewhere. Again, we use system expansion to deal with the higher yield and run the production of crops elsewhere in reverse.

For the acidification scenario, we need more chemicals than for the reference one. First, we need a strong acid – sulfuric acid (H_2SO_4) in this example – to decrease slurry pH from around 7 to 5.5. Second, we need additional lime to counteract soil acidification when acidified slurry is field applied, as compared to the need for liming when conventionally managed, non-acidified slurry is applied. We need to calculate the amount of these chemicals that is needed per functional unit.

A simple example of how to adjust processes to the functional unit is illustrated in Figure 7.2.2. Data were collected on storage of slurry in the pig housing and in an outdoor storage tank. The input to the pig housing was 3.0 ton slurry, 5 kg of a slurry constituent (e.g. N) was emitted, and the remainder was the output. Storage in an outdoor storage tank had an input of 7.0 kg slurry, 0.022 kg was emitted, and the remainder was the output. We now want to investigate the resources needed for and emissions associated with the treatment of 1 ton of pig slurry, our functional unit. Therefore, we first scale down the slurry flow for the housing, so that 1.0 ton enters, 1.75 kg is emitted, and the remainder is the output. Then, the slurry storage process is scaled up so that the input is equal to the output of the housing process. This means that 3.14 kg is emitted and approximately 1.0 ton is the output of the storage process. Emissions to the environment from the processes exemplified in Figure 7.2.1 are given in Table 7.2.1. In reality, there will be more emissions, but they have been left out for reasons of simplicity.

The global warming potentials (GWPs) for the complete conventional and acidification scenarios are calculated in Table 7.2.2. Evidently, the impact potential within the impact category GWP is larger for the conventional manure management scenario.

The three impact categories that are analyzed here are GWP, acidification potential (AP), and eutrophication potential (EP). The impact potentials for acidification and eutrophication are calculated in the same way as illustrated in Table 7.2.2 for global warming. The acidification scenario shows lower impact potentials for global warming and acidification.

Table 7.2.1 *Emissions during life cycle stages associated with one functional unit.*

	Environmental emissions				
	NH_3	N_2O	NO_3^-	CH_4	CO_2
Housing					
Conventional	1.111	0.020	–	0.428	1.59
Acidification	0.336	0.020	–	0.206	34.77
Storage					
Conventional	0.053	–	–	0.640	3.23
Acidification	0.010	–	–	0.031	0.77
Field application					
Conventional	0.850	0.095	0.014	–	94.97
Acidification	0.305	0.140	1.696	–	66.64
Avoided N fertilizer					
Conventional	–0.022	–0.066	–0.004	–0.033	–16.89
Acidification	–0.022	–0.066	–0.004	–0.033	–16.89
Avoided crop production					
Conventional	–	–	–	–	–
Acidification	–0.027	–0.022	–0.894	–0.006	–33.21

Table 7.2.2 *Global warming potentials (GWPs) of the conventional scenario and the acidification scenario.*

	Conventional scenario			Acidification scenario		
	Q_i (kg)	CF (CO_2 eq.)	IP (kg CO_2 eq.)	Q_i (kg)	CF (CO_2 eq.)	IP (kg CO_2 eq.)
CO_2	82.9	1	82.9	52.1	1	52.1
CH_4	1.04	25	25.9	0.20	25	5.0
N_2O	0.05	298	14.6	0.07	298	21.6
Total			123.4			78.7

Q_i, emission of substance i; CF, characterization factor; IP, impact potential.

The differences range from 36% for GWP to 70% for AP. However, the EP is more or less equal in the two scenarios. When looking at GWP in more detail, we can observe that higher emissions are associated with the acidification process, but these are off–set by avoided crop production, due to higher fertilizer substitution in the acidification scenario.

Table 7.2.3 shows the impact potentials, followed by the normalized impact potentials and weighted impact potentials, as well as the overall weighted impact potentials, for the example with conventional and acidification technologies for the handling of manure illustrated in Figure 7.2.2. These potentials have been calculated according to the normalization and weighting principles discussed in Chapter 7.1.

On a global scale, an average person makes a yearly contribution to global warming of 6890 kg CO_2 eq. The contribution of handling 1 ton of slurry in the conventional scenario

Table 7.2.3 *Impact potentials for the conventional scenario and the acidification scenario.*

Impact potentials			
Scenario	GWP (kg CO_2 eq.)	AP (kg SO_2 eq.)	EP (kg N eq.)
Conventional	123.4	4.9	−0.04
Acidification	78.7	1.5	0.00
Normalized impact potentials			
	GWP (PE)	AP (PE)	EP (PE)
Conventional	0.016	0.665	−0.001
Acidification	0.011	0.201	0.000
Weighted impact potentials			
	GWP (PET)	AP (PET)	EP (PET)
Conventional	0.002	0.109	0.000
Acidification	0.001	0.033	0.000
Total weighted impact potential (PET)			
Conventional	0.111		
Acidification	0.034		

GWP, global warming potential; AP, acidification potential; EP, eutrophication potential; PE, person equivalents; PET, targeted person equivalents.

Normalization references:

Global warming potential	6890	kg CO_2	Impact assessment method: ReCiPe
Acidification	7.34	kg SO_2	Impact assessment method: ReCiPe
Eutrophication	38.2	kg N	Impact assessment method: ReCiPe

Weighing factors:

Global warming potential	9.3	Impact assessment method: ReCiPe
Acidification	6.1	Impact assessment method: ReCiPe
Eutrophication	6.6	Impact assessment method: ReCiPe

is 123.4 kg CO_2 eq. The normalized global warming impact is 123.4/6890 = 0.018 person equivalents (PEs). In Table 7.2.3, we can furthermore see that the normalized AP is the highest of the three impact categories analyzed, for both scenarios.

For weighting, we need to multiply the normalized results by a weighting factor. We can see that acidification has the highest weighted impact potential. The total weighted impact potentials for the conventional and acidification scenarios indicate that the acidification technology is associated with smaller environmental impacts. This means that acidification technology could be an option worth considering for the handling of manure, from an environmental point of view.

7.2.2 Conclusion

Application of the LCA approach to compare different manure treatment options includes many assumptions, data sources, and interpretation choices, all of which will influence the outcome of the comparison. We have deliberately kept the example discussed in this chapter relatively simple, so the different assumptions and interpretations are transparent and easy to follow. Real–world cases of manure management are often much more complex. Transparency and clear description of assumptions made and methodologies applied in the reporting of the outcomes are imperative to justify conclusions about the most environmentally friendly options. Examples of such more complex, real–life cases including slurry acidification can be found in ten Hoeve et al. [2, 3], who examine the reduction in environmental impacts from slurry acidification, applied either in the animal house or during field application and in combination with mechanical separation. Other, more complex slurry and digestate treatment examples can be found in ten Hoeve et al. [4] and in Chapter 7.3.

Acknowledgments

This chapter was written with financial support from the People Programme (Marie Curie Actions) of the European Union's Seventh Framework Programme under the *ReUseWaste* Initial Training Network project (REA grant agreement no. 289887) as well from the Danish Council for Strategic Research, under the *CleanWaste* (no. 2104–09–0056) and *BioChain* projects (no. 12–132631).

References

1. Fangueiro, D., Hjorth, M., and Gioelli, F. (2015). Acidification of animal slurry – a review. *Journal of Environmental Management* **149**: 46–56.
2. ten Hoeve, M., Nyord, T., Peters, G.M. et al. (2016). A life cycle perspective of slurry acidification strategies under different nitrogen regulations. *Journal of Cleaner Production* **127**: 591–599.
3. ten Hoeve, M., Gómez–Muñoz, B., Jensen, L.S., and Bruun, S. (2016). Environmental impacts of combining pig slurry acidification and separation under different regulatory regimes – a life cycle assessment. *Journal of Environmental Management* **181**: 710–720.
4. ten Hoeve, M., Hutchings, N.J., Peters, G. et al. (2014). Life cycle assessment of pig slurry treatment technologies for nutrient redistribution in Denmark. *Journal of Environmental Management* **132**: 60–70.

7.3

Case Study: Composting and Drying & Pelletizing of Biogas Digestate

Katarzyna Golkowska[1], Ian Vázquez-Rowe[1,2,3], Daniel Koster[1], Viooltje Lebuf[4], Enrico Benetto[1], Céline Vaneekhaute[5], and Erik Meers[5]

[1] *Luxembourg Institute of Science and Technology (LIST)Network, Environmental Research and Innovation (ERIN), Belvaux, Luxembourg*

[2] *Pontifical Catholic University of Peru, Department of Engineering, Peruvian LCA Network, San Miguel, Peru*

[3] *Department of Chemical Engineering, University of Santiago de Compostela, Santiago de Compostela, Spain*

[4] *Flemish Coordination Centre for Manure Processing, Ghent, Belgium*

[5] *Faculty of Bioscience Engineering, Laboratory of Analytical and Applied Ecochemistry, University of Ghent, Ghent, Belgium*

7.3.1 Introduction

Operation of biogas plants on farming sites has become a common practice in many countries in Europe [1, 2]. Concentration of biogas plants digesting energy crops and organic waste in the regions of intensive livestock farming has contributed to a local surplus of digestate [3–5]. In such areas, a substantial part of digestate cannot be applied on fields without causing serious environmental impacts due to exceeding environmentally acceptable annual nutrient spreading limits per hectare.

The region of Flanders (Belgium) is one of those dealing with the nutrient surplus issue. Flemish authorities have introduced treatment of digestate as an obligatory processing

Biorefinery of Inorganics: Recovering Mineral Nutrients from Biomass and Organic Waste, First Edition.
Edited by Erik Meers, Gerard Velthof, Evi Michels and René Rietra.
© 2020 John Wiley & Sons Ltd. Published 2020 by John Wiley & Sons Ltd.

step [6]. In this way, the water content of digestate can be reduced, making it easy to transport, and the streams of nitrogen (N), phosphorous (P), and potassium (K) can be separated and steered by targeted use of diverse treatment products. Based on their nutritional characteristics, the treatment products can be selectively used for fertilizing or – due to their low water content – transported to other locations, if not needed in the region of their production. Additionally, digestate treatment contributes to reduction of the direct emissions related to the spreading of untreated digestate [7–9] and gains on importance in the fertilizing industry. Fertilizer producers are strongly influenced by the unstable prices of fossil fuels and are increasingly affected by the depletion of natural P resources [10].

In this context, the aim of the analysis presented in this chapter was to support policy makers and other involved parties in the search for sustainable solutions to the problem of excess biogas digestate. For this purpose, we analyzed two plants in Flanders, one using composting and the other using drying and pelletizing processes to treat biogas digestates.

7.3.2 Tunnel Composting *vs* Baseline Scenario

In this case study, we took as a baseline the common practice of spreading digestate without any pre-treatment on agricultural land. We compared this with treatment of digestate of exactly the same characteristics through a tunnel composting plant and spreading of the product on agricultural land. Figure 7.3.1 presents a simplified flow chart showing both processes.

The tunnel composting plant processes the solid fraction (SF) of digestate with a dry matter (DM) content of ca. 26% together with other biomass streams (i.e. poultry manure, organic waste, SF of pig slurry). After entering the plant, the digestate undergoes regular composting treatment. Pre-mixed biomass is placed in the aerated tunnels. Electric energy produced through photovoltaic panels is used to heat the biomass to 70 °C for hygienization purposes. After hygienizing for 1 hour, the air is blown out to the acid washers to avoid

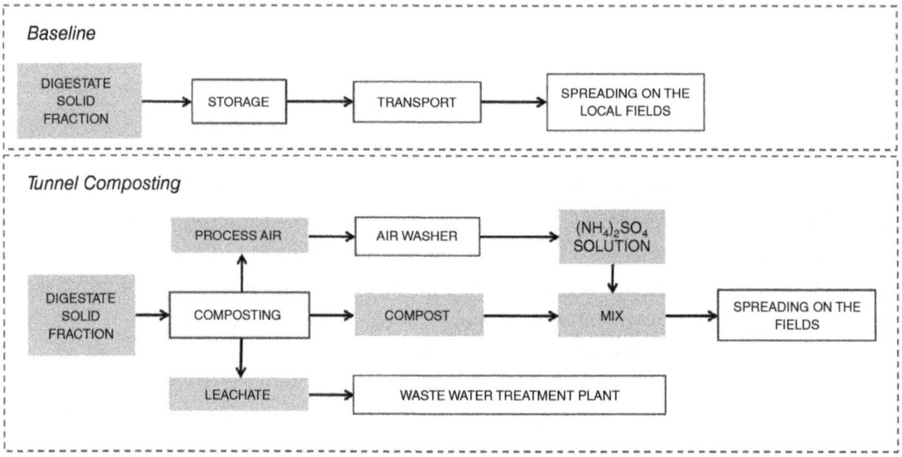

Figure 7.3.1 Simplified flow charts of the tunnel composting process and linked baseline scenario.

overheating and self-ignition of the biomass. The composting time in the tunnels varies a lot depending on the season and the demand – in summer, the compost can sometimes only be kept in the tunnels for 3 days, while in winter it can be kept for up to 6 weeks. The aeration time may therefore range between 1 hour in summer and 1 week in winter. The gaseous emissions from input storage, composting, and product storage are all treated with the acid washer followed by biofilters. Use of sulfuric acid allows for the removal of up to 98% of the ammonia from the emissions. Three intermediate products are created during the treatment: compost, ammonia sulfate solution, and leachate. However, only the first two are combined into final product, while leachate is treated in a wastewater treatment plant (WWTP). Both compost and ammonia sulfate solution can be considered as market-ready products. The rationale behind mixing them into one final product is linked to the current market situation or to client demand.

7.3.3 Drying and Pelletizing *vs* Baseline Scenario

As a second digestate treatment process, we analyzed drying and pelletizing. Again, we took spreading of digestate without any pre-treatment on agricultural land as our baseline scenario. Figure 7.3.2 presents simplified flow charts of the two processes.

The drying and pelletizing plant treats the SF of different digestate types with an average DM content of ca. 26%. The input digestate is stored in a silo that incorporates air washer treatment of gaseous emissions. In the first step of the process, the input digestate is mixed with dry digestate that has a DM content of ca. 90%. In this way, an input stream with an average DM content of ca. 56% is produced, which can further be treated in a fluidized bed drier. The dried digestate is mixed with ammonium sulfate – the product coming from the acid air washer – and subsequently pelletized, cooled, and stored in the silo. Half of the electricity needed in the plant is delivered from the Flemish electric grid, while the other half, together with the heat, is produced by the plant's own combined heat and power (CHP) unit based on the combustion of natural gas. The gaseous emissions from input storage, drying, and pelletizing are all treated with air washers. The intermediate product created

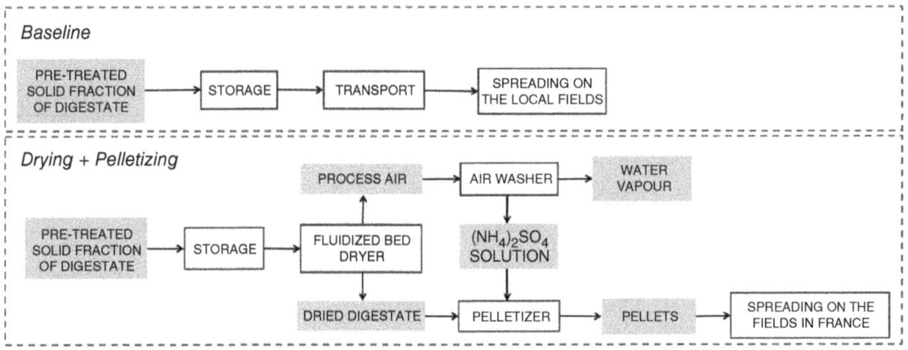

Figure 7.3.2 *Simplified flow charts of the drying and pelletizing process and linked baseline scenario.*

during the air treatment process, the ammonia sulfate solution, is mixed with dried digestate and pelletized, though it could also be used as a separate product if there were a market for it.

7.3.4 Assumptions and Calculations Related to Biomass Flow

7.3.4.1 Characteristics of the Input and Output Streams

The N, P, and K streams were calculated based on the typical characteristics of biogas digestate from Flanders. Information about the nutrient fractions assigned to the intermediate streams (e.g. SFs or liquid fractions [LFs]) was either delivered by the plant operators or extracted from [11]. Table 7.3.1 summarizes the characteristics of the input stream analyzed in both treatment processes and in their respective baseline scenarios. The calculated N, P, and K contents of the final products are provided in Table 7.3.2. More technical details regarding the assumptions and calculations can be found in [12].

Table 7.3.1 *Characteristics of the input streams and energy supply of digestate treatment plants.*

Applied technology	Composting	Drying and pelletizing
Input volume (t/a)	124 000	108 000
Input DM content	26%	56%
Input characteristics	SF of digestate	SF (50%) and dried digestate (50%)
Total N (kg/t_{FM})	6.00	15.80
P_2O_5 (kg/t_{FM})	10.2	21.30
K_2O (kg/t_{FM})	4.20	16.20
Electricity consumption (kWh_{el}/t input)	21	30
Source of electric energy	Fossil fuels + solar energy	Natural gas
Thermal energy (kWh_{th}/t input)	—	142
Source of thermal energy	—	Natural gas combustion

Table 7.3.2 *Characteristics of the selected final products of digestate treatment plants.*

Applied technology	Composting	Drying and pelletizing
Output volume (t/a)	50 000	79 000
Mixed intermediates	Compost + $(NH_4)_2SO_4$– solution	Dried digestate + $(NH_4)_2SO_4$– solution
Current spreading scenario	Export to France (transporting distance of 350–500 km)	Transcontinental exports (Asia, Africa, and South-America)
DM content	50–60%	90%
Total N (kg/t)	9.5	22.8
P_2O_5 (kg/t)	15.9	24.4
K_2O (kg/t)	6.4	31.9

7.3.4.2 Storage, Transport, and Spreading

In the baseline scenario, we assumed that digestate is stored for 3 months before it is brought to the fields. This assumption is based on common agricultural practices in the Netherlands. For the treatment scenarios, we based our calculations on shorter storage periods of 5–7 days for composting and 4–5 days for drying and pelletizing, according to information delivered by the plant operators. The emissions produced during the storage of intermediate products were modeled based on data retrieved from [13, 14]. Further details can be found in [12].

In the baseline scenario, the digestate is spread on local fields (within 100 km distance). The final use of treated digestate varies depending on the legal status and availability of land for spreading and of spreading needs and restrictions. Some treatment plants manage to distribute their treatment products locally, while others are forced (or prone) to transport over long distances, to locations where farmers are willing to acquire their products and are allowed to use them on agricultural land. An overview of the current transport situation for the analyzed treatment plants is provided in Table 7.3.2. Since the individual product distribution scenarios differ strongly, for the sake of comparability it was assumed that treated products were exported to northern regions of France. In order to reach a better transferability of the results, additional transporting distances (+25% and +50%) were included into the environmental assessment. In the baseline scenarios, digestate was assumed to be spread locally on Flemish fields within 100 km distance from the plants.

Spreading techniques and patterns were unknown and probably vary substantially depending on the end customer. However, the integration of spreading into environmental analysis was of great importance, since in this step the treatment products (or untreated digestate) for the first time have direct interactions with the natural environment, which results in direct emissions to soil, water, and air. Thus, in the analyzed case studies, a similar spreading approach was assumed for the baseline and treatment scenarios, considering the specifications (nutrient content) of the different products as well as fertilizing restrictions and common fertilizing practices.

All products were assumed to be spread on fields with winter wheat cultivations and a secondary summer crop. A sandy loam texture was chosen following the typical local soil characteristics. For this type of soil and crop, a maximum amount of 170 kg N per ha^{-1} and 75 kg P per ha is allowed to be spread in Flanders according to current legislation [15]. Additionally, different nitrogen uptake efficiency (NUE) rates were included in the study [16], following the Flemish legislation.

Outputs, in both the baseline and the treatment scenarios, are stackable, and in the analysis are assumed to be spread on fields with a hydraulic spreader. An additional sensitivity analysis was included in the life cycle assessment (LCA) study to capture the possible changes due to use of an alternative spreading technique, i.e. surface incorporation *via* harrowing

7.3.4.3 Supporting Data

The data linked to the background processes were obtained from the ecoinvent® database [17]. While the operational inputs and flow streams for different treatment systems were relatively well characterized based on the plant data provided by the managers, no primary data was available for emissions to air, soil, and water. In most cases, no further information was available with regard to the spreading of the final products on agricultural land. These data were retrieved from the literature: [14, 18–20] for the storage emissions, [19, 21, 22] for emissions related to the treatment process, and ecoinvent® guidelines [23] for spreading on fields. More details can be found in [12].

7.3.5 Goal, Scope, and Assessment Methods

The environmental evaluation of the different digestate treatment scenarios was conducted using the LCA method. Details on the LCA methodology can be found in Chapter 7.1.

The main goal of this LCA study was to assess the environmental impacts related to the treatment of digestate through tunnel composting and drying and pelletizing, including the subsequent application of the products on the fields, in comparison to the impacts attributed to direct spreading of untreated digestate on fields. We chose an attributional perspective, which considers a steady-state assessment of technologies given the interest in the environmental profile of the production system from an operational point of view [24]. A consequential approach, which would evaluate the possible effects of the full-scale introduction of digestate treatment products into the market, was out of scope for this particular study.

The selected functional unit was fixed as 1 ton of input digestate ready to be treated (in the treatment scenarios) or directly spread on the fields without treatment (in the baseline scenarios). The selection of functional unit was based on a technology comparison perspective. Such choice allowed us to identify the potential benefits and constrains of digestate treatment versus spreading of untreated stackable digestate on local or distant fields.

The data were computed using the ReCiPe assessment method [25]. Impacts linked to toxicity were analyzed with the USEtox assessment method [26]. Both environmental values – the mid-point and the single score (for details, see Chapter 7.1) – were determined and interpreted. In this way, we could analyze the influence of different treatment steps or phases on diverse environmental processes and compartments.

The soil organic matter (OM) impact category was not included in the study due to the absence of specific soil data linked to the spreading areas. More detail regarding the assessment methods can be found in [12].

The complete analyses included an assessment of the environmental impacts in the 18 different categories: climate change, ozone depletion, terrestrial acidification, freshwater eutrophication, marine eutrophication, human toxicity, photochemical oxidant formation, particulate matter formation, terrestrial ecotoxicity, freshwater ecotoxicity, marine ecotoxicity, ionizing radiation, agricultural land occupation, urban land occupation, natural land transformation, water depletion, metal depletion, and fossil depletion.

7.3.6 Results

Potential effects of implementing different scenarios (see Figures 7.3.1 and 7.3.2) in each unique impact category are presented in Figure 7.3.3. The results after applying normalization and weighting procedures in order to express the outcome of the analysis as a single score (easier to understand for non-scientific stakeholders, but includes more uncertainties and subjectivity) are shown in Figure 7.3.4.

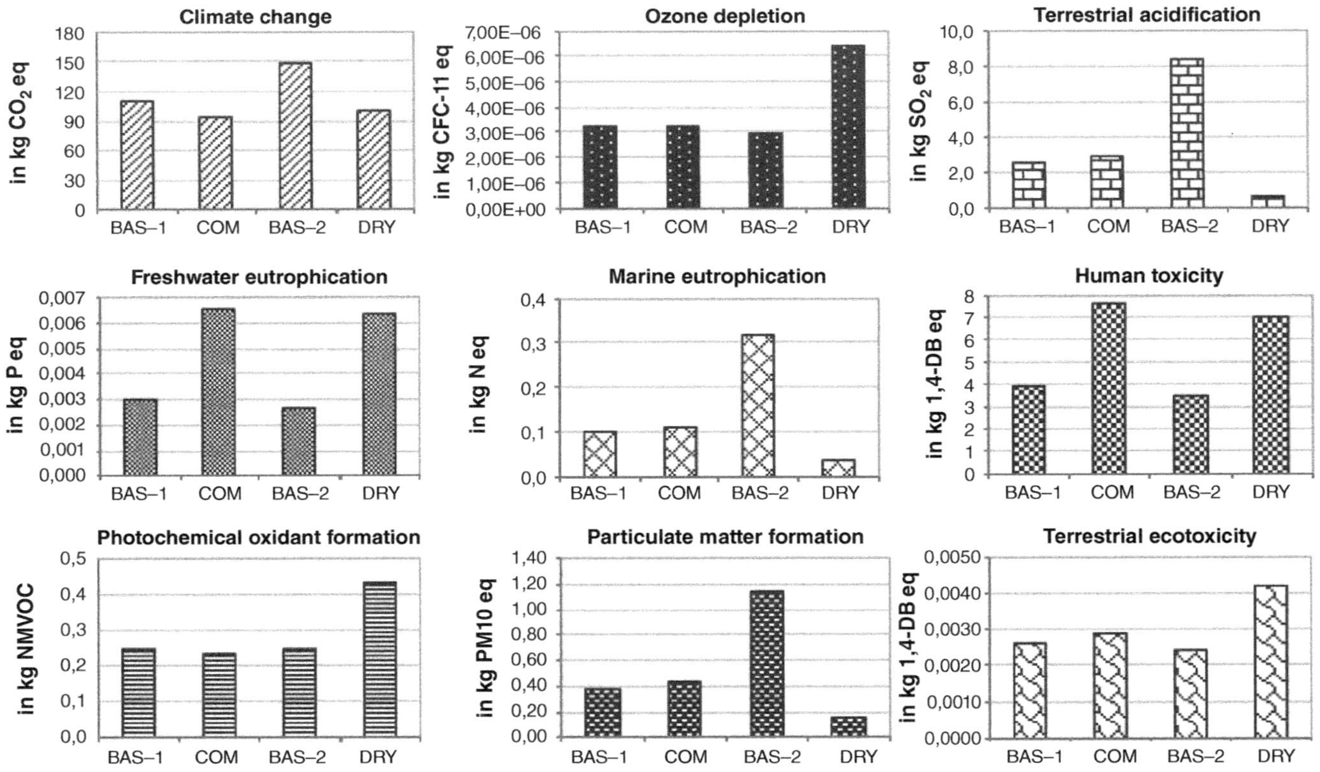

Figure 7.3.3 *Results of the mid-point assessment for the analyzed treatment systems and their respective baseline scenarios. CFC, chlorofluorocarbons; 1,4-DB, 1,4 dichlorobenzene; NMVOCs, non-methane volatile organic compounds; PM10, particulate matter 10 μm or less in diameter.*

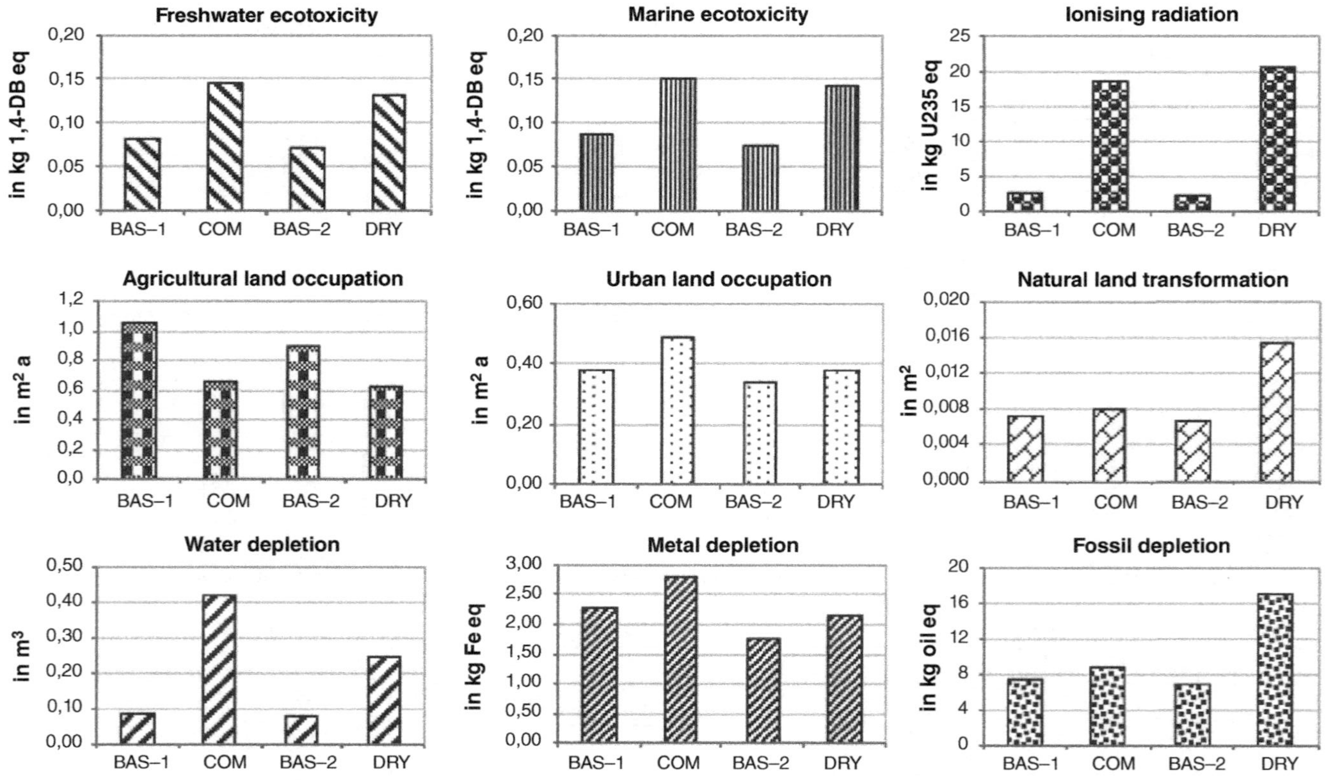

Figure 7.3.3 *(Continued)*

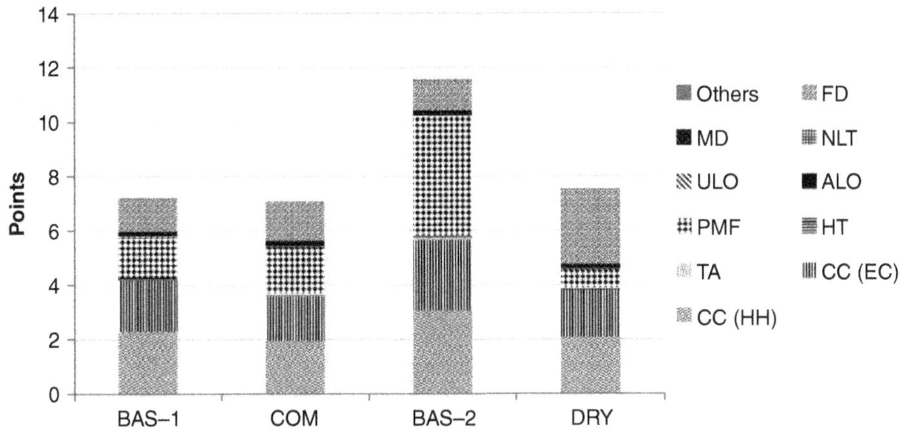

Figure 7.3.4 *Environmental impacts of the analyzed treatment systems and their respective baseline scenarios expressed as a weighted single score. FD, fossil depletion; MD, metal depletion; NLT, natural land transformation; ULO, urban land transformation; ALO, agricultural land transformation; PMF, particulate matter formation; HT, human toxicity; TA, terrestrial acidification; CC(EC), climate change ecosystem; CC(HH), climate change human health; COM, tunnel composting scenario; BAS-1, baseline scenario for tunnel composting; DRY, drying and pelletizing scenario; BAS-2, baseline scenario for drying and pelletizing.*

7.3.6.1 Tunnel Composting

If the unweighted impacts are compared, in the majority of categories a slight to clear increase in environmental burdens can be observed (see Figure 7.3.3) for tunnel composting as compared to the direct spreading of digestate on the fields (BAS-1). At the level of a single-score comparison (see Figure 7.3.4), the reduction of weighted impacts linked to climate change influencing human health is counterbalanced by increased particulate matter formation, which has a stronger effect in the same damage category, and by increased fossil depletion. Based on these results, treatment of digestate through tunnel composting brings about no significant environmental benefits as compared to the spreading of untreated digestate. Composting mainly allows for the emissions from the field works in the baseline scenario (BAS-1) to be shifted towards storage, processing, and air treatment in the treatment scenario – the steps requiring additional use of resources (e.g. water, fossil fuels, etc.).

7.3.6.2 Drying and Pelletizing

At the mid-point level (see Figure 7.3.3), we can observe increased impacts in 13 out of the 18 impact categories (e.g. ozone depletion, ionizing radiation, metal depletion) and a clear impact reduction in only 5 (e.g. climate change, marine eutrophication, terrestrial acidification). However, from a single-score perspective, drying and pelletizing represents a better solution for dealing with the stackable fraction of digestate than spreading without further treatment (see Figure 7.3.4). The total weighted environmental impacts are up to 35% lower than for the baseline scenario (BAS-2). The final single-score result is mostly influenced by reductions of impacts in the category of climate change (linked both to ecosystem quality and human health) and by lower impacts for particulate matter formation – all linked to an important reduction in ammonia emissions through treatment. On the other hand, drying and pelletizing requires use of considerable amounts of fossil fuel-based energy and quadruples the impacts of fossil depletion at the single-score level. The processes that generated the majority of the emissions and contributed most to the major impacts in the study are: storage of the input material, drying, and field operations.

7.3.6.3 Ecosystem Quality

The single-score indicator, through its specific weighting, puts increased focus on resources and human health damage. Some impacts and benefits linked to the ecosystem quality may be overlooked when comparing only single-score results. Analysis of the mid-point results revealed that for both treatment systems, much higher impacts could be observed on human toxicity, freshwater and marine ecotoxicity (less evident for terrestrial ecotoxicity), and freshwater eutrophication than with direct spreading of digestate on fields. These impacts are mainly linked to increased use of fossil energy. On the other hand, marine eutrophication and terrestrial acidification potentials can be reduced through drying and pelletizing of the digestate. This positive effect is reached though a reduction in ammonia emissions via the treatment.

7.3.6.4 Energy, Transport, and Spreading

The increase in energy intensity when introducing conversion technologies appears either to have a marginal impact on the overall (single-score) results or to be counterbalanced by impact reductions in other environmental dimensions. Impacts linked to energy use may be substantially reduced in the drying and pelletizing process through changes in the energy source from fossil fuel-based to renewable.

Independent of the treatment scenario, the simulated increase of 50% in transporting distances did not cause any significant changes in the impacts profile. For treated products, there is no need to alter spreading techniques, since the study shows that no improvement can be reached by this means.

7.3.7 Conclusion

If treatment of digestate is regionally pursued to reduce environmental impacts, it is very important already at the investment planning stage or on the development of a regional environmental strategy to investigate the environmental impacts of different technological options. For example, from an environmental perspective, the choice of locally produced heat and power from renewable resources for use in treatment plants may bring substantial reductions in environmental impacts.

Since the reduction of certain environmental impacts through treatment is often counterbalance by an increase in other burdens (mainly linked to the processing steps), it has to be clearly decided which environmental burdens are of the greatest importance for the region and should hence be reduced through such treatment in the first stage.

Treatment of biogas digestate through drying and pelletizing can be implemented to reduce environmental impacts if the main aim is to reduce the impact related to climate change or particle formation. If ecosystem quality (i.e. on water ecotoxicity, eutrophication) or human health (i.e. on human toxicity, ionizing radiation) effects are to be reduced or if damage to the ozone layer or the use of fossil fuels is to be avoided, then treatment of digestate does not represent a good solution, at least not in the scenario analyzed in this study.

Treatment of the SF of digestate through composting is not recommended, since the additional economic effort is not counterbalanced by any additional environmental benefit.

Acknowledgments

This study is an output of the ARBOR project. It was financially supported by the European Union within the framework of the INTERREG IVB NWE program. The authors wish to express their gratitude to J. de Vries and P. Hoekma (Wageningen University), T. Rehl and J. Müller (University of Hohenheim), V. Hallau (Störk GmbH), A. Meier (Amandus Kahl GmbH & Co. KG), and B. Ryckaert (Inagro) for valuable scientific exchange as well as data support. Dr. Ian Vázquez-Rowe would also like to thank the Galician Government for financial support (I2C postdoctoral student grants program).

References

1. Umweltbundesamt (2009) Characteristics of renewable energies, numbers and values. Available from: http://www.umweltbundesamt-daten-zur-umwelt.de (accessed December 23, 2019).
2. Fabbri, C., Soldano, M., and Piccinini, S. (2010). The farmer believes in biogas and the number bears this out. *Informatore Agrario* **66** (30): 63–67.
3. Prapaspongsa, T., Christensen, P., Schmidt, J.H., and Thrane, M. (2010). LCA of comprehensive pig manure management incorporating integrated technology systems. *Journal of Cleaner Production* **18** (14): 1413–1422.
4. Rehl, T. and Müller, J. (2011). Life cycle assessment of biogas digestate processing technologies. *Resources, Conservation and Recycling* **56** (1): 92–104.
5. Brouwer, F., Hellegers, P., Hoogeveen, M., and Luesink, H. (1999). *Managing Nitrogen Pollution from Intensive Livestock Production in the EU*, 23–42. The Hague: Agricultural Economics Research Institute (LEI).
6. Lebuf, V., Accoe, F., Vaneeckhaute, C., et al. (2012) Nutrient recovery from digestates: techniques and end-products. Fourth International Symposium on Energy from Biogas and Waste, Venice.
7. Holm-Nielsen, J.B., Al Seadi, T., and Oleskowicz-Popiel, P. (2009). The future of anaerobic digestion and biogas utilization. *Bioresource Technology* **100** (22): 5478–5484.
8. Golkowska, K., Vázquez-Rowe, I., Benetto, E., Koster, D. (2012) Life cycle assessment of ammonia stripping treatment of biogas digestate. Fourth International Symposium on Energy from Biomass and Waste, Venice.
9. Vázquez-Rowe, I., Golkowska, K., Lebuf, V., et al. (2013) To treat or not to treat? Environmental assessment of digestate drying technologies using LCA methodology. 13th World Congress on Anaerobic Digestion. Recovering (bio) Resources for the World, Santiago de Compostela.
10. Fixen, P.E. and Johnston, A.M. (2012). World fertilizer nutrient reserves: a view to the future. *Journal of the Science of Food and Agriculture* **92** (5): 1001–1005.
11. Bakx, T., Membrez, Y., Mottet, A., et al. (2009) État de l'art des méthodes (rentables) pour l'élimination, la concentration ou la transformation de l'azote pour les installations de biogaz agricoles de taille petite/moyenne. Bern: Federal Department for the Environment, Transport, Energy and Communication (DETEC) .
12. Vázquez-Rowe, I., Golkowska, K., Lebuf, V. et al. (2015). Environmental assessment of digestate treatment technologies using LCA methodology. *Waste Management* **43**: 442–459.

13. Oenema, O., Velthof, G. L., Verdoes, N., et al. (2000) Forfaitaire warden voor gasvormige stikstofverliezen uit stallen en mestopslagen. Wageningen: Alterra.
14. International Panel on Climate Change (2013) Climate Change 2013. The Physical Science Basis. Working Group I. Contribution to the Fifth Assessment Report of the Intergovernmental Panel on Climate Change. Intergovernmental Panel on Climate Change.
15. Flemish Manure Agency (2012) VLM fertilisation restrictions on grassland and arable land. Brussels: Flemish Manure Agency.
16. Food and Agriculture Organization of the United Nations (2011) Decree Concerning the Protection of Water against Nitrate Pollution from Agricultural Sources. May 13, 2011. Decree No.: BS13.05.2011-MAP4.
17. Frischknecht, R., Jungbluth, N., Althaus, H.J., et al. (2007) Implementation of Life Cycle Impact Assessment Methods. Ecoinvent report No. 3, v2.0. Dübendorf: Swiss Centre for Life Cycle Inventories.
18. De Mol, R.M., Hilhorst, M.A. (2003) Methaan-, lachgas- en ammoniakemissies bij productie, opslag en transport van mest. IMAG Rapport 2003-03.
19. Amon, T., Amon, B., Kryvoruchko, V. et al. (2006). Optimising methane yield from anaerobic digestion of manure: effects of dairy systems and of glycerine supplementation. *International Congress Series* **1293**: 217–220.
20. De Vries, J.W., Groenestein, C.M., and de Boer, I.J.M. (2012). Environmental consequences of processing manure to produce mineral fertilizer and bio-energy. *Journal of Environmental Management* **102**: 173–183.
21. Rehl, T. (2012) Drying technologies for digestate treatment. Emissions of ammonia and nitrous oxides during drying. Unpublished. Stuttgart.
22. Deckx, J., En Deboosere, S. (2005) Manure treatment according to the Trevi concept. Available from: https://www.trevi-env.com/files/pub/146.pdf (accessed December 23, 2019).
23. Nemecek, T., Kägi, T., Blaser, S. (2007) Life cycle inventories of agricultural production systems. Final report ecoinvent v2.0 No. 15. Swiss Centre for Life Cycle Inventories, Dübendorf, Switzerland.
24. International Organization for Standardization (2006) Environmental management – life cycle assessment – principles and framework. ISO 14040-44.
25. Goedkoop, M., Heijungs, R., Huijbregts, M., et al. (2009) ReCiPe 2008. A life cycle impact assessment method which comprises harmonised category indicators at the midpoint and the endpoint level. Report I: Characterisation, Ministry of Housing, Spatial Planning and Environment (VROM).
26. European Commission (2010). *International Reference Life Cycle Data System (ILCD) Handbook – General Guide for Life Cycle Assessment*. Brussels: Publications Office of the European Union.

Section VIII

Modeling and Optimization of Nutrient Recovery from Wastes: Advances and Limitations

8.1

Modeling and Optimization of Nutrient Recovery from Wastes: Advances and Limitations

Céline Vaneeckhaute[1], Erik Meers[2], Evangelina Belia[3], and Peter Vanrolleghem[4]

[1] *BioEngine, Research Team on Green Process Engineering and Biorefineries, Chemical Engineering Department, Laval University, Quebec, QC, Canada*

[2] *Ecochem: Laboratory of Analytical and Applied Ecochemistry, Ghent University, Ghent, Belgium*

[3] *Primodal Inc., Quebec, QC, Canada*

[4] *modelEAU, Department of Civil and Water Engineering, Laval University, Quebec, QC, Canada*

8.1.1 Introduction

Driven by economic, ecological, and community considerations, wastewater treatment plants (WWTPs) are increasingly being transformed into water resource recovery facilities (WRRFs). Successful recovery of water and energy is already widely applied. Attention is now increasingly being paid to the extraction of other valuable products from waste(water)s, in particular nutrients. Though to date many processes for the recovery of nutrients from waste(water) have been proposed and applied to varying degrees [1, 2], challenges remain in improving their operational performance, decreasing their economic costs, and recovering nutrients as pure marketable products with added value for the agricultural sector.

Previous studies provide evidence of the agronomic value of recovered products [3–6]. However, a prerequisite for marketing and recognition in environmental legislations is that these biobased fertilizers can compete with conventional fertilizer quality specifications.

Biorefinery of Inorganics: Recovering Mineral Nutrients from Biomass and Organic Waste, First Edition.
Edited by Erik Meers, Gerard Velthof, Evi Michels and René Rietra.
© 2020 John Wiley & Sons Ltd. Published 2020 by John Wiley & Sons Ltd.

The fact that WRRFs aim at delivering high-value products that can partially replace those produced by other means (e.g. chemical mineral nitrogen [N] production through the Haber–Bosch process) leads to a paradigm shift in specifications of the outputs of the facilities: no longer will these be treated wastewaters and biosolids (i.e. organic thick fractions), but rather products that must compete with what is already on the market. Previous research [3–5] demonstrated that there are still some qualitative bottlenecks for product reuse requiring further optimization. Moreover, a problem still exists in the variability of digestate (and manure) composition over time. Hence, in order to move toward more sustainable fertilization practices, it is crucial that farmers and operators are able to predict the macronutrient content (mainly N, phosphorus [P], potassium [K], and sulfur [S]) of their end products.

From the literature [2], the techniques available or under development for nutrient recovery from digestate and their recovered fertilizer products (in parenthesis) are: (i) chemical precipitation/crystallization (struvite, calcium phosphates), (ii) gas stripping (ammonia, NH_3) and absorption (ammonium sulfate [AmS] solution), (iii) acidic air scrubbing (AmS-solution), (iv) membrane separation (N/K-concentrates), (v) ammonia sorption (N-zeolites), and (vi) biomass production and harvest (biomass). Hence, in contrast to the traditional biological nutrient removal technologies used in WWTPs, such as the activated sludge (AS) system for N removal, the main unit processes considered in WRRFs rely on (changes in) the physicochemical properties of the solution. Important properties are, for example, ion activities, the chemical redox state, and the degree of solution supersaturation, in order to effectively perform precipitation, extraction, stripping, phase separation, crystallization, sorption, and filtration processes for recovery. In turn, these fundamental properties are determined by the underlying chemical solution speciation, which is the detailed distribution of total component amounts between the ionic species physically present in the system. Consequently, the production of a pure and marketable fertilizer from a complex waste matrix is challenging.

Resource-recovery treatment trains are being conceived to maximize the recovery of interesting products from waste streams (WWTP sludge, manure, etc.) at minimal cost and environmental impact. A state-of-the-art example is given in Figure 8.1.1, and further examples can be found in Verstraete and Vlaeminck [7]. However, finding the appropriate combination of technologies for a particular waste flow and the optimal operational conditions for the unit processes in the overall treatment train are key concerns.

Mathematical models have become very important tools for technology design, performance optimization, and process troubleshooting of treatment systems as they are both time- and cost-efficient [8]. Moreover, models can fill the gap between lab/pilot-scale experiments and commercial-scale operation [9]. Though a number of models of treatment facilities have been developed and extensively applied [8, 10], they focus on biological processes for the removal of nutrients (N, P) and chemical oxygen demand (COD). Fundamental physicochemical properties of the solution are not or only insufficiently accounted for, though they clearly play a major role in resource recovery.

In order to integrate nutrient and energy recovery processes in the existing mathematical model libraries (e.g. the standard activated sludge model [ASM] library provided by the International Water Association) or to predict the physicochemical waste input properties (sludge, manure, etc.) for a nutrient recovery treatment train (Figure 8.1.1), existing models must be extended so that they allow for physicochemical transformations to

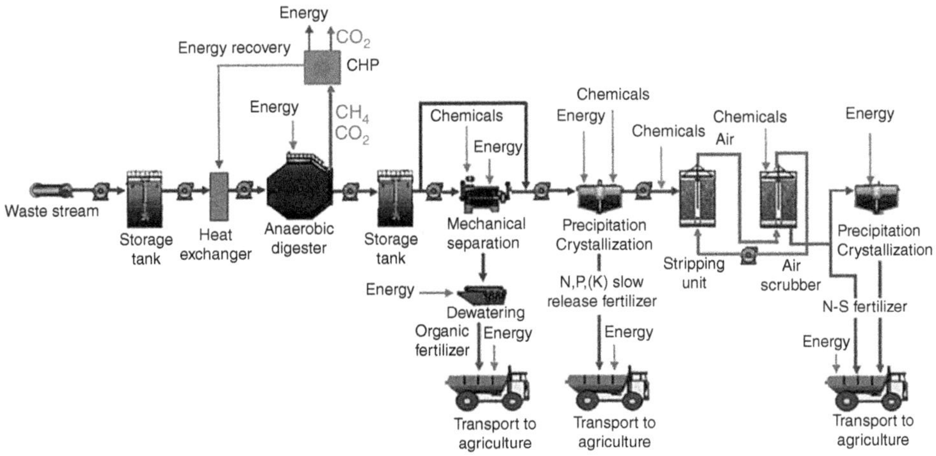

Figure 8.1.1 *Treatment train for the recovery of energy, organic fertilizer (settled solids), ammonium sulfate fertilizer, and NPK slow-release (struvite) fertilizer from a waste stream. CHP, combined heat and power generation.*

occur. Critical elements to be added include accurate descriptions of acid–base reactions, slow precipitation kinetics, liquid–gas exchange, sorption/desorption, and interactions with organics in the complex mixture of chemicals dealt with by the systems in place [11]. Moreover, WRRF models should provide information on the physicochemical characteristics (e.g. macronutrient content, particle diameter, density, etc.) of the recovered products in order to determine and control their fertilizer properties [12]. Hence, considerable research is required before integrated models become available that allow for the design and optimization of WRRFs.

Initial important steps led toward the creation a physicochemical modeling framework compatible with the current more biological process-oriented modeling frameworks [11, 13–17]. However, these modeling studies focus on the integration of simplified physicochemical models in the existing biological nutrient removal models, such as the P precipitation model in the ASM2d [15]. Moreover, their scope stops at the anaerobic digestion (AD) of WWTP sludge, where it generally aims at the prediction of uncontrolled struvite precipitation during digestion. No work has been done on the development of generic (i.e. widely applicable) models for the controlled nutrient recovery treatment train following the digester (Figure 8.1.1). Consequently, models to adequately put together a nutrient recovery treatment train of unit processes and their operating conditions in order to maximize resource recovery and fertilizer quality in a sustainable and cost-effective way are lacking, though the need for them clearly exists.

This chapter first gives a brief overview of the most important fertilizer quality specifications in order to compile essential model outputs with a focus on fertilizer commercialization, before elaborating on the limitations and advances in nutrient recovery process modeling and optimization. The focus is on AD and the most established nutrient recovery technologies available to date, as selected by Vaneeckhaute et al. [2]. Finally, based on the findings of the review, it discusses objectives and ongoing and future work in terms of nutrient recovery model (NRM) development and implementation.

8.1.2 Fertilizer Quality Specifications

8.1.2.1 Generic Fertilizer Quality Requirements

For efficient use in the agricultural sector, recovered nutrient products must have the following characteristics:

a) Consistent chemical nutrient composition and uniform distribution compatible with fossil reserve-based chemical fertilizers:

The three principal macronutrients in fertilizer mixes, so called because they are required in the largest quantities, are N, P, and K. The most common fertilizers in current use are mixtures of compounds containing these three components, conventionally expressed in terms of the relative percentages of N, P_2O_5, and K_2O by weight [18]. The nutrient ratio to be used in mixed fertilizers depends on crop requirements and soil characteristics; for example, a 1 : 1 : 1 (N : P_2O_5 : K_2O) ratio is the base fertilizer for grain crops, sugar beets, potatoes, and vegetables on soddy podzols, gray forests, and chernozems, while a 1 : 1.5 : 1 ratio is applied at planting time for grains, vegetables, and industrial crops [18].

Other important macronutrients include sulfur (S), calcium (Ca), magnesium (Mg), hydrogen (H), oxygen (O), and carbon (C) [18]. Nine additional elements are essential nutrients for many plants, albeit in small quantities. Hence, they are called micronutrients or trace elements. These are boron (B), chlorine (Cl), cobalt (Co), copper (Cu), iron (Fe), manganese (Mn), molybdenum (Mo), nickel (Ni), and zinc (Zn) [18]. They can be applied separately as micronutrient fertilizer, but are often incorporated in mixed fertilizers (ratios depend on crop and soil conditions).

b) A low salinity, sodicity, and a pH close to neutral:

The term "soil salinity" refers to the presence of electrolytic mineral solutes, most commonly Na^+, K^+, Ca^{2+}, Mg^{2+}, Cl^-, SO_4^{2-}, NO_3^-, HCO_3^-, and CO_3^{2-}, in concentrations that are harmful to many plants in the soil and in the aqueous solution within it. Salinity is usually expressed in terms of total dissolved solids (TDSs) or electrical conductivity (EC). "Soil sodicity" generally refers to the dispersion of clay resulting in deterioration of soil structure by clogging of large pores. This occurs when the sodium (Na) ion predominates in the exchange complex of the soil. Hence, the sodium adsorption ratio (SAR) – the ratio of monovalent Na over divalent Ca and Mg – is an important parameter to evaluate. Fertilizers can also affect a soil's pH. When acidity levels are too high (the is pH too low), essential minerals and nutrients may be prevented from reaching a plant's root system, the concentration of potentially toxic metal ions may increase, and the activity of soil microorganisms may be inhibited [18]. Moreover, strongly acidic or basic fertilizers may cause plant burning, while basic fertilizers may also favor NH_3 volatilization [19].

c) Desirable physical characteristics:

The important physical properties of liquid fertilizers are density and viscosity. The strength of the gelling agent (i.e. the thickener) is also critical. It should be strong enough to keep the solids in suspension, but not so strong that the liquid is too thick to be pumped and poured. Important physical characteristics of solid fertilizers are particle size, density, granule hardness, and moisture content [20–25]. The effects of these parameters and typical values for conventional chemical mineral fertilizers are presented in Table 8.1.1.

Table 8.1.1 *Desirable physical characteristics of solid fertilizers: parameter, process affected, impact, and typical values.*

Parameter	Process affected	Impact	Typical values
Particle size and distribution	Fertilizer effectiveness	• size ↓ → dissolution in H_2O ↑ → rate of nutrient release ↑ → nutrient leaching ↑	1–4 mm
	Further processing	• size ↑ → ease of washing ↑, filtering ↑, transportation ↑ and storing ↑	
	Purity	• size ↑ → surface area to volume ratio ↓ → purity ↑ • uniformity ↑ → purity ↑	
	Occupational health and safety	• granulation ↑ → fertilizer powder ↓	
	Environmental aspects	• granulation ↑ → dust formation ↓ → nutrient leaching ↓	
Density	Storing	• density ↑ → packing volume ↓	700–1570 $kg\,m^{-3}$
	Calibrating machinery		
Hardness	Handling and storing	• hardness ↑ → resistance to crushing forces, abrasion, and impacts ↑ • hardness ↑ → thermal stability ↑	crushing strength: 0.5–7.5 $kg\,cm^{-2}$
	Environmental aspects	• hardness ↑ → fertilizer dustiness ↓	
Moisture content	Handling and storing	• CRH ↑ → ease of handling and storing in wet environments ↑ • CRH ↓ → clump formation ↑, ease of spreading ↓, ease of storing ↓ (should be prevented from getting wet) • surface area ↑ → water absorption ↑	CRH: 72–92%

CRH, critical relative humidity = relative humidity of the surrounding atmosphere (at a certain temperature) at which the material begins to absorb moisture from the atmosphere and below which it will not absorb atmospheric moisture.
Source: Compiled from [20–25].

It should be noted that granule hardness also depends on the chemical composition of the fertilizer, the shape of the particles, and how much moisture the fertilizer contains. Moisture absorption in turn depends on the chemical composition of the fertilizer, environmental conditions, and the shape and size of the particles.

d) No/minimal pathogen content:

Depending on the temperature of the process, AD as a pre-treatment step before nutrient recovery can provide partial or complete pasteurization of the waste material [26, 27]. In European legislation, a product is considered pasteurized if it has been subjected to 1 hour's heating at 70 °C or an equivalent treatment (regulation EG 1069/2009 or former 1774/2002 [28, 29]), whereas in the United States and Canada the requirement to obtain class A biosolids (= potential use at home gardens, lawns, etc.) is at least 30 minutes' heating at 70 °C or an equivalent thereof [27, 30].

e) No/minimal odor:

Anaerobic (co-)digestion of organic wastes results in odor reduction [26, 27]. However, to meet regulatory standards for odor and greenhouse gas emissions, air scrubbers are required at most waste processing facilities.

f) Be authorized for registration and application in accordance with regulatory standards:

In the European Union (EU), new fertilizers should obtain a European Commission (EC) conformity certificate (= conform to the revised European Fertilizer Regulation criteria) in order to be sold throughout Europe. The revision of the EU Fertilizer Regulation 2003/2003 widens its scope to include inorganic, organomineral, and organic fertilizers, organic soil improvers, liming products, growing media, and plant biostimulant and agronomic fertilizer additives. This will considerably facilitate the introduction to the market of both organic products containing recycled nutrients (e.g. processed biosolids, digestates, composts, biochars) and inorganic recovered products (e.g. struvite, phosphates recovered from sewage sludge, incineration ash, etc.).

8.1.2.2 Points of Attention for Biobased Products

The most important physicochemical qualitative fertilizer properties that deserve attention when using biobased products in agriculture can be derived from the agronomic results presented by Vaneeckhaute et al. [3–5]. These concern (i) the pH, (ii) the salt content, (iii) the SAR, (iv) the macronutrient (N, P, K, S, Ca, Mg) and (organic) carbon content, (v) the macronutrient use efficiency, and (vi) impurities (e.g. iron [Fe] and aluminum [Al] compounds). Important factors determining a product's economic value, next to its nutrient content, are its density (for liquid fertilizers) and particle size (for granular fertilizers). Hence, WRRF models should allow the accurate prediction of these product characteristics under variable operating conditions and variable input compositions.

8.1.3 Modeling and Optimization: Advances and Limitations

This section describes in more detail the advances and limitations in modeling and optimization of AD and the most established nutrient recovery technologies selected by Vaneeckhaute et al. [2]: P precipitation/crystallization, NH_3 stripping and absorption, and acidic air scrubbing.

8.1.3.1 Anaerobic Digestion

The reaction system in AD is complex, with a number of sequential and parallel steps. These reactions can be divided into biochemical reactions, which act on the pool of biologically available organic components, and physicochemical reactions, which are not biologically mediated and encompass liquid–liquid reactions (i.e. ion association/dissociation), gas–liquid exchange (i.e. gas transfer), and liquid–solid transformation (i.e. precipitation and dissolution of ions). AD is affected by several operating conditions, such as the specific characteristics of the waste stream, temperature, pH, macro- and micronutrients, inhibition (NH_3, volatile fatty acids [VFAs], shock loading [i.e. sudden and drastic increase of load]), toxicity (e.g. H_2S), retention time, mixing conditions, and feeding strategy [31–34]. Monitoring VFAs and alkalinity during digestion is particularly important for efficient digester process control because the acid/alkalinity ratio will change before the pH begins to drop, which is fatal for methanogenic bacteria [35]. Several authors underline the importance of modeling the physicochemical system in anaerobic processes. The following arguments are used:

a) A number of biological inhibition factors can be expressed physicochemically, such as pH, free acids and bases, and dissolved gas concentrations [36].
b) Major performance variables such as gas flow and carbonate alkalinity are dependent on the correct estimation of physicochemical transformations [36].
c) Often, pH control with a strong acid or base is the major operating cost. In this case, the control setpoint (pH) must be calculated from the physicochemical state [36, 37].
d) The acid–base subsystem is vitally important to calculating gas transfer (lots of gases are also acids or bases), while gas transfer has a significant impact on the acid–base subsystem through its effect on pH [11].
e) Chemical speciation of major solutes in digestion is required to, for example, understand the toxicity of NH_3 and VFAs and to mitigate uncontrolled struvite precipitation in the reactor, piping, and equipment [32, 33].
f) Precipitation processes are critical in modern waste(water) treatment, as they describe the behavior of P in all stages (e.g. struvite formation), especially during digestion of sludge from enhanced biological phosphorous removal (EBPR) [38], precipitation of Fe and Al with P after addition of Fe/Al-salts [39], and Ca and Mg scaling [11, 40–42].
g) The presence of precipitates provides a slow buffer to changes in pH, and emerging processes such as P recovery are highly dependent on metal ion precipitation [40, 43].

The lack of ion activity correction (at low conductivity), ion pairing (at high conductivity), precipitation, and P modeling are assumed to be the main limitations of Anaerobic Digestion Model No. 1 (ADM1), which is the generic AD model currently provided by the International Water Association [11, 37]. Due to a lack of activity corrections, ADM1 also fails to predict pH correctly. pH is vital for proper precipitation prediction [44]. In recent years, some attempts have been made to improve the predictability of AD by integrating physicochemistry, mainly in order to overcome the nuisance problem of struvite precipitation [17, 33, 40, 42, 45–47]. However, to date, no generic approach has been agreed upon in order to incorporate solution speciation based on ion activity and the kinetics of precipitation of multiple minerals that share common ions, as well as competing reactions such as ion pairing, in an anaerobic digester for organic waste treatment.

Furthermore, the ADM1 application has practical problems related to the characterization of the digester feedstock and the associated model definition of the enzymatic disintegration and hydrolysis steps. As biological wastes are heterogeneous and dynamically changing in composition, it is difficult to find unique parameter values that are applicable to all possible combinations and ratios of wastes together with decaying anaerobic biomass. Since ADM1 was published, several methods have been developed to overcome such parameter estimation and substrate fractionation problems. These are based on elemental analysis [48–50], physicochemical analysis [51], the conversion of other model outputs (e.g. ASM outputs) to state variables used in ADM1 [52–54], and anaerobic respirometry [55, 56]. A first step toward a more advanced dynamic interface to ADM1 in order to simulate the digestion of a combination of waste streams by evaluating their independent hydrolysis rates is the integrated solid waste co-digestion (GISCOD) modeling tool developed by Zaher et al. [50].

8.1.3.2 Phosphorus Precipitation/Crystallization

The ability to predict the P precipitation potential from a waste(water) flow is an important consideration for designers and operators in order to determine the feasibility and economics of nutrient recovery (e.g. as struvite [$MgNH_4PO_4 : 6H_2O$]) and for the subsequent design and operation of reactors for P crystallization. For good product quality control, it is essential to know the best conditions under which the target precipitation reaction is likely to occur. Based on experimental studies conducted to date, P recovery through crystallization has been found to be mainly affected by the following operating factors: solution supersaturation [57, 58], pH [59–61], Mg : P molar ratio in the case of struvite [60–64], crystal retention time [62], recycle ratio (i.e. the ratio between the feed flow and the recycle flow) [62], reactor seeding [58, 65], temperature [32, 66], and turbulence and mixing [58, 67]. The crucial value to control is the supersaturation value, followed by the total crystal surface, retention time, and flow pattern. To optimize the size of recovered crystals, researchers have often tested crystallization on to seed materials such as sand [68] and preformed crystals [69]. Seeding clearly impacts the final particle size distribution (PSD).

Struvite solubility in particular is widely studied. However, the conditions reported to be optimal for struvite crystallization vary from publication to publication [70–72], and to date the purity of the product precipitated cannot be guaranteed due to the availability of foreign ions and co-precipitation. Schneider et al. [73] underlined the importance of modeling solution thermodynamics and the presence of foreign ions in nutrient recovery systems, as the constituent species concentrations, the solution pH, and ionic strength directly determine the generation of supersaturation.

Though a substantial number of models have been developed for P precipitation and crystallization, these mostly focus on the precipitation of one target compound (e.g. struvite) [74–78]; a couple of examples are $Ca_5(PO_4)_3OH$ [79] and $FePO_4$ [15, 39], which account only for the solubility product and supersaturation ratio of the target species (e.g. for struvite based on the three main constituents, Mg^{2+}, PO_4^{3-}, and NH_4^+). Besides the equilibrium model developed by Lee et al. [80] for struvite formation with simultaneous Ca precipitation, no other models have been described in the literature for simulation of P recovery as a pure target product (e.g. struvite) under the competitive inhibition of other ions (e.g. Ca and Fe ions). Moreover, no models applied for nutrient recovery from waste(water) by means of precipitation/crystallization account for supersaturation ratios and the solubility products of multiple competing precipitation reactions based on a detailed chemical solution speciation, including for examplePO_4^{3-}, HPO_4^{2-}, Ca^{2+}, Fe^{3+}, and Al^{3+} ionic species, and the time-dependent behavior of supersaturation. As a result, given the complex nature of (digested) waste and the multiple competing processes (complex formation, ion exchange, co-precipitation, etc.), the current models often overestimate phosphorus removal efficiencies [81, 82] or underpredict the precipitate formation potential [83, 84]. Hence, pilot testing is still indispensable for proper design and process performance evaluation. Besides, a product's fertilizer potential is affected by concurrent precipitation. For example, Vaneeckhaute et al. [5] observed that $FePO_4$ sludge is not an interesting product in terms of P release for agricultural crop growth because of its high P-binding capacity, in contrast to the valuable slow-release struvite fertilizer.

Next to the simplified thermodynamic approach of the current models for P precipitation/crystallization, another limitation is that most of the present studies focus on the development of a thermodynamic chemical equilibrium model. Much less work has been carried out to couple solution thermodynamics to the fundamental kinetics of P crystallization, which involves the relatively slow processes of nucleation, crystal growth, agglomeration, and breakage, and hence should be modeled dynamically [64, 81, 85–89]. The available studies focus on the precipitation of N-struvite only and are limited to crystal growth and occasionally nucleation, described by empirical power laws. Only Le Corre et al. [64, 89] have studied aggregation, and they suggest that aggregation without the addition of coagulants is not significant in the case of struvite. However, Galbraith et al. [87] proposed a generic model for P crystallization driven by the three key mechanisms of nucleation, growth, and aggregation. Nevertheless, this model is likely too complex for direct application to real-world, large-scale nutrient recovery systems, since it employs complex crystal population dynamics. To date, it has only been tested on synthetic solutions containing principal reactants, while it is expected that the presence of competing ions and suspended particles in the complex matrix of (digested) waste flows will significantly influence the rate of crystal formation and purity [90]. Hence, as in all other kinetic modeling efforts described so far, a significant problem of this model is that the underlying kinetic equations are all driven by solution supersaturation, which is not adequately accounted for due to the simplified thermodynamic approach (even though the kinetics are described in a very complex way).

Ideally, a more easily applicable generic modeling approach for real waste flows would build up a detailed chemical speciation model to correctly predict supersaturation of multiple precipitates, coupled to a simplified classical kinetic model (e.g. [91]) to describe the main slow mechanisms involved in crystal formation. As such, the model can provide accurate information on product quantity and quality, including purity and particle size, which is essential to obtaining marketable end-products acceptable for agricultural use.

8.1.3.3 Ammonia Stripping and Absorption

The operational pH and temperature are the most important factors in the NH_3-NH_4^+ equilibrium. Above pH 10, NH_3 predominates in the solution and an increased temperature enhances NH_3 stripping [92]. However, as with P precipitation, the optimal conditions for NH_3 removal reported in the literature are very variable. Lemmens et al. [93] state that for optimal NH_3 removal, the pH of the liquid fraction (LF) should be around 10 and the temperature around 70 °C. However, Liao et al. [94, 95] find no appreciable improvement in NH_3 removal from changing the temperature at high pH (10.5–11.5), while other studies show that "complete" removal without chemical addition is possible at a temperature of 80 °C [96] or 60 °C [97]. The latter would be caused by the transformation of bicarbonate (HCO_3^-) to carbonate (CO_3^{2-}) at higher temperatures, resulting in a pH increase through carbon dioxide (CO_2) stripping. Hence, the alkalinity of the feed flow is also very important in determining optimal process conditions and costs. This underlines the essence of modeling treatment trains for resource recovery, as alkalinity is also a key operational factor for AD. Hence, process optimization of these systems is interconnected.

The rate of mass transfer of a compound is proportional to the contact area, which is determined by the specific surface area ($m^2\ m^{-3}$) and the degree of wetness of the packing material (if there is any), which, in turn, is affected by the means of wetting, such as trickling, spraying, or submerging, and the liquid flow rate [98, 99]. In practice, air stripping in packed towers typically leads to scaling and fouling of the packing material due to reactions between CO_2 in the air and some metal ions in the waste(water). Slaked lime is therefore often added to adjust the pH of the waste flow and reduce the carbonate content before it enters the stripping tower [100, 101]. Alternatively, calcium carbonate ($CaCO_3$) can simultaneously be recovered in the stripping column [102]. In case of high buffering capacity, an additional CO_2 stripper before the NH_3 stripping process might be economical (RVT Process Equipment, Steinwiesen, Germany, pers. comm.).

In order to improve mass transfer and eliminate scaling problems, some new gas–liquid contactors without packing have been used for gas–liquid operation in recent years, such as the water-sparged aerocyclone [103–105] and bubble column reactor (BCR) [98, 106]. In these systems, the gas film resistance is decreased and the gas–liquid contact area increased. This accelerates the mass transfer of NH_3 (which has a low Henry coefficient or very high solubility) from the liquid to the gas phase [106, 107].

Mathematical models are particularly important for process optimization and scale-up of stripping systems, in order to fill the information gap between lab/pilot and full scale [9, 98]. For example, most laboratory stripping experiments use blowers instead of fans because there are no appropriate fans available for lab-scale NH_3 stripping experiments. At a commercial scale, though, it is possible to achieve a large volume of airflow with low air-pressure fans, which require minimal electrical consumption.

Previously reported theoretical modeling studies are mostly based on empirical methods, such as mass transfer correlation of the volatile compound under study, and assume that the flow in stripping towers is homogenous (i.e. spatially independent) [108]. Collivignerelli et al. [98] and Powers et al. [106] developed steady-state models for BCRs and accounted for the entrainment of NH_3 in bubbles, as well as for the temperature and pH dependency of the Henry coefficient. Yu et al. [9] showed that the distribution of temperature and liquid volume fraction in a packed tower is not homogenous. They underlined the importance of considering liquid residence time in one's design. Little effort has been made to connect mass and heat transfer with the chemical reactions (other than NH_3-NH_4^+ equilibrium) occurring in stripping towers for N recovery. For instance, $CaCO_3$ precipitation, which mainly causes the previously mentioned blocking and scaling problems, may occur and should be quantified. Also, the simultaneous removal of other volatile compounds, such as amines, from the waste flow should be considered in view of odor emission control. Moreover, a series of strippers could be implemented for combined N and S recovery [109]. Yet, again, because of the complex nature of (digested) waste material, modeling the chemical speciation and solution thermodynamics – next to an appropriate rate-based mass transfer model – is essential to improving process design, operational performance, and recovered product quality.

8.1.3.4 Acidic Air Scrubbing

The equipment most often used for acidic air scrubbing to capture NH_3 is packed towers and venturi scrubbers. The pH is controlled, usually at a value below 4, by addition of acid (typically sulfuric acid, H_2SO_4) to the recirculation water in order to enhance the mass transfer and absorption processes. A minimum water discharge rate is required to prevent precipitation of ammonium sulfate ($(NH_4)_2SO_4$, AmS) on the packing column, which causes blocking and clogging of the system. At an NH_3 removal efficiency of 95%, the discharge water production is about 0.2 m^3 kg^{-1} NH_3 [99]. Melse and Ogink [99] reported that the AmS concentration of acidic air scrubbers is usually controlled at a level of ±150 g AmS l^{-1}, which is roughly 40% of the maximum solubility at pH 4. However, commercial processes combining stripping and NH_3 absorption seem to achieve more concentrated solutions of 25% AmS (*ANAStrip*, GNS, Halle, Germany, pers. comm.), 38% AmS (RVT Process Equipment, Steinwiesen, Germany, pers. comm.), or 40% AmS (*Amfer*, Colsen, Hulst, the Netherlands, pers. comm.). From a review of the literature and contact with technology providers, process control with pH measurement and automatic water discharge appears to be sufficient to guarantee adequate NH_3 recovery. In order to simultaneously capture sour reacting components (e.g. hydrogen sulfide [H_2S] and CO_2), a two-stage scrubber is often used in the field of gas purification, capturing these compounds in an alkaline aqueous solution such as sodium hydroxide (NaOH) [110]. Using a subsequent bioreactor, elemental S can simultaneously be recovered after biological oxidation [111]. Furthermore, a recent study [112] demonstrated the economic viability of integrating NH_3 stripping, absorption, and biogas purification (H_2S and CO_2 absorption) for both N and S recovery.

The rate of mass transfer of a compound to the liquid phase is proportional to the concentration gradient between the gas and liquid phases. The concentration in the liquid phase is determined by the component solubility, the rate of water discharge and fresh water supply, the pH, and, if applicable, the transformation of NH_3 (or H_2S, etc.) into other compounds [113]. Furthermore, the empty-bed air residence time, which can be calculated by dividing the reactor volume by the air flow rate, determines the total mass transfer and depends on the solubility. Simplified models have been developed in the literature to predict the performance of a counter-current gas absorption tower based on the preceding mass transfer principles [113–115]. Manuzon et al. [113] designed and optimized a prototype acid spray wet scrubber for single- and multi-stage NH_3 absorption, while commercial-scale H_2S stripping columns have also been modeled [116]. Nevertheless, no generic absorber model with the purpose of nutrient recovery that takes into account ion activity and other physicochemical reactions than the targeted gas–liquid equilibrium (e.g. NH_3-NH_4^+) has been proposed. For instance, precipitation of AmS and simultaneous absorption/volatilization of multiple compounds is not considered.

Finally, several technologies are commercially available for the crystallization of AmS. These involve both evaporative crystallizers for undersaturated solutions (external heat is required to obtain supersaturation) and reaction crystallizers for concentrated reactants (use of dissolution and reaction heat only). Examples are the Oslo-type and Draft Tube Baffled (DTB)-type crystallizers [117]. Currently 80–90% of AmS crystallizers operate in evaporative mode and DTB is recommended [118], resulting in crystal sizes of 2.0–2.4 mm. Mathematical models of FBR crystallizers for AmS production from NH_3 and H_2SO_4 have been reported in literature [119–121], because of their long record of industrial application. Usually, the process is described in terms of dynamic heat and mass balances, combined with a dynamic population balance, which describes the crystal size distribution (CSD). The nucleation rate is traditionally described using an empirical power law based on supersaturation, which is the main driving force [120]. However, as supersaturation is difficult to measure, it is often replaced by the growth rate, which is also a function of supersaturation, but can be estimated experimentally. Because such experiments are expensive and time-consuming, models that are able to adequately predict solution supersaturation or sensors that allow online monitoring of this control parameter would be valuable tools for crystallization and product quality optimization (see earlier).

8.1.4 Modeling Objectives and Further Research

8.1.4.1 Definition of Modeling Objectives

From the preceding literature review, it is clear that a first important research objective is the development of generic integrated biological-physicochemical three-phase process models for AD and for the most established nutrient recovery systems available to date. These models should include an accurate chemical solution speciation, as well as reaction dynamics, though a balance must be found between model accuracy, complexity, and simulation times (see later).

Second, the generic model library should be applicable as a tool for process optimization of single-nutrient recovery systems, as well as for determination of optimal unit process combinations, in order to maximize resource recovery (nutrients, energy) from a particular waste stream and minimize energy and chemical requirements. Modeling of treatment trains is important, as a combination of suboptimal unit processes may lead to an overall optimal output. Moreover, modeling of treatment trains can help identify bottlenecks in operational strategies and treatment processes upstream. For example, the use of Fe or Al salts to improve separation/dewatering has a huge impact on the bioavailability of P [5] and limits the potential for its recovery as a valuable fertilizer product downstream.

In summary, a common base for modeling of nutrient recovery processes is required, which should not only facilitate process and treatment train implementation, but also serve as a generic framework that makes outcomes more comparable and compatible.

8.1.4.2 Toward a Generic Nutrient Recovery Model Library

In order to meet these needs and objectives, the development and implementation of a generic NRM library has been described by Vaneeckhaute et al. [122]. The generic models included in this library are based on mass balances to describe physicochemical and biochemical transformation and transport processes, as well as on accurate calculations of

Table 8.1.2 Recommended model outputs for each unit process.

Model output	AD	Prec	Strip	Scrub
Biogas volume and composition	X	—	—	—
Consumables (air, chemicals, heat, etc.)	X	X	X	X
Fertilizer quantity (mass/volume)	X	X	X	X
Fertilizer (and/or effluent) quality				
Dry weight content/density	X	X	(X)[a]	X
Macronutrient content (N, P, K, S, Ca, Mg)	X	X	X	X
Macronutrient use efficiency (N, P, K, S)	X	X	(X)[a]	X
Micronutrient content	X	X	(X)[a]	(X)[b]
Organic carbon content	X	X	(X)[a]	—
Particle size	—	X	—	—
pH	X	X	(X)[a]	X
Purity	—	X	X	X
Salt content	X	X	(X)[a]	X
Scaling potential	X	X	X	X

AD, anaerobic digestion; Prec, precipitation/crystallization; Strip, stripping; Scrub, scrubbing.
[a]Values in parenthesis refer only to the effluent quality from the stripping unit, not to the stripped gas.
[b]Depending on the origin of the acid used in the air scrubber, micronutrients can be taken into account or ignored.

water chemistry in order to correctly define solution speciation and the driving forces for component transformation (e.g. supersaturation). A dynamic modeling approach (one that accounts for time-dependent changes in the state of the system) is used in the NRM library, because the models must be applicable to real-time situations and variable operating conditions, such as (i) periodical load variations (e.g. absence of operators at weekends/evenings, seasonal variations, etc.), (ii) individual disturbances (e.g. rain events and incorrect control manipulations), and (iii) systems that are operated intermittently or cyclically, as with multiple-nutrient recovery processes (e.g. intermittent aeration in stripping systems) and (semi-)batch processes to obtain target fertilizer specifications (e.g. a target AmS concentration via acidic air scrubbing).

The waste treatment system to be described consists of interactions between three phases: liquid, solid, and gas. Both heterogeneous transfer reactions that occur between phases (gas transfer, liquid–solid transfer) and homogenous transformation reactions that occur within a single phase (biodegradation, acid–base chemistry, ion pairing) are taken into account in the NRM models. Model outputs involve fertilizer quality and quantity measurements. Based on the literature review, desired outputs for each resource recovery process discussed are compiled in Table 8.1.2.

Through the interrelated chemical and biological processes, the mass balance- and continuity-based process models quantitatively fix the relationship between all components considered in the system so that the system's output is governed completely by the input waste stream characteristics and the applied process conditions. Factors that are expected to influence the process outputs and hence which are included as NRM inputs are presented in Table 8.1.3.

As such, the mathematical NRMs can show how a change in the preceding model inputs will impact on the value of the process outputs. Thus, the models can be used as a valuable tool for process optimization.

Table 8.1.3 *Factors that potentially influence the model outputs per unit process and references for corresponding equations.*

Influencing factor	AD	Prec	Strip	Scrub	References
Aeration (air flow rate)	–	(X)[a]	X	X	[27]
Alkalinity (addition/removal)	X	–	X	–	[123]
Bubble size	–	–	X	X	[124]
Chemical pH adjustment (acid/base dose)	X	X	X	X	[123]
Feed composition	X	X	X	X	–
Feed flow rate	X	X	X	X	–
Heating (temperature)	X	X	X	X	[27]
Mixing ($\overline{G}$ value)[b]	X	X	–	–	[123]
Reactor seeding	–	X	–	–	[73]
Reactor height	–	–	X	X	[124]
Residence time (liquid, air, crystals)	X	X	X	X	[27]

AD, anaerobic digestion; Prec, precipitation/crystallization; Strip, stripping; Scrub, scrubbing.
[a]Values in parenthesis represent the use of air instead of chemicals for pH increase.
[b]$\overline{G}$ is the root mean square velocity gradient [Time^{-1}], which depends on the power input [125].

8.1.4.3 Numerical Solution

The numerical solution of model equations is a critical step in combining biological and chemical reactions because of the stiffness that arises when considering reactions with very different conversion rates (i.e. where the range of time constants is large) [16]. Because they are much more rapid, homogenous physicochemical reactions can be assumed at equilibrium, as compared to the time scale of heterogeneous physicochemical reactions and biological reactions [11]. Therefore, for fast reactions, the steady-state solutions are perfectly adequate and a thermodynamic equilibrium approach may be applied. However, for slower reactions, a kinetic approach must be used because we are interested in the time-variable or dynamic variation of the constituents. This makes the simulation of such a system challenging, and in order to avoid excessively long simulation times, one needs to be somewhat creative when implementing the model.

In the case of dynamic models, two possible solution procedures have been applied to date for stiff systems:

1. The ordinary differential equation (ODE) approach: All reactions are calculated simultaneously using ODEs, as in Musvoto et al. [46, 126] and Sotemann et al. [47].
2. The differential algebraic equation (DAE) approach: Slower reactions are represented by ODE and fast reactions are calculated using algebraic equations (AEs) at each iteration step, as in Batstone et al. [11], Brouckaert et al. [40], Volcke et al. [127], and Rosen et al. [128]. The modeler can choose between using a tailored code to solve the water chemistry and using an external software tool such as PHREEQC [129] or Visual MINTEQ [130] at each iteration step.

The use of an external geochemical software tool with designated thermodynamic databases to calculate chemical speciation and pH is interesting. Tools such as PHREEQC and MINTEQ are generally accepted for use in equilibrium water quality modeling and have a dedicated and proven solver for chemical speciation calculations. However, simulation times using the full PHREEQC/MINTEQ thermodynamic databases for chemical speciation may be longer than when using an integrated code [16]. On the other hand, the latter may be less flexible and complete. Hence, an important challenge exists in the development of an efficient methodology for solving the (stiff) equations in NRMs. A compromise should be found between model accuracy and simulation time.

Vaneeckhaute et al. [122] developed an efficient procedure for calling PHREEQC-selected chemical speciation outputs from Modelica coded kinetic transformation models using the Tornado/WEST software kernel. A reduction of execution time was established at two critical points during model simulations: (i) during uploading and reading of the database and input files (through PHREEQC model reduction); and (ii) during the transfer of data between PHREEQC and Tornado/WEST (through tight model coupling). A three- to fivefold improvement of model simulation speeds was obtained using the developed reduced models as compared to the full PHREEQC and MINTEQ databases. Hence, the proposed procedure provides an efficient methodology for the numerical solution of future NRM applications.

8.1.5 Conclusion

The construction of a generic NRM library based on detailed chemical solution speciation and reaction dynamics and aiming at fertilizer quantity and quality as model outputs is of preliminary importance in facilitating future nutrient recovery processes and treatment train implementations. The library should be applicable as a generic tool for process optimization of single-nutrient recovery systems, as well as for determination of optimal unit process combinations, in order to maximize resource recovery (nutrients, energy) from a particular waste stream and minimize energy and chemical requirements. A numerical solution should be regarded as a critical step in resource recovery modeling. Starting from the modeling objectives outlined in the present chapter, the development and implementation of a generic NRM library, including an efficient procedure for numerical solution, has been proposed by Vaneeckhaute et al. [122].

Acknowledgments

This work has been funded by the Natural Science and Engineering Research Council of Canada (NSERC) through the award of an NSERC Discovery Grant (RGPIN-2017-04838) and by NSERC, the Fonds de Recherche sur la Nature et les Technologies (FRQNT), and Primodal Inc. through the award of a BMP Innovation scholarship (BMP Innovation 178263) to the first author. Peter Vanrolleghem holds the Canada Research Chair in Water Quality Modeling.

References

1. Lebuf V, Vaneeckhaute C, Accoe F, et al. Nutrient Recovery from Digestate in North West Europe. 2013.
2. Vaneeckhaute, C., Lebuf, V., Michels, E. et al. (2017). Nutrient recovery from bio-digestion waste: systematic technology review and product classification. *Waste and Biomass Valorization* **8**: 21–40.

3. Vaneeckhaute, C., Meers, E., Michels, E. et al. (2013). Closing the nutrient cycle by using bio-digestion waste derivatives as synthetic fertilizer substitutes: a field experiment. *Biomass and Bioenergy* **55**: 175–189.
4. Vaneeckhaute, C., Ghekiere, G., Michels, E. et al. (2014). Assessing nutrient use efficiency and environmental pressure of macro-nutrients in bio-based mineral fertilizers: a review of recent advances and best practices at field scale. *Advances in Agronomy* **128**: 137–180.
5. Vaneeckhaute, C., Janda, J., Meers, E., and Tack, F.M.G. (2016). Phosphorus use efficiency in bio-based fertilizers: bioavailability and fractionation. *Pedosphere* **26**: 310–325.
6. Velthof GL. Mineral Concentrate from Processed Manure as Fertiliser. 2015. Available from: https://edepot.wur.nl/352930 (accessed December 23, 2019).
7. Verstraete, W. and Vlaeminck, S.E. (2011). ZeroWasteWater: short-cycling of wastewater resources for sustainable cities of the future. *International Journal of Sustainable Development and World Ecology* **18**: 253–264.
8. Rieger, L., Gillot, S., Langergraber, G. et al. (2012). *Guidelines for Using Activated Sludge Models*. London: International Water Association.
9. Yu, L., Zhao, Q., Jiang, A., and Chen, S. (2011). Analysis and optimization of ammonia stripping using multi-fluid model. *Water Science and Technology* **63**: 1143–1152.
10. Henze, M., Gujer, W., Mino, T., and van Loosdrecht, M.C.M. (2000). *Activated Sludge Models ASM1, ASM2, ASM2d and ASM3*. London: International Water Association.
11. Batstone, D.J., Amerlinck, Y., Ekama, G. et al. (2012). Towards a generalized physicochemical framework. *Water Science and Technology* **66**: 1147–1161.
12. Vanrolleghem, P.A. and Vaneeckhaute, C. (2014). Resource recovery from waste water and sludge: modelling and control challenges. In: *IWA Specialist Conference on Global Challenges: Sustainable Wastewater Treatment and Resource Recovery, Conference Proceedings*. Kathmandu: International Water Association.
13. Fernández-Arévalo, T., Lizarralde, I., Grau, P., and Ayesa, E. (2014). New systematic methodology for incorporating dynamic heat transfer modelling in multi-phase biochemical reactors. *Water Research* **60**: 141–155.
14. Grau, P., de Gracia, M., Vanrolleghem, P.A., and Ayesa, E. (2007). A new plant-wide modelling methodology for WWTPs. *Water Research* **41**: 4357–4372.
15. Hauduc H, Takacs I, Smith S, et al. A dynamic model for physicochemical phosphorus removal: validation and integration in ADM2d. In: 4th IWA/WEF Wastewater Treatment Modelling Seminar (WWTmod2014), Conference Proceedings, Spa; 2014.
16. Lizarralde I, Brouckaert CJ, Vanrolleghem PA, et al. Incorporating water chemistry into wastewater treatment process models: Critical review of different approaches for numerical resolution. In: 4th IWA/WEF Wastewater Treatment Modelling Seminar (WWTmod2014), Conference Proceedings, Spa; 2014.
17. Takacs, I., Murthy, S., Smith, S., and McGrath, M. (2006). Chemical phosphorus removal to extremely low levels: experience of two plants in the Washington, DC area. *Water Science and Technology* **53**: 21–28.
18. Hillel, D. (2008). *Soil in the Environment. Crucible of Terrestrial Life*. New York: Academia Press.
19. Sommer, S.G., Schjoerring, J.K., and Denmead, O.T. (2004). Ammonia emission from mineral fertilizers and fertilized crops. *Advances in Agronomy* **82**: 557–622.
20. Barnes B, Fortune T. Blending and spreading fertilizer: physical properties. Available from: https://www.fertilizer-assoc.ie/wp-content/uploads/2014/10/Blending_Spreading-Physical_Characteristics_B_Barnes.pdf (accessed December 23, 2019).
21. Dombalov, I., Pelovski, Y., and Petkova, V. (1999). Thermal stability and properties of new NPS-fertilisers. *Journal of Thermal Analysis and Calorimetry* **56**: 87–94.
22. Fittmar, H. (2009). Fertilizers 4. In: *Ullmann's Encyclopedia of Industrial Chemistry* (ed. M.E. Trenkel). Weinheim: Wiley-VCH.

23. Haby, V.A., Backer, M.L., and Feagley, S. (2003). Solids and fertilizers. In: *Vegetable Resources*. College Station, TX: Agrilife Extension Service.
24. McCauley, A., Jones, C., Jacobsen, J., and Module, N. (2009). Commercial fertilizers and soil amendments. In: *Nutrient Management Course*. Bozeman, MT: Montana State University.
25. Sahoy H. Fertilizers and their use. Available from: https://extension.tennessee.edu/publications/Documents/PB1637.pdf (accessed December 23, 2019).
26. Bond, T., Brouckaert, C.J., Foxon, K.M., and Buckley, C.A. (2012). A critical review of experimental and predicted methane generation from anaerobic codigestion. *Water Science and Technology* **65**: 183–189.
27. Tchobanoglous, G., Burton, F., and Stensel, H.D. (2003). *Wastewater Engineering: Treatment and Reuse*. New York: McGraw Hill.
28. European Commission. Regulation (EC) No. 1774/2002 of 3 October 2002 to laying down health rules concerning animal by-products not intended for human consumption. Brussels: European Commission; 2002.
29. European Commission. Regulation (EC) No. 1069/2009 of 21 October 2009 to laying down health rules concerning animal by-products not intended for human consumption and repealing Regulation (EC) No 1774/2002. Brussels: European Commission; 2009.
30. MDDEFP. Guide sur le recyclage des matières résiduelles fertilisantes: Critères de référence et normes réglementaires. 2012.
31. Astals, S., Esteban-Gutierrez, M., Fernandez-Arevalo, T. et al. (2013). Anaerobic digestion of seven different sewage sludges: a biodegradability and modelling study. *Water Research* **47**: 6033–6043.
32. Bhuiyan, M.I.H., Mavinic, D.S., and Beckie, R.D. (2009). Determination of temperature dependence of electrical conductivity and its relationship with ionic strength of anaerobic digester supernatant, for struvite formation. *Journal of Environmental Engineering and Science* **135**: 1221–1226.
33. Hafner, S.D. and Bisogni, J.J. (2009). Modeling of ammonia speciation in anaerobic digesters. *Water Research* **43**: 4105–4114.
34. Zhang, L. and Jahng, D. (2010). Enhanced anaerobic digestion of piggery wastewater by ammonia stripping: effects of alkali types. *Journal of Hazardous Materials* **182**: 536–543.
35. Vanrolleghem, P.A. and Lee, D.S. (2003). On-line monitoring equipment for wastewater treatment processes: state of the art. *Water Science and Technology* **47**: 1–34.
36. Batstone, D.J., Keller, J., Angelidaki, I. et al. (2002). The IWA anaerobic digestion model no. 1 (ADM1). *Water Science and Technology* **45**: 65–73.
37. Lauwers, J., Appels, L., Thompson, I.P. et al. (2013). Mathematical modelling of anaerobic digestion of biomass and waste: power and limitations. *Progress in Energy and Combustion Science* **39**: 383–402.
38. Ikumi DS. The development of a three phase plant-wide mathematical model for sewage treatment. PhD Thesis: University of Cape Town; 2011.
39. Hauduc H, Takacs I, Smith S, et al. A dynamic physicochemical model for phosphorus removal. IWA Nutrient Removal and Recovery 2013, Conference Proceedings, Vancouver; 2013.
40. Brouckaert CJ, Ikumi DS, Ekama GA. A three phase anaerobic digestion model. 12th IWA Anaerobic Digestion Conference (AD12), Conference Proceedings, Guadalajara, 2010.
41. Harding TH. A steady state stoichiometric model describing the anaerobic digestion of BEPR WAS. Thesis: University of Cape Town; 2009.
42. Van Rensburg, P., Musvoto, E.V., Wentzel, M.C., and Ekama, G.A. (2003). Modelling multiple mineral precipitation in anaerobic digester liquor. *Water Research* **37**: 3087–3097.
43. Kim, B.U., Lee, W.H., Lee, H.J., and Rim, J.M. (2004). Ammonium nitrogen removal from slurry type swine wastewater by pretreatment using struvite crystallization for nitrogen control of anaerobic digestion. *Water Science and Technology* **49**: 215–222.
44. Ganigue, R., Volcke, E.I.P., Puig, S. et al. (2010). Systematic model development for partial nitrification of landfill leachate in a SBR. *Water Science and Technology* **61**: 2199–2210.

45. Lizarralde I, Brouckaert CJ, Ekama GA, Grau P. Incorporating water chemistry into the steady-state models for wastewater treatment processes: Case study anaerobic reactor in the SANI process. 13th World Congress on Anaerobic Digestion (AD-13): Recovering (Bio) Resources for the World, Conference Proceedings, Santiago de Compostela; 2013.
46. Musvoto, E.V., Wentzel, M.C., and Ekama, G.A. (2000). Integrated chemical-physical processes modelling – II. Simulating aeration treatment of anaerobic digester supernatants. *Water Research* **34**: 1868–1880.
47. Sotemann, S.W., Musvoto, E.V., Wentzel, M.C., and Ekama, G.A. (2005). Integrated biological, chemical and physical processes kinetic modelling Part 1 – Anoxic-aerobic C and N removal in the activated sludge system. *Water SA* **31**: 529–544.
48. Grau, P., Beltrán, S., de Gracia, M., and Ayesa, E. (2007). New mathematical procedure for the automatic estimation of influent characteristics in WWTP's. *Water Science and Technology* **56**: 95–106.
49. Kleerebezem, R. and Van Loosdrecht, M.C.M. (2006). Waste characterization for implementation in ADM1. *Water Science and Technology* **54**: 157–174.
50. Zaher, U., Li, R., Jeppsson, U. et al. (2009). GISCOD: general integrated solid waste co-digestion model. *Water Research* **43**: 2717–2727.
51. Batstone, D.J., Tait, S., and Starrenburg, D. (2009). Estimation of hydrolysis parameters in full-scale anaerobic digesters. *Biotechnology & Bioengineering* **102**: 1513–1520.
52. Copp JB, Belia E, Peerbolte A, et al. Integrating anaerobic digestion into plant-wide wastewater treatment modeling. WEFTEC, Conference Proceedings, New Orleans. Los Angeles; 2004.
53. Vanrolleghem, P.A., Rosen, C., Zaher, U. et al. (2005). Continuity-based interfacing of models for wastewater systems described by Petersen matrices. *Water Science and Technology* **52**: 493–500.
54. Zaher, U., Grau, P., Benedetti, L. et al. (2007). Transformers for interfacing anaerobic digestion models to pre- and post-treatment processes in a plant-wide modelling context. *Environmental Modelling & Software* **22**: 40–58.
55. Girault, R., Bridoux, G., Nauleau, F. et al. (2012). A waste characterisation procedure for ADM1 implementation based on degradation kinetics. *Water Research* **46**: 4099–4110.
56. Zaher, U., Buffière, P., Steyer, J.-P., and Chen, S. (2009). A procedure to estimate proximate analysis of mixed organic wastes. *Water Environment Research* **81**: 407–415.
57. Bouropoulos, N.C. and Koutsoukos, P.G. (2000). Spontaneous precipitation of struvite from aqueous solutions. *Journal of Crystal Growth* **213**: 381–388.
58. Ohlinger, K.N., Young, T.M., and Schroeder, E.D. (1998). Predicting struvite formation in digestion. *Water Research* **32**: 3607–3614.
59. Doyle, J.D., Oldring, K., Churchley, J., and Parsons, S.A. (2002). Struvite formation and the fouling propensity of different materials. *Water Research* **36**: 3971–3978.
60. Münch, E.V. and Barr, K. (2001). Controlled struvite crystallisation for removing phosphorus from anaerobic digester sidestreams. *Water Research* **35**: 151–159.
61. Nelson, N.O., Mikkelsen, R.L., and Hesterberg, D.L. (2003). Struvite precipitation in anaerobic swine lagoon liquid: effect of pH and Mg : P ratio and determination of rate constant. *Bioresource Technology* **89**: 229–236.
62. Adnan, A., Mavinic, D.S., and Koch, F.A. (2003). Pilot-scale study of phosphorus recovery through struvite crystallization – examining the process feasibility. *Journal of Environmental Engineering and Science* **2**: 315–324.
63. Jeong, Y.K. and Hwang, S.J. (2005). Optimum doses of Mg and P salts for precipitating ammonia into struvite crystals in aerobic composting. *Bioresource Technology* **96**: 1–6.
64. Le Corre, K.S., Valsami-Jones, E., Hobbs, P., and Parsons, S.A. (2007). Kinetics of struvite precipitation: effect of the magnesium dose on induction times and precipitation rates. *Environmental Technology* **28**: 1317–1324.
65. Wu, Q.Z. and Bishop, P.L. (2004). Enhancing struvite crystallization from anaerobic supernatant. *Journal of Environmental Engineering and Science* **3**: 21–29.

66. Doyle, J.D., Philp, R., Churchley, J., and Parsons, S.A. (2000). Analysis of struvite precipitation in real and synthetic liquors. *Process Safety and Environmental Protection* **78**: 480–488.
67. Regy S, Mangin D, Klein JP, Thronton C. Phosphate recovery by struvite precipitation in a stirred reactor. Available from: https://www.phosphorusplatform.eu/images/download/Regy-Mangin-Lagep-Report-struvite-precipitation-2001.pdf (accessed December 23, 2019).
68. Battistoni, P., De Angelis, A., Prisciandaro, M. et al. (2002). P removal from anaerobic supernatants by struvite crystallization: long term validation and process modeling. *Water Research* **36**: 1927–1938.
69. Shimamura, K., Tanaka, T., Miura, Y., and Ishikawa, H. (2003). Development of a high efficiency phosphorus recovery method using a fluidised-bed crystallized phosphorus removal system. *Water Science and Technology* **48**: 163–170.
70. Andrade, A. and Schuling, R.D. (2001). The chemistry of struvite crystallization. *Mineralogical Journal* **23**: 37–46.
71. Huang, H., Song, Q., and Xu, C. (2011). The mechanism and influence factors of struvite precipitation for the removal of ammonium nitrogen. *Manufacturing Process Technology* **189–193**: 2613–2620.
72. Le Corre, K.S., Valsami-Jones, E., Hobbs, P., and Parsons, S.A. (2009). Phosphorus recovery from wastewater by struvite crystallization: a review. *Critical Reviews in Environmental Science and Technology* **39**: 433–477.
73. Schneider, P.A., Wallace, J.W., and Tickle, J.C. (2013). Modelling and dynamic simulation of struvite precipitation from source-separated urine. *Water Science and Technology* **67**: 2724–2732.
74. Bhuiyan, M.I.H., Mavinic, D.S., and Beckie, R.D. (2007). A solubility and thermodynamic study of struvite. *Environmental Technology* **28**: 1015–1026.
75. Celen, I., Buchanan, J.R., Burns, R.T. et al. (2007). Using a chemical equilibrium model to predict amendments required to precipitate phosphorus as struvite in liquid swine manure. *Water Research* **41**: 1689–1696.
76. Miles AE, Ellis TG. Recovery of nitrogen and phosphorus from anaerobically wastes using struvite precipitation. National Conference on Environmental Engineering, Waste resources and the Urban Environment, Conference Proceedings, Chicago, IL; 1998.
77. Mohan, G.R., Gadekar, S., and Pratap, P. (2011). Development of a process model for recovery of nutrients from wastewater by precipitation as struvite. *Florida Water Resources Journal* **1**: 17–22.
78. Wu, Y. and Zhou, S. (2012). Improving the prediction of ammonium nitrogen removal through struvite precipitation. *Environmental Science and Pollution Research* **19**: 347–360.
79. Maurer, M., Abramovich, D., Siegrist, H., and Gujer, W. (1999). Kinetics of biologically induced phosphorus precipitation in wastewater treatment. *Water Research* **33**: 484–493.
80. Lee, S.-H., Yoo, B.-H., Lim, S.J. et al. (2013). Development and validation of an equilibrium model for struvite formation with Ca co-precipitation. *Journal of Crystal Growth* **372**: 129–137.
81. Rahaman, M.S., Mavinic, D.S., and Ellis, N. (2008). Phosphorus recovery from anaerobic digester supernatant by struvite crystallization: model-based evaluation of a fluidized bed reactor. *Water Science and Technology* **58**: 1321–1327.
82. Rahaman, M.S., Mavinic, D.S., Meikleham, A., and Ellis, N. (2014). Modelling phosphorus removal and recovery from anaerobic digester supernatants through struvite crystallization in a fluidized bed reactor. *Water Research* **51**: 1–10.
83. Doyle, J.D. and Parsons, S.A. (2002). Struvite formation, control and recovery. *Water Research* **36**: 3925–3940.
84. Parsons, S.A., Wall, F., Doyle, J. et al. (2001). Assessing the potential for struvite recovery at sewage treatment works. *Environmental Technology* **22**: 1279–1286.
85. Ali, M.I. and Schneider, P.A. (2008). An approach of estimating struvite growth kinetic incorporating thermodynamic and solution chemistry, kinetic and process description. *Chemical Engineering Science* **63**: 3514–3525.
86. Bhuiyan, M.I.H., Mavinic, D.S., and Beckie, R.D. (2008). Nucleation and growth kinetics of struvite in a fluidized bed reactor. *Journal of Crystal Growth* **310**: 1187–1194.

87. Galbraith, S.C., Schneider, P.A., and Flood, A.E. (2014). Model-driven experimental evaluation of struvite nucleation, growth and aggregation kinetics. *Water Research* **56**: 122–132.
88. Harrison ML, Johns MR, White ET, Mehta CM. Growth rate kinetics for struvite crystallisation. 14th International Conference on Process Integration, Modelling and Optimisation for Energy Saving and Pollution Reduction, Pts 1 and 2; 2011.
89. Le Corre, K.S., Valsami-Jones, E., Hobbs, P., and Parsons, S.A. (2007). Impact of reactor operation on succes of struvite precipitation from synthetic liquors. *Environmental Technology* **28**: 1245–1256.
90. Quintana, M., Sanchez, E., Colmenarejo, M.F. et al. (2005). Kinetics of phosphorus removal and struvite formation by the utilization of by-product of magnesium oxide production. *Chemical Engineering Journal* **111**: 45–52.
91. Perez, M., Dumont, M., and Acevedo-Reyes, D. (2008). Implementation of classical nucleation and growth theories for precipitation. *Acta Materialia* **56**: 1219–2132.
92. Saracco, G. and Genon, G. (1994). High-temperature ammonia stripping and recovery from process liquid wastes. *Journal of Hazardous Materials* **37**: 191–206.
93. Lemmens B, Ceulemans C, Elslander H, et al. Beste Beschikbare Technieken (BBT) voor mestverwerking. Ghent; 2007.
94. Liao, P.H., Chen, A., and Lo, K.V. (1995). Removal of nitrogen from swine manure wastewaters by ammonia stripping. *Bioresource Technology* **54**: 17–20.
95. Gustin, S. and Marinsek-Logar, R. (2011). Effect of pH, temperature and air flow rate on the continuous ammonia stripping of the anaerobic digestion effluent. *Process Safety and Environmental Protection* **89**: 61–66.
96. Bonmati, A. and Flotats, X. (2003). Air stripping of ammonia from pig slurry: characterisation and feasibility as a pre- or post-treatment to mesophilic anaerobic digestion. *Waste Management* **23**: 261–272.
97. Campos, J.C., Moura, D., Costa, A.P. et al. (2013). Evaluation of pH, alkalinity and temperature during air stripping process for ammonia removal from landfill leachate. *Journal of Environmental Science and Health. Part A – Toxic/Hazardous Substances & Environmental Engineering* **48**: 1105–1113.
98. Collivignarelli, C., Bertanza, G., Baldi, M., and Avezzu, F. (1998). Ammonia stripping from MSW landfill leachate in bubble reactors: process modeling and optimization. *Waste Management & Research* **16**: 455–466.
99. Melse, R.W. and Ogink, N.W.M. (2005). Air scrubbing techniques for ammonia and odor reduction at livestock operations: review of on-farm research in the Netherlands. *Transactions of the Asae* **48**: 2303–2313.
100. Alitalo, A., Kyro, A., and Aura, E. (2012). Ammonia stripping of biologically treated liquid manure. *Journal of Environmental Quality* **41**: 273–280.
101. Environmental Protection Agency. Wastewater Technology Fact Sheet: Ammonia Stripping. Available from: https://nepis.epa.gov/Exe/ZyNET.exe/P10099PH.TXT?ZyActionD=ZyDocument&Client=EPA&Index=2000+Thru+2005&Docs=&Query=&Time=&EndTime=&SearchMethod=1&TocRestrict=n&Toc=&TocEntry=&QField=&QFieldYear=&QFieldMonth=&QFieldDay=&IntQFieldOp=0&ExtQFieldOp=0&XmlQuery=&File=D%3A%5Czyfiles%5CIndex%20Data%5C00thru05%5CTxt%5C00000025%5CP10099PH.txt&User=ANONYMOUS&Password=anonymous&SortMethod=h%7C-&MaximumDocuments=1&FuzzyDegree=0&ImageQuality=r75g8/r75g8/x150y150g16/i425&Display=hpfr&DefSeekPage=x&SearchBack=ZyActionL&Back=ZyActionS&BackDesc=Results%20page&MaximumPages=1&ZyEntry=1&SeekPage=x&ZyPURL (accessed December 23, 2019).
102. GNS. Proven Method for the Removal of Ammonia Nitrogen from Digestate since 2008. Technical fact sheet. Available from: https://www.gns-halle.de/downloads/info_digestate_treatement.pdf (accessed December 23, 2019).
103. Bokotko, R.P., Hupka, J., and Miller, J.D. (2005). Flue gas treatment for SO_2 removal with air-sparged hydrocyclone technology. *Environmental Science and Technology* **39**: 1184–1189.

104. Quan, X., Wang, F., Zhao, Q. et al. (2009). Air stripping of ammonia in a water-sparged aerocyclone reactor. *Journal of Hazardous Materials* **170**: 983–988.
105. Quan, X., Ye, C., Xiong, Y. et al. (2010). Simultaneous removal of ammonia, P and COD from anaerobically digested piggery wastewater using an integrated process of chemical precipitation and air stripping. *Journal of Hazardous Materials* **178**: 326–332.
106. Powers, S.E., Collins, A.G., Edzwald, J.K., and Dietrich, J.M. (1987). Modeling an aerated bubble ammonia stripping process. *Journal of the Water Pollution Control Federation* **59**: 92–100.
107. Mattermuller, C., Gujer, W., and Giger, W. (1981). Transfer of volatile substances from water to the atmosphere. *Water Research* **15**: 1271–1279.
108. Arogo, J., Zhang, R.H., Riskowski, G.L. et al. (1999). Mass transfer coefficient of ammonia in liquid swine manure and aqueous solutions. *Journal of Agricultural Engineering Research* **73**: 77–86.
109. Lee, D., Lee, J.-M., Lee, S.-Y., and Lee, L.-B. (2002). Dynamic simulation of the sour water stripping process and modified structure for effective pressure control. *Chemical Engineering Research and Design* **80**: 167–177.
110. Brettschneider, O., Thiele, R., Faber, R. et al. (2004). Experimental investigation and simulation of the chemical absorption in a packed column for the system NH_3-CO_2-H_2S-NaOH-H_2O. *Separation and Purification Technology* **39**: 139–159.
111. Lens, P.N.L., Kennes, C., Le Cloirec, P., and Dehusses, M.A. (2006). *Waste Gas Treatment for Resource Recovery*. London: IWA Publishing.
112. Jiang, A., Zhang, T., Zhao, Q.-B. et al. (2014). Evaluation of an integrated ammonia stripping, recovery, and biogas scrubbing system for use with anaerobiclly digested dairy manure. *Biosystems Engineering* **119**: 117–126.
113. Manuzon, R.B., Zhao, L.Y., Keener, H.M., and Darr, M.J. (2007). A prototype acid spray scrubber for absorbing ammonia emissions from exhaust fans of animal buildings. *Transactions of the Asabe* **50**: 1395–1407.
114. Calvert, S. and Englund, H. (1984). *Handbook of Air Pollution Technology*. New York: Wiley.
115. Fair, J.R., Steinmeyer, D.E., Penney, W.R., and Crocker, B.B. (1997). Gas absorption and gas-liquid system design. In: *Perry's Chemical Engineer's Handbook* (eds. R.H. Perry, D.W. Green and D.O. Maloney). New York: McGraw Hill.
116. Taylor R, Wilkinson P, Kooijman H. Rate-based modeling of two commercial scale H2S stripping columns. Distillation Absorption 2010, Conference Proceedings, Eindhoven; 2010.
117. Hofmann G, Paroli F, Van Esch J. Crystallization of ammonium sulphate: state of the art and new developments. Icheap-9, 9th International Conference on Chemical and Process Engineering; 2009.
118. Gea-Messo PT. Ammoniumsulfate crystallization. Available from: https://www.gea.com/en/applications/chemicals/agrochemicals/agrochemicals_solution-crystelization.jsp (accessed December 23, 2019).
119. Belcu, M. and Turtoi, D. (1996). Simulation of the fluidized-bed crystallizers.1. Influence of parameters. *Crystal Research and Technology* **31**: 1015–1023.
120. Daudey PJ. Crystallization of ammonium sulfate: secondary nucleation and growth kinetics in suspension. Available from: http://resolver.tudelft.nl/uuid:b2331fb2-5e02-41f0-bbaa-a83b71f45973 (accessed December 23, 2019).
121. Kubota, N. and Onosawa, M. (2009). Seeded batch crystallization of ammonium aluminum sulfate from aqueous solution. *Journal of Crystal Growth* **311**: 4525–4529.
122. Vaneeckhaute, C., Claeys, F.H.A., Tack, F. et al. (2018). Development, implementation, and validation of a generic nutrient recovery model (NRM) library. *Environmental Modelling and Software* **99**: 170–209.
123. Crittenden, J.C., Trussell, R.R., Hand, D.W. et al. (2012). *MWHs Water Treatment: Principles and Design*. New York: Wiley.
124. Gujer, W. (2008). *Systems Analysis for Water Technology*. Berlin: Springer Verlag.

125. Camp, T.R. and Stein, P.C. (1943). Velocity gradients and hydraulic work in fluid motion. *Journal of the Boston Society of Civil Engineers* **30**: 203–221.
126. Musvoto, E.V., Wentzel, M.C., Loewenthal, R.E., and Ekama, G.A. (2000). Integrated chemical-physical processes modelling – I. Development of a kinetic-based model for mixed weak acid/base systems. *Water Research* **34**: 1857–1867.
127. Volcke EIP, Van Hulle S, Deksissa T, et al. Calculation of pH and concentration of equilibrium components during dynamic simulation by means of a charge balance. Coupure Links. 65362. 2005.
128. Rosen, C., Vrecko, D., Gerneay, K.V. et al. (2006). Implementing ADM1 for plant-wide benchmark simulations in Matlab/Simulink. *Water Science and Technology* **40**: 11–19.
129. Parkhurst, D.L. and Appelo, C.A.J. (2013). *PHREEQC Version 3 – A Computer Program for Speciation, Batch-Reaction, One-Dimensional Transport, and Inverse Geochemical Calculations*. Denver, CO: United States Geological Survey.
130. Allison JD, Brown DS, Novo Gradac KJ. MINTEQA2/PRODEFA2. A Geochemical Assessment Model for Environmental Systems: Version 3.0. 1991.

8.2

Soil Dynamic Models: Predicting the Behavior of Fertilizers in the Soil

Marius Heinen, Falentijn Assinck, Piet Groenendijk, and Oscar Schoumans

Wageningen Environmental Research, Wageningen University & Research, Wageningen, The Netherlands

8.2.1 Introduction

Crops are grown on arable soils to feed both people and animals and to provide fibers and biofuels. Besides carbon, hydrogen, and oxygen, plants need nutrients to grow. For growth and proper functioning, they need to take up macronutrients (including nitrogen [N], phosphorus [P], and potassium [K]) and micronutrients (including iron [Fe] and zinc [Zn]) [1]. Most or all of these nutrients are indigenously present in soil; however, they are not always easily available, or available in the quantities required by the crop. Therefore, nutrients should be applied by adding (in)organic fertilizers.

The main purpose of adding synthetic fertilizers, animal manure, compost, and biobased derivatives (such as products resulting from manure digestives, algae digestives, and ashes) is to supply nutrients to the root zone so that they can easily be taken up by the crop via their roots. The efficiency of nutrient application highly depends on the synlocalization and synchronization of their availability and the demand for them in the soil root zone [2, 3]. When nutrients reside in the soil, besides uptake by the roots, they are subject to other processes: they can be transported away from the root system with draining water (leaching), used by microbes (oxygen donor in the denitrification process; immobilization), adsorbed

---

Biorefinery of Inorganics: Recovering Mineral Nutrients from Biomass and Organic Waste, First Edition.
Edited by Erik Meers, Gerard Velthof, Evi Michels and René Rietra.
© 2020 John Wiley & Sons Ltd. Published 2020 by John Wiley & Sons Ltd.

by the adsorption complex (reversible), fixed (mostly irreversible), or precipitated. In order to understand these processes, their importance, and their magnitude, simulation models have been used at large scale. There is no such model that can be regarded as the ultimate best one. Each has its own specific purpose. Inter-model comparisons have been carried out in recent decades [4–15]. Some examples of high-quality models will be presented later in the chapter.

The simplest approach to quantifying the nutrient surplus on the soil balance is via the calculation of the difference between the nutrient supply to the field (fertilizers, manure, deposition, biobased derivatives) and the crop offtake. The nutrient surplus on the soil balance is a measure of potential pollution of soil, air, and water by agricultural practices; the lower the surplus, the better it is for the environment. This approach assumes that no changes in nutrient storage in the soil occur. More sophisticated approaches (models, see later) try to handle (all) the main processes regarding nutrients in the soil, and come up with predictions for crop uptake, losses to the environment (leaching to groundwater and open surface water; gaseous emission), changes in storage in the soil, and intermediate process state variables.

The aim of this chapter is to give an overview of the most dominant processes for the situation where organic biobased derivatives are added to the soil, with emphasis on decomposition, production of mineral nutrients (which can be taken up by the root system), and losses to the environment. Though organic fertilizers may contain several nutrients, the main focus will be on N and P.

8.2.2 Soil N and P Processes

8.2.2.1 Main Dynamic Processes

The main soil nutrient – typically N and P – dynamic processes can be divided into sources and sinks (Table 8.2.1); for N, we distinguish between nitrate (NO_3) and ammonium (NH_4).

Nutrient transport in the soil is generally described by the convective dispersion–diffusion equation given by (in the vertical dimension only):

$$\frac{\partial Q}{\partial t} = -\frac{\partial q_s}{\partial z} + P - S = -\frac{\partial qc}{\partial z} + \frac{\partial}{\partial z}\left(\theta D \frac{\partial c}{\partial z}\right) + P - S \qquad (8.2.1)$$

where Q is the total amount of the nutrient present in the soil (g cm^{-3} [soil]), t is the time (d), q_s is the solute flux density (mg cm^{-2} d^{-1}), c is the concentration of the nutrient in the soil solution (g cm^{-3} [solution]), q is the water flux density (cm^3 cm^{-2} d^{-1}), z is the vertical coordinate (cm), θ is the volumetric soil water content (cm^3 cm^{-3}), D is the combined dispersion–diffusion coefficient (cm^2 d^{-1}), P represents the sources (produced) (g cm^{-3} d^{-1}), and S represents the sinks (g cm^{-3} d^{-1}). Equation (8.2.1) applies to any nutrient, including N (NO_3, NH_4), P, and dissolved organic components like dissolved organic nitrogen (DON) and dissolved organic phosphorus (DOP). The rate of change of Q is due to four subprocesses (four terms on the right-hand side of Eq. (8.2.1): convection or transportation with the main water flow, dispersion–diffusion due to concentration gradients, production, and sink terms. The production and sink terms represent local processes like mineralization and root uptake, but they can also represent lateral inputs or outputs mimicking multidimensional transport; for example, transport toward drain pipes, ditches, canals,

Table 8.2.1 Summary of sources and sinks.

Sources	Sinks
(Sub)surface application: fertilizer, manure, slurry, biobased derivatives, atmospheric deposition (wet, dry), incorporation of (dead) plant material; often, surface-applied sources are mixed in the top soil layer, e.g. via plowing	Volatilization
Mineralization from decomposing organic material	Immobilization
Nitrification (NO_3 produced from NH_4)	Nitrification (NH_4 transformed into NO_3) Denitrification (NO_3 lost as N_2O or N_2) Leaching (NO_3, NH_4, N_2O, P)
Desorption or dissolution (NH_4, P)	Adsorption, fixation, precipitation (NH_4, P) Plant uptake

or other open water systems. In this chapter, we focus on the main processes (mainly *P*, *S*) for the situation of applied biobased derivatives: organic matter (OM) decomposition, nitrification, denitrification, volatilization, leaching, and sorption.

8.2.3 Other Related State and Rate Variables

8.2.3.1 Water Flow

Part of the solute transport is convection with the water flow. Therefore, water fluxes must be known, which can be calculated by considering water movement, for example using the state-of-the-art Richards [16] equation, combined with the Darcy [17]–Buckingham [18] expression for the water flux density, such as is used in the models SWAP [19], HYDRUS [20], and FUSSIM [21, 22]. For this, the highly non-linear relationships between pressure head (h, cm), volumetric water content (θ, cm^3 cm^{-3}), and hydraulic conductivity (K, cm d^{-1}) must be known. In some cases, mostly large-scale studies, soil water movement is considered by a simple tipping bucket (capacity) water-balance approach [23–25].

8.2.3.2 Soil Water Content

Most of the processes mentioned in Table 8.2.1 are dependent on the soil water content θ (cm^3 cm^{-3}) or the degree of saturation S_e (dimensionless). Often, the rate (coefficient) of the process is reduced by a dimensionless factor between zero and one. In some cases, the pressure head is used as the determining variable.

8.2.3.3 Soil Temperature

Many of the processes mentioned in Table 8.2.1 are dependent on soil temperature T (°C). The rate (coefficient) of the process is reduced by a dimensionless factor, typically by a Q_{10}, Arrhenius, or exponential relationship. In extreme cases, modelers approximate the soil temperature by the ambient air temperature [26]. Otherwise, soil temperature can be modeled as a diffusion process in the soil, determined by the volumetric heat capacity and thermal conductivity.

8.2.3.4 Soil pH

Some processes mentioned in Table 8.2.1 are dependent on soil pH. Their rate (coefficient) is reduced by a dimensionless pH factor between zero and one.

8.2.3.5 Gas Transport

During denitrification, gases are formed. Most models assume that these gases instantaneously disappear into the atmosphere, and no gas transport in the soil is considered. If gas transport were to be considered, one could take into account the residence time of the gases in the soil, which might then undergo further transformation. We will not discuss gas transport in this chapter.

8.2.3.6 Crop Growth and Nutrient Demand

Uptake of water and uptake of nutrients are sinks in soil process models. Therefore, soil process models need information on crop development, either through coupling with a sophisticated crop growth model, via internal crop growth modules, or from predefined demand functions (input tables). In the cycles of N and P in agronomic situations, crop uptake is by far the most dominant item in the N and P balances.

The primary aim of adding fertilizers is to make nutrients available for uptake. Therefore, it is necessary to know how much nutrient (typically N and P) is taken up by the crop. This appears to be highly variable, depending on the climate, cropping circumstances, water shortage, pest and diseases, soil type, and so on. Thus, for each situation, crop uptake must be determined before biobased derivatives are added to the soil as fertilizers. The large variation in N yields (N removed via the harvested crop) for different crops in 27 EU countries as observed by Velthof et al. [27] is caused by several factors, among them interaction with the availability of other nutrients, including micronutrients. A few examples are provided in Table 8.2.2, including calculated P yields (as P_2O_5).

If biobased derivatives are used, it should also be investigated whether these materials influence rooting patterns or are influenced by, for example, root exudates. If biobased derivatives cause acidification or salinization, roots may not penetrate the zones where they are located. On the other hand, if root exudates influence the dissolution or decomposition of the derivatives, this may influence their availability for uptake and losses to the environment.

8.2.3.7 Dynamic Simulation

The fate of N and P is determined by the convective dispersion–diffusion equation (Eq. (8.2.1)), which is a non-linear partial differential equation that must be solved numerically. For this, the soil is divided into several layers, for each of which the state and rate variables are simulated. If multidimensional modeling is required, the soil is divided into several

Table 8.2.2 *Statistics (minimum, maximum, average, median, and coefficient of variation, cv) on N and P_2O_5 yields in selected crops.*

Crop	N ($kg\,ha^{-1}$)					P_2O_5 ($kg\,ha^{-1}$)			
	Min.	Max.	Avg.	Median	cv (%)	Min.	Max.	Avg.	Median
Apple	1	26	8.3	5	81.8	0.3	7.0	2.2	1.4
Barley	11	106	50.4	44	57.1	5.8	55.8	26.5	23.2
Grass	70	241	104.9	85.5	46.1	25.9	89.3	38.9	31.7
Fodder maize	16	165	75.0	60.5	53.6	10.0	103.1	46.9	37.8
Grain maize	15	153	74.4	65	53.8	8.3	85.0	41.3	36.1
Oat	12	106	41.7	28.5	67.3	6.3	55.8	21.9	15.0
Other cereals	17	89	48.5	50	38.8	7.4	38.7	21.1	21.7
Potato	18	160	71.1	59	53.8	7.2	64.0	28.5	23.6
Rape	26	127	59.5	55	47.1	14.4	70.6	33.0	30.6
Rye	10	68	33.7	25.5	55.4	5.3	35.8	17.7	13.4
Soya	23	153	69.4	56.5	52.1	6.4	42.5	19.3	15.7
Sugar beet	13	111	65.6	64	40.7	7.2	61.7	36.4	35.6
Soft wheat	21	167	72.7	51	60.0	8.4	66.8	29.1	20.4
Tomato	4	433	94.5	39	115.3	1.4	149.3	32.6	13.4

Data for P_2O_5 calculated using provided N:P_2O_5 ratios in harvested crops.
Source: Calculated from the non-zero data for 27 EU countries provided by Velthof et al. [27].

2D or 3D elements. In each layer (element), the different processes are considered, and solutes may be transported from one to another layer (element). Most models include (daily) weather data to feed the associated soil water and temperature modules (see later).

8.2.4 Organic Matter

Organic compounds in the soil (OM, organic fertilizers, biobased derivatives, soil biomass) are subject to (micro)biological conversion reactions [28]. Decomposition of OM usually occurs relatively quickly as long as enough molecular oxygen is present. OM is ultimately transformed into carbon dioxide (CO_2) and water and partly incorporated in biomass, thereby releasing other components that were present in the organic material, such as mineral and dissolved organic forms of N and P. When OM decomposes, part of the total conversion results in CO_2 production (dissimilation) and the remaining part results in another carbon pool or biomass (assimilation). The ratio between soil biomass formation and total amount of decomposition is known as the assimilation efficiency. Anaerobic decomposition also exists, but this occurs at a much slower (factor 0.01–0.001) rate [28, 29].

Two major models can be distinguished in modeling mineralization of OM: multi-pool models with a constant decomposition rate factor and models with a time-dependent decomposition rate factor. De Willigen et al. [30] compared seven models describing the decomposition and accumulation of OM in soil. Two of these (MINIPa, MINIPb) belong to the class with a time-dependent decomposition rate factor, while the other five (CESAR, RECAFS/CNGRAS, ANIMO, NUCSAM, CENTURY) belong to the class of multi-pool models with a constant decomposition rate factor. In another study, the performances of nine soil OM models (ROTHC, CANDY, DNDC, CENTURY, DAISY, NCSOIL, SOMM, ITE, and Verberne) were compared using datasets from seven long-term experiments [31].

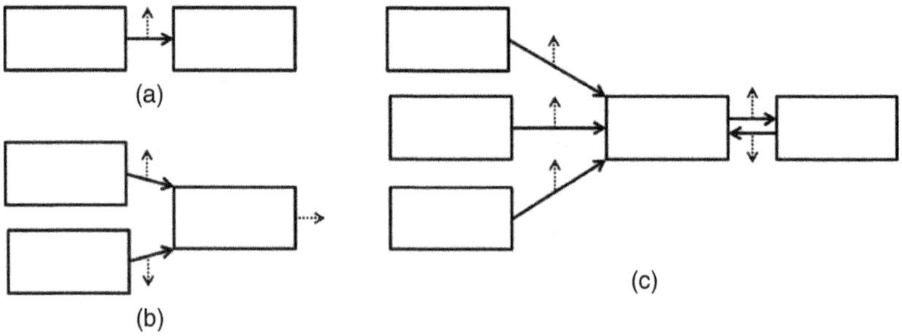

Figure 8.2.1 Examples of multi-pool C dynamics models: (a) two; (b) three; and (c) five pools. Solid arrows indicate transformations (assimilation) and dashed lines represent CO_2 production (dissimilation).

Despite differences in model structure, in most cases the predicted soil organic carbon content (and corresponding model errors) did not differ significantly from one another. The user, or model, has to decide in what form the mineral N is released: partly as NH_4 and the remainder as NO_3.

8.2.4.1 Multi-Pool Models with Constant Decomposition Rate Factor

In many cases, the decomposition of OM is considered to occur at a rate proportional to the amount present. This is the classical first-order decay model, given by:

$$\frac{\mathrm{d}C}{\mathrm{d}t} = -kC \tag{8.2.2}$$

where C is the size of the organic carbon (C) pool (kg ha^{-1}), t is the time (d), and k is the constant decay rate coefficient (d^{-1}).

Often, the decay results in intermediate (new) pools, such that a series of pools needs to be considered that shows decay and formation. Each pool has its own constant C:N ratio and its own efficiency that indicates how much of the C released from one pool is incorporated into another. The remainder is then equal to the CO_2 production. Figure 8.2.1 shows some examples of multi-pool models.

Theoretically, for M pools, $M \times M$ transformations would be possible. In reality, only a few will be present and considered. The C transformation of pool C_i is given by [30, 32]:

$$\frac{\mathrm{d}C_i}{\mathrm{d}t} = E_{1,i}k_1C_1 + \ldots - k_iC_i + \ldots + E_{M,i}k_MC_M \tag{8.2.3}$$

where $E_{j,i}$ is the efficiency of the decomposition from pool j to pool i. Setting up Eq. (8.2.3) for all pools results mathematically in a matrix problem, which can be solved with the help of the Laplace transformation [30] or by an integration procedure, such as the Runge–Kutta method as used in MOTOR [33].

The total N mineralization follows from the C transformation and the C:N ratio of the pool:

$$\frac{\mathrm{d}N}{\mathrm{d}t} = \sum_{i=1}^{M} \frac{\mathrm{d}C_i/\mathrm{d}t}{(C:N)_i} \tag{8.2.4}$$

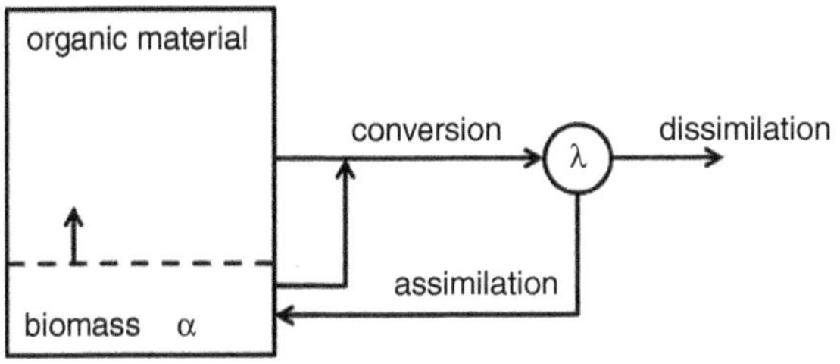

Figure 8.2.2 Schematic representation of the single-pool model consisting of the time variable components of applied organic material and biomass.

where C:N is the C to N ratio of the pool, which remains constant in time. As such, this approach cannot be used to predict whether immobilization will occur. For that, an additional pool or description is required that represents the incorporation of C and N into microbial biomass [32, 33].

This approach is flexible and allows the user to define different pools and the transformations between them. Biomass can be considered as a separate pool. Additions of new organic materials are made by increasing some of the existing pools, provided that the total additions of C (and other nutrients) are guaranteed.

Similar to N, the mineralization for other nutrients such as P can be modeled, provided that corresponding parameter values are used for E and C:P ratio.

8.2.4.2 Models with a Time-Dependent Decomposition Rate Factor

Mineralization from a single pool with a time-dependent decay rate coefficient is given by:

$$\frac{\mathrm{d}C}{\mathrm{d}t} = -k(t)C(t) \tag{8.2.5}$$

For example, $k(t)$ can be defined as [34]:

$$k(t) = bm(a + t)^{-m-1} \tag{8.2.6}$$

The time course of total C follows from integration of Eq. (8.2.5) using Eq. (8.2.6) and is given by [34, 35]:

$$C_{\mathrm{tot}}(t) = C_0 \exp[b(a + t)^{\mathrm{m}} - ba^{\mathrm{m}}] \tag{8.2.7}$$

where C_{tot} is the total C content of the organic material (pool) considered (mg kg^{-1}), C_0 is the initial amount (mg kg^{-1}), t is time (d), a is the so-called (apparent) initial age of the organic material (d), and b and m are dimensionless empirical constants: $b = 4.7$ and $m = -0.6$. Examples of a for some organic materials are presented in Table 8.2.3. Rapidly decomposing materials have low values of a; for example, materials with $a < 1.2$ degrade in a few years.

In this model, the carbon in the pool consists of C from the applied material and C incorporated in biomass (Figure 8.2.2). Part of the carbon decomposed is produced as CO_2 (dissimilation), and the other part is incorporated in the biomass (assimilation).

Table 8.2.3 *Initial age for some organic materials, sorted from low to high.*

Organic material	Initial age, *a* (yr)	References
Glucose[a]	0.31–0.38	[36]
Hemicellulose[a]	0.42–1.04	[36]
Cellulose[a]	0.64–0.83	[36]
Sunn hemp residues	0.84	[37]
Oat shoots	0.87–1.54	[36]
Maize straw	0.96–1.21	[36]
Foliage	0.99	[28]
Crop residues (aboveground)	0.99	[38]
Pinto peanut residues	0.99	[39]
Stylo residues	0.99	[39]
Brachiaria residues	0.99	[37]
Ryegrass	1.08–1.59	[36]
Velvet bean residues	1.11	[39]
Barley straw	1.12–2.65	[36]
Green manure	1.27	[38]
Manure, pig	1.36	[38]
Manure, chicken (laying hens)	1.36	[38]
Manure, chicken litter	1.36	[38]
Slurry, pig	1.36	[38]
Slurry, poultry	1.36	[38]
Liquid manure, pig	1.36	[38]
Calopo residues	1.41	[39]
Inga residues	1.41	[37]
Straw	1.41; 1.50	[28, 34, 38]
Manure, young meat chicken	1.45	[38]
Root and stubble residues	1.57	[38]
Mature wheat straw	1.70–2.09	[36]
Mushroom compost	1.96	[38]
Leave residues	2.18	[38]
Litter deciduous trees	2.28	[28, 34]
Farmyard manure	2.45	[28, 34, 38]
Manure, cattle	3.16	[38]
Slurry, cattle	3.16	[38]
Liquid manure, cattle	3.16	[38]
Litter spruce trees	3.34	[28]
Fir needles	3.36	[34]
Sawdust	3.65; 3.69	[28, 34, 38]
GFT compost	3.68	[38]
Spruce needles	3.69	[38]
Compost	3.7	[38]
Peat-1	3.95	[28, 34]
Peat-2	5.47	[28, 34]
Peat mixture	5.47	[38]
Peat-3	7.87	[28, 34]
Peat-4	13.62	[28, 34]
Stable soil organic matter	14–28	[40]

[a]Prepared from barley straw.

The corresponding time course of total N in the same pool can be given by [32]:

$$N_{tot}(t) = N_0 \exp[b(a+t)^m - ba^m(1+\lambda)] \times \frac{\beta_0 \exp[b\lambda a^m] + (\alpha - \beta_0)\exp[b\lambda(a+t)^m]}{\alpha} \quad (8.2.8)$$

where N_{tot} is the total N content of the organic material (pool) considered (mg kg^{-1}), N_0 is the initial amount (mg kg^{-1}), λ is the ratio between assimilation and dissimilation (assumed constant, e.g. $\lambda = 0.5$), α is the C:N ratio of the microbial biomass (mg C mg^{-1} N; assumed constant, e.g. $\alpha = 10$), and β_0 is the initial C:N ratio of the organic material (mg C mg^{-1} N). Note that λ and E (as used earlier) are related according to $\lambda = E/(1-E)$.

The time course of the C:N ratio of the OM pool, $\beta(t)$, follows from the division of Eq. (8.2.7) by Eq. (8.2.8). It can be shown that for $\beta(t) > \alpha(1 + 1/\lambda)$, this leads to N immobilization. For example, for $\alpha = 10$ and $\lambda = 0.5$, immobilization occurs as long as $\beta(t) > 30$. In other words, if an organic material with $\beta_0 > 30$ is supplied to the soil, this will lead initially to N immobilization. As time progresses, $\beta(t)$ decreases, and as soon as $\beta(t)$ becomes less than 30, N is released (mineralized).

Similar to N, the mineralization for other nutrients such as P can be modeled, provided that corresponding parameter values are used for α, λ, and C:P ratio.

The attractiveness of the time-dependent Janssen model is that it is a one-parameter model [34, 35]. Unlike with the traditional first-order models, the Janssen model results in finite C_{tot} and N_{tot} contents when time goes to infinity; these finite contents are dependent on the initial age of the material (and for N_{tot} also on α, λ, and β_0) [32].

Because of the time-dependent decay coefficient in the Janssen model, it is not possible to raise a pool with newly added organic material, even if it is the same material that was supplied before. Thus, for each material and each addition, a separate pool must be considered.

Though the Janssen model was derived based on experimental data for peat soils and other organic materials, experience has shown that it is not always able to predict well the early-stage fast decay that is often observed [36, 41, 42].

The Janssen model can be approximated by a two- or three-pools concept with constant decomposition rate constants [32, 43]. The correspondence is good for a limited period of time (several years).

8.2.4.3 Environmental Response Factors

As many microbial processes are temperature-dependent, so too will be OM decomposition. Therefore, most models will make use of a temperature correction factor. This temperature dependency is modeled by either a Q_{10} (van 't Hoff) function, a linear function, a power function, an Arrhenius-type function, an S-shaped function, or a heat sum concept [44]. Often, the Q_{10} (the increase factor for an increase in temperature of 10 °C) or Arrhenius temperature reduction function f_T is used, here given by:

$$f_T = Q_{10}^{(T-T_{ref})/10} \quad (8.2.9)$$

and:

$$f_T = A^{(T-T_{ref})} \quad (8.2.10)$$

where T is the soil temperature (°C) and T_{ref} is the reference temperature where $f_T = 1$ (°C). The increase factors Q_{10} and A are related by $Q_{10} = A^{10}$. From the thermodynamics underlying the Arrhenius equation, it follows that A (and thus also Q_{10}) is temperature-dependent; this is generally neglected, and constant values for A and Q_{10} are used [11].

The decomposition is also considered to be dependent on oxygen availability and soil water content. In practice, this is mostly modeled by a dependency on soil water content (or degree of saturation) only, where extreme wet conditions are a proxy for lack of oxygen. In general, this dependency states that the decomposition rate is optimal for a certain range of water content and decreases (linear or curvilinear) at lower or higher water contents.

For the first-order decay models with constant decay rate coefficient, the temperature T (°C) and water content S (dimensionless degree of saturation) correction factors f_T and f_S added to Eq. (8.2.2) result in:

$$\frac{dC}{dt} = -f_T f_S kC \tag{8.2.11}$$

For the models with time-varying decay rate coefficient, the addition of correction factors f_T and f_S to Eq. (8.2.7) results in:

$$C_{tot}(t) = C_0 \exp[b(a + f_T f_S t)^m - ba^m] \tag{8.2.12}$$

Note that the time-varying decay coefficient is not linearly dependent on f_T and f_S. This means that $C(t)$ cannot be calculated directly and must be approximated numerically, for example via Euler integration [32].

8.2.5 Nitrogen

Nitrogen in soils can be present as ammonium (NH_4), nitrate (NO_3), nitrite (NO_2), nitric oxide (NO), nitrous oxide (N_2O; gas or dissolved in water), di-nitrogen (N_2; gas), DON, or incorporated in OM. The major pools are NH_4, NO_3, and DON. Production of the greenhouse gas N_2O during denitrification needs attention.

Adding organic biobased derivatives to the soil may increases the N pools NH_4, NO_3, and DON, as well as N bound in the OM pool. The latter will be released in time, as described earlier. The mineral N (NH_4, NO_3) will increase the indigenous amounts already present in the soil. It will then take part in the conversion and transportation processes that act on these pools. The main processes considered here are adsorption (NH_4), nitrification ($NH_4 \rightarrow NO_3$), denitrification ($NO_3 \rightarrow N_2O + N_2$), volatilization ($NH_4 \rightarrow NH_3$), and leaching.

8.2.5.1 Adsorption and Desorption

Cations in soil solution are subject to sorption to the adsorption complex (negative sites of clay particles, OM). This process is generally described by a linear, a Langmuir, or a Freundlich adsorption isotherm. Such an isotherm gives the relationship between the amount in the soil solution and the amount bound on the adsorption complex. The expressions for the total amount of NH_4 will then be given by:

Linear:

$$Q = \theta c + kc \tag{8.2.13}$$

where k is the linear adsorption coefficient (ml cm^{-3}).
Langmuir:

$$Q = \theta c + \sum_{i=1}^{M} \frac{a_i b_i c}{1 + b_i c} \tag{8.2.14}$$

where M represents the number of Langmuir isotherms to consider, a is the maximum of the isotherm (mg cm^{-3}), and b is a parameter (ml mg^{-1}).
Freundlich:

$$Q = \theta c + mc^n \tag{8.2.15}$$

where m (mg cm^{-3}) and n (dimensionless) are the Freundlich parameters.

In general, the affinity for NH_4 at the adsorption complex will be low. According to soil chemistry, the preference at the adsorption complex follows the lyotropic series: $Ca^{2+} \geq Mg^{2+} \gg K^+ \cong NH_4{}^+ > Na^+$ [45]

Sorption is influenced by the volumetric water content θ, as is obvious from the preceding equations. Adsorption is usually assumed independent of soil temperature.

8.2.5.2 Nitrification

Nitrification is a microbial process of nitrogen compound oxidation where NH_4 is transformed in NO_3:

$$NH_4^+ + 2O_2 \rightarrow NO_3^- + H_2O + 2H^+ \tag{8.2.16}$$

Intermediate products only exist for extremely short times and are thus neglected. By-products of nitrification are the release of NO and N_2O, in case of incomplete oxidation (also known as nitrifying denitrification).

Nitrification is dependent on oxygen availability (or its proxy, the degree of saturation). Often, a similar dependency is used as for the mineralization process: for intermediate degree of saturation, the process occurs at the optimum rate, and at lower and higher degrees of saturation the rate decreases (curvi)linearly.

Nitrification is also dependent on soil temperature as microbial activity is temperature-dependent. Often, an Arrhenius function or a van 't Hoff Q_{10} function is used for temperature dependency [46].

8.2.5.3 Denitrification

Denitrification is an anaerobic process where microbes use nitrate as an oxygen donor when decomposing OM. The complete denitrification process can be expressed as a redox reaction:

$$2NO_3^- + 10e^- + 12H^+ \rightarrow N_2 + 6H_2O \tag{8.2.17}$$

Denitrification reduces the nitrate content, so that less nitrate can leach downward and root uptake can be hindered. Denitrification generally proceeds through some combination of the following intermediate forms:

$$NO_3^- \rightarrow NO_2^- \rightarrow NO + N_2O \rightarrow N_2 \tag{8.2.18}$$

Full reduction will lead to N_2 (gas) emission to the atmosphere, which is harmless to the environment. However, incomplete reduction will result in emission of the greenhouse gas N_2O. N_2O may also dissolve in water and can then be subject to leaching.

This process depends on soil environmental conditions such as oxygen content (or its proxy, the degree of saturation), soil temperature, and soil pH. Based on an inventory of 59 models, denitrification can often be modeled by the following universal equation [11]:

$$D_a = \alpha \cdot f[\mathrm{NO_3}, S_e, T, \mathrm{pH}] \tag{8.2.19}$$

where D_a is the actual denitrification rate (kg N ha^{-1} d^{-1}), α is either the potential denitrification rate (kg N ha^{-1} d^{-1}) or a first-order decay rate coefficient (d^{-1}), $f[]$ represents some functional dependency of its arguments (dimensionless, or kg ha^{-1} in the case of first-order decay), NO_3 is the nitrate content (kg N ha^{-1}), S_e is the degree of saturation (dimensionless), T is the soil temperature (°C), and pH is the soil pH (dimensionless).

Aerobic denitrification also occurs because some microorganisms simultaneously use both oxygen and nitrate as oxidizing agents [47]; this process has a higher potential to create the harmful by-product N_2O [48].

8.2.5.4 Leaching

Leaching is the loss of nutrients via (mostly downward or through artificial drains) draining of water as a result of excess rainfall or irrigation. It is of environmental concern as it contributes to the contamination of the groundwater and surface-water systems. Nutrients can be considered as lost when they are no longer available for uptake by plants. In some cases, typically in situations with shallow groundwater, nutrients present in the upper groundwater may be transported upward (via capillary rise, diffusion) and then can become available again for root uptake. Thus, it is not straightforward to uniquely define leaching as a permanent loss of nutrients.

In general, leaching can be considered as the net loss of nutrients below a certain depth (e.g. the maximum rooting depth or the groundwater level):

$$L_{\mathrm{N},z=z_\mathrm{L}} = \int_0^t (q_\mathrm{s})_{z=z_\mathrm{L}} \mathrm{d}t = \int_0^t \left(qc - \theta D \frac{\partial c}{\partial z} \right)_{z=z_\mathrm{L}} \mathrm{d}t \tag{8.2.20}$$

where $L_{\mathrm{N},z=z\mathrm{L}}$ is the net, time-integrated loss of N across depth $z = z_\mathrm{L}$ (mg cm^{-2}), where z_L is the user-defined leaching depth (cm).

8.2.5.5 Gaseous N Losses

Ammonia (NH_3) volatilization from field-applied animal manure represents a major source of atmospheric pollution and reduces the nitrogen fertilizer value of the manure [49]. Loss rates are generally high directly after application and decrease as a function of time. The loss is related to the total concentration of ammonia (NH_3) and ammonium (NH_4) and the pH, which determines the equilibrium between NH_4 and NH_3. In the ALFAM model, total volatilization is described by a Michaelis–Menten-type function [50]:

$$N_{\mathrm{Vol}}(t) = N_{\max} \frac{t}{t + K_{\mathrm{m,Vol}}} \tag{8.2.21}$$

where N_{Vol} is the cumulative amount of N lost by volatilization (kg ha^{-1}), t is time (d), N_{max} is the total loss as $t \rightarrow \infty$ (kg ha^{-1}), and $K_{m,Vol}$ is the time where $N_{Vol}(t) = N_{max}/2$. The parameters N_{max} and $K_{m,Vol}$ were modeled as functions of climate (air temperature, wind speed), slurry composition (dry matter [DM] content), soil conditions (soil water content), and application method (broadcast, band spreading, injection).

Alternatively, volatilization of NH_4 may be modeled as a constant weight fraction f_{Vol} (kg kg^{-1}) of the added material. In that case, the actual addition of NH_4 to the soil equals $(1-f_{Vol})$ times the total NH_4 content in the added material [51]. This is similar to a situation where no volatilization is modeled and the addition is taken such that the corrected amount of NH_4 is added to the soil [40].

During both the nitrification and the denitrification processes, intermediate gases NO and N_2O are formed. These can be emitted to the atmosphere, contributing to greenhouse gas emissions, or they can further be reduced to the harmless gas N_2. Modeling detailed (de)nitrification and gaseous losses is complex. For example, the DNDC model considers the growth and death rates of (de)nitrifying microbes and NO, N_2O, and N_2 diffusion in the soil [52]. In most agroecological models, however, (de)nitrification is only modeled at the scale of losses from the NH_4 pool and of production and losses from the NO_3 pool. A simple model can then be used to determine, for example, the ratio of N_2/N_2O production during the denitrification process that is a function of the water-filled pore space [53]. Chen et al. [54] give an overview of models ranging from attempts to comprehensively simulate all soil processes in relation to N_2O emission to more empirical approaches requiring minimal input data.

8.2.6 Phosphorus

Phosphorus in soils is present in mineral (70–90%) and organic (10–30%) forms. Mineral P can be present bound to the solid phase, in precipitates of primary and secondary crystals, or as soluble inorganic P in solution. In solution, the soluble mineral form present depends on pH ($H_2PO_4^-$, HPO_4^{2-}, PO_4^{3-}). The amount of organic P depends on the turnover (mineralization/immobilization) or decomposition of OM and applied organic materials (e.g. manure), and part of the OM may be soluble (designated DOP). Particulate movement of P is often associated with soil erosion, and it is a physical process for diffuse transfer of P from agriculture to water systems [55]. Particulate transfer of P is (arbitrarily) assumed to occur with material >0.45 μm (or sometimes >0.4 μm, >0.2 μm). Categorization of P into dissolved or particulate forms is, therefore, an analytical convenience and may have limited environmental relevance [55]. In practice, P can also be associated to <0.45 μm-sized colloids in solution and is difficult to distinguish from dissolved inorganic and organic forms.

Often, the modeling of P has been done less intensively than for N. Keating et al. [56] give the following argumentation for this: "Unlike the management of N, there has been little need for detailed crop models to evaluate alternative strategies for management of P … Particularly in high input agricultural systems there are few prospects for improving management of P beyond recommendations for amount and method of application of fertiliser. Empirical relationships between yield and soil P tests have been adequate to gain insights into crop responsiveness to alternative fertiliser P sources and their residual effects." Sattari [57] lists several models describing or considering phosphorus cycling in soil–plant systems, including WOFOST, PHOSMOD, APSIM, CENTURY/DAYCENT, DSSAT, FIELD, ECOSYS, EPIC, PARJIB, QUEFTS, and DPPS. Schoumans et al. [14] describe nine nutrient-loss models, mainly focusing on P losses to surface water: NL-CAT (a combination of the models ANIMO/SWAP/SWQN/SWQL), REALTA, NLES-CAT, MONERIS, TRK (a combination of the models SOILNDB/HBV-N), SWAT, EveNFlow, NOPOLU, and a source apportionment (SA) approach. These models differ in the complexity of their process description, data requirement, temporal scale (daily, monthly, seasonal, year), and system boundaries. The models with a clear hydrological component are able to pay attention to specific pathways such as leaching from the root zone, subsurface runoff, surface runoff, and erosion.

For a typical growing season, about 10–20% of the applied fertilizer P is taken up by the crop; the remainder accumulates in soil [58]. The residual effect of P application is larger than this 10–20%, as in later growing seasons the accumulated P (provided it is not lost by leaching, runoff, or erosion) can be taken up by the crops.

Adding biobased derivatives (fertilization) means addition of mineral P, organically bound P (solid or dissolved), or incorporated P in OM: mineral pools can be raised, and P release from OM decomposition follows from C mineralization (see earlier) provided proper C:P ratios are used.

8.2.6.1 Adsorption, Desorption, Fixation, and Precipitation

Inorganic P can be assumed to be present in solution, adsorbed to soil particles (strongly adsorbed and slowly reversible bound P), and precipitated. The phosphate sorption to soil particles exists as a fast reversible adsorption reaction (Q_A) and a time-dependent (slow) diffusion/precipitation or "absorption" reaction (Q_F) that is strongly bound and can be considered as slowly reversible bound P. Inorganic P can be available as natural minerals (e.g. hydroxyapatite) or as secondary formed precipitates (e.g. dicalciumphosphate). If the soil concentration is supersaturated with phosphate, several secondary minerals will be formed, which most of the time will come easily into solution as the P concentration in soil solution drops down again. These are mainly mono- and di-calcium phosphates ($Ca(H_2PO_4)_2$, $CaHPO_4$, respectively) and partly mono- and di-magnesium phosphates.

The total amount of inorganic P in the soil (Q, as used in Eq. (8.2.1)) is given by:

$$Q = \theta c + Q_A + Q_F + Q_P = Q_{lab} + Q_F + Q_P \tag{8.2.22}$$

where c is the P concentration in solution ($mg\,ml^{-1}$), Q_A is the amount of adsorbed P ($mg\,cm^{-3}$), Q_F is the amount of strongly bound P ($mg\,cm^{-3}$), Q_P is the amount of chemically precipitated P ($mg\,cm^{-3}$), and Q_{lab} is the labile amount of P with $Q_{lab} = \theta c + Q_A$ ($mg\,cm^{-3}$).

Similar to ammonium, P-adsorption can be described by adsorption isotherms:

Langmuir (see Eq. (8.2.14)):

$$Q_A = \theta c + \sum_{i=1}^{M} \frac{a_i b_i c}{1 + b_i c} \tag{8.2.23}$$

where M represents the number of Langmuir isotherms to consider, a is the maximum of the isotherm ($mg\,cm^{-3}$), and b is a parameter ($ml\,mg^{-1}$).

Freundlich (see Eq. (8.2.15)):

$$Q_A = \theta c + mc^n \tag{8.2.24}$$

where m (mg cm^{-3}) and n (dimensionless) are the Freundlich parameters.

For P-fixation, a three-pool Freundlich-type model can be used [59]. In this model, the rate of change for the fixated pool is given by:

$$\frac{\partial Q_F}{\partial t} = \sum_{i=1}^{3} \alpha_i (K_{F,i} c^{N_i} - Q_{s,i}) \tag{8.2.25}$$

where Q_F is the total amount of P fixated with $Q_F = Q_{s,1} + Q_{s,2} + Q_{s,3}$ (mg cm^{-3}), $Q_{s,i}$ is the amount of P fixated in pool i (i = 1, 2, 3; mg cm^{-3}), α_i is an extinction coefficient for pool i (d^{-1}), N_i is an exponent coefficient for pool i (dimensionless), and $K_{F,i}$ is a fixation coefficient for pool i (mg cm^{-3} [mg ml^{-1}]$^{-N}$), given by:

$$K_F = b[\text{Al} + \text{Fe}]\rho_d 31 \cdot 10^{-3} \tag{8.2.26}$$

where b is a fixation parameter ([mg ml^{-1}]$^{-N}$), [Al + Fe] is the oxalate extractable amount of aluminum (Al) and iron (Fe) (mmol kg^{-1}), ρ_d is the dry bulk density (g cm^{-3}), 31 is the molecular weight of P (mg mmol^{-1}), and 10^{-3} is a conversion factor (kg g^{-1}).

At given c and known quantities of Q_s (i.e. from a previous time step), the rate of change of Q_F can be calculated. The new quantities Q_s follow from the individual time rate of changes through integration.

Chemical P-precipitation occurs when the concentration in the soil solution reaches the equilibrium concentration (c_{eq}), which is sometimes regarded as pH-dependent. This process is often modeled as a reversibly instantaneous reaction. The concentration of P in solution will never exceed c_{eq}. For proper modeling, the time when c_{eq} is reached must be discovered during the modeling process.

8.2.6.2 Calculation of Soil-Available P

Several soil tests exist to determine plant-available P, including the Olsen, Morgan, Bray, Mehlich, and calcium acetate lactate (CAL) tests. However, no models are known that can predict these quantities, because it is unknown which inorganic or organic fractions are released during extraction (often at a pH different from the soil pH). For very weak extraction methods at the pH of the soil (e.g. water extractions like the P_w method [60] or low-salt-concentration extractions like P-$CaCl_2$ [61]), the P concentration can be assessed based on the chemical reactions in the soil (desorption and dissolution). Such an approach has been proposed for P_w [62, 63].

8.2.6.3 Leaching

Leaching of P is calculated similarly to that for N; see Eq. (8.2.20).

8.2.7 Indices of Nutrient Use Efficiency

There are several methods for assessing the uptake or fertilizer efficiency of a nutrient applied to a soil. It goes beyond the scope of this chapter to discuss all possible measures, including their advantages and disadvantages; for that, the reader is referred to Syers et al. [58], Cassman et al. [64], and Dobermann [65]. In general, it is advisable to include several indices to quantify uptake or fertilizer efficiency [65]. Different indices cannot be interchanged, as they have different meanings. Some are intended to quantify the fertilizer recovery efficiency (RE), others to qualify the nutrient use efficiency at the level of a cropping system, where not only the nutrient removed with the crop but also the amounts contained in recycled crop residues and incorporated into soil OM and inorganic pools are taken into account [64].

Roughly two approaches can be considered [58]. The *difference method* is applicable to nutrients that are mobile and thus easily available for uptake and of which at the end of the growing season very little is left in the soil to be beneficial for subsequent crops. This is typically the case for N. Examples are the apparent crop recovery efficiency (*RE*) and agronomic efficiency (*AE*) (Table 8.2.4). The *mass balance method* is typically useful for nutrients for which large residues remain in the soil after the end of the growing season, which can be used by subsequent crops for a number of years thereafter. An example is the nutrient recovery (*NR*) (Table 8.2.4).

In quantifying the effect of manure or biobased derivatives, it is often compared to that of a default mineral fertilizer (e.g. ammonium nitrate), based either on uptake or on DM production. Examples are the nutrient activity coefficient (*NAC*) and the nutrient fertilizer replacement value (*NFRV*) (Table 8.2.4). For equal uptake, $(U - U_0)$ is equal for the treatments with mineral fertilizer and with biobased derivative, so that $NAC = NFRV_a$. If *NAC* or $NFRV < 1$ (or $< 100\%$), the remaining nutrient can cause burden to the environment (leaching, denitrification). The *NAC* values are sometimes used in policy making and decisions on whether or not a certain biobased derivative is allowed to be used as fertilizer. Depending on the research goal, nutrient uptake U may refer to the harvested part of the crop, the aboveground biomass, or the total biomass; this should be mentioned when reporting the indices.

8.2.8 Other Nutrients

In an agronomic sense, the majority of the agroecosystem models are restricted to N and sometimes P. These nutrients are considered separately, and no interactions are dealt with. If one wishes to consider multiple ions, interactions may no longer be neglected and a comprehensive chemical description of the processes in soil needs to be considered. A quantitative, thermodynamic description of the system is mathematically difficult and an arduous task because of the large number of possible chemical reactions [68].

All soil chemical models, regardless of purpose or scale, consist of three main components [69]: (i) a description of water movement, which determines advective transport; (ii) a description of the equilibrium between precipitated, exchange, solution, and gas phases; and (iii) a description of the transport of the dissolved chemical species. Parker et al. [70] consider in detail the utilization of chemical equilibrium models in plant mineral nutrition research. Some specific chemical equilibrium models include (all separately described by Loeppert et al. [68]): MINTEQA2/PRODEFA2, GEOCHEM-PC, SOILCHEM, C-SALT,

***Table 8.2.4** Commonly used definitions of nutrient recovery indices and indices for quantifying the effect of one fertilizer type relative to another [58, 64–67].*

Apparent crop recovery efficiency [58, 64–67] (= apparent nutrient recovery, Chapter 5.1)	$RE = \dfrac{U - U_0}{F}$	(8.2.27)
Physiological efficiency [58, 64, 65]	$PE = \dfrac{Y - Y_0}{U - U_0}$	(8.2.28)
Agronomic efficiency [58, 64, 65, 67] (= apparent nutrient efficiency, Chapter 5.1)	$AE = RE \times PE = \dfrac{Y - Y_0}{F}$	(8.2.29)
Partial factor productivity [58, 64, 65]	$PFP = \dfrac{Y}{F}$	(8.2.30)
Nutrient recovery [58]	$NR = \dfrac{U}{F}$	(8.2.31)
Nutrient activity coefficient [67]	$NAC = \dfrac{RE_{\text{bio}}}{RE_{\text{inorg}}}$	(8.2.32)
Nutrient fertilizer replacement value [66]	$NFRV\|_{U_{\text{bio}}=U_{\text{inorg}}} = NAC\|_{U_{\text{bio}}=U_{\text{inorg}}} = \left.\dfrac{F_{inorg}}{F_{bio}}\right\|_{U_{\text{bio}}=U_{\text{inorg}}}$	(8.2.33)
	$NFRV\|_{Y_{\text{bio}}=Y_{\text{inorg}}} = NAC\|_{Y_{\text{bio}}=Y_{\text{inorg}}} = \left.\dfrac{F_{inorg}}{F_{bio}}\right\|_{Y_{\text{bio}}=Y_{\text{inorg}}}$	(8.2.34)

All measures result in (weight) fractions and can be given in percentages by multiplication with 100%.

F	Amount of nutrient applied (kg or kg ha^{-1}).
U, U_0	Total crop nutrient uptake in aboveground biomass at maturity in treatment with and without (0) nutrient application (kg or kg ha^{-1}).
Y, Y_0	Crop yield in treatment with and without (0) nutrient application (kg or kg ha^{-1}).
bio	Subscript referring to biobased derivative.
inorg	Subscript referring to a default inorganic fertilizer (e.g. ammonium nitrate).

ALCHEMI, CHESS, GMIN, and LEACHM. Other examples are EPIDIM [71] and ORCHESTRA [72].

8.2.9 Overview of Processes in Selected Soil Dynamics Models

Table 8.2.5 presents an overview of processes for some selected soil dynamic models:

- ANIMO [43, 51];
- COUPMODEL [73, 74];

Table 8.2.5 *Overview of processes considered (+, yes; −, no) in some selected agroecological models.*

	ANIMO	COUPMODEL	DAISY	DAYCENT	DNDC	EU-Rotate_N	NDICEA	PASTIS	RothC	SOILN	STICS	SUNDIAL	
General processes													
Water flow	−	+	+	+	+	+	+	+	+	−	+	+	+
Solute transport	+	+	+	+	+	+	+	+	−	+	+	+	+
Gas transport	+	+	−	+	+	−	−	−	−	−	−	−	−
Heat flow	+	+	+	+	+	−	−	+	−	−	−	−	+
Crop uptake	+	+	+	+	+	+	+	−	−	+	+	+	+
Fertilization	+	+	+	+	+	+	+	−	+	+	+	+	+
Soil tillage	+	+	+	+	+	+	+	−	−	+	+	−	+
Crop rotations	+	+	+	+	+	+	+	−	+	+	+	+	+
Time step[a]	d	h	h	d	d	d	w	d	m	d	d	w	d
OM dynamics													
Decomposition rate type[b]	c	c	c	c	c	c	t	c	c	c	c	c	c
Number of pools[c]	pf	pf	p	pf	pf	p	f	p	p	p	pf	pf	pf
Carbon													
Mineralization	+	+	+	+	+	+	+	+	+	+	+	+	+
DOC	+	+	+	+	+	−	−	−	−	−	−	−	−
Nitrogen													
Nitrate	+	+	+	+	+	+	+	+	−	+	+	+	+
Ammonium	+	+	+	+	+	+	−	+	−	+	+	+	+
Mineralization	+	+	+	+	+	+	+	+	−	+	+	+	+
Immobilization	+	+	+	+	+	+	+	+	−	+	+	+	+
Nitrification	+	+	+	+	+	+	−	+	−	+	+	+	+
Denitrification	+	+	+	+	+	+	+	−	−	+	+	+	+
Ad- and desorption	+	+	+	−	+	−	−	+	−	−	−	−	+
Volatilization	+	+	+	+	+	+	+	−	−	+	+	+	+
Leaching	+	+	+	+	+	+	+	+	−	+	+	+	+
Gaseous losses	+	+	+	+	+	+	−	−	−	−	+	+	+
Fixation	−	+	+	+	+	+	+	−	−	−	+	+	+
Deposition	+	+	+	+	+	−	+	−	−	+	−	+	+
DON	+	+	−	+	−	−	−	−	−	−	−	−	+
Phosphorus													
PO4	+	−	−	−	−	−	−	−	−	−	−	−	+
Mineralization	+	−	−	+	−	−	−	−	−	−	−	−	+
Immobilization	+	−	−	+	−	−	−	−	−	−	−	−	+

Table 8.2.5 *(continued)*

	ANIMO	COUPMODEL	DAISY	DAYCENT	DNDC	EU-Rotate_N	NDICEA	PASTIS	RothC	SOILN	STICS	SUNDIAL	
Ad- and desorption	+	−	−	+	−	−	−	−	−	−	−	−	+
Fixation	+	−	−	+	−	−	−	−	−	−	−	−	+
Precipitation	+	−	−	−	−	−	−	−	−	−	−	−	−
Leaching	+	−	−	+	−	−	−	−	−	−	−	−	+
DOP	+	−	−	+	−	−	−	−	−	−	−	−	+
Environmental response factors													
Moisture content	+	+	+	+	+	+	+	+	+	+	+	+	+
(An)aerobic	+	+	−	+	+	−	−	−	−	−	+	−	−
Temperature	+	+	+	+	+	+	+	+	+	+	+	+	+
pH	+	+	−	+	+	+	+	−	−	+	−	−	−
Texture	−	−	+	+	+	+	+	−	+	−	+	+	−
Other interesting aspects													
Plant exudates	+	+	+	−	−	−	−	−	−	−	−	+	−
^{14}C and/or ^{15}N	−	−	−	+	−	−	−	−	+	−	−	+	−
Greenhouse gases	+	+	+	+	+	−	−	−	−	−	−	+	−

References for the models are given in the main text.
[a]h, hour; d, day; w, week; m, month.
[b]c, constant decomposition rate; t, time-dependent decomposition rate.
[c]p, predefined pools; f, flexible pools; pf, some predefined and some flexible pools.

- DAISY [75, 76];
- DAYCENT [77–80];
- DNDC [81, 82];
- EU-Rotate_N [83–85];
- NDICEA [40, 86];
- PASTIS [87, 88];
- RothC [26, 89, 90];
- SOILN [91–93];
- STICS [94–96];
- SUNDIAL [97, 98]; and
- SWAT [99, 100].

All models describe C dynamics, and, except for RothC, all consider N dynamics as well. The P cycle is much less taken into account in agroecosystem models. Only ANIMO, DAYCENT, and SWAT consider it and dissolved organic nutrients (N, P). ANIMO and SOILN do not simulate water flow themselves; this should be provided by an external hydrological model. NDICEA is the only model that uses a time-dependent decomposition rate factor; all others use constant decomposition rate factors. None of the considered models are capable of taking into account the effects on pH and water retention of the addition of a biobased derivative to the soil. These observations are based on a limited set of models, so cannot be generalized. It should be obvious that when selecting a model to use in evaluating the influence of biobased derivatives on the C, N, and P dynamics in soils, care must be taken regarding what processes are expected to be important and what outcome is desired.

8.2.10 Model Parameterization of Biobased Fertilizers

At the scale of the laboratory of mesocosm experiments it has often been demonstrated that the addition of biobased derivatives or processed wastes to natural soils can have a great effect on, for example, CO_2 evolution rates, percentage mineralization, N immobilization, denitrification, N_2O emission [101–103], and P availability [104]. Thus, addition of biobased derivatives may lead to an increase of leaching and gaseous emissions after soil application, or to a decrease of nutrient availability to plants. However, in a study by Vaneeckhaute et al. [105], addition of biobased derivatives sometimes resulted in fewer greenhouse gas emissions. Despite their great potential, agroecosystem models have only been used to a very limited degree to assess the environmental impacts of applying biobased derivatives or composted municipal solid waste (MSW) on agricultural land [106]. According to the ReUseWaste project (www.reusewaste.eu), best agronomic practices may be developed by calibrating simulation models on experimental data (laboratory and field) for new manure products or biobased derivatives developed therefrom. With the calibrated simulation models, new scenarios for manure nutrients and OM can be run [106]. In this section, a few examples in which agroecosystem models have been used will be briefly provided.

Dalemo et al. [107] studied the effects on N emissions from soil in an environmental systems analysis of waste management strategies. In an existing organic waste research model (ORWARE [108]), only direct emissions were included, and a detailed description of flows in the soil was lacking. Dalemo et al. used parts of the SOILN [91] model to simulate N turnover and emissions. Their results indicated the vital importance of considering N emissions from soil when comparing biological waste management systems with other waste management methods, especially with regard to eutrophication effects. Moreover, soil emissions are also important when comparing the environmental impacts of anaerobic digestion and composting systems.

Gerke et al. [109] modeled long-term compost application effects on nitrate leaching, soil OM content, and crop production using the DAISY model [75, 76]. For a 50-year period, they concluded from their simulation scenarios that the differences between compost types (non-matured and matured) were small compared to site-specific properties (soil temperature and water balance) and management practices (crop rotation). With respect to nitrate leaching, the effect of different compost application scenarios was highly sensitive to sand and relatively insensitive to loam. Relatively high crop yields and acceptably low nitrate concentrations in the drainage water were obtained, but further optimizations may still be possible.

Bruun et al. [106] used DAISY to study the application of processed organic municipal solid waste (MSW) on agricultural land. New organic materials added to the soil, such as

MSW, were split over two addition pools and one soil OM pool [75, 76]. Each of these pools decomposed according to first-order kinetics using a fixed C:N ratio to define the N-dynamics. It was the distribution of the added materials between the pools, their turnover rates, and the C:N ratios that determined the C and N mineralization dynamics of the added material. A range of different scenarios representing different geographical areas and different farm and soil types under Danish conditions and legislation were considered. Compared to standard scenarios, the application of processed organic MSW resulted in increased but highly variable emissions (leaching, tile drainage, N_2O emission, ammonia volatilization). A sensitivity analysis showed that the estimates were not very sensitive to the mineralization dynamics of the processed organic MSW.

Vaneeckhaute et al. [110, 111] assessed nutrient balances and evaluated the physicochemical soil fertility and quality in field tests in which biodigestion waste derivatives were used as a replacement for synthetic fertilizers or animal manure. For N, they made use of the NDICEA model [40] to obtain estimates of different N outputs from the soil, including volatilization, denitrification, and leaching. Based on a 2-year test, they concluded that the use of the biodigestion waste derivative caused a small, non-significant improvement in maize yield, physicochemical soil fertility, and soil quality. In addition, the energetic potential per hectare of harvested energy maize was slightly higher, and the economic and ecological benefits significantly higher, when digestate derivatives were used, as compared to animal manure supplied with synthetic fertilizers.

Many agroecological models have been calibrated and validated for synthetic (inorganic) and common organic fertilizers (e.g. manure, slurry, compost, crop residues). In order to determine whether biobased derivatives to be used as fertilizers can be used in existing models, their composition needs to be known, and preferably it should be tested whether the model descriptions are valid for the material considered. Here, we assume that the materials under consideration have already passed legislation.

The properties of the biobased derivatives must pertain to the specific model requirements. Some basic information might include:

- total content in solids: C, N, P;
- mineral contents in solids: NH_4, NO_3, ortho-P; and
- solubility of mineral N and P (e.g. at three pH values, say 5.0, 6.5, and 8.0).

As an illustration, Table 8.2.6 provides some of these basic properties for different biobased derivatives. The C:N ratios vary from 0 (struvite) to very low (anaerobically digested MSW) to very high (biochar). The percentage P varies from 1.1 (liquid fraction of pig slurry) to 12.6 (struvite).

If the biobased derivative is of organic type, it is likely that one will want to model its decomposition via the OM module present in the model (see earlier). Basically, there are two approaches: (i) use the pre-defined OM pools and their properties; or (ii) (if possible) define new pools with their own decomposition rates, efficiencies, and C, N, and P contents. For both cases, the validity (and parameterization) needs to be performed beforehand, for example via a decomposition experiment in the laboratory.

The decomposition of C from added organic material or soil plus added organic material can often be predicted accurately with first-order decay models with or without time-dependent decay coefficients (see earlier), as shown by Luxhøi et al. [112] and Marcelis et al. [113] (see Figure 8.2.3). However, N mineralization appears to be more

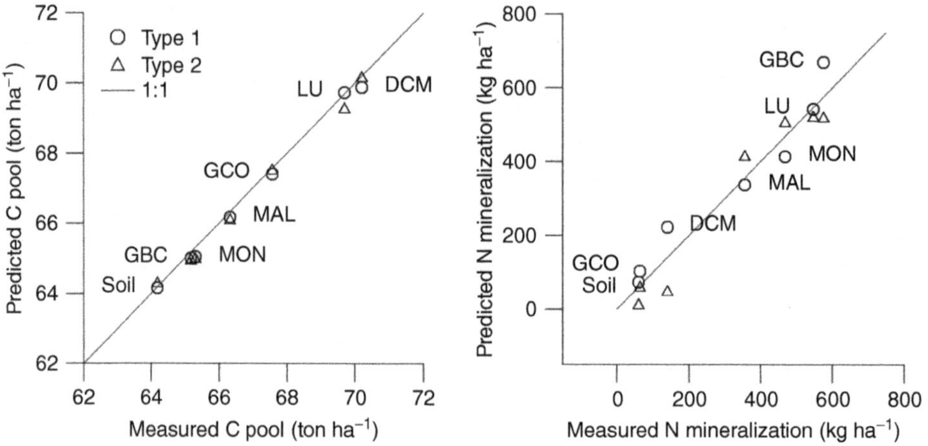

Figure 8.2.3 *Measured C pool and N mineralization after 84 days of incubation versus predicted quantities using two types of OM decay models: Type 1 with a constant decay rate coefficient and five pools and Type 2 with a time-varying decay coefficient. Data refer to soil (sand with 4.4% OM content in the top 10 cm) and soil amended with digested cattle manure (DCM), grinded blood clot (GBC), maltaflor (MAL), monterra (MON), lucerne (LU), and green compost (GCO). Source: Data from Marcelis et al. [113].*

difficult to predict (e.g. Figure 8.2.3); for example, Luxhøi et al. [112] had to use modifications of their originally chosen simple description of N mineralization.

Decomposition of pure minerals in biobased derivatives does not need to be considered. For example, struvite (magnesium-ammonium-phosphate, $(NH_4)Mg(PO_4)\cdot 6H_2O$) is a mineral containing NH_4 and PO_4 (Table 8.2.6) that can be produced from sludge or animal manure. When struvite is added to the soil, it is only necessary to have a proper description of the dissolution of the mineral. The dissolution in water is slow and increases at lower pH values [115]. If biobased derivatives contain mineral N and mineral P in their liquid components, these amounts can be directly added to the total or liquid phase in the model.

Other important qualifications of biobased derivatives are the humification coefficient and the effective organic carbon content. The humification coefficient (hc) is the amount of the added organic fertilizer that remains in the soil after 1 year [116]. For example, if the decomposition of the organic fertilizer is given by Eq. (8.2.7) then hc is given by $hc = C_{tot}(1)/C_0 = \exp[b(a+1)^m - ba^m]$, and thus can follow directly from the initial age parameter a (see examples in Table 8.2.5; Figure 8.2.4). The higher the effective organic carbon content (C_{eff}), the less OM is degraded in the soil [34, 117]. The C_{eff} is substantially higher for biowaste compost than for animal manures [117].

8.2.11 Conclusion

Agronomy is facing conflicting issues regarding agronomic and environmental goals. Bridging the gap between them is a true challenge. Increasing OM contents in soils is desirable for sustainability, improving soil quality or soil health, and adding (organic) nutrients for crop uptake. On the other hand, adding material to the soil may also lead to

Table 8.2.6 *Some properties of different biobased derivatives as obtained from indicated references.*

	[110]			[106]		[112]	[109]	[114]		[104]	[115]					
	Pig slurry	Mixture (50/50) of digestate and liquid fraction of digestate	Liquid fraction of digestate	Composted MSW[a]	Anaerobically digested MSW[a]	Composted MSW[a]	Anaerobically digested MSW[a]	Non-matured compost	Matured compost	Birchwood biochar	Solids from manure digestate (SOL)	SOL + straw	SOL + wood chip char	Composted SOL + straw	Composted SOL + wood chip char	Struvite
DW (%[a])	10	6.2	2.5	50	1[b]	31	63[c]	60	66	—	30.8	36.4	38.2	35.9	38.4	—
OC (%[a])	42	38	25	40	38	42	27	37.5	26.5	81	40.8	42.0	51.7	37.7	44.9	—
OC (g kg FW^{-1})	42	23.6	6.3	200	3.8	130	170	225	175	—	126	153	197	135	172	—
N total (g kg FW^{-1})	8.1	4.7	3.6	19	10	4.9	16.2	12	14.2	—	7.7	7.1	7.0	7.6	8.4	—
N total (%[a])	8.1	7.6	14.4	3.8	100	1.59	2.57	2	2.2	0.24	2.49	1.96	1.82	2.13	2.2	5.7
C:N	5.19	5.01	1.74	10.53	0.38	26.42	10.51	18.75	12.32	338	16.4	21.4	28.4	17.7	20.4	—
NH_4-N (g kg FW^{-1})	5.6	3.1	2.8	—	—	—	—	—	—	—	—	—	—	—	—	—
NO_3-N (g kg FW^{-1})	0.011	0.019	0.02	—	—	—	—	—	—	—	—	—	—	—	—	—
N_{org} (g kg FW^{-1})	2.5	1.6	0.82	—	—	4.65	13.23	—	—	—	—	—	—	—	—	—
NH_4-N (g g^{-1})	0.69	0.66	0.78	0.107	0.5	—	—	—	—	—	—	—	—	—	—	0.057
P total (g kg FW^{-1})	2.4	0.9	0.27	—	—	—	—	—	—	—	4.7	4.2	4.4	4.6	5.2	—
P total (%[a])	2.4	1.45	1.08	—	—	—	—	—	—	—	1.52	1.15	1.16	1.29	1.36	12.6
C:P	17.5	26.2	23.1	—	—	—	—	—	—	—	26.9	36.5	44.6	29.2	33.0	—

MSW, municipal solid waste; DW, dry weight; OC, organic carbon; FW, fresh weight; Norg, organic nitrogen; —, not provided or not applicable.
[a]Percentages based on DW.
[b]Value confirmed by S. Bruun (pers. comm.).
[c]High according to S. Bruun (pers. comm.), and may have been 6.3.

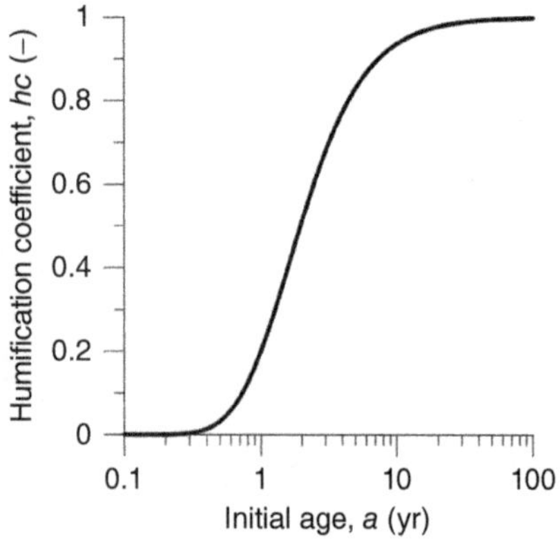

Figure 8.2.4 *The humification coefficient as a function of the initial age for the time-dependent decomposition model Eq. (8.2.7).*

environmental burden, as it may cause an increase in leaching or greenhouse gas emissions. The situation becomes even more complicated as the climate changes, influencing rate processes in the soil.

For situations where one wants to use an existing agroecosystem model to simulate the effects of a certain biobased fertilizer on soil properties, crop uptake, and losses to the environment, two import aspects need to be considered:

1. Has the model under consideration been validated for use with the biobased fertilizer that is under research? Often, models have been used for animal manure, but the transition to biobased fertilizers can only be made once the fertilizers have been properly characterized (e.g. humification coefficient, nutrient use efficiency).
2. Has the model under consideration been validated for climate change conditions?

The latter is more difficult to achieve, as climate change refers to the future, meaning validation is not possible.

Positive aspects of adding biobased derivatives to agricultural cropping systems include [111]:

- observed higher nutrient use efficiencies;
- reduced leaching;
- small (yet not always statistically significant) increases in yield; and
- addition of organic carbon and other nutrients (such as Ca, Mg, S).

Further long-term field experiments are recommended.

To what extent biobased derivatives contribute to environmental burden (i.e. denitrification and leaching, typically with respect to N) is dependent on how much mineral nutrient is added directly to the mineral pools; depending on the increase of the mineral pool, the current demand of the crop for uptake of that particular mineral, and the excess of water (rain, irrigation) in the period following the time of application, leaching may occur and can then be partly attributed to the addition of the biobased derivative.

Similarly, but depending on the anaerobicity in the soil, the addition may give an increase in denitrification. Depending on the decomposability of the biobased derivative, the release of nutrients can cause an increase in the mineral pool during the season, which may result in an increase of leaching and denitrification in situations where the rate of release is not in agreement with the rate of crop demand for uptake. Climate change may enhance these effects. Further, biobased fertilizers may contain other nutrients and contaminants (heavy metals, organic contaminants).

Indirectly, addition of biobased derivatives may influence transformation and transport rates. For example, increased OM contents may influence water holding capacity and thereby anaerobicity and soil temperature. Where the soil pH is influenced, this may affect denitrification, ammonia volatilization, and the solubility of phosphorus.

References

1. Marschner, H. (1995). *Mineral Nutrition of Higher Plants*, 2e. Cambridge, MA: Academic Press.
2. De Willigen, P., van Noordwijk, M. (1987) Roots, plant production and nutrient use efficiency. PhD Thesis: Wageningen Agricultural University.
3. Oenema, O. and Pietrzak, S. (2002). Nutrient management in food production: achieving agronomic and environmental targets. *Ambio* **31**: 159–168.
4. De Willigen, P. and Neeteson, J.J. (1985). Comparison of six simulation models for the nitrogen cycle in the soil. *Fertilizer Research* **8**: 157–171.
5. De Willigen, P. (1991). Nitrogen turnover in the soil-crop system; comparison of fourteen simulation models. *Fertilizer Research* **27**: 141–149.
6. Vereecken, H., Jansen, E.J., Hack-ten Broeke, M.J.D. et al., eds. (1991). Comparison of simulation results of five nitrogen models using different data sets. Soil and Groundwater Research Report II. Nitrate in Soils. Luxembourg: Commission of the European Communities.
7. Diekkrüger, B., Söndgerath, D., Kersebaum, K.C., and McVoy, C.W. (1995). Validity of agroecosystem models. A comparison of results of different models applied to the same data set. *Ecological Modelling* **81**: 3–29.
8. McGill, W.B. (1996). Review and classification of ten soil organic matter (SOM) models. In: *Evaluation of Soil Organic Matter Models*, NATO ASI Series, vol. **138** (eds. D.S. Powlson, P. Smith and J.U. Smith), 111–132. Berlin: Springer-Verlag.
9. Wu, L. and McGechan, M.B. (1998). A review of carbon and nitrogen processes in four soil nitrogen dynamics models. *Journal of Agricultural Engineering Research* **69**: 279–305.
10. McGechan, M.B. and Wu, L. (2001). A review of carbon and nitrogen processes in European soil nitrogen dynamic models. In: *Modeling Carbon and Nitrogen Dynamics for Soil Management* (eds. M.J. Shaffer, L. Ma and S. Hansen), 103–167. Boca Raton, FL: Lewis.
11. Heinen, M. (2006). Simplified denitrification models: overview and properties. *Geoderma* **133**: 444–463.
12. Herbst, M., Fialkiewicz, W., Chen, T. et al. (2005). Intercomparison of flow and transport models applied to vertical drainage in cropped lysimeters. *Vadose Zone Journal* **4**: 240–254.
13. Kersebaum, K.C., Hecker, J.M., Mirschel, W., and Wegehenkel, M. (2007). Modelling water and nutrient dynamics in soil–crop systems: a comparison of simulation models applied on common data sets. In: *Modelling Water and Nutrient Dynamics in Soil–Crop Systems* (eds. K.C. Kersebaum, J.M. Hecker, W. Mirschel and M. Wegehenkel), 1–17. New York: Springer.
14. Schoumans, O.F., Silgram, M., Groenendijk, P. et al. (2009). Description of nine nutrient loss models: capabilities and suitability based on their characteristics. *Journal of Environmental Monitoring* **11**: 506–514.
15. Groenendijk, P., Heinen, M., Klammler, G. et al. (2014). Performance assessment of nitrate leaching models for highly vulnerable soils used in low-input farming based on lysimeter data. *Science of the Total Environment* **499**: 463–480.

16. Richards, L.A. (1931). Capillary conduction of liquids through porous mediums. *Physics* **1**: 318–333.
17. Darcy, H. (1856). *Les Fontaines Publique de la Ville de Dijon*. Paris: Dalmont.
18. Buckingham, E. (1907). *Studies on the Movement of Soil Moisture*. Bulletin 38. US Bureau of Soils. Washington, DC: US Department of Agriculture, National Agricultural Library.
19. Kroes, J.G., van Dam, J.C., Groenendijk, P., et al. (2008) SWAP Version 3.2. Theory Description and User Manual. Alterra Report 1649. Wageningen: Alterra. Available from: http://edepot.wur.nl/176385 (accessed December 23, 2019).
20. Šimůnek, J., van Genuchten, M.T., and Šejna, M. (2012). *The HYDRUS Software Package for Simulating the Two- and Three-Dimensional Movement of Water, Heat, and Multiple Solutes in Variably-Saturated Porous Media, Technical Manual, Version 2.0*. Prague: PC Progress Available from: http://www.pc-progress.com (accessed December 23, 2019).
21. Heinen, M. (2001). FUSSIM2: brief description of the simulation model and application to fertigation scenarios. *Agronomie* **21**: 285–296.
22. Heinen, M. and de Willigen, P. (1998). *FUSSIM2 a Two-Dimensional Simulation Model for Water Flow, Solute Transport and Root Uptake of Water and Nutrients in Partly Unsaturated Porous Media*. Quantitative Approaches in Systems Analysis No. 20. Wageningen: DLO Research Institute for Agrobiology and Soil Fertility and the C.T. de Wit Graduate School for Production Ecology Available from: http://edepot.wur.nl/4408 (accessed December 23, 2019).
23. Milly, P.C.D. (1994). Climate, soil water storage, and the average annual water balance. *Water Resources Research* **30**: 2143–2156.
24. Atkinson, S.E., Woods, R.A., and Sivalpan, M. (2002). Climate and landscape controls on water balance model complexity over changing timescales. *Water Resources Research* **38**: 1314.
25. Ruud, N., Harter, T., and Naugle, A. (2004). Estimation of groundwater pumping as closure to the water balance of a semi-arid, irrigated agricultural basin. *Journal of Hydrology* **297**: 51–73.
26. Coleman, K., Jenkinson, D.S. (2014) RothC – a model for the turnover of carbon in soil. Model description and users guide. Harpenden: Rothamsted Research. Available from: https://www.rothamsted.ac.uk/sites/default/files/RothC_guide_WIN.pdf (accessed December 23, 2019).
27. Velthof, G.L., Oudendag, D.A., Oenema, O. (2007) Development and application of the integrated nitrogen model MITERRA-EUROPE. Alterra Report 1663.1. Wageningen: Alterra. Available from: http://edepot.wur.nl/30636 (accessed December 23, 2019).
28. Rijtema, P.E., Groenendijk, P., and Kroes, J.G. (eds.) (1995). *Environmental Impact of Land Use in Rural Regions. The Development, Validation and Application of Model Tools for Management and Policy Analysis*, Series on Environmental Science and Management, vol. **1**, 131–165. London: Imperial College Press.
29. Hämäläinen, M. (1991). *Principal Variations in the Chemical Composition of Peat*. Uppsala: Swedish University Agricultural Science Department Chemistry.
30. De Willigen, P., Janssen, B.H., Heesmans, H.I.M., et al. (2008) Decomposition and accumulation of organic matter in soil: comparison of some models. Alterra Report 1726. Wageningen: Alterra. Available from: http://edepot.wur.nl/15401 (accessed December 23, 2019).
31. Smith, P., Smith, J.U., Powlson, D.S. et al. (1997). A comparison of the performance of nine soil organic matter models using datasets from seven long-term experiments. *Geoderma* **81**: 153–225.
32. Heinen, M., de Willigen, P. (2005) Comparison of organic matter models MOTOR and MINIP [Dutch]. Alterra Report 1260. Wageningen: Alterra. Available from: http://www2.alterra.wur.nl/Webdocs/PDFFiles/Alterrarapporten/AlterraRapport1260.pdf (accessed December 23, 2019).
33. Assinck, F.B.T., Rappoldt, C. (2004) MOTOR 2.0: Module for transformation of organic matter and nutrients in soil. User Guide and Technical Documentation. Alterra Report 933. Wageningen: Alterra. Available from: http://edepot.wur.nl/20815 (accessed December 23, 2019).
34. Janssen, B.H. (1984). A simple method for calculating decomposition and accumulation of "young" soil organic matter. *Plant and Soil* **76**: 297–304.

35. Janssen, B.H. (1986). Nitrogen mineralization in relation to C:N ratio and decomposability of organic materials. *Plant and Soil* **181**: 39–45.
36. Yang, H.S. (1996) Modelling organic matter mineralization and exploring options for organic matter management in arable farming in Northern China. PhD Thesis: Wageningen Agricultural University. Available from: http://edepot.wur.nl/202123 (accessed December 23, 2019).
37. do Nascimento, A.F., de Sá Mendonça, E., Carvalho Leite, L.F. et al. (2012). Calibration and validation of models for short-term decomposition and N mineralization of plant residues in the tropics. *Scientia Agricola* **69**: 393–401.
38. Pronk, A.A. (2007) Organic matter management on sandy soil with special attention to dune sand. Literature study [Dutch]. Note 487. Wageningen: Plant Research International. Available from: http://edepot.wur.nl/155342 (accessed December 23, 2019).
39. do Nascimento, A.F., de Sá Mendonça, E., Carvalho Leite, L.F., and Lima Neves, J.C. (2011). Calibration of the century, Apsim and NDICEA models of decomposition and N mineralization of plant residues in the humid tropics. *Revista Brasileira de Ciencia do Solo* **35**: 917–928.
40. van der Burgt, G.J.H.M., Oomen, G.J.M., Habets, A.S.J., and Rossing, W.A.H. (2006). The NDICEA model, a tool to improve nitrogen use efficiency in cropping systems. *Nutrient Cycling in Agroecosystems* **74**: 275–294.
41. Yang, H.S. and Janssen, B.H. (2000). A mono-component model of carbon mineralization with a dynamic rate constant. *European Journal of Soil Science* **51**: 517–529.
42. Yang, H.S. and Janssen, B.H. (2002). Relationship between substrate initial reactivity and residues ageing speed in carbon mineralization. *Plant and Soil* **239**: 215–224.
43. Groenendijk, P., Renaud, L., Roelsma, J. (2005) Prediction of nitrogen and phosphorus leaching to groundwater and surface waters: process descriptions of the ANIMO 4.0 model [Dutch]. Alterra Report 983. Wageningen: Alterra. Available from: http://edepot.wur.nl/35121 (accessed December 23, 2019).
44. Kätterer, T., Reichstein, M., Andrén, O., and Lomander, A. (1998). Temperature dependence of organic matter decomposition: a critical review using literature data analyzed with different models. *Biology Fertility of Soils* **27**: 258–262.
45. Bolt, G.H., Bruggenwert, M.G.M., and Kamphorst, A. (1978). Adsorption of cations in soil. In: *Soil Chemistry. A. Basic Elements* (eds. G.H. Bolt and M.G.M. Bruggenwert), 54–90. Amsterdam: Elsevier Scientific Publishing Company.
46. Rodrigo, A., Recous, S., Neel, C., and Mary, B. (1997). Modelling temperature and moisture effects on C–N transformations in soils: comparison of nine models. *Ecological Modelling* **102**: 325–339.
47. Robertson, L.A. and Kuenen, J.G. (1984). Aerobic denitrification: a controversy revived. *Archives of Microbiology* **139**: 351–354.
48. Lloyd, D. (1993). Aerobic denitrification in soils and sediments: from fallacies to facts. *Trends in Ecology & Evolution* **8**: 352–356.
49. Jarvis, S.C. and Pain, B.F. (1990). Ammonia volatilization from agricultural land. In: *Proceedings of the Fertiliser Society No. 298*. Peterborough: Greenhill House.
50. Søgaard, H.T., Sommer, S.G., Hutchings, N.J. et al. (2002). Ammonia volatilization from field-applied animal slurry – the ALFAM model. *Atmospheric Environment* **36**: 3309–3319.
51. Renaud, L.V., Roelsma, J., Groenendijk, P. (2005) ANIMO 4.0: user's guide of the ANIMO 4.0 nutrient leaching model. Alterra Report 224. Wageningen: Alterra. Available from: http://www2.alterra.wur.nl/Webdocs/PDFFiles/Alterrarapporten/AlterraRapport224.pdf (accessed December 23, 2019).
52. Li, C.S. (2000). Modeling trace gas emissions from agricultural ecosystems. *Nutrient Cycling in Agroecosystems* **58**: 259–276.
53. Parton, W.J., Mosier, A.R., Ojima, D.S. et al. (1996). Generalized model for N2 and N_2O production from nitrification and denitrification. *Global Biogeochemical Cycles* **10**: 401–412.
54. Chen, D., Li, Y., Grace, P., and Mosier, A.R. (2008). N_2O emissions from agricultural lands: a synthesis of simulation approaches. *Plant and Soil* **309**: 169–189.

55. Haygarth, P.M. and Jarvis, S.C. (1999). Transfer of phosphorus from agricultural soils. *Advances in Agronomy* **66**: 195–249.
56. Keating, B.A., Carberry, P.S., Hammer, G.L. et al. (2003). An overview of APSIM, a model designed for farming systems simulation. *European Journal of Agronomy* **18**: 267–288.
57. Sattari, S.Z. (2014) The legacy of phosphorus. Agriculture and food security. PhD Thesis: Wageningen University. Available from: http://edepot.wur.nl/316019 (accessed December 23, 2019).
58. Syers, J.K., Johnston, A.E., and Curtin, D. (2008). *Efficiency of Soil and Fertilizer Phosphorus Use. Reconciling Changing Concepts of Soil Phosphorus Behaviour with Agronomic Information.* FAO Fertilizer and Plant Nutrition Bulletin 18. Rome: FAO.
59. Schoumans, O.F. (1995) Description and validation of the process formulation of abiotic phosphate reactions in non-calcareous sandy soils [Dutch]. Report 381. Wageningen: DLO-Staring Centre. Available from: http://edepot.wur.nl/301272 (accessed December 23, 2019).
60. Sissingh, H.A. (1969). Die Lösung der Bodenphosphorsäure bei wässriger Extraktion in Verbindung mit der Entwicklung einer neuen P-Wasser-Methode. *Landwirtschaftliche Forschungen Sonderheft* **23** (11): 110–120.
61. Houba, V.J.G., Novozamsky, I., Lexmond, T.M., and van der Lee, J.J. (1990). Applicability of 0.01 M CaCl2 as a single extraction solution for the assessment of the nutrient status of soils and other diagnostic purposes. *Communications in Soil Science and Plant Analysis* **21** (19–20): 2281–2290.
62. van Noordwijk, M., de Willigen, P., and Ehlert, P.A.I. (1990). A simple model of P uptake by crops as a possible basis for P fertilizer recommendations. *Netherlands Journal of Agricultural Science* **38** (3A): 317–322.
63. Schoumans, O.F. and Groenendijk, P. (2000). Modeling soil phosphorus levels and phosphorus leaching from agricultural land in the Netherlands. *Journal of Environmental Quality* **29**: 111–116.
64. Cassman, K.G., Peng, S., Olk, D.C. et al. (1998). Opportunities for increased nitrogen-use efficiency from improved resource management in irrigated rice systems. *Field Crops Research* **56**: 7–39.
65. Dobermann, A. (2007). Nutrient use efficiency – measurement and management [Dutch]. Fertilizer Best Management Practices. General Principles, Strategy for their Adoption, and Voluntary Initiatives vs Regulation, 1–28. Papers presented at the IFA International Workshop on Fertilizer Best Management Practices 7-9 March 2007, Brussels, Belgium. Paris: International Fertilizer Industry Association.
66. Lalor, S.T.J., Schröder, J.J., Lantinga, E.A. et al. (2011). Nitrogen fertilizer replacement value of cattle slurry in grassland as affected by method and timing of application. *Journal of Environmental Quality* **40**: 362–373.
67. Ehlert, P.A.I., Nelemans, J., Velthof, G.L. (2012) Stikstofwerking van mineralenconcentraten. Stikstofwerkingscoëfficiënten en verliezen door denitrificatie en stikstofimmobilisatie bepaald onder gecontroleerde omstandigheden. Alterra Report 2314. Wageningen: Alterra. Available from: http://edepot.wur.nl/235503 (accessed December 23, 2019).
68. Loeppert, R.H., Schwab, A.P., and Goldberg, S. (1995). *Chemical Equilibrium and Reaction Models.* SSSA Special Publication Number 42. Madison, WI: Soil Science Society of America.
69. Hutson, J.L. and Wagenet, R.J. (1995). The application of chemical equilibrium in solute transport models. In: *Chemical Equilibrium and Reaction Models.* SSSA Special Publication Number 42 (eds. R.H. Loeppert, A.P. Schwab and S. Goldberg), 97–112. Madison, WI: Soil Science Society of America.
70. Parker, D.R., Chaney, R.L., and Norvell, W.A. (1995). Chemical equilibrium models: application to plant nutrition research. In: *Chemical Equilibrium and Reaction Models.* SSSA Special Publication Number 42 (eds. R.H. Loeppert, A.P. Schwab and S. Goldberg), 163–200. Madison, WI: Soil Science Society of America.
71. Groenendijk, P.G. (1987). *Onderzoek naar de effecten van wateraanvoer en peilveranderingen in agrarische gebieden op de waterkwaliteit in natuurgebieden. deel 3: epidim, een chemisch evenwichtsmodel met adsorptie, verwering en neerslag van calciet*, ICW Nota, vol. **1774**. Wageningen:

Instituut voor Cultuurtechniek en Waterhuishouding Available from: http://edepot.wur.nl/217806 (accessed December 23, 2019).

72. Meeussen, J.C.L. (2003). ORCHESTRA: an object-oriented framework for implementing chemical equilibrium models. *Environmental Science and Technology* **37**: 1175–1182.
73. Jansson, P.E., Karlberg, L. (2010) COUP Manual. Coupled Heat and Mass Transfer Model for Soil–Plant–Atmosphere Systems. Available from: https://drive.google.com/file/d/0B0-WJKp0fmYCZ0JVeVgzRWFIbUk/view (accessed December 23, 2019).
74. Jansson, P.E. (2012). COUPMODEL: model use, calibration, and validation. *Transactions of the ASABE* **55** (4): 1335–1344.
75. Hansen, S., Jensen, H.E., Nielsen, N.E., and Svendsen, H. (1990). *DAISY – Soil Plant Atmosphere System Model.* Copenhagen: Royal Veterinary and Agricultural University Copenhagen.
76. Hansen, S., Abrahamsen, P., Petersen, C.T., and Styczen, M. (2012). DAISY: model use, calibration and validation. *Transactions of the ASABE* **55** (4): 1315–1333.
77. Parton, W.J., Hartman, M., Ojima, D., and Schimel, D. (1998). DAYCENT and its land surface submodel: description and testing. *Global and Planetary Change* **19**: 35–48.
78. Parton, W.J., Holland, E.A., Del Grosso, S.J. et al. (2001). Generalized model for NO_x and N_2O emissions from soils. *Journal of Geophysical Research* **106**: 17403–17419.
79. Del Grosso, S.J., Parton, W.J., Mosier, A.R. et al. (2001). Simulated interaction of carbon dynamics and nitrogen trace gas fluxes using the DAYCENT model. In: *Modeling Carbon and Nitrogen Dynamics for Soil Management* (eds. M.J. Shaffer, L. Ma and S. Hansen), 303–332. Boca Raton, FL: Lewis.
80. Del Grosso, S.J., Mosier, A.R., Parton, W.J., and Ojima, D.S. (2005). DAYCENT model analysis of past and contemporary soil N_2O and net greenhouse gas flux for major crops in the USA. *Soil & Tillage Research* **83**: 9–24.
81. Li, C., Frolking, S., and Frolking, T.A. (1992). A model of nitrous oxide evolution from soil driven by rainfall events: I. Model structure and sensitivity. *Journal of Geophysical Research* **97**: 9759–9776.
82. Li, C., Frolking, S., Harriss, R., et al. (2012) User's Guide for the DNDC Model. Version 9.5. Institute for the Study of Earth, Oceans and Space, University of New Hampshire. Available from: http://www.dndc.sr.unh.edu/model/GuideDNDC95.pdf (accessed December 23, 2019).
83. Rahn, C.R., Zhang, K., Lillywhite, R. et al. (2007) Brief Description of the EU-ROTATE_N Model. The Integrated Model – Description of the Model Subroutines. University of Warwick. Available from: http://www2.warwick.ac.uk/fac/sci/lifesci/wcc/research/nutrition/eurotaten (accessed December 23, 2019).
84. Rahn, C., Zhang, K., Lillywhite, R. (2009) User Manual EU-ROTATE_N. University of Warwick. Available from: http://www2.warwick.ac.uk/fac/sci/lifesci/wcc/research/nutrition/eurotaten (accessed December 23, 2019).
85. Rahn, C.R., Zhang, K., Lillywhite, R. et al. (2010). EU-Rotate_N – a decision support system – to predict environmental and economic consequences of the management of nitrogen fertiliser in crop rotations. *European Journal of Horticultural Science* **75**: 20–32.
86. Koopmans, C.J. and Bokhorst, J. (2002). Nitrogen mineralisation in organic farming systems: a test of the NDICEA model. *Agronomie* **22**: 855–862.
87. Lafolie, F. (1991). Modelling water flow, nitrogen transport and root uptake including physical non-equilibrium and optimization of the root water potential. *Fertilizer Research* **27**: 215–231.
88. Garnier, P., Néel, C., Mary, B., and Lafolie, F. (2001). Evaluation of a nitrogen transport and transformation model in a bare soil. *European Journal of Soil Science* **52**: 253–268.
89. Coleman, K., Jenkinson, D.S., Crocker, G.J. et al. (1997). Simulating trends in soil organic carbon in long-term experiments using RothC-26.3. *Geoderma* **81**: 29–44.
90. Coleman, K. and Jenkinson, D.S. (1996). RothC-26.3 – a model for the turnover of carbon in soils. In: *Evaluation of Soil Organic Matter Models*, NATO ASI Series, vol. **138** (eds. D.S. Powlson, P. Smith and J.U. Smith), 237–246. Berlin: Springer-Verlag.

91. Johnsson, H., Bergström, L., and Jansson, P.E. (1987). Simulated nitrogen dynamics and losses in a layered agricultural soil. *Agriculture, Ecosystems and Environment* **18**: 333–356.
92. Johnsson, H., Larsson, M., Mårtensson, K., and Hoffmann, M. (2002). SOILNDB: a decision support tool for assessing nitrogen leaching losses from arable land. *Environmental Modelling & Software* **17**: 505–517.
93. Bergström, L., Johnsson, H., and Torstensson, G. (1991). Simulation of soil nitrogen dynamics using the SOILN model. *Fertilizer Research* **27**: 181–188.
94. Brisson, N., Mary, B., Ripoche, D. et al. (1998). STICS: a generic model for the simulation of crops and their water and nitrogen balances. 1. Theory and parameterization applied to wheat and corn. *Agronomie* **18**: 311–346.
95. Brisson, N., Ruget, F., Gate, P. et al. (2002). STICS: a generic model for simulating crops and their water and nitrogen balances. II. Model validation for wheat and maize. *Agronomie* **22**: 69–92.
96. Brisson, N., Gary, C., Justes, E. et al. (2003). An overview of the crop model STICS. *European Journal of Agronomy* **18**: 309–332.
97. Bradbury, N.J., Whitmore, A.P., Hart, P.B.S., and Jenkinson, D.S. (1993). Modelling the fate of nitrogen in crop and soil in the years following application of ^{15}N-labelled fertilizer to winter wheat. *Journal of Agricultural Science* **121**: 363–379.
98. Smith, J.U., Bradbury, N.J., and Addiscott, T.M. (1996). SUNDIAL: a PC-based system for simulating nitrogen dynamics in arable land. *Agronomy Journal* **88**: 38–43.
99. Neitsch, S.L., Arnold, J.G., Kiniry, J.R., Williams, J.R. (2011) Soil and Water Assessment Tool. Theoretical Documentation Version 2009. Texas Water Resources Institute Technical Report Nr. 406. College Station, TX. Available from: http://swat.tamu.edu/media/99192/swat2009-theory.pdf (accessed December 23, 2019).
100. Arnold, J.G., Kiniry, J.R., Srinivasan, R., et al. (2012) Soil & Water Assessment Tool. Input/Output Documentation Version 2012. Texas Water Resources Institute TR-439. Available from: http://swat.tamu.edu/media/69296/SWAT-IO-Documentation-2012.pdf (accessed December 23, 2019).
101. Busby, R.R., Torbert, H.A., and Gebhart, D.L. (2007). Carbon and nitrogen mineralization of non-composted and composted municipal solid waste in sandy soils. *Soil Biology & Biochemistry* **39**: 1277–1283.
102. Fangueiro, D., Pereira, J., Chadwick, D. et al. (2008). Laboratory assessment of the effect of cattle slurry pre-treatment on organic N degradation after soil application and N_2O and N_2 emissions. *Nutrient Cycling and Agroecosystems* **80**: 107–120.
103. Bertora, C., Alluvione, F., Zavattaro, L. et al. (2008). Pig slurry treatment modifies slurry composition, N_2O, and CO_2 emissions after soil incorporation. *Soil Biology & Biochemistry* **40**: 1999–2006.
104. Christel, W., Bruun, S., Magid, J., and Jensen, L.S. (2014). Phosphorus availability from the solid fraction of pig slurry is altered by composting or thermal treatment. *Bioresource Technology* **169**: 543–551.
105. Vaneeckhaute, C., Meers, E., Michels, E. et al. (2013b). Ecological and economic benefits of the application of bio-based mineral fertilizers in modern agriculture. *Biomass and Bioenergy* **49**: 239–248.
106. Bruun, S., Hansen, T.L., Christensen, T.H. et al. (2006). Application of processed organic municipal solid waste on agricultural land – a scenario analysis. *Environmental Modeling and Assessment* **11**: 251–265.
107. Dalemo, M., Sonesson, U., Jönsson, H., and Björklund, A. (1998). Effects of including nitrogen emissions from soil in environmental systems analysis of waste management strategies. *Resources, Conservation and Recycling* **24**: 363–381.
108. Dalemo, M., Sonesson, U., Björklund, A. et al. (1997). ORWARE – a simulation model for organic waste handling systems. Part 1: Model description. *Resources, Conservation and Recycling* **21**: 17–37.
109. Gerke, H.H., Arning, M., and Stöppler-Zimmer, H. (1999). Modeling long-term compost application effects on nitrate leaching. *Plant and Soil* **213**: 75–92.

110. Vaneeckhaute, C., Meers, E., Michels, E. et al. (2013a). Closing the nutrient cycle by using bio-digestion waste derivatives as synthetic fertilizer substitutes: a field experiment. *Biomass and Bioenergy* **55**: 175–189.
111. Vaneeckhaute, C., Ghekiere, G., Michels, E. et al. (2014). Assessing nutrient use efficiency and environmental pressure of macronutrients in biobased mineral fertilizers: a review of recent advances and best practices at field scale. *Advances in Agronomy* **128**: 137–180.
112. Luxhøi, J., Bruun, S., Jensen, L.S. et al. (2007). Modelling C and N mineralization during decomposition of anaerobically digested and composted municipal solid waste. *Waste Management Research* **25**: 170–176.
113. Marcelis, L.F.M., Voogt, W., de Visser, P.H.B., et al. (2003) Organic matter management in biological greenhouse culture. Chrysanthemums 2002 [Dutch]. Report 70. Wageningen: Plant Research International. Available from: http://edepot.wur.nl/32198 (accessed December 23, 2019).
114. Sun, Z., Moldrup, P., Elsgaard, L. et al. (2013). Direct and indirect short-term effects of biochar on physical characteristics of an arable sandy loam. *Soil Science* **178**: 465–473.
115. Manning, D.A.C. (2008). Phosphate minerals, environmental pollution and sustainable agriculture. *Elements* **4**: 105–108.
116. van Dijk, W., van Dam, A.M., van Middelkoop, J.C., et al. (2005) Onderbouwing N-werkingscoëfficiënt overige organische meststoffen [Dutch]. Report PPO 343. Lelystad: Praktijkonderzoek Plant en Omgeving B.V. Available from: http://edepot.wur.nl/93807 (accessed December 23, 2019).
117. Veeken, A. and Hamelers, B. (2002). Sources of Cd, Cu, Pb and Zn in biowaste. *The Science of the Total Environment* **300**: 87–98.

Index

Biorefinery of Inorganics: Recovering Mineral Nutrients from Biomass and Organic Waste, First Edition.
Edited by Erik Meers, Gerard Velthof, Evi Michels and René Rietra.
© 2020 John Wiley & Sons Ltd. Published 2020 by John Wiley & Sons Ltd.